2nd International Conference on Measurement Control and Instrumentation (MCAI 2022)

Journal of Physics: Conference Series Volume 2366

Guangzhou, China
22 – 24 July 2022

ISBN: 978-1-7138-6474-5
ISSN: 1742-6588

Printed from e-media with permission by:

Curran Associates, Inc.
57 Morehouse Lane
Red Hook, NY 12571

proceedings.com

Some format issues inherent in the e-media version may also appear in this print version.

Institute of Physics
Dirac House, Temple Back
Bristol BS1 6BE UK

Phone: 44 1 17 929 7481
Fax: 44 1 17 920 0979

techtracking@iop.org

Additional copies of this publication are available from:

Curran Associates, Inc.
57 Morehouse Lane
Red Hook, NY 12571 USA
Phone: 845-758-0400
Fax: 845-758-2633
Email: curran@proceedings.com
Web: www.proceedings.com

TABLE OF CONTENTS

Preface

The 2022 2nd International Conference on Measurement Control and Instrumentation (MCAI 2022) was held in Guangzhou, China from July 22 to 24, 2022. In view of the current COVID-19 pandemic, it was chosen to be held online. MCAI 2022 promoted research and developmental activities in the fields of "Measurement Control" and "Instrumentation" as well as scientific information interchange between researchers, engineers, and practitioners working all around the world.

In the conference, we had the honor of having invited both Prof. Guixiong Liu from South China University of Technology, China and Prof. Zhen Zhou, Guangzhou Hexin Instrument Co., Ltd., China to serve as our Conference General Chairs. The conference consisted of keynote speeches, oral presentations and online Q&A discussion, attracting 100 individuals and institutions worldwide. Primarily, keynote speakers were each allocated 30-45 minutes to hold their speeches. Then in the oral presentations, the excellent papers we had selected were presented by their authors by turns.

We were delighted and honored to invite three prominent professors as keynote speakers to deliver their speeches. For a start, Prof. Zongmin Ma from North University of China showed us his research: Non-contact Atomic Force Microscopy Instruments and Atomic Scale Precision Measurements. By developing the ultra-high vacuum non-contact atomic force microscope (NC-AFM), he solved the key technologies of probe vibration control and precise detection, force spectrum measurement, and realized atomic scale imaging characterization in terms of non-destructive imaging of magnetic information and uncoupled measurement of surface charge. Besides, Assoc. Prof. Zhiwei Shi from Guangdong University of Technology, China made a report on Multi-sensor Information Fusion Technology for Intelligent Networked Vehicles. Aiming at the development of intelligent networked vehicles, he proposed a vehicle headlight far-light system based on multi-sensor information fusion, which mainly includes three major parts: information fusion, drive control and optical design, ensuring in a maximum degree the intelligent lighting functions and simple structure, low cost. The wonderful speeches of each keynote speaker had triggered heated discussion. Besides, every participant praised this conference for disseminating useful and insightful knowledge.

The proceedings of MCAI 2022 is a compilation of the accepted papers and represent an interesting outcome of the conference. These papers feature but are not limited to the following topics: Virtual Instrument, Mechanical and Electrical Integration, Precision Assessment, Automatic Control Principle, etc. All the papers have been checked through rigorous review and processes to meet the requirements of publication.

We would like to express our sincere appreciation to all of those who supported MCAI 2022. In particular, we would like to thank the Journal of Physics: Conference Series, for the endeavor of all the colleagues in publishing this paper volume. The help and contribution of each individual and institution was instrumental in the success of the conference and with the strong support, we are sure MCAI 2022 has been beneficial to all the attendees.

Committee of MCAI 2022

Committee member

Conference General Chairs
Prof. Guixiong Liu, South China University of Technology, China
Prof. Zhen Zhou, Guangzhou Hexin Instrument Co., Ltd., China

Organizing Committee Chairs
Prof. Jiaxiang Xue, South China University of Technology, China
Prof. Pingyu Zhu, Guangzhou University, China
Prof. Minghua Chen, South China National Centre of Metrology, China

Technical Program Committee Chair
Prof. Hamid Reza Karimi, Politecnico di Milano, Italy

Publication Chair
Assoc. Prof. P. C. Srinivasa Rao, Koneru Lakshmaiah University (KLU), India

Academic Committee
Researcher Yaotang Li, Guangzhou Electronic Technology Co., Ltd.
Researcher Taobo Cheng, Guangdong Institute of Intelligent Manufacturing, GDAS., China
Prof. Xiaobin Hong, South China University of Technology, China
Prof. Yaohua Deng, Guangdong University of Technology, China
Prof. Lifu Li, South China University of Technology, China
Prof. Peng Han, South China Normal University, China
Prof. Yanming Quan, South China University of Technology, China
Prof. Liming Wu, Guangdong University of Technology, China
Prof. Luxin Tang, Guangdong University of Technology, China
Prof. Weixin Wang, South China Agricultural University, China
Prof. Buhai Shi, South China Agricultural University, China
Prof. Ping Yao, Guangdong Polytechnic Normal University, China
Prof. Kai Wu, South China University of Technology, China
Prof. Zhen Li, South China Agricultural University, China
Prof. Chen Xu, Guangdong Artificial Intelligence and Digital Economy Laboratory (Guangzhou), China
Prof. Shuqiang Huang, Jinan University, China
Prof. Xiaohong Huang, Guangdong Industry Polytechnic, China
Prof. Yong You, Beijing Institute of Technology, Zhuhai, China
Prof. Xuefeng Zhou, Guangdong Institute of Intelligent Manufacturing, GDAS., China
Prof. Guojian Huang, Guangdong Mechanical and Electrical Polytechnic (GDMEP) , China
Prof. Chuangmian Huang, China Ceprei Laboratories, China
Prof. Zuohua Huang, South China Normal University, China
Prof. Jianxing Cheng, Zhongkai University of Agriculture and Engineering, China
Prof. Xufeng Li, Guangdong Institute of Special Equipment Inspection and Research, China
Prof. Dachuan Cai, China National Analytical Center, Guangzhou, China
Prof. Dongsheng He, China National Quality Supervision and Testing Center for Smart, China
Prof. Feijiang Huang, Guangzhou Maritime University, China
Prof. Xiaoyong Luo, Xinhui KangYu Control Systems Engineering Inc., China
Prof. Xiangbao Meng, Guangdong Agricultural Machinery Research Institute, China
Prof. Junbin Xu, South China National Centre of Metrology, China
Prof. Rongbin Yu, Guangdong Testing Institute of Product Quality Supervision, China
Researcher Mingzhe Yuan, Shenyang Institute of Automation, Chinese Academy of Sciences, China

Peer Review Statement

All papers published in this volume have been reviewed through processes administered by the Editors. Reviews were conducted by expert referees to the professional and scientific standards expected of a proceedings journal published by IOP Publishing Publishing.

- **Type of peer review:** Single Anonymous
- **Conference submission management system:** Morressier
- **Number of submissions received:** 97
- **Number of submissions sent for review:** 89
- **Number of submissions accepted:** 55
- **Acceptance Rate (Submissions Accepted / Submissions Received × 100):** 56.7
- **Average number of reviews per paper:** 2
- **Total number of reviewers involved:** 20
- **Contact person for queries:**
  Name: Xuexia Ye
  Email: xx.ye@keoaeic.org
  Affiliation: AEIC Academic Exchange Information Centre

Design and calibration of the DC electric-field strength measuring instrument by horizontal vibration

Fengchun Liu[1,a], Shan Tu[1], Yifeng Zhang[1], Yuhuan Liu[2]

[1]State Grid Shanxi Electric Power Company Electric Power Research Institute, Taiyuan, China

[2]State Grid Electric Power Research Institute Co. LTD, Nanjing, China

[a]842341737@qq.com

Abstract—The synthesis electric field is an important indicator to assess electromagnetic environment of DC transmission line ,and field mill is often used in practical engineering.In order to solve the problem that the shielding plate in the field mill have unreliable grounding due to rotation and the electric field is distorted by accumulated charge, the structure of static shielding plate and moving induction plate are designed in this paper. On this basis, a complete synthetic electric field measuring device is developed.The instrument is calibrated under laboratory conditions, and the results show that calibration curve has good linear characteristic.

1. Introduction

During operation of uHVDC(ultra High Voltage Direct Current) transmission lines, corona may occur, generating electric charges near the wires. Some of the charge is applied to the electric field, moving towards the ground, and this charge changes the nominal electric field (the electric field generated by a live wire)[1-3]. The electric field generated by the space charge is superimposed on the nominal electric field to form a synthetic electric field. In this regard, the electric field limit under the HVDC transmission line is given in China. In theory, the synthetic electric field intensity can be determined by calculation and empirical prediction. However, due to the complexity of the actual situation, many factors need to be considered. In practical engineering, measurement instruments are often used for direct measurement[4-6].

According to the principle of measuring sensors, the existing electric field measuring devices can be divided into two categories: electrical measurement and optical measurement. Optical measuring device has high sensitivity, but strict requirements on the use environment, easy to be affected by temperature, and can not meet the needs of long time measurement, so in practical engineering, field mill measuring device based on electrical measuring principle has been widely used. According to the reaction of the front-line personnel, sometimes there is still a large error in the actual use of the field grinding device. The analysis shows that the ground of the rotating fan of the field grinding device is not reliable, and the space charge accumulated by the rotating fan cannot be released in time during operation, leading to the error of the results[7-8].

In order to improve the structure of the field mill, the shielding electrode was changed from moving plate to static one to avoid the influence of unreliable grounding. In this paper, a rotation-vibration motion conversion structure is designed to drive the induction pole to vibrate horizontally when the motor rotates. After a complete measuring device is developed, it is calibrated under laboratory conditions.

2. Principle of Field Mill

Without consideration of complex factors such as weather, synthetic electric field for the time-varying field, in order to obtain the electric field signal, time-varying capacitance in the synthesis of measuring electric field construction, the external resistance on both ends of the capacitor, forming a closed loop, the loop will generate an electric current, then can obtain the voltage, resistance to voltage signal modulation, receive for chip processing of signal, Finally, the electric field intensity can be calculated[9-10].

Schematic diagram of differential field mill is shown in Figure 1.

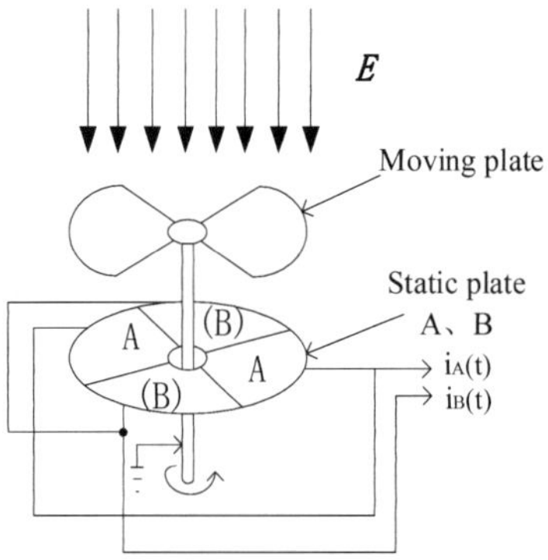

Figure 1 Schematic diagram of differential field mill

The device forms a capacitor between the moving plate and the fixed plate, and the area of the capacitor is changing, and the charge Q accumulated at both ends of the capacitor is

$$Q = \varepsilon E S(t) \tag{1}$$

Where

ε, the dielectric constant of the gas between the moving and stationary plates

E, the synthetic electric field

S(t), the area exposed to the air by the stationary plate, which changes with time

Further, the induced current i(t) on each fixed plate is

$$i(t) = \frac{dQ}{dt} = \varepsilon E \frac{dS(t)}{dt} \tag{2}$$

In the differential field mill measuring device, the interference of the DC component of the circuit signal can be eliminated by the difference between the positive induction current $i_A(t)$ and the negative induction current $i_B(t)$. The DC component is the current formed by the ions in the air entering the induction pole.

3. Horizontal vibration measuring device

3.1. The Difference With The Field Mill

The principle of the horizontal vibration measuring device is the same as that of the field grinding measuring device, but considering that the shielding electrode of the field grinding device has unreliable grounding, the shielding electrode is designed as a fixed metal plate, in which empty gaps are left to pass the power supply load. In order to induce the charge on the induction plate and make the induction plate vibrate horizontally, the schematic diagram is shown in Figure 2.

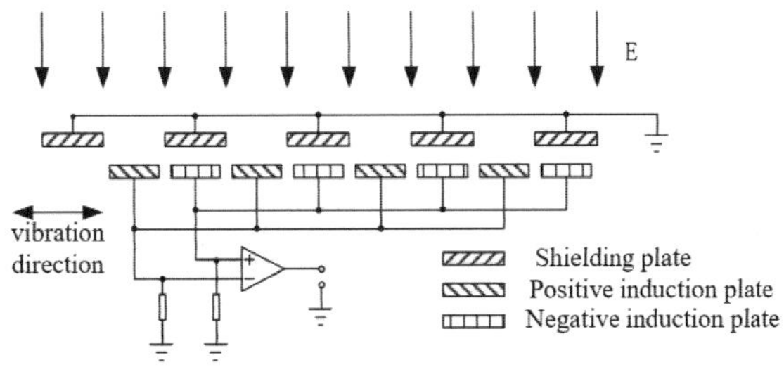

Figure 2 Schematic diagram of horizontal vibration sensor

3.2. The Realization of Vibration

Theoretically speaking, the motion mode is linear motor -- voice coil motor, more suitable for the design of the structure, and voice coil motor has the characteristics of high acceleration, fast response. However, the voice coil motor is expensive, and the rotary motor can also drive the induction plate to vibrate after adjustment. Therefore, this paper selects the rotary motor to drive from the perspective of economy, and designs a motion conversion structure that turns rotation into vibration as shown in FIG. 3.

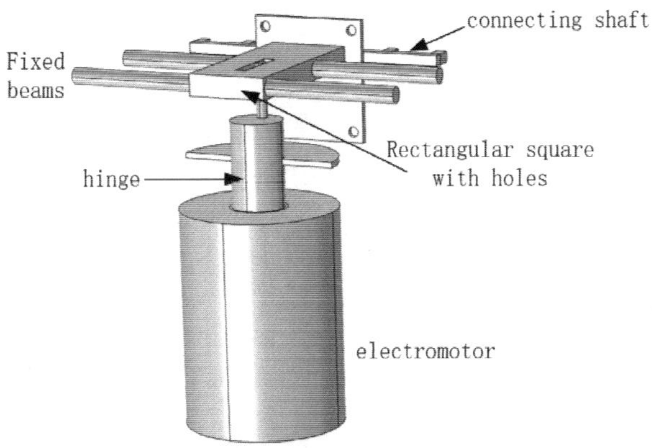

Figure 3 Schematic diagram of motion transformation structure

There is a rod on the upper end of the motor shaft, extending into the rectangular box. Rectangular blocks can move about on the fixed beam, and the contact surface is smooth. In addition, the rectangular block is connected to a shaft connecting the sensor pole. In the measuring device, the motor is placed horizontally in a line parallel to the rotating shaft, and a schematic diagram looking toward the bottom of the rotating shaft is shown in FIG. 4.

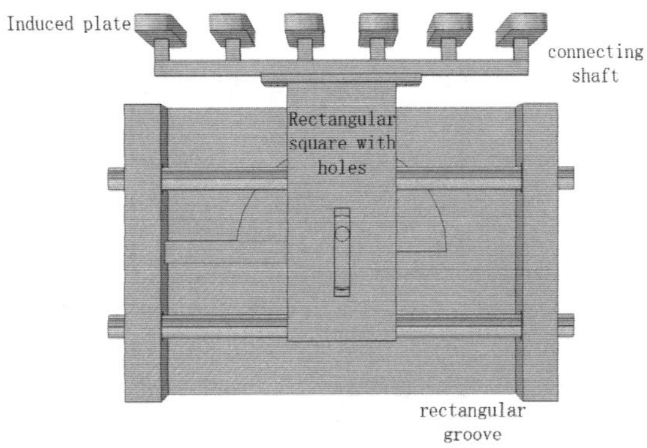

Figure 4 schematic diagram looking toward the bottom of the rotating shaft

In FIG. 4, the motion conversion structure is fixed in the groove and placed horizontally. When the motor rotates, the rotating shaft drives the rectangular block with holes to move left and right on the fixed beam, and the induction electrode moves horizontally together with the connecting shaft.

3.3. Shape design of shielding metal hollow slot

In order to analyze the current waveform of the induction electrode, the motion process of the sensor is analyzed first. The profile structure of the sensor is shown in Figure 5.

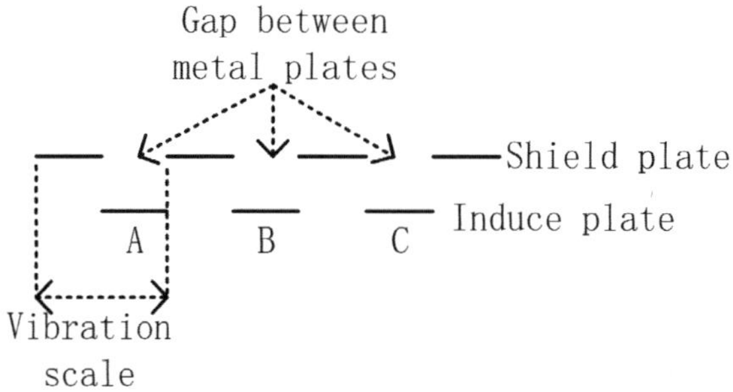

Figure 5 Schematic diagram of vibration sensor profile

As shown in figure 5, the shielding metal(shield plate) is fixed, and the space is left in the metal sheet, and the induction electrode is controlled by the motor in the vibration range, so the effective area of the charge will depend on the part of the induction electrode exposed to the gap. In order to constrain the induction current, the current waveform is positively string, the derivative of the area is sinusoidal, and it needs to be optimized for the boundary of the metal gap or the induction electrode boundary. This section is designed for the boundary of the gap, and the induction electrode can fully receive the particles that cross the gap at the bottom of the gap, in which case the induction current waveform will depend only on the boundary of the gap.

When the time t is 0, the induction electrode is located on the left side of the vibration range, and the electrode is completely blocked and the induction current is 0. The space gap center is the origin and the coordinate system is established. Based on the sensor as the vibration type, the selection coordinate system is suitable for cartesian coordinate system when determining the boundary equation of the gap structure. Considering the characteristics of the X-axis of the X-axis, the equation of the above part of the X-axis is

$$y = y(x) \tag{3}$$

The motion process is shown in FIG. 6.

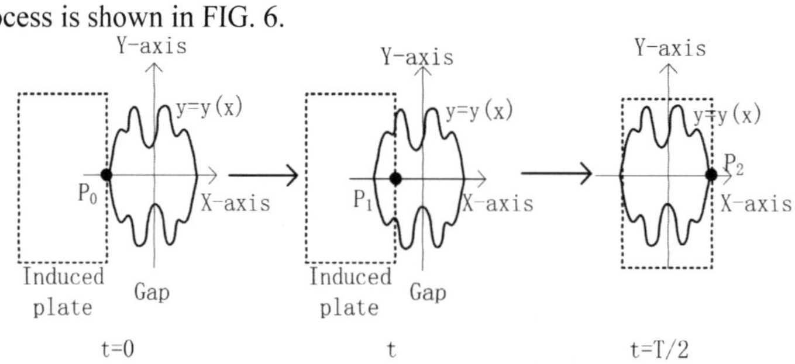

Figure 6 Schematic diagram of relative motion process in half period

As shown in Figure 6, when t=0, the induction pole is completely shielded and the induction current is 0. At this time, the intersection point P of the induction pole and X-axis is at P0. As the electrode moves, P moves to P1 and P2 moves to T = t /2. As the electrode movement is controlled by the motor, when the motor rotation frequency is F, the abscissa of point P changes with time as

$$x = -R\cos 2\pi f t \tag{4}$$

Where R is half of the gap width.
So there is

$$\frac{dx}{dt} = 2\pi f R \sin 2\pi f t \tag{5}$$

The purpose of the constraint on y is to satisfy the area of the time t

$$\frac{dS(t)}{dt} = C\sin 2\pi f t \tag{6}$$

Where C is any non-zero constant, which can be obtained by combining with Equation (5)

$$\begin{aligned}\frac{dS}{dx} &= \frac{dS}{dt}\cdot\frac{dt}{dx} \\ &= C\sin(2\pi f t)\frac{1}{2\pi f R\sin 2\pi f t} \\ &= \frac{C}{2\pi f R}\end{aligned} \tag{7}$$

In half a cycle, the expression of the area exposed to the gap by the induction motor when the sensor vibrates is

$$S(x) = 2\int_{-R}^{x} y(x)dx \tag{8}$$

Equation (9) can be obtained from Equation (8)

$$\frac{dS}{dx} = 2y(x) \tag{9}$$

By comparing equation (7) with Equation (9), it can be obtained

$$y = \frac{C}{4\pi f R} = C_0 \tag{10}$$

According to equation (10), when the void is rectangular, its length is arbitrary constant and its width is 2R, the induced current waveform is sinusoidal. The physical picture of the shell is shown in Figure 7.

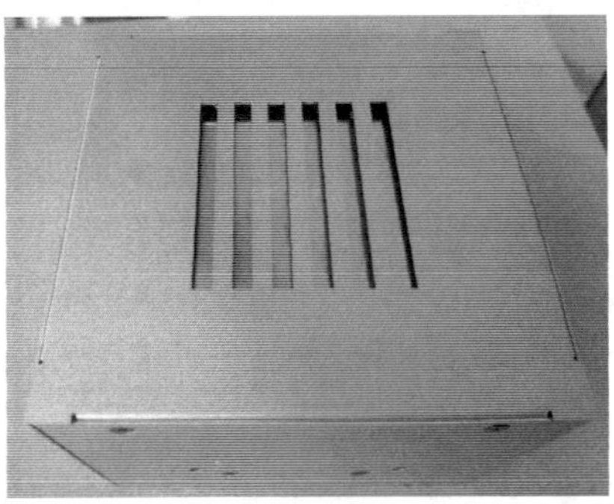

Figure 7 Practicality picture of the shell

4. PROTOTYPE CALIBRATION

The calibration curves of the synthetic field measuring device are a family of straight lines passing through the origin, and different straight lines correspond to the measurement curves under different ion current densities. In this paper, the curve with ion current density of 0 was firstly calibrated. The schematic diagram of the calibration device[11] is shown in FIG. 8. Where V is the voltage applied by the plates, d is the distance between the plates, and the electric field intensity within the plates is E=V/d. When the distance between the plates is less than the radius of the plates, the electric field in the middle region between the two plates can be regarded as a uniform field.

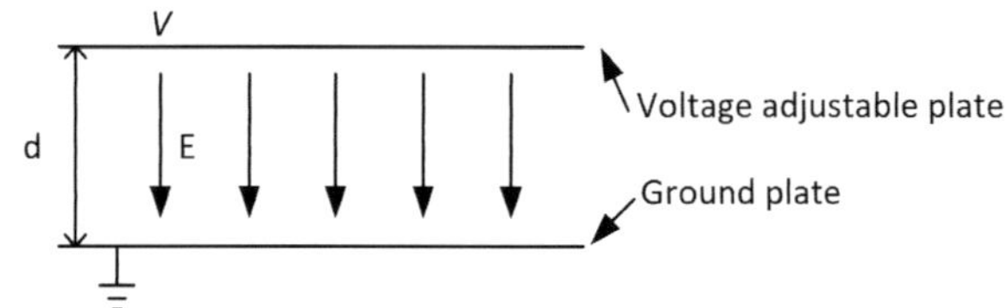

Figure 8 Schematic diagram of the calibration device

There is a groove in the center of the grounding plate. During calibration, the synthetic field measuring device is placed in the groove. By applying different voltages to the pressurized plate, the comparison curve between the data of the measuring device and the actual field intensity is obtained, and the measurement results are shown in Fig.9.

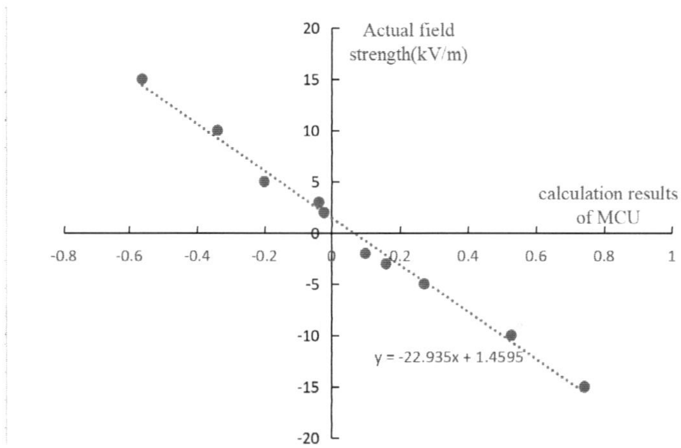

Figure 9 Calibration result curve

On the measured surface, the linear characteristics of vibration sensor are good.

5. ELECTRIC FIELD MEASUREMENT

Field measurement was carried out on Mincun-Dike road of ±1100kV Jiquan line. The lines were arranged horizontally, the number of splits was 8, pole spacing was 24m, wire height to ground was 23m, air relative humidity was 33% ~ 58%, temperature was 5.5℃ ~ 13.8℃, wind speed was 0 ~ 0.4m/s, the origin of measurement was 50m outside the dc positive pole line. Make a measurement in the direction of the negative transmission line. The equipment used for measurement includes the original field mill device and the vibration measuring device developed in this paper. 45 sets of the two devices are placed under the line according to the law of remote thinning and close density. 100 sets of data are measured for each device during measurement. According to the unified national standard GB 39220-2020, the measurement data of dc electric field is analyzed using E_n, which means percentile level of Total Electric Field(PLTEF). The continuous measurement data of the synthetic electric field at the monitoring point (sampling values at equal time intervals) are sorted according to the absolute value from small to large. The nth % data is called the cumulative percentage synthetic electric field value En, which means that the absolute value of n% measurement data is less than or equal to E_n within the measurement time. For example, E_{95} and E_{80} indicate that the absolute value of measurement data is less than or equal to E_{95} and E_{80} in 95% and 80% of the measurement time, respectively. In this paper, E_{80} and E_{95} of the two devices were analyzed, and the error curve was drawn as shown in FIG. 10

$$\eta = \frac{E_{95} - E_{80}}{E_{95}} \times 100\% \tag{11}$$

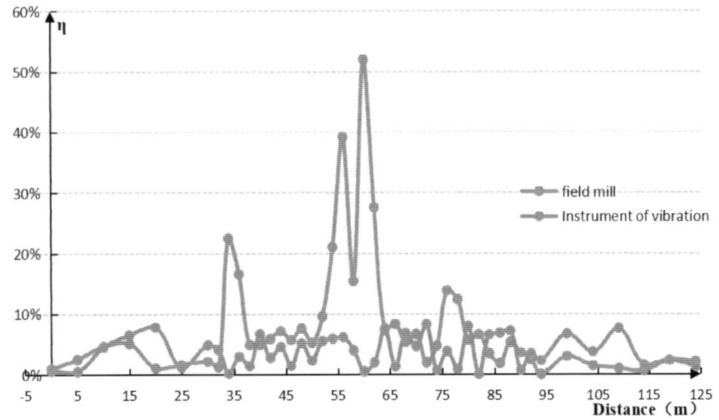

Figure 10 The distribution of η

It can be seen from FIG. 10 that electric field distortion exists in the field grinding measurement equipment during measurement, and the relative error of E80 and E95 exceeds 10% in 10 of the 45 equipment, while the vibration measurement device can control the error of both relatively well. In theory, during the measurement time, the field strength does not change with time, and the error between E80 and E95 should be close to 0. In fact, the slight difference between E80 and E95 may be caused by the influence of wind, or the load change changes the transmission line current, which slightly affects the distribution of ion flow.

6. Conclusion

(a) In view of the unreliable grounding of the shielding pole of the field grinding measuring device, a structure of fixed shielding pole and vibration instead of rotation is proposed in this paper, and a synthetic electric field measuring device is successively developed. The device was calibrated in laboratory environment and the results show that the device has good linearity.

(b) The relative error between the existing field grinding device and vibration measurement E95 and E80 is analyzed through the actual measurement under the DC transmission line. It can be seen that the vibration measurement device avoids the electric field mutation caused by the unreliable grounding of the shield pole and limits the error within 10%.

Acknowledgment

Thanks to the sponsorship of State Grid Corporation, as well as the great support of colleagues.

Thanks to sponsor: State Grid Corporation of China headquarters management science and technology project support: Research on synthetic electric field measurement device and application based on vibration sensing, 5500-202115370a-0-0-00.

Management Technology Project of The Headquarters of State Grid Co., LTD.Research on synthetic electric field measuring Device based on vibration sensor and its application.5500-202115370A-0-0-00

References

[1] Q. Tashi,; Song Yue; Bujie Duo; Lei Gao; Deji Tenzin; Liugang Li,et al., "The Measurement and Analysis of Electromagnetic Environment of DC Transmission Line in High Altitude

Areas," 2021 3rd Asia Energy and Electrical Engineering Symposium (AEEES), 2021, pp. 147-150, doi: 10.1109/AEEES51875.2021.9403109.
[2] Zhang Guangzhou. Calculation and Measurement of Hybrid Electric Field from AC/DC Lines[D]. Wuhuan.University, 2012.
[3] Yang F, Liu Z, Luo H et al. Calculation of Ionized Field of HVDC Transmission Lines by the Meshless Method[J]. IEEE Transactions on Magnetics, 2014, 50(7):1-6.
[4] Zhen Y, Cui X, Lu Tet al. High Efficiency FEM Calculation of the Ionized Field Under HVDC Transmission Lines [J]. IEEE Transactions on Magnetics, 2012, 48(48):743-746.
[5] Zhang Guangzhou. Research on Electromagnetic Environment of HVDC Transmission Engineering [D]. Huazhong University of Science and Technology, 2006.
[6] Qin Ho - ho. Research on Numerical Calculation method of Ground Synthetic Electric Field for UHVDC Transmission Line [D]. Chongqing University, 2014.
[7] ZHANG Hongchuan, YUAN Haiwen, LU Jiayu et al. Electric Power System Technology, 2012, 36(12):182-188.
[8] Yang Yong, LU Jiayu, LEI Yinzhao. Calculation of Ground Synthetic Electric Field for Dual-circuit DC line in the same corridor [J]. Power grid technology, 2009,33 (19) : 181-185.
[9] Yang Fan, DAI Feng, LUO Hanwu, et al. Distribution characteristics and Influencing Factors of ion Flow field in HVDC Transmission Line under Fog and Haze [J]. Transactions of China Electrotechnical Society, 2016, 31(12):49-57.
[10] Zhang Yemao, Lu Yao, Wang Yanzhao, et al. Design of synchronous measurement system for DC Synthetic Field and Ion Current Density Based on Wireless Communication [J]. High Voltage Technology, 2019, 45(01):336-342.
[11] Yang F, Liu Z, Luo H et al. Calculation of Ionized Field of HVDC Transmission Lines by the Meshless Method[J]. IEEE Transactions on Magnetics, 2014, 50(7):1-6.
[12] Zhen Y, Cui X, Lu Tet al. High Efficiency FEM Calculation of the Ionized Field Under HVDC Transmission Lines [J]. IEEE Transactions on Magnetics, 2012, 48(48):743-746.
[13] Research on Electric Field Measurement Technology for AC and DC Power Transmission Engineering [R], State Grid Electric Power Research Institute, 2009.
[14] Design and application of DC synthetic field Measurement Device [R], State Grid Electric Power Research Institute, 2010.

Design of Testing Device and Precision Assessment for Temperature Control Blanket in Medical Use

Yang Xu[a], Tingting Ren[*a], Wei Gong[a], Han Luo[a], Tao Zhang[a]

[a] Chongqing Academy of Metrology and Quality Inspection, Chongqing, China

[*]Corresponding author: 755245093@qq.com

ABSTRACT: The temperature control blanket in medical use is a device that can help regulate the temperature of the human body by controlling the temperature of the circulating liquid in the blanket and physically warming or cooling the human body. This article presents a novel design of precision assessment method of temperature control blanket by studying the essential metrological parameters of the device and designing a testing device for load simulation. The experimental results show that the measurement method and the design of the simulated load presented in this article can establish the metrological traceability system of temperature control blanket in medical use in China in order to ensure the accuracy and reliability of the device, which is of great importance for the quality control of the device.

1. Introduction

The temperature control blanket in medical use (hereinafter referred to as blanket) is a device that can help regulate the temperature of the human body by controlling the temperature of the circulating liquid in the blanket and physically warming or cooling the human body. The blanket can be divided into cooling type, heating type, and cooling & heating type according to its application[1]. The structure of the blanket mainly comprises refrigeration or heating system, temperature control system, water circulation control system, connection pipe, water blanket, and skin temperature sensor.

The working principle of the blanket is to use a compressor (or semiconductor) or a heater to provide a cold or a heat source to cool or heat the water in the water tank. The temperature control system controls the water temperature, and the water of required temperature is output to the water blanket by the water circulation system. The water blanket regulates the patient's body temperature by the heat exchange with the patient's body.

The cooling type blanket is mainly used for the hypothermia therapy, which is a physical method to reduce the patient's body temperature to the expected level to achieve the specific disease treating purpose. Hypothermia therapy has been widely used in cardiac surgeries and neurosurgeries[3]. It has been found that hypothermia therapy can regulate cerebral blood flow, reduce cerebral oxygen metabolic rate, improve cellular energy metabolism, reduce the release of excitatory amino acids, reduce the production of oxygen free radicals, reduce intracellular calcium overload, increase the synthesis of neuronal ubiquitin, reduce neuronal necrosis, promote the recovery of intercellular signal transduction, reduce the area of cerebral infarction, and reduce cerebral edema and intracranial pressure[2].

The heating type is mainly used for heating and heat preservation of patients in perioperative period. Perioperative hypothermia can lead to a variety of complications, such as slow metabolism of narcotic drugs (enhanced toxic and side effects of drugs), prolonged anesthesia recovery time, coagulation disorders, decreased immune function, abnormal cardiovascular function, postoperative bleeding, high wound infection rate, systemic inflammatory response syndrome, organ function disorder, and postoperative chills. The application of heating blanket can maintain the normal body temperature of patients during perioperative period, which can effectively alleviate the hypothermia syndrome during perioperative period.

However, until now no national metrological specifications applicable to temperature control blanket has been issued in China, and the corresponding traceability system of the device has not been established yet. In reference of YY 0952-2015 *Temperature Control Blanket for Medical use*[4] and JJF (Chongqing) 029-2018 *Calibration Specification for Temperature Control Blanket of Medical Use*[5], this article studies the influencing factors of the key technical parameters of the device such as "temperature indication error of circulating liquid ", "temperature indication error of skin temperature sensor", "heating rate" and "cooling rate"[4], and design a novel measurement system for metrological calibration of the device, in order to ensure its accuracy and reliability, which will ultimately improve medical treatment quality of China.

2. Measurement method

2.1 *Temperature indication error of circulating liquid*

Put the standard platinum resistance thermometer at the output port of circulating liquid to measure the temperature of circulating liquid. Set the temperature of the circulating liquid at the upper limit, middle point and lower limit of the temperature range of the blanket. When the temperature of circulating liquid is stable, record the indication temperature of the blanket t_y and the indication temperature of standard thermometer t_{sy}, and the temperature indication error of circulating liquid Δt_y is calculated as follows according to (1)[5].

$$\Delta t_y = t_y - t_{sy} \tag{1}$$

2.2 *Temperature indication error of skin temperature sensor*

The temperature indication error of skin temperature sensor is measured at the temperature of 35.0 ℃, 37.0 ℃, 39.0 ℃ and 41.0 ℃. Keep the temperature of the standard thermostatic water bath constant at the measuring point. Insert the standard platinum resistance thermometer and the skin temperature sensor vertically into the thermostatic bath at the same time. Record the indication temperature of the skin temperature sensor t_c and the indication temperature of standard thermometer t_{sc}, and the temperature indication error of skin temperature sensor Δt_c is calculated as follows according to (2)[5].

$$\Delta t_c = t_c - t_{sc} \tag{2}$$

2.3 *Average cooling rate under no-load condition*

Set the temperature of circulating liquid at 30 ℃, and start refrigeration function of the blanket after the temperature of circulating liquid is stable. Use a stopwatch to measure the time T during the temperature drop from 25℃ to 10℃, and the average cooling rate under no-load condition T_S is calculated according to (3)[5]:

$$T_S = \frac{25-10}{T} \tag{3}$$

2.4 *Average heating rate under no-load condition*

Set the temperature of circulating liquid at 25 ℃, and start heating function of the blanket after the temperature of circulating liquid is stable. Use a stopwatch to measure the time T during the temperature rise from 30℃ to 40℃, and the average heating rate under no-load condition T_S is calculated according to (4)[5]:

$$T_S = \frac{40-30}{T} \tag{4}$$

2.5 Average cooling rate of simulated load

The blanket is loaded with simulated load, and the water inside the simulated load is preheated to 36.5℃±0.5℃. See chapter 3 for the fabrication and application method of the simulated load. Set the temperature of circulating liquid of the blanket at the lower limit of the temperature range of the blanket, and start refrigeration function of the blanket. The measurement begins as soon as the temperature of the circulating water reaches the lower limit of the temperature range of the blanket, and the measurement time T_F should be no less than 4 hours. The average cooling rate of simulated load T_{SF} is calculated according to (5)[5]:

$$T_{SF} = \frac{\overline{t_{0F}} - \overline{t_{iF}}}{T_F} \tag{5}$$

Where $\overline{t_{0F}}$ is average temperature of five temperature measurement points in the water bath of the simulated load at the beginning of the measurement, while $\overline{t_{iF}}$ is average temperature of five temperature measurement points in the water bath of the simulated load in the end of the measurement.

2.6 Average heating rate of simulated load

The blanket is loaded with simulated load, and the water inside the simulated load is preheated to 30.0℃±0.5℃. See chapter 3 for the fabrication and application method of the simulated load. Set the temperature of circulating liquid of the blanket at the upper limit of the temperature range of the blanket, and start heating function of the blanket. The measurement begins as soon as the temperature of the circulating water upper the lower limit of the temperature range of the blanket, and the measurement time T_F should be no less than 4 hours. The average heating rate of simulated load T_{SF} is calculated according to (6)[5]:

$$T_{SF} = \frac{\overline{t_{iF}} - \overline{t_{0F}}}{T_F} \tag{6}$$

3. Fabrication and application method of simulated load

3.1 Fabrication of simulated load

The simulated load is composed of metal water tank, heating device, water circulating device and temperature measuring device. The dimension of metal water tank cavity is 610 mm (L) ×520 mm (W) ×200 mm (H). The thickness of the box is no more than 3 mm, the external surface of the box bottom is flat, and all metal surfaces except the bottom are covered with thermal insulation material, as demonstrated in Figure 1. The water temperature can maintain at 36.5 ℃ ± 1.0 ℃ for at least 4 hours. The water circulating velocity shall be more than 5 l/min. The schematic diagram of simulated load is demonstrated in Figure 2.

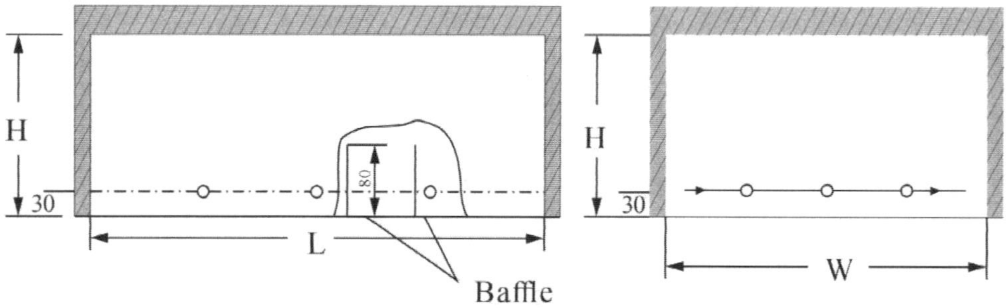

Figure 1. The dimension of metal water tank cavity

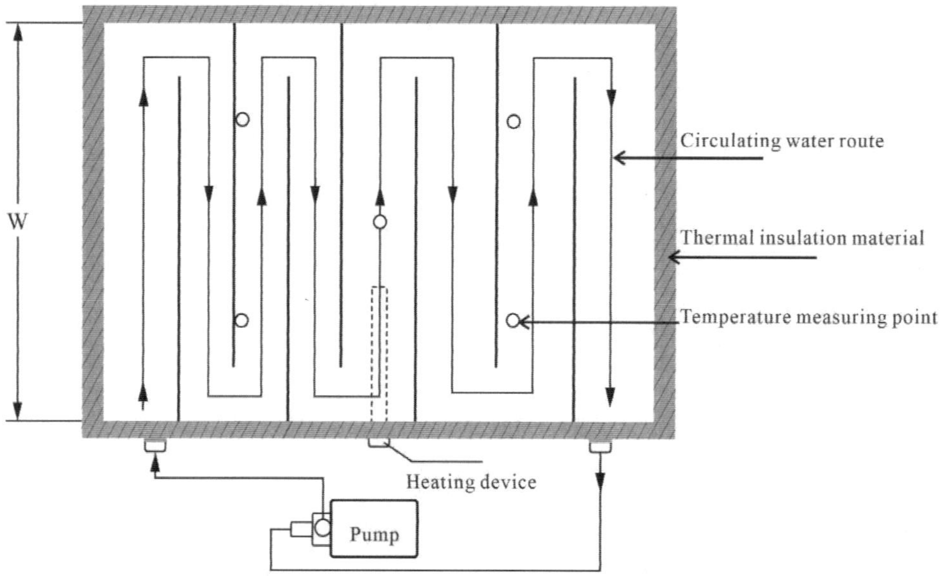

Figure 2. The schematic diagram of simulated load

3.2 Application method of simulated load

The blanket shall be placed on a pine board with an area of not less than the blanket surface and a thickness of not less than 20 mm when measuring the cooling or heating rate of simulated load. The simulated load shall be placed on the blanket and fully contacted with blanket. The mass of the water injected into the tank shall be 50 kg ± 1 kg. The temperature measuring point is 30 mm away from the bottom of the water tank, and 5 digital thermometers are used for simultaneous measurement. The location of 5 measurement points is shown in Figure 3.

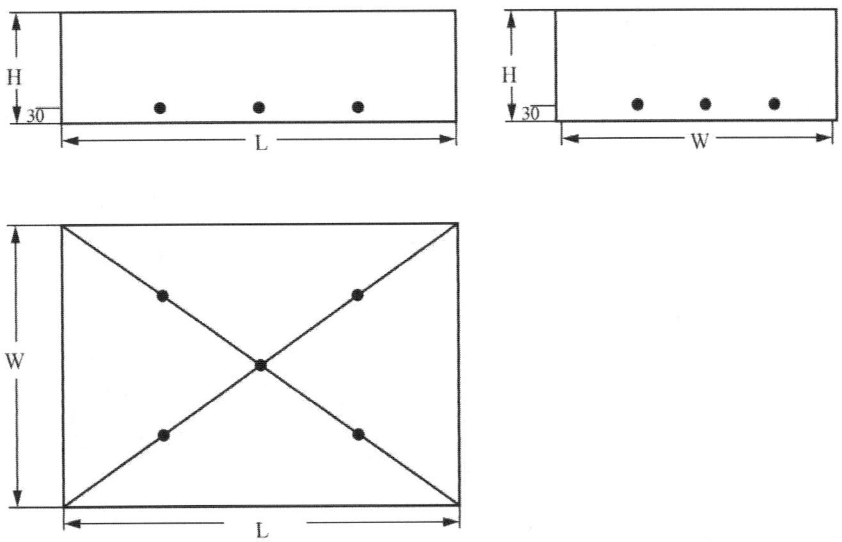

Figure 3. The location of 5 measurement point

4. Experimental results

Select 3 different types of temperature control blanket manufactured by different manufacturers as the measurement subjects for verification experiment. The measurement results are shown in the Table 1.

Table 1. The sizes and correction values of matched gauge blocks

Calibration item	Type		
	DWK- II	RC-2000 II	FY-1008
Temperature indication error of circulating liquid	1.2℃	1.3℃	-1.2
Temperature indication error of skin temperature sensor	-0.1℃	-0.1℃	0.1℃
Average cooling rate under no-load condition	1.0℃/min	1.1℃/min	1.2℃/min
Average heating rate under no-load condition	2.0℃/min	2.2℃/min	2.1℃/min
Average cooling rate of simulated load	1.5℃/h	1.7℃/h	1.6℃/h
Average heating rate of simulated load	2.2℃/h	2.1℃/h	2.2℃/h

In reference of YY 0952-2015 *Temperature Control Blanket for Medical use*[4], JJF (Chongqing) 029-2018 *Calibration Specification for Temperature Control Blanket of Medical Use*[5], and operation instructions of different types of blankets, the recommended metrological characteristic indexes of temperature control blanket in medical use are shown in Table 2.

Table 2. Recommended metrological characteristic index

Metrological characteristic parameter	Recommended index
Temperature indication error of circulating liquid	MPE:±1.5℃
Temperature indication error of skin temperature sensor	MPE:±0.2℃
Average cooling rate under no-load condition	≥1.0℃/min
Average heating rate under no-load condition	≥2.0℃/min
Average cooling rate of simulated load	≥1.2℃/h
Average heating rate of simulated load	≥1.5℃/h

It can be seen from the figures demonstrated in Table 1 that the measurement results can meet the technical requirement of the temperature control blanket in medical use. Therefore, the measurement method and the design of simulated load presented in this article can full satisfy the measurement need of temperature control blanket in medical use.

5. Conclusion

This article puts forward the measurement method for temperature control blanket in medical use which can comprehensively evaluates the metrological performance of the device, including temperature indication error of circulating liquid, temperature indication error of skin temperature sensor, average cooling rate under no-load condition, average heating rate under no-load condition, average cooling rate of simulated load, and average heating rate of simulated load, based on the application requirements of the blanket. The experimental results has proven that the measurement method and the design of the simulated load presented in this article are scientific and applicable, which provides a reference for the for the calibration of temperature control blanket in medical use.

References

[1] Ren, T. T., et al (2019) Study on temperature calibration method of temperature control blanket in medical use. J. Measurement Technique, 51-53.
[2] Zhang, X. D., et al. (2020) The inspection method of some items in YY 0952-2015 "Temperature Control Blanket for Medical Use" . J. Medical Equipment, 33(7): 37-38.
[3] Zheng, X. E., et al. (2010) Application of medical temperature control blanket in mild hypothermia therapy for patients with craniocerebral injury. J. Modern Nurse, 02: 81-82.
[4] YY 0952-2015 Temperature Control Blanket for Medical Use. China Quality and Standards Publishing & Media Co., Ltd., Beijing.
[5] JJF (Chongqing) 029, "Calibration Specification for Temperature Control Blanket of Medical Use," Chongqing Market Regulation Bureau, Chongqing, (2018).

Structural Strength Analysis and Stress Testing of Portal Crane based on Finite Element Method

Min Chen[1*], Shan He[1]

[1]Guangzhou Academy of Special Equipment Inspection &Testing

[*]502414769@qq.com

Abstract. In order to ensure the safety of the long-term service shipbuilding portal crane, this paper analyzes the stress distribution of the crane metal structure by using the finite element analysis method, which plays an auxiliary role in the crane safety assessment. The working condition at the most dangerous working position of the crane is selected as the stress test condition for analysis, and compared with the simulation calculation. The finite element calculation results show that the stress distribution of the whole machine is relatively reasonable. The overall stress is basically less than 100 MPa, and the maximum stress value is 131 MPa, which is less than the allowable stress value of the material. And through the comparison with the actual stress testing data, they are also basically consistent, and the structure of the crane can be used safely.

1. Introduction

The crane is used to carry out the rag, transportation, installation, or uninstallation of materials, which can greatly improve labor efficiency and promote production. It is widely used in various departments of the national economy [1]. Portal crane has the advantages of efficient operating capacity, low cost construction costs, strong versatility and universality. Due to the long-term work environment such as open -air and corrosion, its metal structure is prone to cracks, corrosion and other defects such as cracks such as cracks and corrosion. Due to the long -term work environment such as open -air and corrosiveness, the metal structure of the main loading parts is prone to defects such as cracks, corrosion, etc. Once the crane metal structure has defects, it will cause major safety hazards to the production of shipbuilding operations, and may cause major economic losses and casualties. For cranes that have been in service for more than 15 years, the State Administration of market supervision proposed that safety assessment should be carried out to ensure the safe operation of cranes [2-3]. Therefore, analyzing the strength of the metal structure of the crane and testing the stress at the key position of the metal structure can improve the safety of the operation and reduce the probability of potential safety hazards [4-5]. In this paper, the finite element analysis method is used to analyze the strength of a shipbuilding portal crane, and at the same time, it is compared with stress testing data of the safety evaluation to ensure the safe and reliable operation of the crane.

2. Overview of portal crane

A portal crane has been in service for more than 50 years in a shipyard, as shown in Figure 1. The rated lifting capacity of the crane is 12 tons, and the maximum working range is 32 meters. The specific technical parameters are shown in Table 1. The metal structure of the crane includes combined boom, A-frame, balance beam, turntable structure and gantry structure. The combined boom adopts the cable type combined boom, which is not common at present [6]. Due to the long working time of the crane, the metal structure has certain damage that needs to be considered, and the thickness of the material has some reduction. The actual measured data are used in the calculation.

Table 1. Technical Parameter

Parameter	Value
Rated capacity/t	12
Radius/m	9-32
Load-lifting height/m	32.09
Lifting speed/(m/min)	13
Derricking speed/(m/min)	20
Main structural materials	Q235

Figure 1. Portal crane

3. Establishment of finite element model

In order to obtain the stress distribution and strength analysis results of the crane, it is necessary to establish a finite element model.

3.1. Selection of cell grid element

According to the structural and mechanical characteristics of the crane, the main metal structure of the crane can be simplified into a spatial bar structure, and the finite element analysis software is used to model and calculate the crane. In the modeling process, three cell grid types element are used. The first element type is beam188. It is adopted for boom, elephant nose beam, A-frame, balance beam, rotary table and gantry of main structure. The second element type is link180. It is a three-dimensional rod element, which is used for the simulation of flexible cables and gantry links. The third element type is mass21. The counterweight, pulley block and other mechanisms can be regarded as centralized mass, which can be simulated with this element type.

3.2. Set material properties and boundary conditions

The metal structure material of the portal crane is Q235 carbon structural steel, with a density of $7.85*10^{-6}$kg/mm^3, a yield limit of 235mpa, an elastic modulus of $2.1*10^5$mpa and a Poisson's ratio of 0.3.

The bottom of the gantry structure of the crane is rigidly connected with the foundation, that is, all of the six degrees of freedom are applied to the four legs of the gantry. In addition, the connection between components should also be accurately simulated. The connections between the boom and the elephant nose beam, the turntable, the A-frame and the balance beam are hinged, and they are simulated by the node degrees of freedom at the coupling connection point in the finite element model. They can all rotate around the hinge axis, so the degree of freedom constraints in five directions other than this rotation are imposed. The finite element model of the crane is shown in Figure 2.

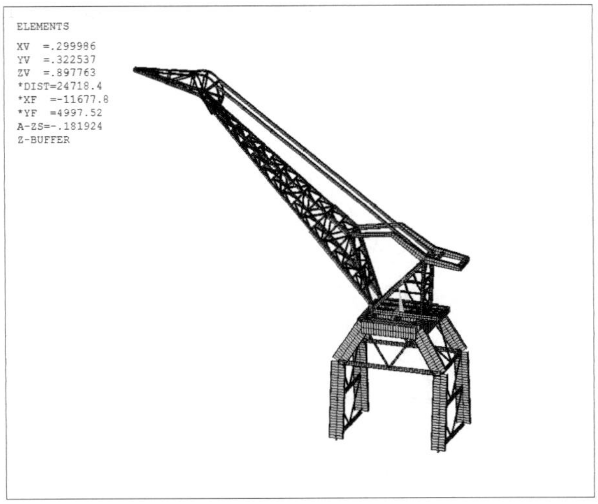

Figure 2. The finite element model of the crane

3.3. Develop load combinations

According to the specifications and the actual needs of the crane safety assessment, the most unfavorable operating conditions of the crane are selected for calculation and analysis. According to the stress characteristics, the most dangerous working condition of the portal crane is that the amplitude is located at the maximum working range of 32 meters, and the lifting load is 10 tons. Therefore, it is necessary to analyze the stress and strain distribution of the crane at this position. Since the crane is basically in the windless state during the safety assessment stress testing, the finite element analysis adopts the windless condition. Table 2 shows the calculated load combination of crane

Table 2. Load combination of the crane

Working condition	Amplitude	Lifting capacity	Load combination
Windless	32meter	10tons	I

4. Analysis of finite element calculation

Through finite element analysis and calculation, it is obtained that the maximum equivalent stress on the metal structure of the crane appears at the part where the balance beam is connected with the boom, which is 131.3mPa. The stress distribution of the crane and components are shown in Figure 3, figure 4 and figure 5. It can be seen from Figure 3 that the overall stress of the crane is not large, and the stress value is generally below 100mPa. The places with high stress appear in the stress concentration at the discontinuity of the structure, as well as on the nose beam and balance beam. Because the crane has been in service for a long time, and some stiffeners were not considered in the process of model simplification, the maximum stress of the finite element calculation result is larger than the actual stress.

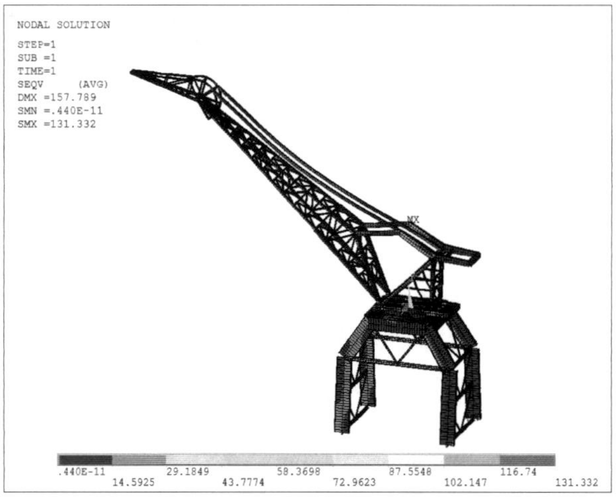

Figure 3. equivalent stress diagram of the crane

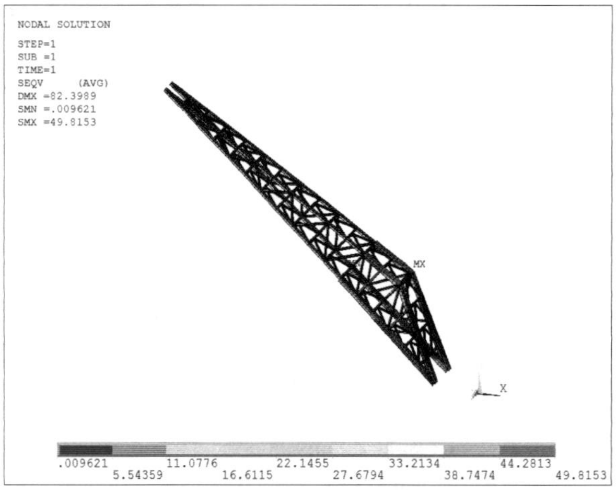

Figure 4. equivalent stress diagram of boom

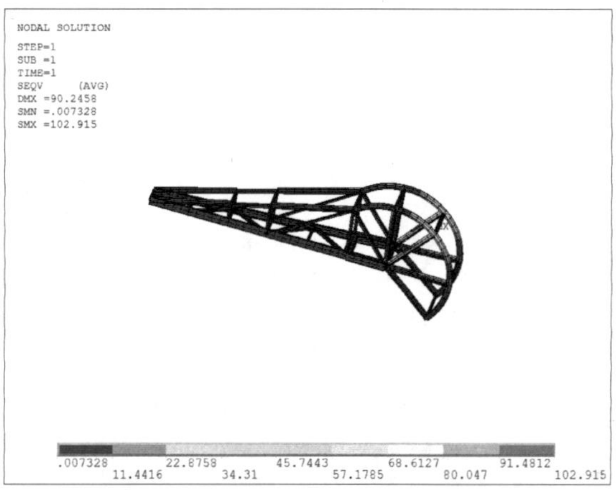

Figure 5. equivalent stress diagram of elephant-nose bridge

5. Stress testing of metal structure

According to the results of finite element calculation, the location section with large equivalent stress is selected as the location of the stress testing point. Then, according to the evaluation needs, set the test conditions and carry out the field stress testing on the crane [7]. The testing results do not include self-weight stress. The test results of load cases consistent with the finite element calculation are shown in Table 3.

Table 3. Stress testing condition

Serial number	Working condition
1	Amplitude:9 meters, lifting capacity:10 tons, lifting brake is applied
2	Luffing to 32 meters, descent braking
3	Rotate 45 ° to the North
4	Rotate 90 ° to the West
5	Rotate 45 ° to the North
6	Boom luffing to 9meters
7	Uninstall

According to the testing results, the maximum stress value of the crane appears on the I-beam at the tension side of the A-frame, which is 86.1mpa. Except for the A-frame stress value, the stress values of other testing points are relatively small. Since the measured stress value on site does not consider the self-weight stress of the crane metal structure, the actual testing result is lower than the finite element calculation value, but the overall stress distribution trend of the metal structure is consistent. The testing results under this working condition are shown in Table 4.

Table 4. Testing results of each measuring point

Measuring position	Maximum stress value /mPa
top of boom	17.4
lifting point of boom	21.8
I-beam at compression side of A-frame	65.7
I-beam at Tension side of A-frame	86.1
Turntable cross beam	11.2
Gantry I-beam	11.5
I-beam at the bottom of gantry	9.9

6. Concluding remarks

Through the finite element analysis of the metal structure of the shipbuilding portal crane and the comparison with the stress testing results, it can be seen that the distribution trend of the two stress values is relatively consistent. Through finite element calculation, the maximum equivalent stress of the crane appears at the position where the boom lifting point is connected with the balance beam, and the maximum value is 131.3mpa. The stress distribution of the crane and the A-frame is also reasonable, which is consistent with the trend of the 5. Stress testing results. All calculation results and test results are within the allowable stress of the material.

Acknowledgments
This work is supported by project Guangdong special equipment safety technology Collaborative Innovation Center (2021b1111600002), Guangdong market supervision and Administration Bu. (2021CT07).

References

[1] Chendi Huang. 2012 Current situation and development trend of portal crane. (Science and technology communication vol 6) chapter 1 p57-58.
[2] Weizhang Chen. 1988 Crane metal structure. Beijing: People's Communications Press. p46.
[3] GB/T 3811-2008. 2008 Crane design specification [s] Beijing: China Standards Press.
[4] Zhongcheng Han. 2019 Full automatic finite element analysis of portal crane structure. (Lifting and transportation machinery vol19) p81-84.
[5] Fang Su. 2020 Finite element simulation calculation of steel structure of portal crane. (Construction machinery vol7) p 50-54.
[6] Yu Zhang. 2012 Research on safety evaluation method of Portal Crane. (China water transportation vol6) p34-35.
[7] Fusen Li. 2020 Estimation and analysis of residual life of metal mechanism of portal crane based on stress test. (Lifting and transportation vol5) p 45–48.

Development and Application of Vibration Comfort Detection Device for Escalator Based on MEMS Acceleration Sensor

Jianxun Chen[1*], Zhengwu Qi[1], Ningxiang Yang[1], Weili Luo[2]

[1.] Zhuhai Branch, Guangdong Institute of Special Equipment Inspection and Research, Zhuhai, Guangdong Province 519000

[2.] Zhuhai Anyue Technology Co., Ltd., Zhuhai, Guangdong Province 519002

[*]Affiliation: Zhuhai Branch, Guangdong Institute of Special Equipment Inspection and Research, 133 Renmin West Road, Zhuhai, Guangdong Province 519002, P. R. China. Tel.: +86 0756 2661879;

[*]jianxun_chen@126.com (Dr. JX. Chen)

Abstract. Vibration comfort is an important indicator for evaluating escalator riding experience. The measuring principle of vibration comfort of escalator step and armbands was analyzed. The calculation methods of vibration acceleration extraction, frequency meter weight, root mean square (RMS) and vector sum were summarized. A wireless triaxial acceleration data acquisition module was developed based on micro-electromechanical system (MEMS) acceleration sensor, MCU and high-speed WIFI module, and a vibration comfort detection application (APP) based on the Android operating system was developed. The calibrated detection device was applied to escalator field test. The results showed that the maximum vibration RMS and average RMS of the step and the armband on both sides can be quickly calculated, and the vibration comfort level can be divided, which provided important data reference for equipment maintenance.

1. Introduction

Escalator is widely used in large shopping malls, rail transit stations, airports, tourist attractions, and other crowded places. Equipment failure or improper operation often causes serious casualties and adverse social impacts [1-4] in the accidents such as falling, trampling, and shearing. As a transportation tool closely related to people's production and life, the safety performance has been widely concerned. At the same time, as passengers focus more on the riding experience, more and more attention has been paid to the quality of escalator rides with vibration as the main performance. Especially for the elderly, children, and some patients, the abnormal vibration of the escalator will cause psychological fear and even physical reaction, which indirectly affects the safety of the escalator. In view of the current status of escalator ride comfort testing in China, the State Administration for Market Regulation and the National Standardization Administration have released GB/T 24474.2-2020 *Measurement of Passenger Quality - Part 2: Escalator and Moving Sidewalks*, which has been officially implemented since April 2021. The standard specifies the escalator or moving sidewalk carrier and handrail vibration measurement method [5].

At present, there are many theoretical studies on the condition monitoring and fault diagnosis of escalators based on vibration analysis [6-9]. However, the research and development of instruments that can measure and evaluate the vibration comfort of escalators according to the measurement standards are still relatively few, with a lack of special testing equipment compatible with escalator step vibration and handrail vibration comfort analysis.

The vibration state is mainly characterized by vibration acceleration. Compared with traditional acceleration sensors, MEMS acceleration sensors have the advantages of small size, low cost, high integration, and data accuracy, which have been widely used in natural disaster monitoring [10], inclination detection [11], displacement or velocity measurement [12, 13], medical health monitoring [14], industrial vibration monitoring [15] and other fields. In the field of vertical elevator inspection and detection, the sensor has been used in the elevator running characteristic detection [16], car vibration detection [17], elevator attitude evaluation [18], and brake performance safety detection [19]. In this study, an escalator vibration comfort detection device is developed based on a MEMS triaxial acceleration sensor, which can realize the vibration measurement and evaluation of escalator steps and handrails for online detection.

2. Detection principle

The functional area division of the escalator and the equipment coordinate system are shown in Figure 1. The moving part can be divided into upper horizontal section, lower horizontal section, and inclined section by the area. GB/T 24474.2-2020 specifies the coordinate system when measuring escalator transportation quality, in which the X-axis and the Y-axis are the horizontal directions, and the Z-axis is the vertical direction. The X-axis points to the horizontal motion direction of the cascade when the escalator is downward, the Z-axis is vertical to the cascade, and the three-axis direction conforms to the right-hand rule of the coordinate. The vibration comfort measurement is only for the process of the steps and the handrails on both sides moving to the inclined section of the equipment. The steps need to analyze the vibration acceleration in the three-axis direction of the X-axis, Y-axis, and Z-axis, and the handrail only needs to analyze the vibration acceleration in the X_h direction along the handrail.

When collecting vibration data, the escalator should be in a normal operation state, and the vibration source that may affect the measurement of vibration results outside the equipment should be shut down. During the step vibration test, the data coordinate system of the three-axis acceleration data acquisition module should correspond to the escalator step coordinate system, and in the handrail vibration test, the Y data axis of the three-axis acceleration data acquisition module should be consistent with the X_h direction. In addition, the test shall include at least triaxial acceleration data for the entire inclined segment.

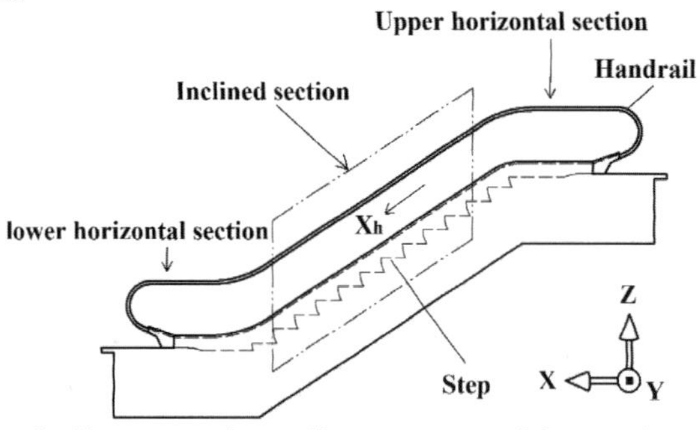

Figure 1. Schematic diagram and coordinate system of the escalator functional area

3. Data analysis method

3.1. Extraction of vibration acceleration

Vibration is the phenomenon in which the acceleration value changes alternately with time relative to the average acceleration without oscillation, and the premise of vibration analysis is to obtain the vibration acceleration in the direction of the specified coordinate axis.

On the one hand, due to the comprehensive influence of the welding position error of the MEMS chip on the circuit board, the assembly error of the circuit board in the data module, and the placement posture of the data module on the surface of the step or handrail, when collecting the step vibration acceleration, the three-axis directions of the MEMS acceleration sensor cannot completely correspond to the escalator coordinate system. To improve the accuracy of subsequent data analysis, it is necessary to correct the declination angle of the three-axis acceleration data through coordinate transformation, and obtain the acceleration data in the absolute horizontal and vertical directions:

$$\boldsymbol{A_1} = \boldsymbol{CA_0} \tag{1}$$

$\boldsymbol{A_0}$ refers to the coordinate vector of the original triaxial acceleration data obtained during the step vibration test or the handrail vibration test:

$$\boldsymbol{A_0} = \left[a_{x0}, a_{y0}, a_{z0}\right]^T \tag{2}$$

$\boldsymbol{A_1}$ is the coordinate vector of the three-axis acceleration data after declination correction:

$$\boldsymbol{A_1} = \left[a_{x1}, a_{y1}, a_{z1}\right]^T \tag{3}$$

$\boldsymbol{C}$ is the coordinate transformation matrix:

$$\boldsymbol{C} = \begin{bmatrix} 1 & 0 & -\frac{\sin\alpha}{\sin\gamma} \\ 0 & 1 & -\frac{\sin\beta}{\sin\gamma} \\ 0 & 0 & \frac{1}{\sin\gamma} \end{bmatrix} \tag{4}$$

In matrix $\boldsymbol{C}$, α, β and γ are the angle radians between the sensor X data axis, Y data axis, and Z data axis and the horizontal plane, respectively. During the step vibration test, α and β are approximately zero, and γ is approximately 0.5π. During the vibration test of the handrail, α is approximately zero, β is approximately the inclination angle of the escalator, and γ is approximately the supplementary angle of the inclination angle.

On the other hand, due to the influence of the earth's gravity, the obtained acceleration data already contains the DC component of the gravitational acceleration, which needs to be deducted to obtain the vibration acceleration component:

$$\boldsymbol{A} = \boldsymbol{A_1} - \boldsymbol{G} \tag{5}$$

In the formula, $\boldsymbol{A}$ refers to the coordinate vector of three-axis vibration acceleration when X and Y axes are in horizontal positions.

$$\boldsymbol{A} = \left[a_x, a_y, a_z\right]^T \tag{6}$$

$\boldsymbol{G}$ is the gravitational acceleration correction vector:

$$\boldsymbol{G} = [0, 0, g]^T \tag{7}$$

In the formula, g is the gravitational acceleration coefficient, which is approximately 9.8m/s^2.

Through the projection of a_y and a_z in the downward direction of the handrail in Formula (6), the vibration acceleration a_h of the handrail in the direction of the X_h coordinate axis can be obtained:

$$a_h = a_y \cos\beta - a_z \sin\beta \tag{8}$$

$\boldsymbol{A}$ is used to analyze the step triaxial vibration, and a_h is applied to analyzing the handrail vibration.

3.2. Frequency Weighting

Affected by the physiological structure of the human body, different vibration frequencies will lead to differences in the perception of vibration amplitude by passengers. To make the vibration test results more in line with the human body's feeling of ride comfort, it is necessary to perform frequency weighting on the vibration acceleration. According to GB/T 23716-2009 *Human Body Response to Vibration - Measuring Instruments*, frequency weighting can be decomposed into four processes: high-pass and low-pass second-order Butterworth filtering, a-v transform filtering, and high-order filtering. The total frequency weighting function is the product of the above four filter functions. The Z-transform transfer function of the digital filter in each filtering process is:

$$H(z) = \frac{\boldsymbol{aZ}}{\boldsymbol{bZ}} \tag{9}$$

Among them, $H(z)$ is the Z-transform output value, and a refers to the digital filter numerator vector:

$$\boldsymbol{a} = [1, a_2, a_3] \tag{10}$$

$\boldsymbol{b}$ is the denominator vector of the digital filter:

$$\boldsymbol{b} = [b_1, b_2, b_3] \tag{11}$$

The $\boldsymbol{Z}$ vector is:

$$\boldsymbol{Z} = [1, z^{-1}, z^{-2}]^T \tag{12}$$

Different frequency weighting methods are suitable for different vibration sensing parts, vibration directions, and human body postures. The vertical whole-body vibration in standing posture adopts frequency weighting W_d, the vertical whole-body vibration in standing posture adopts frequency weighting W_k, and the hand transmission vibration adopts frequency weighting W_h. Different filter vectors $\boldsymbol{a}$ and $\boldsymbol{b}$ are used in different filtering processes under different frequency weighting methods. The recursive calculation formula for digital filtering of the m-th acceleration data is:

$$y(m) = b_1x(m) + b_2x(m-1) + b_3x(m-2) - a_2y(m-1) - a_3y(m-2) \tag{13}$$

In the formula, x is the data sequence before filtering, y is the data sequence after filtering, and m is the number of the filtered data in the sequence.

3.3. Calculation of RMS value

The *RMS* value is an effective means to evaluate the strength of vibration energy. For step vibration, the *RMS* value of the three-axis vibration acceleration needs to be calculated separately. For the vibration of the handrail, it is only necessary to calculate the *RMS* value of the vibration acceleration of the handrail in the downward X_h direction. The time length of the calculation window is 1s. The calculation formula of the N-th *RMS* data is:

$$RMS = \sqrt{\frac{\sum_{i=1}^{n} a_{N+i-1}^2}{n}} \tag{14}$$

In the formula, n is the number of acceleration data within a time constant of 1s. For steps of 500 and handrails of 3000, a is the frequency-weighted acceleration value.

3.4. Calculation of the vibration vector sum

For step vibration, the RMS value vector also needs to be calculated to estimate the overall vibration energy:

$$a_{xyz} = \sqrt{a_x^2 + a_y^2 + a_z^2} \tag{15}$$

Where, a_{xyz} is the sum of vibration vectors, a_x is the *RMS* value of X-axis vibration, a_y is the *RMS* value of Y-axis vibration, and a_z is the *RMS* value of Z-axis vibration.

4. Development of detection device

4.1. Hardware Design

The MEMS sensor of the ADXL three-axis acceleration digital chip of Anolog Devices was adopted, and the main controller of the STM32F407 microcontroller from Italian and France Semiconductor Company was used. The hardware system is powered by DC 3.7V lithium-ion battery. To meet the vibration comfort measurement, the three-axis acceleration range was set to ±2g, and the data sampling rate was changed according to the measurement mode. The printed circuit board of the three-axis acceleration data acquisition module is shown in Figure 2. The developed data acquisition module has been calibrated and measured by a third-party organization before application. The error of the three-axis acceleration measurement is less than ±1.5%.

Figure 2. Circuit board of three-axis acceleration data acquisition module

4.2. Software Development of Lower Computer

The Kei MDK5.0 development tool was used for embedded development of single-chip microcomputers, which realizes MEMS chip parameter setting, acceleration data acquisition, data frame packaging, and data transmission. Besides, acceleration data is stored in front of the floating-point quad byte and the frame rate is 10 frames per second. The data transmission protocol is shown in Table 1. The frame header "AA" indicates that the lower computer sends data to the outside, and the RX address is the address of the intelligent terminal. The TX address is the address of the acceleration module, and the number of bytes is the number of bytes contained in the acceleration data segment. During the step vibration test, each frame contains 50 sets of X, Y, and Z-axis data (600 bytes). During the handrail vibration test, each frame contains 300 sets of X, Y, and Z-axis data (3600 bytes). The check bit is the low byte of the data sum in the data frame except the frame header.

Table 1. Data transmission protocol

Frame header	RX address	TX address	function code	number of bytes	data	check bit
0X AA	0X 01	0X 02	0X FE	0X 04 B0	Triaxial data	0X XX

4.3. Software development

The application program (APP) applied to the Android system was developed by the JAVA language, whose functional architecture is shown in Figure 3, including five functional units.

The unit of data acquisition and export obtain the three-axis vibration data of the escalator carrier or the handrail according to the sampling parameters, and the original data of the table file can be exported for analysis by other platforms. The unit of vibration comfort analysis is used to analyze the step vibration and handrail vibration comfort parameters. The unit of the result query and report can export the test report, and the data can be stored in the cloud for remote viewing, downloading, printing or verification. The unit of the acceleration module calibration is used to collect the vibration data of the standard vibration source in the third-party metrological calibration mechanism. The measurement error is calculated, and the latest triaxial acceleration correction coefficient is stored in the acquisition hardware to achieve calibration. The unit of module status monitoring is used to diagnose the running status of the mail, monitor the power information, and read the hardware ID number and other parameters after networking, which is convenient for the instrument to trace the test results.

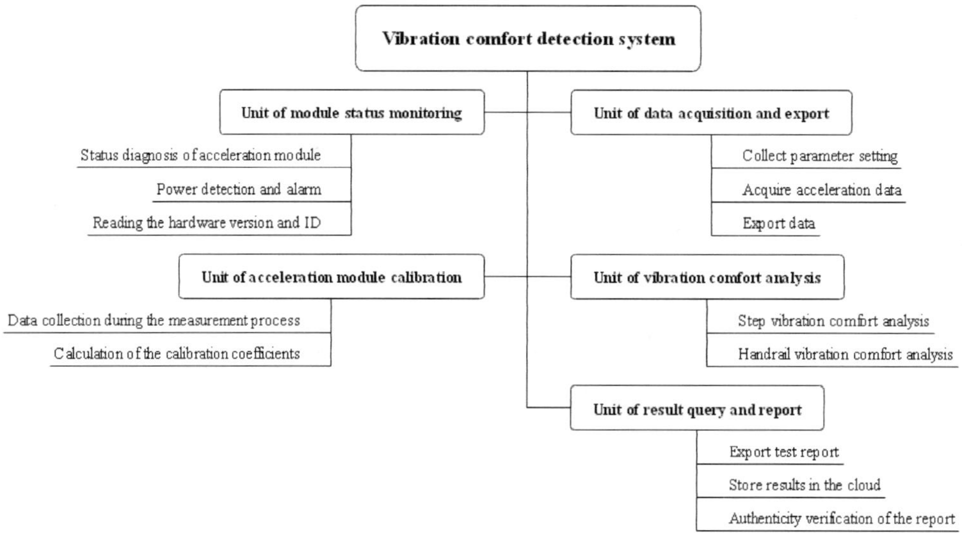

Figure 3. Functional architecture of the vibration comfort detection system APP

5. Field detection

5.1. Step vibration detection

As shown in Figure 4, taking the vibration comfort measurement of a descending escalator in a large shopping mall as an example, the step test process is as follows: 1) When the escalator is running normally, it is needed to operate the APP at the entrance and exit of the escalator to start data collection. At the same time, quickly place the three-axis acceleration data collection module in the center of a step; 2) The tester stands on the next step of the measured step and runs with the step; 3) The three-axis acceleration data acquisition module is removed when the measured step is approaching the comb-tooth plate of the upper entrance and exit, and the APP is operated to end the data acquisition and the elevator ride; 4) The device automatically analyzes the maximum RMS value and the average RMS value of the stepped three-axis vibration acceleration.

5.2. Handrail vibration detection

As shown in Figure 5, the vibration test process of the handrail when going down is as follows: 1) At the entrance and exit of the escalator, the three-axis acceleration data acquisition module is held and placed on the handrail to ensure that the module and the measured position of the handrail remain relatively static; 2) The tester goes down with the escalator and operates the APP to start data collection at the latest time before the escalator runs to the start of the inclined section; 3) When the handrail runs to the lower level of the escalator, the three-axis acceleration data acquisition module is removed to operate the APP to end the data acquisition and elevator ride; 4) The device automatically analyzes the maximum vibration RMS value and the average vibration RMS value in the X_h direction of the escalator.

Figure 4. Vibration test process of escalator steps

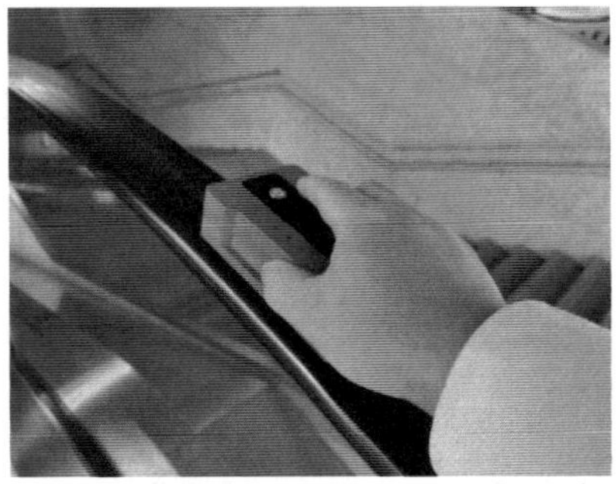

Figure 5. Vibration test process of escalator handrails

6. Results and analysis

6.1. Step vibration

The curve of the stepped three-axis original acceleration data after declination correction is shown in Figure 6. The X-axis and Y-axis acceleration data fluctuate up and down near the zero value, and the Z-axis data contains the DC component of the gravitational acceleration, which fluctuates up and down near the gravitational acceleration. When the escalator is running, the vibration amplitude of the steps in the left and right direction (Y-axis) and the up and down direction (Z-axis) is significantly higher than that in the front and rear direction (X-axis).

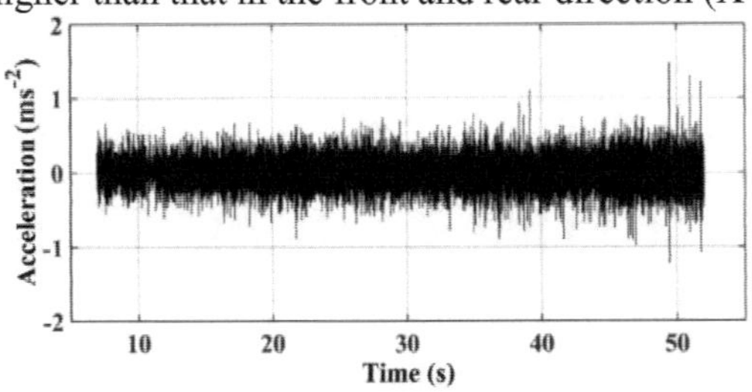

(a) X axis

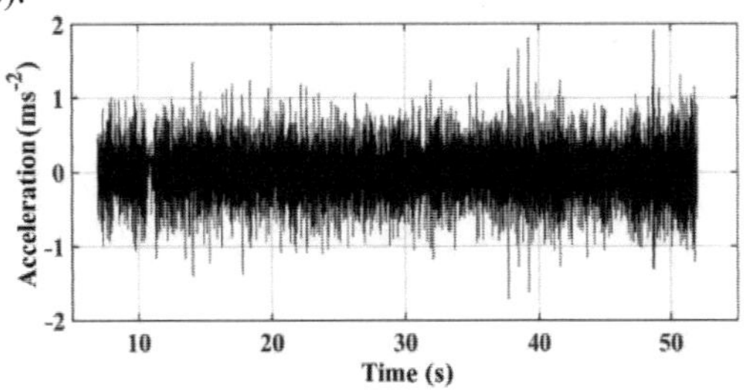

(b) Y-axis

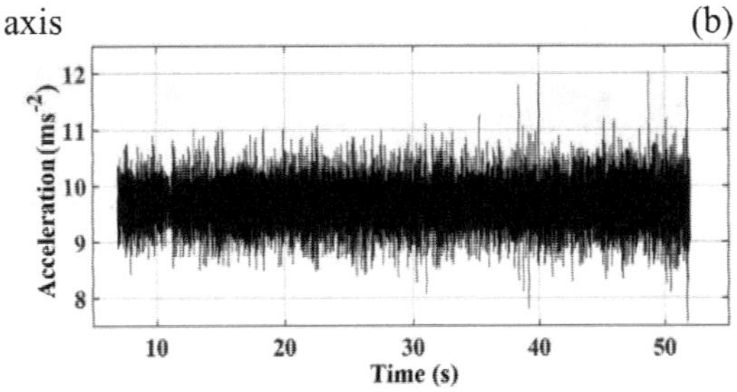

(c) Z-axis

Figure 6. Original data curve of the stepped three-axis acceleration

The data curve of the triaxial vibration acceleration after frequency weighting and RMS calculating is shown in Figure 7. It can be seen that the acceleration vibration intensity of the Z-axis is the largest, and its RMS value is twice greater than that of the X-axis or the Y-axis. In addition, when riding the ladder, the step vibration mainly comes from the up and down vibration, and the vibration intensity is the weakest before and after the step.

The RMS vector sum curve of stepped triaxial vibration acceleration is shown in Figure 8. The maximum value of the vector sum is 0.137 m / s^2, which occurs before the end of the step down to the inclined section, and the average value is 0.100 m / s^2. Appendix C.2.3 of GB/T 13441.1-2007 *Mechanical Vibration and Shock - Evaluation of Human Body Exposure to Whole Body Vibration Part*

1: General Requirements proposes the comfort response to the vibration environment [20]. According to the standard, when the total value of comprehensive vibration in public transport is less than 0.315 m / s^2, the human body cannot feel uncomfortable. Therefore, the vibration comfort of the measured escalator steps is "good", and the mechanical transmission parts related to step operation do not have obvious wear or failure.

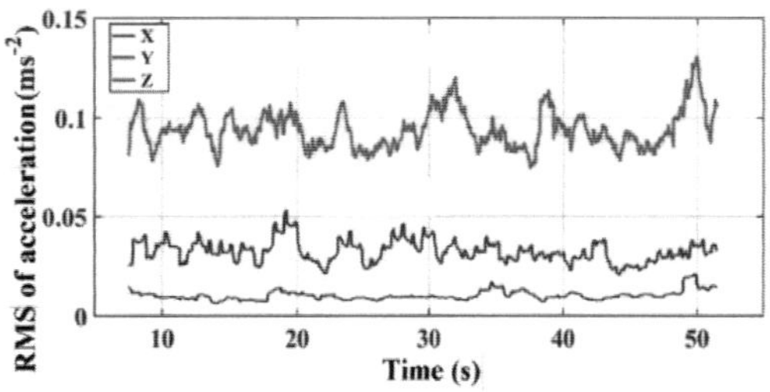

Figure 7. RMS curve of stepped three-axis vibration acceleration

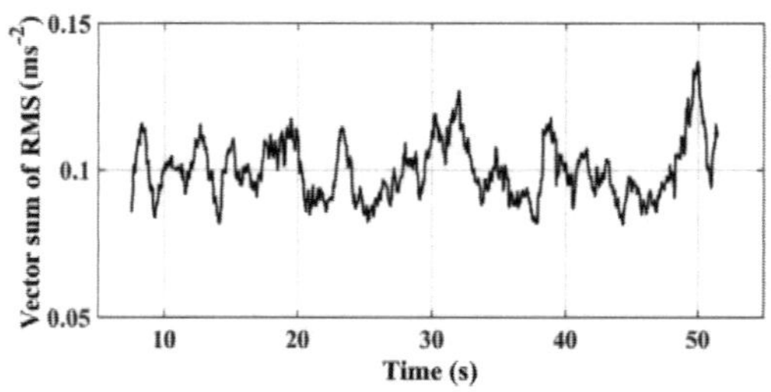

Figure 8. Vector sum curve of stepped three-axis vibration acceleration

6.2. Handrail vibration
The original acceleration curve of the handrail along the X_h coordinate axis of the escalator is shown in Figure 9. When running to the upper horizontal section and the lower horizontal section, the acceleration fluctuates up and down near the zero value. During the operation of the inclined section for nearly 50 seconds, the acceleration fluctuates up and down near the projected value of the gravitational acceleration along the X_h axis.

The vibration acceleration of the left and right handrails at the inclined section was extracted to calculate the RMS value. The curve is shown in Figure 10. The maximum RMS value of vibration on the left handrail is 0.441 m/s^2, and the average RMS value is 0.342 m/s^2. The maximum RMS value of the vibration on the right handrail is 0.447 m/s^2, and the average RMS value is 0.315 m/s^2.

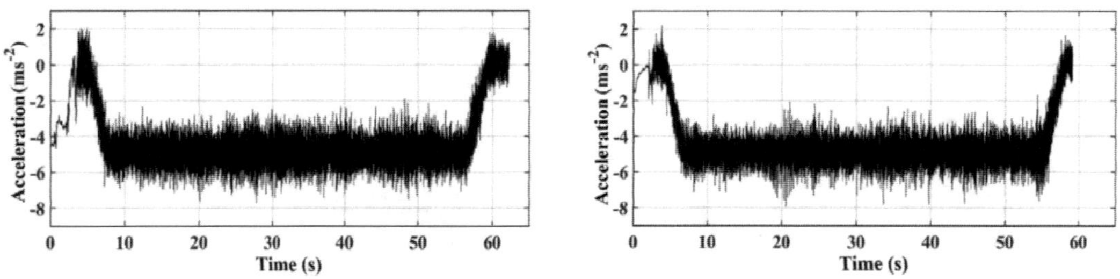

(a) Left handrail (b) Right handrail
Figure 9. Original acceleration curve of handrail running direction

Since GB/T 24474.2-2020 does not give the quantitative range of handrail vibration comfort, and referring to DB35/T 1936-2020 *Escalator and Moving Walkway Operation Vibration Measurement and Evaluation*, when the handrail vibration RMS value is less than or equal to 0.35 m /s^2, the operating vibration evaluation level is "good", and the passengers feel comfortable. When the vibration RMS value of the handrail is greater than 0.35 m/s^2 and less than or equal to 0.5 m/s^2, the operating vibration evaluation level is "medium", and passengers at this level do not feel uncomfortable. Therefore, it can be seen that passengers feel comfortable with the hand-transmitted vibration of the handrails on both sides of the escalator when the escalator is running. The total time when the vibration acceleration RMS value of the right-hand rail is greater than 0.35 m/s^2 is shorter than that of the right handrail, indicating that the right-hand rail of the ladder runs relatively more smoothly.

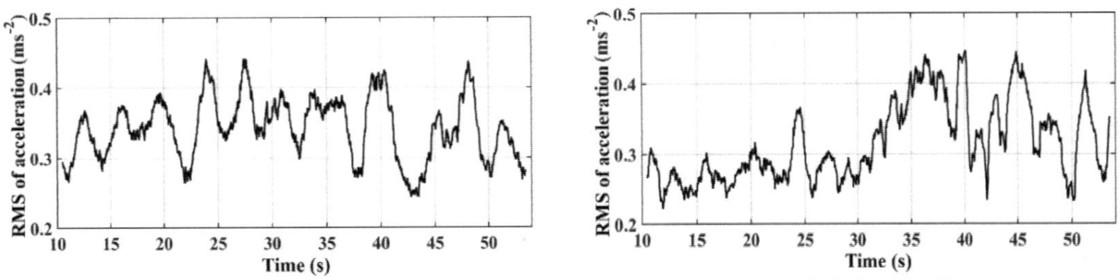

(a) handrail (b)Right handrail

Figure 10. RMS curve of handrail vibration acceleration

7. Conclusion

The three-axis acceleration sensor was used to collect the acceleration data of the inclined section steps and the handrail during the operation of the escalator, and the vibration acceleration was extracted after removing the reading component of the gravity acceleration. After frequency weighting, vibration RMS value and triaxial vector sum calculation, the maximum vibration RMS value, and the average vibration RMS value can be counted so as to divide the vibration comfort level. Moreover, a three-axis acceleration data acquisition module was developed based on a MEMS acceleration sensor and high-performance single-chip microcomputer, and an APP was developed for an Android based vibration comfort detection system with the help of the data analysis method. The calibrated detection device can meet the requirements of detecting vibration comfort of escalator steps and handrails online, providing important data reference for evaluating the escalator transportation quality.

Acknowledgments

This work has been supported by Guangdong Administration for Market Regulation (project nos. 2022CT04) and Zhuhai Science and Technology Innovation Bureau (project nos. ZH22036201210036PWC).

References

[1] Beards P J, Frost G, Healey N, et al. (2022) A new Human factors incident taxonomy for members of the public (HFIT-MP): An investigation of escalator incidents[J]. Safety Science, 147(2):105597.

[2] Liu Z. L, Lu R. F, Chen J. (2021) A case study of hidden dangers in the monitoring switch of the escalator cover plate[J]. China Elevator, 32(01):17-19.

[3] Lin S. F. (2021) Development of escalator synchronization rate measurement system based on nRF24L01+[J]. Industrial Instrumentation and Automation Device, (06):30-33.

[4] Su X. F, Huang Z. L. (2019) Case analysis of personal injury caused by slipping of escalator handrails [J]. China Elevator, 30(23):54-55.

[5] GB/T 24474.2-2020, Ride quality measurement - Part 2: Escalators and moving walks [S].

[6] Liang M. J, Peng X. J. (2021) Design and implementation of escalator mechanical vibration data acquisition system based on Labview [J] Industrial Instrumentation and Automation Device, (02):54-58.

[7] Zhang Y, Li X. (2021) Research on escalator condition monitoring and early warning system based on vibration analysis[J]. Urban Express Rail Transit, 34(06):149-154.

[8] Liu D.Y, Liang M. J, Yang N. X, et al. (2021) Research on vibration signal feature extraction and classification method of escalator sprocket bearing [J]. Application of Single Chip Microcomputer and Embedded System, 21(10):36-41.

[9] Zhang J. H, Shi X, Gao X, et al. (2017) Experimental analysis of lateral vibration of escalator chain drive[J]. Mechanical Design and Research, 33(05):168-172.

[10] Wang W. X, Yao L, Xu J. (2021) Design of landslide detection system based on MEMS acceleration sensor MPU-6050[J]. Coal and Chemical Industry, 44(02):156-160.

[11] Cao G. Q, Zhao W. S. (2020) Development of high-precision inclination sensor based on MEMS accelerometer[J]. Automation Instruments, 41(03):25-28+35.
[12] Niu W. M, Fang L. Q, Qi Z. Y. et al. (2020) Displacement measurement system based on MEMS acceleration sensor [J]. Instrument Technology and Sensors, (03):62-66+72.
[13] Sun C. M, Xu S. L, Gan W. W., et al. (2019) Research on the application of three-axis accelerometer calibration method in maglev train speed calculation [J]. Locomotive Electric Drive, (04):27-31.
[14] Wu X. Y. (2021) Research on MEMS-driven upper limb rehabilitation posture angle tracking and recognition method for stroke patients [D]. Nanjing University of Information Science and Technology.
[15] Pan Z. M, Wang X, Yuan J., et al. (2021) Design of engineering vibration monitoring system for MEMS accelerometer[J]. Sensors and Microsystems, 40(04):86-89.
[16] Chen J. X, Tai S. L, Yang N. X., et al. (2021) Detection of elevator running characteristics based on MEMS acceleration sensor [J]. Hoisting and Transportation Machinery, (20):84-88.
[17] Wu T. J, Liu P, Yu X. Y. et al. (2021) Research on elevator car vibration sensing network based on MEMS technology [J]. Measurement and Control Technology, 40(12): 26-30.
[18] Guo W, Wu Y. P, Wang T. Y. (2020) MEMS Quaternion Kalman Filter Algorithm for Elevator Attitude Estimation [J]. Computer System Applications, 29(03):246-252.
[19] Li Z. X, Ma H. X, Li G. et al. (2017) Research and Design of Elevator Brake Safety Performance Testing Device [J]. Industrial Safety and Environmental Protection, 43(07): 34-37.
[20] GB/T 13441.1-2007, Mechanical vibration and shock - Evaluation of human exposure to whole-body vibration - Part 1: General requirements [S].

Development and Engineering Application of the Measurement System Based on the Miniaturized Ultrasonic Transducer

YouDi Kong[1,a], JianRui Zhao[1], HaiTao Wang[1], XinPeng Wang[1], GuangJie Wang[1], YuHua Wu[1] and WeiXing Zhang[1]

[1] National Ocean Technology Center, 219 Jieyuanxi Road, Nankai District, Tianjin, China

[a]soul_510270558@126.com

Abstract. The iron terminals are mostly used in the small ultrasonic transducers which are connected to the circuit board by the direct or indirect welding method. Although the method is considered to be firm and reliable, the ultrasonic transducer or the whole system is easy to be damaged when the ultrasonic transducer is replaced or maintained. Moreover, the terminals of the ultrasonic transducer are easy to be oxidized. The miniaturized ultrasonic transducer with the electroplated copper terminals is innovatively designed to enhance the oxidation resistance of its terminals. Then the ultrasonic transducers are installed on the circuit board by the copper claw spring instead of welding. In addition, the waterproof miniaturized ultrasonic transducer is replaced by the fully covered waterproof sound permeable membrane and the non waterproof miniaturized ultrasonic transducer. The experimental results demonstrate that the quality of the signal transmission of the ultrasonic transducer is not affected, and the convenience of replacement and maintenance of the ultrasonic transducer is improved. With this design process, the overall waterproof performance of the system is also significantly enhanced while the system cost is reduced, market future is wide prospect.

1. Introduction

The miniaturized ultrasonic transducers are more and more used in the liquid level control, the obstacle detection, the human body detection, the material level continuity detection, the flow velocity detection and other fields [1]. At present, most of the ultrasonic transducers on the circuit board with the medium or the low frequency are made of the irons, especially the two terminals of the 40KHz ultrasonic transducer on the circuit board. Moreover, they are connected to the circuit board by welding. The connection mode makes the structure firm, and the signal transmission effect is good [2].

There are still several aspects that need to be improved. First and foremost, the iron materials are easy to be oxidized after a long time, and the signal transmission quality is affected by the oxidized iron materials. Besides, as a sensitive element, the internal structure of the ultrasonic transducer is easily damaged by the poor welding process. Last but not least, it is not conducive to the maintenance and replacement of the ultrasonic transducer. In addition, in the aspect of waterproof, most of the non waterproof ultrasonic transducers are made into the waterproof ultrasonic transducers for waterproof and sealing. The main consideration of the design is the waterproof and sealing of the ultrasonic transducer itself, but the waterproof and sealing of the overall measurement structure is still insufficient.

Based on the above problems, the new on-board design method of the miniaturized ultrasonic transducer circuit is proposed in the study. The copper electroplating process is adopted for the terminal to avoid oxidation. The claw spring installed on the circuit board is used to realize the reliable connection with the circuit board by plugging and pulling. The quality of the signal transmission is not affected by the method. The ultrasonic transducer is maintained and replaced conveniently through the method. The waterproof and sealing performance of the overall measurement structure has also been significantly improved through the design method of the waterproof and sound permeable membrane [3].

2. The traditional design method of the small ultrasonic transducer on the circuit board

2.1. Traditional ultrasonic transducer

Taking the 40KHz ultrasonic transducer with two iron terminals as an example, the real object of the common small-scale ultrasonic transducer on the circuit board in the market is shown in Figure 1.

Figure 1. the practicality figure of the traditional miniaturized ultrasonic transducer

2.2. The conventional method of installing the ultrasonic transducer

At present, the on-board connection methods of the small ultrasonic transducers are mostly completed by welding process, which can be divided into direct contact welding and indirect contact welding. The direct contact welding method is shown in Figure 2. The indirect contact welding method is shown in Figure 3.

The direct contact between the ultrasonic transducer and the circuit board is avoided by the indirect contact welding through the decoupling method, which is better than the direct contact welding in protecting the ultrasonic transducer, maintaining and replacing the ultrasonic transducer and ensuring the safety of the overall measurement circuit.

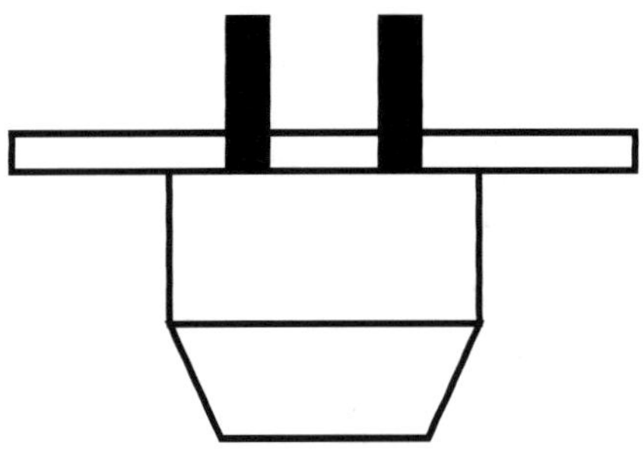

Figure 2. the diagram of the direct contact welding method of the traditional miniaturized ultrasonic transducer

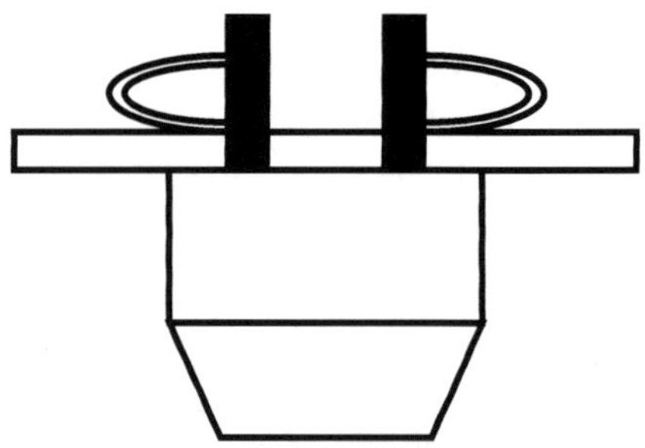

Figure 3. the diagram of the indirect contact welding method of the traditional miniaturized ultrasonic transducer

3. The improved design method of the small ultrasonic transducer on the circuit board

3.1. Improved ultrasonic transducer

While the performance of the improved small ultrasonic transducer is ensured, the two terminals are processed by the copper electroplating process to improve its oxidation resistance. The physical object is shown in Figure 4.

Figure 4. the practicality figure of the traditional miniaturized ultrasonic transducer

3.2. The improved method of installing the ultrasonic transducer

The claw spring is embedded in the corresponding position of the circuit board installed with the ultrasonic transducer. The copper claw spring is a precision machined connector with the copper jack and the gold-plated inner claw spring. The real object of the connector is shown in Figure 5. The ultrasonic transducer is reliably installed on the circuit board by plugging, as shown in Figure 6.

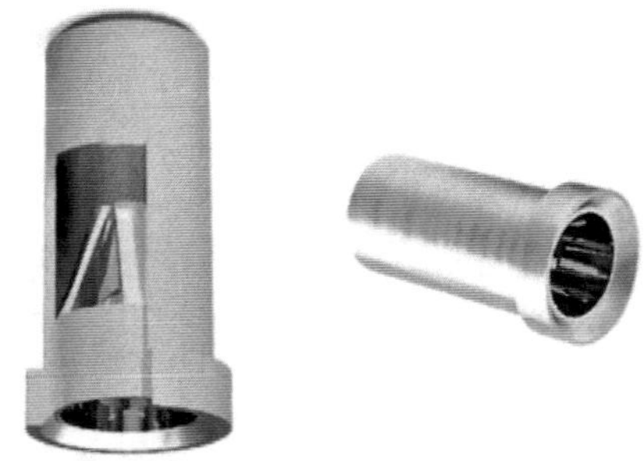

Figure 5. the practicality figure of the copper claw spring

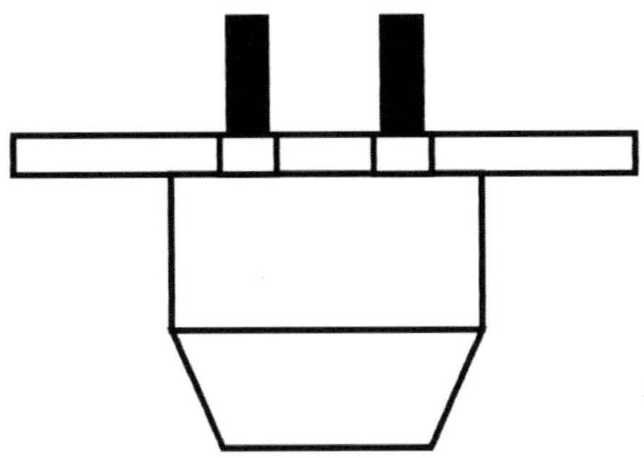

Figure 6. the connection diagram of the plugging mode of the ultrasonic transducer

4. The test scheme design and test analysis

4.1. Design of the test scheme

The test system is designed to test the quality of the signal transmission of the improved on-board ultrasonic transducer. The schematic diagram of the overall structure is shown in Figure 7.

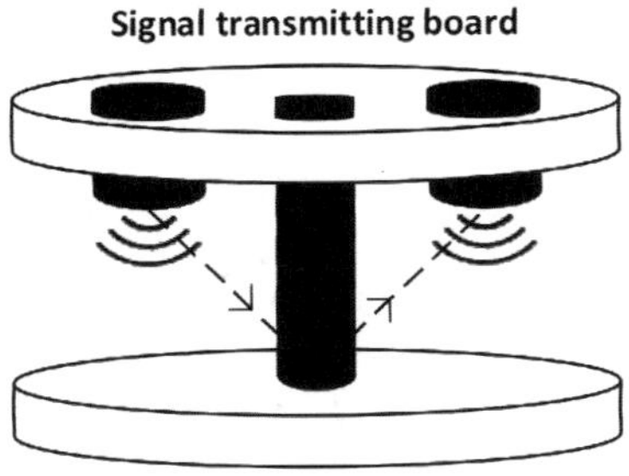

Figure 7. the structural diagram of the test system

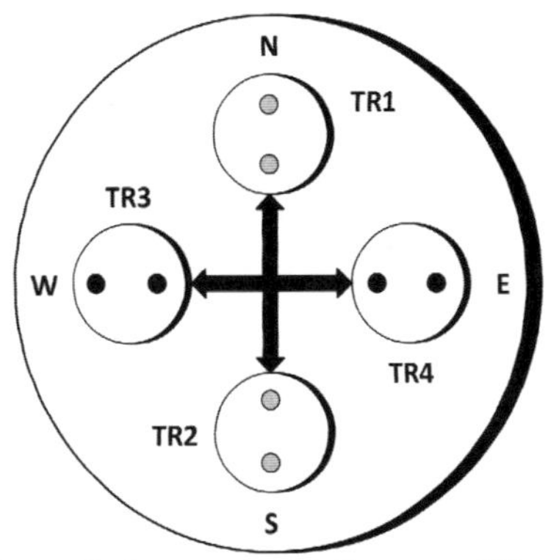

Figure 8. the structural diagram of the distribution of the two-dimensional cross shaped ultrasonic transducer array

The whole system is mainly composed of three parts: the signal transmitting board, the signal reflecting board and the waterproof and sound permeable membrane. The signal transmitting board mainly includes the two-dimensional cruciform ultrasonic transducer array and the signal processing circuit. The structural diagram of the distribution of the two-dimensional cross shaped ultrasonic transducer array is shown in Figure 8. The schematic diagram of the signal processing circuit is shown in Figure 9. The signal reflecting board is mainly used to reflect the transmitted signal of one ultrasonic transducer probe to another ultrasonic transducer probe to receive the echo signal [4].

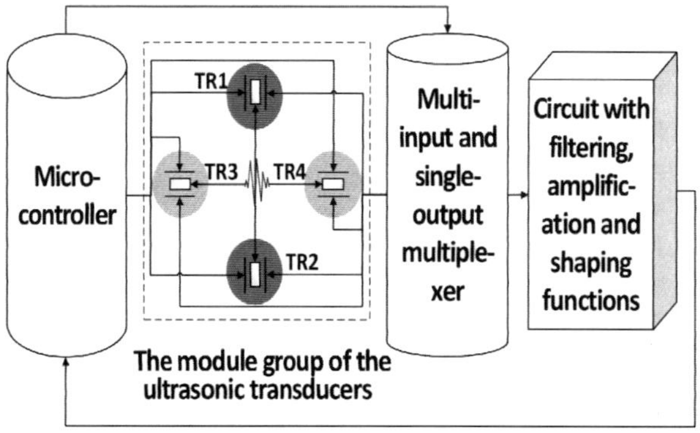

Figure 9. the schematic diagram of the signal processing circuit

TR1 and TR2 are the ultrasonic transducer probes with the electroplated copper terminals. TR3 and TR4 are the ultrasonic transducer probes with the iron terminals. TR1 and TR2 are installed on the circuit board in a plug-in manner through the claw springs.TR3 and TR4 are installed on the circuit board by welding.

The signal processing circuit is composed of four parts: the microcontroller, the module group of the ultrasonic transducers, the multi-input and single-output multiplexer and the circuit with filtering, amplification and shaping functions [5]. The transmitting and receiving signals of the module group of the ultrasonic transducers are controlled by the microcontroller and the multi-input and single-output multiplexer. The received echo signal is returned to the microcontroller for subsequent processing after passing through the circuit with filtering, amplification and shaping functions.

4.2. Analysis of the test

The waveforms of the ultrasonic signals which are transmitted by TR1 and the ultrasonic echo signals which are received by TR2 are shown in Figure 10. The waveforms of the ultrasonic signals which are transmitted by TR3 and the ultrasonic echo signals which are received by TR4 are shown in Figure 11.

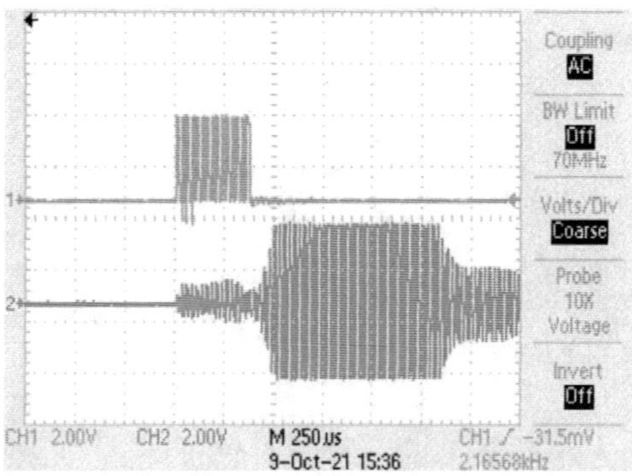

Figure 10. the diagram of the ultrasonic signals transmitted by TR1 and the ultrasonic echo signals received by TR2.

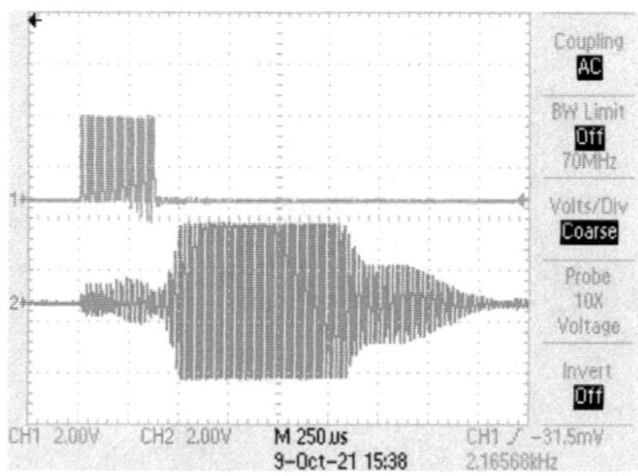

Figure 11. the diagram of the ultrasonic signals transmitted by TR3 and the ultrasonic echo signals received by TR4

Through the waveforms of the ultrasonic signals displayed by the oscilloscope, we can clearly see that the on-board ultrasonic transducer with the electroplated copper terminals installed by the plug-in method and the on-board ultrasonic transducer with the iron terminals installed by the welding process can be identified, sampled and processed by the microcontroller or single chip microcomputer after the signal processing circuit [6].

5. Conclusion

In conclusion, through the design method, the on-board ultrasonic transducer with the electroplated copper terminals installed by plug-in method can be used to realize the function of the on-board ultrasonic transducer with the iron terminals installed by the welding process. The maintainability and the waterproof performance of the improved measurement and control system are better than the traditional measurement and control system.

Acknowledgments

Our research was financed by the National Key Research and Development Program of China(2017YFF0208400) and National Ocean Technology Center Innovation Fund(K5200Z004). We also would like to thank all colleagues who previously provided the technical supports.

References

[1] Zhu K, Chen X, Qu M, Yang D, Hu Land Xie J. 2021 *Non-Contact Ultrasonic Flow Measurement for Small Pipes Based on AlN Piezoelectric Micromachined Ultrasonic Transducer Arrays. Journal of Microelectromechanical Systems,* **30:** 480-487.

[2] Smith L, Li Z and Dixon S. 2022 *Transducer Design for Clamp-on Guided Wave Flow Measurement in Thin-Walled Pipes, IEEE Sensors Journal,* **22:** 10613-10619.

[3] Wang J, Zhang M, Wang Z, Sun S, Ning Y, Yang X and Pang W. 2022 *An Ultra-Low Power, Small Size and High Precision Indoor Localization System Based on MEMS Ultrasonic Transducer Chips. IEEE Transactions on Ultrasonics, Ferroelectrics, and Frequency Control,* **69:** 1469-1477.

[4] Feuillard G, Hue L, Saadaoui N, Nguyen V, Lethiecq M and Saillant J. 2021 *Symmetric Reflector Ultrasonic Transducer Modeling and Characterization: Role of the Matching Layer on Electroacoustic Performance, IEEE Transactions on Ultrasonics, Ferroelectrics, and Frequency Control,* **68:** 3608-3615.

[5] Kang L, Feeney A and Dixon S. 2020 *The High Frequency Flexural Ultrasonic Transducer for Transmitting and Receiving Ultrasound in Air, IEEE Sensors Journal,* **20:** 7653-7660.

[6] Zamora I, Ledesma E, Uranga A and Barniol N. 2022 *Phased Array Based on AlScN Piezoelectric Micromachined Ultrasound Transducers Monolithically Integrated on CMOS, IEEE Electron Device Letters,* **43:** 1113-1116.

Application of On-line Digital Radiographic Inspection for Pipeline with Insulation

Xu-feng Li[1], Jie Sun[1], Shengzi Lu[1]*, Lei Wang[1]

[1]Guangdong Institute of Special Equipment Inspection and Research, Foshan, China 528251, Guangzhou, China 510640

*Corresponding author: raze_z.y@163.com

Abstract. Pipes with insulation are widely used in the industrial field. The traditional nondestructive testing technology for on-line pipes with insulation often requires the removal of a large number of insulation layers and the forcing of the owner to stop for testing. The traditional testing mode not only has low detection efficiency, but also forces the owner to spend a lot of manpower, material resources and financial resources. In recent years, a series of new nondestructive testing (NDT) technologies have different advantages for in-service pipeline testing with coated layers. However, Digital radiographic (DR) testing can satisfy the integrity inspection of on-line pipeline with insulation well. In this paper, the basic principle of DR testing and the configuration of instrument parameters are introduced. Based on the analysis and discussion of the actual testing cases, it is proved that digital ray testing can effectively solve the quality testing of on-line insulation pipeline, and it is a new nondestructive testing technology worthy of further study and popularization.

1. Introduction

Pipes with insulation layers were widely used in petroleum, chemical industry and nuclear power plant. For the integrity inspection of the in-service pipelines covered by insulation layers, ultrasonic testing, radiographic testing, magnetic testing and penetrant testing are the conventional nondestructive testing method we used mainly, but these methods require the owner to stop the production, remove a large number of insulation layers, and polish the weld joint for the testing preparation. It can be seen that the conventional NDT method is time-consuming and laborious, and the detection efficiency is low. For the owner, it is inevitable to waste a lot of money and time to remove the insulation layers before inspection, and reinstall the insulation layers after inspection. Therefore, the need of application of new NDT method without removing the layers is increasing. At present, the emphasis on engineering quality is greatly improved in China, which promotes the rapid development of new technology of NDT, such as ultrasonic guided wave testing[1], pulsed eddy current[2], and digital radiography[3], applied in petrochemical, electric power, chemical and other industrial fields in recent years gradually.

The new NDT methods mentioned above have their own limits. Ultrasonic guided wave detection is a rapid inspection method for pipelines with insulation layers, 100% comprehensive inspection for long-distance pipelines[4] can be carried out after removing the a small part of insulation layers and installing the probe. But ultrasonic guided wave detection cannot determine the defects qualitatively, and the detection sensitivity of local small metal thinning defects is very low, so it can only be used as a preliminary detection method. Pulsed eddy current is a in-situ testing technology which has been gradually applied. It can realize the imaging inspection of the thick-walled pipe without removing the insulation layers, but it cannot detect the defects in weld. Due to the application of DSP (Digital Signal

Processing) in digital radiography, the image quality of DR is much better than the traditional radiographic detection. Meanwhile, film is not needed in the use of DR[5], and the radiation dose of DR is much less. At present, DR is gradually used in the online detection of weld defects and pipe wall thinning.

2. Methodology

2.1 Principle of digital radiographic

Digital radiographic testing is a nondestructive testing technology which converts X-ray directly into digital signal with digital imaging board as the core part. In the detection of the pipe with insulation layers, The X-rays that pass through the pipe wall are captured behind the pipe by an imaging board and turned into image data. After a series of signal amplification and data processing by the computer, the detection image can be shown to reveal the flaw inside the pipe wall and weld[6]. Different from traditional film X-ray detection, DR detection technology does not need to use film and can achieve 100% electronic management, which reduces the management cost. Meanwhile, X-ray has high sensitivity and short penetration time, enabling real-time imaging[7]. So DR testing technology has been widely promoted in various industrial NDT fields. The principle of DR is shown in Figure 1.

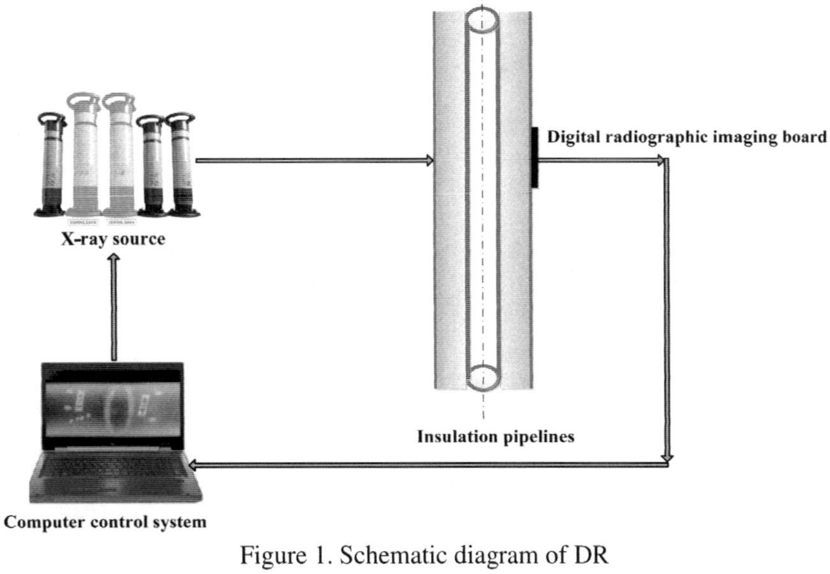

Figure 1. Schematic diagram of DR

2.2 System of digital radiographic

A DR system includes computer control system, pulse X-ray source and digital imaging board.

The computer system controls the excitation of the X-ray source, amplifies and filters the X-ray signal received by the imaging board, and saves the detected signal in real time. The parameters of the computer system adopted in this paper are shown in Table 1.

Table 1. Basic parameters of computer control system

Hardware	Parameters
Screen size	17/15 inch optional
CPU	Intel / core i7-4900mq
Graphics card	NV-K2100m
Graphics card memory capacity	2GB
Hard disk capacity	500GB
Operating system	Windows 7/Windows 10

A portable directional pulse x-ray source suitable for field operation is adopted. The output voltage range of the X-ray machine is 10~225kv, the maximum working current is 4.0 mA, the focus size is 3 mm, the temperature range used is: -30~+60℃, and the maximum penetration tube thickness is 47 mm. Basic parameters of X-ray source is shown in Table 2.

Table 2. Basic parameters of X-ray source

Name	Parameters
X-ray machine	CP225D directional X-ray source
Inherent filtration	0.8mmBe
Maximum penetration	47mm
Operating temperature range	-30~+60℃
Maximum exposure time	99 min 59 sec
Voltage range	10~225KV
Tube current range	1~10mA
Head size	Φ140*725mm
Full load current	4.0mA
Focus size	3.0mm
Total weight	12.1Kg

A digital imaging plate (Bruteck, Germany) is used in this paper. As X-ray source is incentive, the X-ray through heat preservation and the tube body to CsI cesium iodide scintillator, cesium iodide will be issued and reach the plate is proportional to the X-ray imaging visible light, then picked up by the next layer of photoelectric diode visible light and convert them into electrical signals, then through the underlying electrical excitation circuit of each photodiode, convert electrical signals into digital signals. Finally, digital images can be obtained by amplification and filtering by computer system [8-10]. Parameters of the digital imaging plate used are shown in Table 3.

Table 3. Basic parameters of digital imaging plate

Name	Parameters
Scintillation	Cs I
Pixel	139um
Frame rate	2fps
Size	274*184*32.5mm
Acceptable X-ray energy	40-350kVp
The battery number	1
Standby time	≥4.5h
Working temperature	+5℃~+35℃
Quantum testing efficiency	65%
Weight	≤3.5KG
Maximum communication distance	100m

3. Field application

3.1 Pipe with insulation layers in cold storage

In this case, DR was used to inspect the circumferential weld of pipeline with insulation layer in a cold storage. The specification of pipe containing gaseous Freon is Φ89 mm*8.0 mm, and the thickness of insulation layer is 50 mm. The testing voltage of DR was set to 175 kV, and the testing results are shown in Figure 2. It can be seen that an obvious lack of fusion defect at the circumferential weld of the pipeline is shown in the image, and the image quality is good.

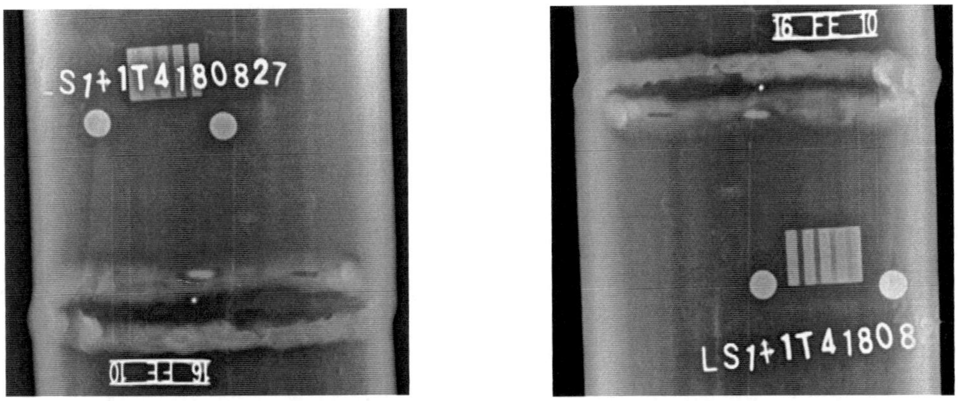

(a) The first image taken (b) The Second image taken

Figure 2. DR testing image of Φ89 mm*8.0 mm pipe

3.2 Ammonia water pipe with insulation layers

In this case, DR was used to inspect the circumferential weld of ammonia water pipeline in a fertilizer plant, as shown in Figure 3. The specification of pipe is Φ219 mm*10.0 mm, and the thickness of insulation layer is 120 mm. The pipe is filled with ammonia water, and 2 circumferential welds (1# and 2#) were inspected by DR. For the 1# welds (Φ219 mm*10.0 mm), the testing voltage of DR was set to 210kV, and the testing results are shown in Figure 4. For the 2# welds (Φ273 mm*12.7 mm), the testing voltage of DR was set to 300kV, and the testing results are shown in Figure 5. As shown in Figure 4, 1# weld is incomplete-penetration, and the crack defect is more obvious by adjusting the gray scale of image. As shown in Figure 5, a serious incomplete-penetration is found in 2# weld by adjusting the gray scale of the blurred image (Figure 5 a) into a clear image (Figure 5 b).

Figure 3. On-site testing of ammonia water pipeline DR

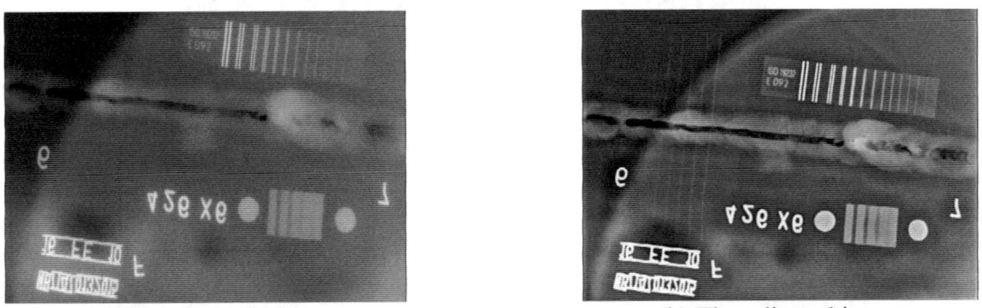

(a) The original image (b) The adjusted image

Figure 4. DR Testing image of 1# weld

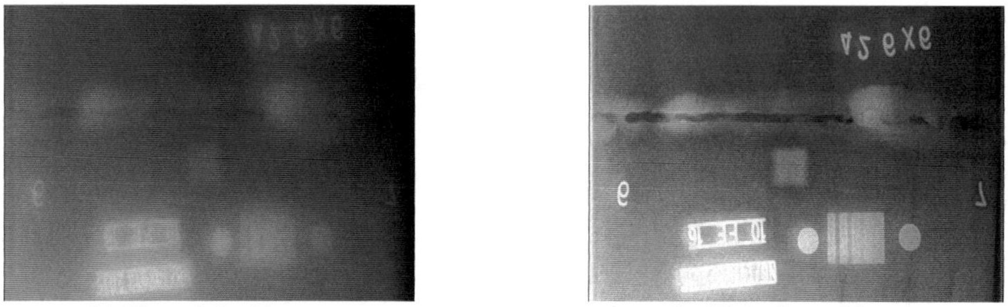

(a) The original image (b) The adjusted image

Figure 5. DR Testing image of 2# weld

3.3 Drive pipe with insulation layers

In this case, DR was used to inspect the butt joint of drive pipe in chemical fiber factory. The specification of the outer pipe and inner pipe is Φ89 mm*8.0 mm and Φ34 mm*3.0 mm, and the thickness of insulation layer is 25 mm. The drive pipe is empty. The testing voltage of DR was set to 150 kV, and the testing results are shown in Figure 6. A serious incomplete-penetration can be found in the weld of outer pipe.

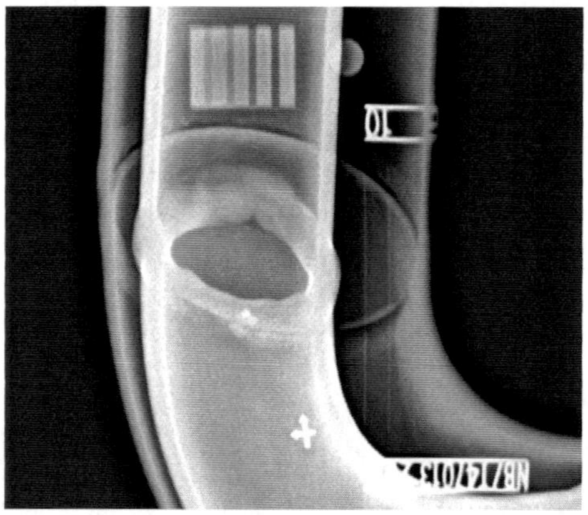

Figure 6. DR testing image of drive pip

4. Analysis and discussion

As can be seen from the above field case, the condition of liquid medium in the pipeline may have a certain influence on the DR inspection result. Besides, the system noise of digital imaging and geometric amplification will affect the DR inspection result.

1) Medium in the tube: if the medium in the tube contains liquid or high density gas-liquid medium, it will generally cause a certain amount of attenuation of photons. Therefore, for DR detection of the pipeline containing such media, filtering and noise reduction should be carried out on the original image.

2) System noise of digital imaging: the main influencing factor that causes the system noise of digital imaging is the light quantum[11]. According to the research of Chen Guang[8] et al., interference can be suppressed by using continuous frame superposition denoising processing in operation.

3) Geometric amplification: DR detection can directly locate and conduct qualitative and quantitative processing of defects through the features of the detection images containing defects obtained. However, in service conditions, the insulation layer has relatively harsh temperature conditions. Limited by the working temperature of the digital imaging board, the distance between the digital imaging board and the coating layer is often kept rather than completely fitted, which is easy to

enlarge the defect image on the side of the ray source, and the change degree may cause misjudgment on the size of the defect.

5. Conclusion

The digital radiographic inspection technology can effectively realize the quality inspection of the in-service pipeline with insulation layer, and avoid the shutdown and removal of a large amount of insulation caused by the conventional nondestructive inspection. On the other hand, the simple operation steps and superior post-processing means of digital radiographic testing make the whole detection process more efficient, the obtained detection image is clearer, the defects are easy to identify, and the digital electronization is easier to manage. Although the medium, imaging plate and coating in the pipe to be tested will affect the results to a certain extent, it can be suppressed by filtering and other means. Therefore, digital radiographic testing is a new nondestructive testing technology for the structural integrity testing of in-service pipeline with coating, which is worthy of in-depth research and promotion.

Acknowledgments

We gratefully acknowledge financial support from the Guangdong Provincial Key R&D Programme (2020B0404010001).

REFERENCES

[1] Zou Ningbo, Chen Haiyun, Liu Quanli, etc Corrosion detection of gas transmission pipeline based on t (0,1) mode ultrasonic guided wave[J] Non destructive testing 2013,35(9): 19-22, 68.
[2] Lu Yang, Li Yongjian, Xu Xiaojia, etc Application of pulsed eddy current based inspection technology for clad pipes[J] Liaoning Chemical Industry Co., Ltd 2017,46(8): 812-814.
[3] Yuan Peixin, Niu Yanping Study on X-ray detection method of corrosion state of pipeline with insulation layer[J] CT Theory and application research, 2007,16 (3): 36-42.
[4] Ding min, yushurong, Hu bin, etc Application of magnetostrictive guided wave technology in pipeline inspection[J] Pipeline technology and equipment, 2013, (1): 19-22.
[5] Liu Quanli, Bai Shiwu, Xue Yan, etc Research on digital radiographic inspection technology of pipeline girth weld[J] Pipeline technology and equipment, 2010, (3): 20-22.
[6] Lei Zhengqiang, Wang Weibin, Wen Yufen, etc Application of radiographic digital imaging inspection technology for girth welds of long-distance pipelines[J] Pipeline technology and equipment, 2021, (1): 52-54,57.
[7] Zhang Wei, sun bona, Dynasty, etc Research on submarine pipeline route detection method[J] Harbor engineering technology, 2015,52 (6): 111-113.
[8] Chen Guang, Shi Kun, Ding Keqin Research on radiographic digital inspection of circumferential weld of pipeline with insulation layer[J] Metal casting, forging and welding technology, 2012,41 (5): 154-155, 158.
[9] Chen Jianxin Application value of digital imaging technology of amorphous silicon X-ray flat panel detector[J] Journal of Fuzhou General Hospital, 2005,12 (1): 57-59.
[10] Chen Guang, Ding Keqin, Liang Lihong Application of portable DR and CR imaging technology in weld testing[J] Nondestructive testing, 2009,31 (6): 494-496.
[11] Xiao Yongshun, Chen Zhiqiang, Zhang Li, et al Experimental study on high energy X-ray imaging based on digital flat panel detector[J] Optical technology, 2003,29 (6): 660-663.

Effects of Impact Hammer Parameters on Mechanical Impedance Measurement

LinXiong Chen[1,a], CaiXia Hu[2,b], Chang Fang[1,c] and Wei Liu[1,d]

[1] Wuhan Second Ship Design and Research Institute, Wuhan, Hubei, China

[2] Weihai Ocean Vocational College, Weihai, Shandong, China

[a]chenlinxiong@foxmail.com, [b]hcx227@163.com, [c]1097328259@qq.com, [d]1134122744@qq.com

Abstract. Base impedance has a direct effect on mechanical equipment's properties, and thus its accurate measurement constitutes an important foundation for the evaluation of these properties. Through theoretical deduction and base impedance measurement, we investigated the influences of impact hammer parameters on mechanical impedance measurement using an impact excitation method. By comparing the effects of hammer head softness and hammer size on base's mechanical impedance measurements, we know that soft and hard hammer heads were suitable for the impedance measurement in low and high frequency regions, respectively, and that hammer size had a little influence on impedance measurements.

1. Introduction

At present, vibration and noise reduction technologies have gained more attention. Vibration isolation device [1] with single-layer or double-layer isolation design is often used to reduce the vibration and noise of mechanical equipment. Their isolation efficiency is usually evaluated with the difference between the vibration level of equipment legs installed on them and that of their base. Base impedance has a direct influence on the measurement of base vibration level, thus directly affecting the isolation efficiency of vibration isolation device [2] [3].

Also, it exerts a direct impact on technical indicators of vibration isolation device. Therefore, it is necessary to impose restrictions when designing the bases of vibration isolation device. On the one hand, too low base impedance may cause high base vibration level, leading to undesirable vibration isolation efficiency. On the other hand, too high base impedance may result in low vibration level. But it'll prevent vibration energy of the vibration isolation device from being transmitted and thus lead to unwanted vibration severity [4]. Relevant international standards set base impedance limit at 100±40dB, and frequency range at 20Hz~2000Hz.

In laboratories impedance is often measured with an impact excitation method according to GB/T 11349.3-2016 *Vibration and Shock - Experimental Determination of Mechanical Mobility - Part 3: Measurement Using Impact Excitation.* As specified in the method, one can calculate a base's mechanical impedance after hammering the base with an impact hammer and collecting impact force signals and the acceleration signals on the base. Given that hammer size and hammer head's softness have a direct effect on the accuracy of experiment-based impedance determination, we aim to investigate the effects of hammer parameters on impedance measurement.

2. Theoretical analysis of base impedance

Today, a large number of pieces of mechanical equipment adopt vibration reduction and isolation design [5]. Their mechanical system's equivalent diagram is as shown in Figure 1, regardless of their vibration isolation model (single-layer or double-layer). Equivalent models of the mechanical system have witnessed remarkable research achievement, and thus they are not the focus of our research. Given this, we directly cited an equivalent model of mechanical system power source and an equivalent vibration isolation network diagram [6], as shown in Figure 2.

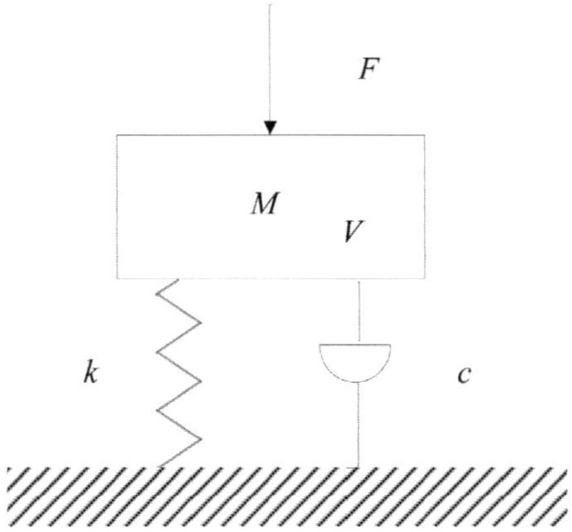

Figure 1. Mechanical system's equivalent diagram

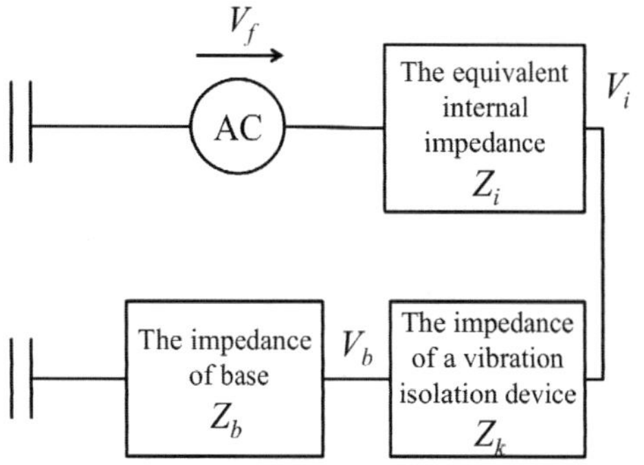

Figure 2. Vibration isolation network equivalent diagram

In Figure 1, F is exciting force, M is the mass of mass block, V is the vibration velocity of mass block, k is the stiffness of spring, and c is damping coefficient. In Figure 2, V_f is an equivalent constant velocity source of mechanical system, V_i is the vibration velocity of mechanical equipment legs, V_b is the vibration velocity of mechanical equipment base, Z_i is the equivalent internal impedance of mechanical system, Z_b is the impedance of base, Z_k is the impedance of a vibration isolation device.

As shown in Figure 2, systems are arranged in series and have identical equivalent exciting force F, and exciting force F, vibration velocity V and mechanical impedance Z follow the relationship shown below:

$$Z = \frac{F}{V} \tag{1}$$

We turned Figure 2 into an electrical series network diagram. Because of identical exciting force F, it corresponds to the electrical current I in electrics. According to Formula (1), vibration velocity V corresponds to the voltage U in electrics, and the reciprocal of mechanical impedance Z is equivalent to the electrical resistance R. Based on this, Formula (1) was transformed into the following form:

$$V = F \bullet \frac{1}{Z} \tag{2}$$

According to electrical knowledge and Formula (2), we know that the mechanical system's internal impedance and base impedance shown in Figure 2 have following relationship:

$$V_i = \frac{\frac{1}{Z_i}}{\frac{1}{Z_k} + \frac{1}{Z_b} + \frac{1}{Z_i}} V_f \tag{3}$$

$$V_b = \frac{\frac{1}{Z_b}}{\frac{1}{Z_k}+\frac{1}{Z_b}+\frac{1}{Z_i}} V_f \tag{4}$$

The isolation efficiency δ was calculated as follows:

$$\delta = 20\lg\left|\frac{V_i}{V_b}\right| = 20\lg\left|\frac{\frac{1}{Z_i}}{\frac{1}{Z_b}}\right| = 20\lg\left|\frac{Z_b}{Z_i}\right| \tag{5}$$

From Formula (3), we understand that equipment legs' vibration velocity V_i will reach its maximum value $\frac{Z_k}{Z_k+Z_i}V_f$ when base impedance Z_b increases towards infinity. In other words, too high base impedance will cause very fast vibration of equipment legs, resulting in the movement of leg vibration level out of specified range. Formula (5) tells us that vibration isolation device will produce the poorest isolation efficiency when base impedance Z_b goes to zero. That is to say, too low base impedance may fail to bring desirable isolation efficiency.

3. Effects of impact hammer parameters on experimental determination of base impedance

Above theoretical deduction shows that the size of base impedance is very important and its experimental determination is also important. In the paper, our focus was on the effects of hammer parameters used in impedance measurement. We planned to investigate the effects of hardware parameters of impact hammer alone, without giving attention to those of software parameters. Thus, we conducted our research on the effects of hammer head material (soft and hard) and additional mass. In our research, we acquired the impact force signal curves of hammers with different parameters, and obtained auto power spectrum of force signal after Fourier transform. By comparing force signal curve and auto power spectrum, we identified how impact hammer parameters affect the experimental determination of base impedance.

3.1. Test results of rubber and aluminum alloy hammer heads

In our test, a 2303-type impact hammer from Endevco was used to hammer the same base. The hammer was first equipped with a soft head made of rubber, and then with a hard head made of aluminum alloy. Time domain signals of force were obtained as shown in Figure 3 and Figure 4.

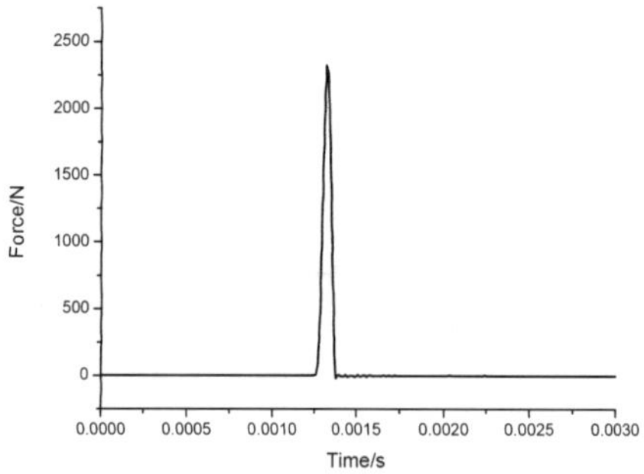

Figure 3. Time domain signals of force by aluminum alloy

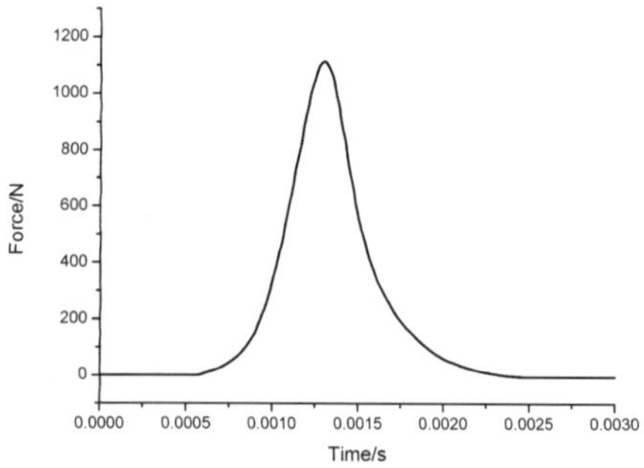

Figure 4. Time domain signals of force by rubber

As shown in these figures:

- The impact force amplitude was about 2400N for the aluminum alloy head, and about 1200N for the rubber head;
- The force impact duration of aluminum alloy head was short, about 0.000125s, compared to 0.0015s for the rubber head;

After Fourier transform of these two force signal curves was conducted, the auto power spectrum of force signal was obtained as shown in Figure 5, with a frequency range of 1Hz~3200Hz.

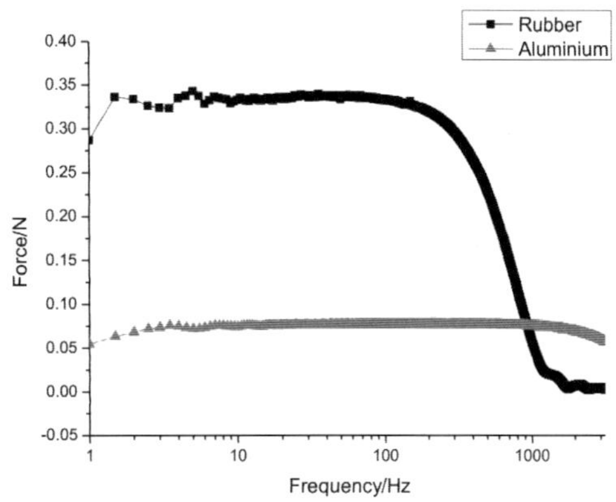

Figure 5. The auto power spectrum of force signal

From the figure, we can see that:

- The impact force amplitude of hammer equipped with a rubber head stood at about 0.34N in the region of 1Hz~100Hz;
- It declined sharply in the region of 100Hz~1000Hz;
- It approached zero when frequency was above 1000Hz;
- In the region of 1Hz~3200Hz, the impact force amplitude of hammer equipped with an aluminum alloy head stayed at about 0.09N.

3.2. Test results of big and small impact hammers

In our test, 2303-type and 2304-type impact hammers from Endevco were used, which are differentiated according to their factory certificates provided by the company. Their parameters are presented in Table 1.

Table 1. Factory parameters of 2303-type and 2304-type impact hammers

Section	Sensitivity	Maximum force	Resonance frequency	Frequency range	Head weight
	mV/N	N	kHz	kHz	g
Model 2303	0.227	35584	50	10	454
Model 2304	0.227	35584	50	10	1362

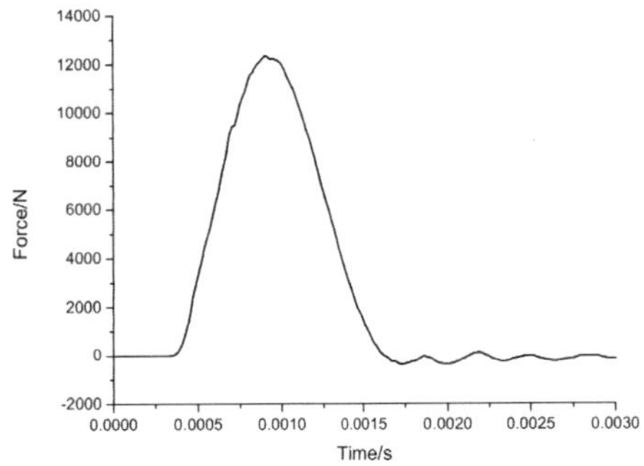

Figure 6. Time domain signals of force by the 2304-type hammer

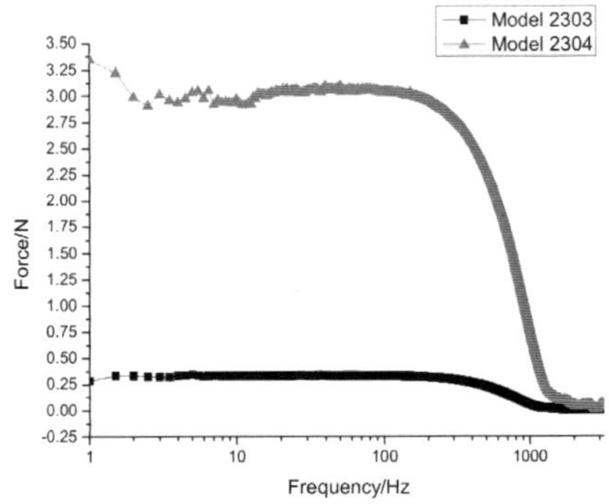

Figure 7. The auto power spectrum of force signal

A soft hammer head made of rubber was used in our test to hammer a same base. Force signal is shown in Figure 4 for the 2303-type hammer, and in Figure 6 for the 2304-type hammer. Figure 7 shows the auto power spectrum obtained after Fourier transform of these force signals, with a frequency range of 1Hz~3200Hz.

As shown in Table 1, with effects of their sizes ignored, the difference between 2303-type and 2304-type hammers lies in their weight alone, which was 454g and 1362g, respectively. It was the weight difference that caused their different impact force. By comparing Figure 4 and Figure 6, we know that:

- The impact force amplitude of 2304-type hammer was about 12000N, compared to 1200N for the 2303-type hammer;
- Both the hammers had impact duration of about 0.0015s.

From Figure 7 which was obtained after Fourier transform of force signals, it can be seen that:

- The overall force amplitude of 2304-type hammer was about 10 times that of 2303-type hammer;
- Their force amplitude didn't show great change in the region of 1Hz~100Hz;
- Their force amplitude decreased significantly in the region of 100Hz~1000Hz;
- In the region of over 1000Hz, their force amplitude almost dropped to zero.

3.3. Summary

Because of the relationship between mechanical impedance Z, hammer impact force F and impact response V as shown in Formula (1), impedance can be calculated by dividing hammer impact force signal by acceleration signal. Impact response is usually measured with a vibration acceleration sensor. Given small assumed background vibration of test environment, and an inherent noise of about 50dB~60dB of such sensor, impact force needs to be high enough to evoke the impact response of base during mechanical impedance measurement, enabling the vibration sensor to reach its signal-to-noise ratio.

In the low frequency region of 1Hz~100Hz, the hard hammer head failed to cause acceleration signal response due to its small impact energy, so that the acceleration sensor didn't meet its signal-to-noise ratio. In contrast, the soft hammer head applied impact for a longer time, and produced higher energy in the low frequency region, which allowed the acceleration sensor to achieve desired signal-to-noise ratio. Therefore, soft hammer head is more suitable for impedance determination in low frequency region.

4. Verification by experiments

In order to verify the conclusion drawn from force signal curves and auto power spectra in Section 3, three groups of impedance measurement were conducted on the same base, with test details described below.

4.1. Test details
We conducted our impedance determination according to GB/T 11349-2006 Vibration and Shock - Experimental Determination of Mechanical Mobility - Part 3: Measurement Using Impact Excitation. Apparatuses used in measurement are provided in Table 2. As required by the standard, we hammered a base for at least three times with an impact hammer, acquired hammer's force signal and the acceleration signal of vibration sensor on the base, and obtained origin mechanical impedance of the base by performing cross-power spectrum calculation on these two types of signals.

Table 2. Apparatuses used in measurement

	Hammer model	Hammer No.	Hammer head Material	Model of Acceleration sensor	No. of Acceleration sensor
Group 1	Model 2303	1091	Aluminum	4514B001	53731, 53670
Group 2	Model 2303	1091	Rubber	4514B001	53731, 53670
Group 3	Model 2304	1481	Rubber	4514B001	53731, 53670

4.2. Test results and analysis
On the test base there were two positions (1# and 2#) for sensor installation, and a vertical upward acceleration sensor was provided on each of them. Origin mechanical impedance tests were conducted on them, respectively, with hammering performed for at least 3~5 times for each group of tests. Calculation was done with a group of data with good correlation, and results are shown in Figure 8 and Figure 9.

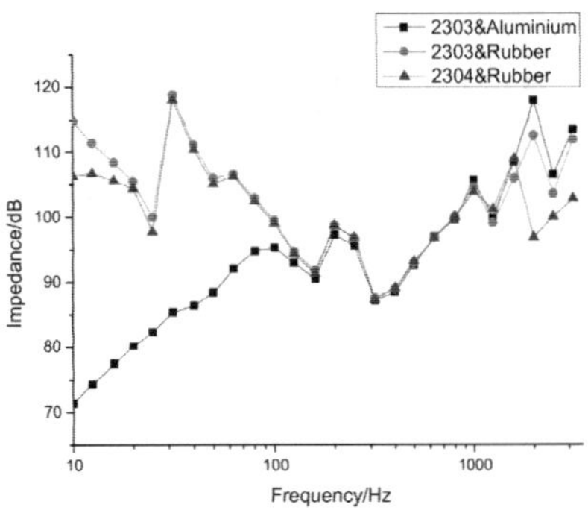

Figure 8. The impedance measurements of 1# position

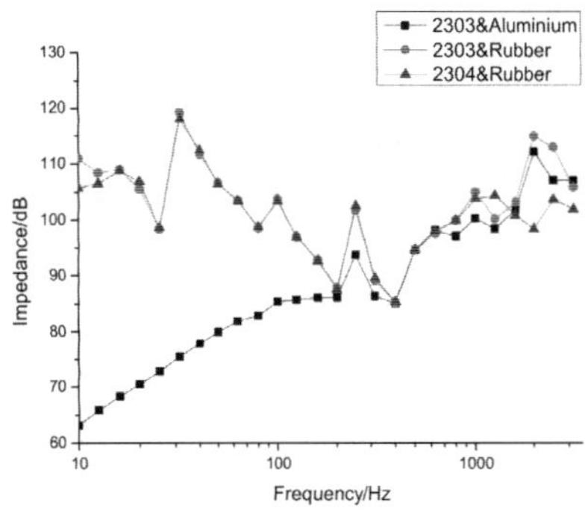

Figure 9. The impedance measurements of 2# position

Figure 8 and Figure 9 show the impedance measurements of test base obtained in the frequency range of 10Hz~3150Hz, with the x-coordinate representing frequency, with an exponential distribution, and the y-coordinate indicating mechanical impedance measurements, with decibel reference value of 1N/(m/s). In these figures, data indicated with a square are the test results of 2303-type hammer equipped with an aluminum alloy head, those indicated with a circle represents the results of 2303-type hammer equipped with a rubber head, and those identified with a triangle are the results of 2304-type hammer provided with a rubber head.

From Figure 8 and Figure 9, we can draw the following conclusion which is approximately in line with that from theoretical analysis:

- The impedance measurements of three groups of test were similar when frequency was above 100Hz;
- For the same model of hammer (2303-type), the aluminum alloy head produced low impedance in the low frequency range of 10Hz~100Hz which was about 20dB lower than that of the rubber head. But these two heads didn't show great difference in their impedance measurements when frequency was above 100Hz;
- Impedance measurements didn't differ greatly when the same rubber head, but different models of hammers were used.

5. Conclusion and further proposal

Based on the theoretical deduction and experimental test conducted in the paper, we can draw the following conclusion. And we put forward follow further proposal for base design and base impedance test:

- The structural impedance of base may cause strong vibration on equipment legs if it is too high, and fail to produce desirable vibration isolation efficiency when it is too low. Therefore, it is recommended to maintain impedance at 100±40dB;
- For impedance test, a hammer head made of soft materials such as rubber is recommended if your focus is on the data in low frequency region, and a hard head made of aluminum alloy may be a good choice if you want the data in high frequency region;
- Above recommendation is applicable to modal test. In other words, a soft hammer head is more suitable to obtain the modal in low frequency region.

At the end of the article, there are still deficiencies in our research. There is a lack of experimental data for impedance testing of bases with different weights. This is also the direction of our follow-up work.

References

[1] Z G Li 2015 Computation and analysis of the isolation performance of floating rafts and double-stage vibration isolation systems *Noise and Vibration Control*

[2] Z X Ye 2019 The influence and optimization of base design to the performance of vibration isolation system *Ship Science and Technology*

[3] C Zhu 2017 Research status and application of mechanical impedance *Science and Technology Innovation Herald* **14**(15) 2

[4] Z Q Xie 2014 Effect of onboard base impedance on vibration isolation efficiency of pump sets floating raft *Journal of Wuhan University of Technology* (*Transportation Science & Engineering*)

[5] L He & W Xu 2013 Naval vessel machinery mounting technology and its recent advances *Shengxue Xuebao / acta Acustica* **38**(2) 128-136

[6] M Y Lu, X Z Chen and X Peng 2008 Effects of the resilient mounting system parameters on the seat vibration level of the machine. *Noise and Vibration Control*

Design of cable pipeline fault location device based on dynamic laser scanning

Zhongxing Li[a], Caihong Zhang[a], Guojian Huang*[a], Haixia Ma[b]

[a]Guangdong Mechanical & Electrical Polytechnic, Guangzhou,China, 510515;

[b]Guangzhou City University of Technology, Guangzhou, China 510800

* Corresponding author: 3136944 @qq.com

ABSTRACT: Aiming at the problem that some of the spare power cable pipelines without cable are damaged, and there is no effective detection equipment to detect the blocked state of the pipeline. This paper proposed a dynamic laser scanning method for obstacles in the pipeline, and designed a power pipeline fault positioning device. The infrared laser ranging technology based on time-of-flight is used to dynamically detect the distance between obstacles in the pipeline and the laser sensor, so as to detect the status and location of pipeline blockage. Taking ARM microcontroller as the core, equipped with VL53L0X laser module, the hardware design of the device was completed, and the software program was completed based on the Keil μVision5 IDE environment. The measurement and error analysis of obstacles of two different materials are carried out. The results show that the longer the measurement distance, the larger the error, and the highest measurement error is 1.47%, which fully meets the needs of engineering measurement.

1. Introduction

At present, the urban distribution network mainly adopts the laying method of underground pipe gallery and pipeline, laying the distribution network underground, without competing with the public space for land resources, and using the form of the underground distribution network can also reduce natural conditions such as ice accretion influence of electrical wiring. When the underground cable passes through the road and other areas, it needs to use the pipe laying method. The pipeline must be laid first, and the cable will pass through the pipeline. In actual engineering, design redundancy will be reserved to prevent the need to re-erect the pipeline due to the increased electricity consumption in the area in the future. Therefore, some spare cable ducts are not laid with cables and are in an idle state. Long-term idle pipelines will cause various problems, such as idle pipelines and cable wells may be damaged due to improper construction by other units (jacking, burial, etc.). Due to the lack of detection methods, the operation and maintenance personnel cannot determine the situation in the pipeline, which may result in the cancellation of the planned project. In emergency repair operations, if the pipeline is blocked and cannot be used, the repair progress and power transmission time will also be extended. Construction workers "penetrating pipe detection" operations on the road are likely to cause traffic congestion and personal traffic accidents. Reference [1] discusses the basic principles and methods of ground penetrating radar, and analyzes the detection of metal pipelines and non-metallic pipelines combined with engineering examples. Reference [2] uses three-dimensional ground penetrating radar to conduct pipeline detection research. The results show that three-dimensional ground penetrating radar can display different directions, different buried depths, and different types of underground pipelines, and it has a significant application effect in urban underground pipeline detection work. Reference [3] uses three-dimensional geological radar to detect underground pipelines in a certain area, and analyzes the detection results and finds that the application effect of three-dimensional geological radar on underground pipeline detection is obvious. The detection of underground pipelines is generally realized by electromagnetic method and ground penetrating radar, and the internal environment monitoring of cable pipelines is mainly realized by camera technology. These two methods cannot well satisfy the problem of collapse, congestion, damage, deformation and other diseases of idle cable ducts. Although the small robot method can solve the problem well, it needs wired power supply, communication, inconvenient to carry, and has the requirements for the pipeline environment, so the application scope is relatively small [4-5]. At present, the market is generally aimed at the detection of underground pipelines and the internal environment monitoring of cable pipelines, but there is no research on the monitoring of idle cable pipelines [6-8]. In order to solve the above problems, improve the work efficiency of operation and maintenance personnel, and reduce the time-consuming, labor-intensive, and high-risk factors, this paper proposes a power cable pipeline fault location detection device based on dynamic laser scanning. It can quickly confirm whether there is a pipeline blockage within a certain distance.

2. The principle of laser measuring distance

Laser measuring uses a laser as a light source, and different methods use different laser materials. The laser of the phase laser measuring is a gas laser, and the gas laser is in a working state of continuous output; the pulsed laser measuring uses a solid laser; the infrared measuring uses a semiconductor laser. Compared with photoelectric measuring, laser measuring can work for a long time, improve measuring accuracy, and reduce weight and power consumption.

Laser measuring can be divided into pulse type and continuous wave phase type. It uses the characteristics of laser unidirectionality and coherence, anti-interference and strong directionality to carry out high-precision measurement and detection. Pulsed laser measuring is based on the calculation of the time-of-flight difference of the laser for ranging. The laser pulse duration is extremely short, the energy is relatively concentrated in time, and the instantaneous power is large. The principle of pulsed laser is similar to ultrasonic measuring. The laser emits a laser signal, which is reflected back to the sensor after encountering an obstacle. Record the round-trip time of the light signal, multiply it by the speed of light (the speed of light is 300,000 km/s), and multiply by 1/2 to get the measured distance. The schematic diagram is shown in Figure 1, and the calculation formula is shown in formula 1:

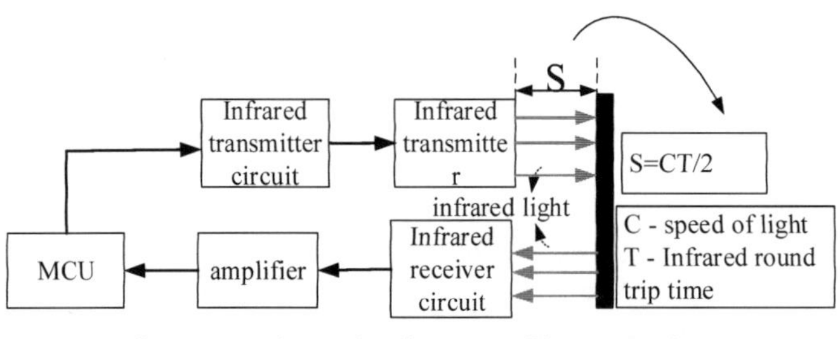

Figure 1. Schematic diagram of laser ranging

$$L = \frac{1}{2} c \cdot t \quad (1)$$

Phase laser measuring is an indirect method to measure the time required for light to travel back and forth through the line. First, use the frequency of the radio band to modulate the amplitude of the laser beam, and measure the phase delay generated by the modulation light going back and forth to the measuring line once, and then calculate the distance represented by this phase delay according to the wavelength of the modulated light. In this paper, the sensor based on the principle of infrared laser measuring is used with high-performance ARM chip to realize fault location of power cable pipeline. The general idea is shown in Figure 2.

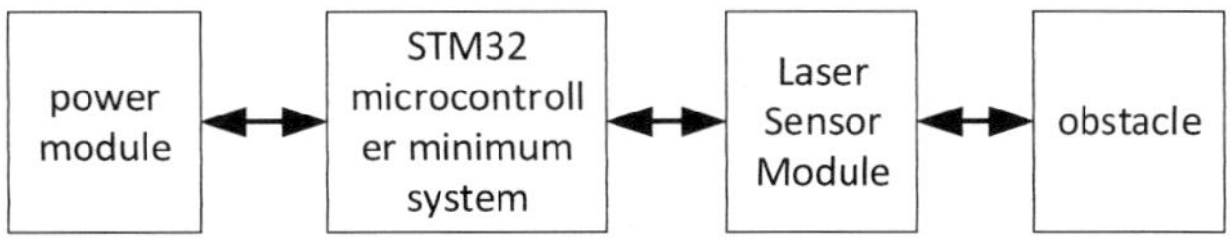

Figure 2. The idea of locating faults in power cable pipelines

3. Design of cable fault location device

3.1 General Design

This design system consists of ARM main control chip, laser sensor, basic circuit module and data processing and command module. The scheme of the system is shown in Figure 3:

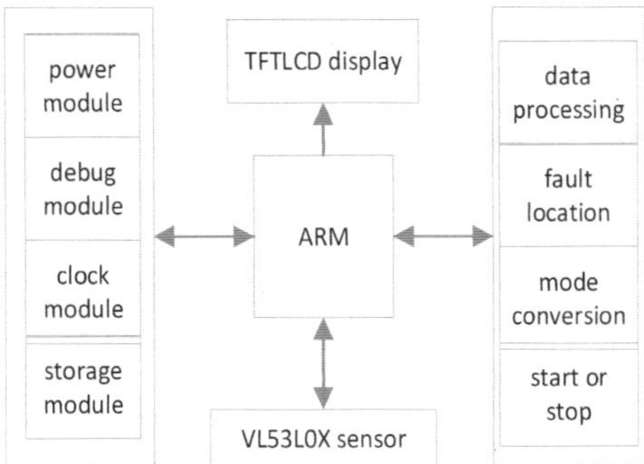

Figure 3. System hardware principle

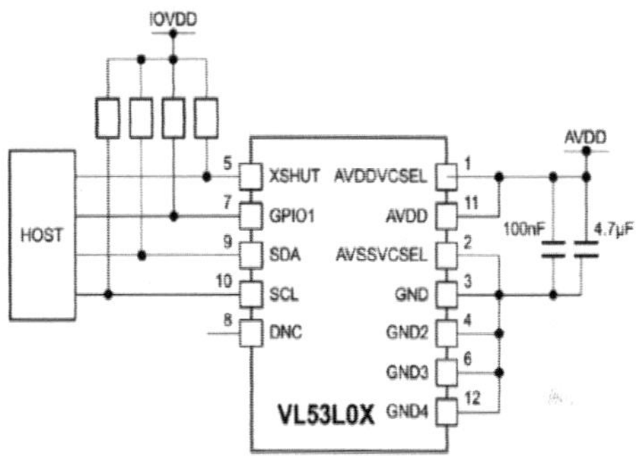

Figure 4. VL53L0X principle Figure

The non-contact VL53L0X laser module is selected as the sensor. Compared with previous generations of laser sensors, the module is small, with strong anti-interference ability, high precision and good stability. The liquid crystal display adopts a low power consumption, color TFTLCD display, which is used to display the distance signal and display the operation mode.

3.2 Hardware selection

3.2.1 ARM processor selection

The ARM processor adopts the STM32 series STM32103RC of STMicroelectronics (ST). It is based on the Cortex-M3 core and combines the 32-bit Harvard micro-architecture and system peripherals that execute Thumb-2 instructions. Reach 1.2DMIPS/MHz processing speed, built-in high-speed memory (up to 128K bytes of Flash and 20K bytes of SRAM). It contains 2 12-bit ADCs, 3 general-purpose 16-bit timers, 1 PWM timer, 2 I2Cs, 2 SPIs, 3 USARTs, 1 USB and 1 CAN, etc. The high integration level of this processor greatly simplifies the peripheral circuits, thereby enhancing the reliability and expansibility of the system and reducing the cost of the monitoring system.

3.2.2 Infrared sensor selection

Infrared sensor is used to detect the position of obstacles so that the corresponding distance can be obtained. According to the position measurement requirements of the device, it can be determined that the distance measurement during the movement needs to reach a sampling frequency of 20 Hz. In the process of measurement, the measurement principle based on infrared is used, which is equivalent to the distance that the laser source is emitted from the device, and is reflected back to the device when it encounters an obstacle. Combined with the actual measurement requirements of the device, the measurement range should reach 2m. Finally, the measurement accuracy of the sensor should be considered to be less than 5%, and the IIC interface or serial interface can be used for communication. After considering the above performance requirements, the non-contact VL53L0X laser is selected as the distance sensor of the device. The schematic diagram of the circuit is shown in Figure 4.

3.3 Software Design

The software of the power cable pipeline fault location detection device mainly completes the system initialization, the opening and closing of the infrared laser, the data collection of the position change during the movement, and the data transmission and display functions. Finally, the obstacle position information is obtained by analyzing, processing and calculating the data. Figure 5 is a software flow of a power cable pipeline fault location detection device.

The software programming is completed under the Keil μVision5 IDE environment of keil company. The main program of the system initializes the system, sends a start command to STM32, selects the mode according to the command to the laser ranging module, and then sends the return data to STM32 through the serial port, STM32 receives the data, processes and calculates, and displays it on the liquid crystal display module. The main program flow chart of the system is shown in Figure 5:

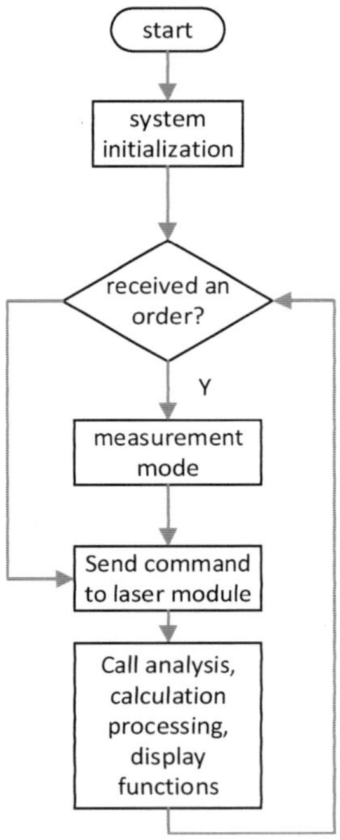

Figure 5. System flow chart

4. Experimental test

The design follows the principles of low cost and portability. When using, just install the device at the opening of the pipe, press the start button, it will automatically fix the posture, and automatically detect whether there is blockage within the detection distance. The Phenomenon Instrument returns the distance from the blockage to the pipe opening. The connection between the laser module and the ARM is shown in Figure 7.

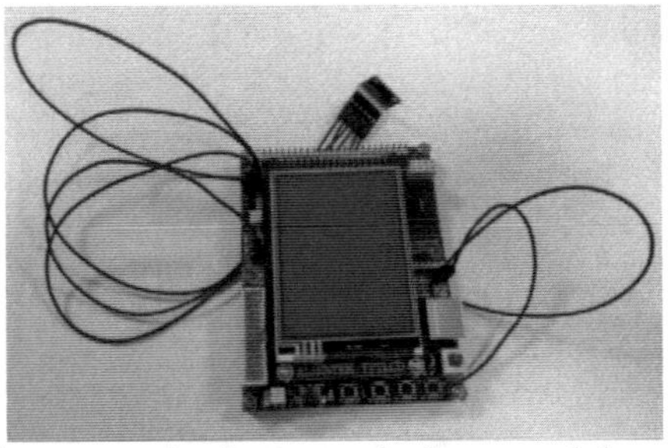

Figure 6. The connection between the laser module and the ARM

The design follows the principles of low cost and portability. The pin connection between the VL53L0X laser module and the main control system STM32F103RCT6 microcontroller is shown in Table 1.

Table 1. VL53L0X and STM32F103RCT6 pin correspondence table

The connection relationship between VL53L0X and STM32F103RCT6						
VL53L0X	VCC	GND	SCL	SDA	INT	XSH
STM32F103RCT6	5V	GND	PA3	PA2	PA4	PA15

The test conditions are as follows: white objects and paper objects are selected for high-precision ordinary distance test and general accuracy long-distance test. The longest distance of high-precision test is 1000mm, and the measurement is performed 10 times. The measurement results of the high-precision measurement mode are shown in Tables 2 and 3, and the measurement results of the long-distance measurement mode are shown in Table 4 and 5:

Table 2. Measurement results of white wall obstacles under high precision

actual distance /mm	measuring distance /mm	error /mm	Relative error
100	100	0	0
200	200	0	0
300	301	1	0.3%
400	401	1	0.25%
500	500	0	0
600	599	1	0.17%
700	700	0	0
800	802	2	0.25%
900	901	1	0.11%
1000	1002	2	0.2%

Table 3. Measurement results of cardboard obstacles under high precision

actual distance /mm	measuring distance /mm	error /mm	Relative error
100	100	0	0
200	200	0	0
300	301	1	0.3%
400	399	1	0.25%
500	501	1	0.2%
600	602	2	0.3%
700	701	1	0.1%
800	802	2	0.25%
900	898	2	0.22%
1000	1000	0	0

As can be seen from Table 2 and 3, in the high-precision measurement mode, the maximum measurement distance is 1000mm. With the increase of the measurement distance, the error does not change much. For different obstacles, the measurement error is basically the same, and the maximum error is only 2mm. This can meet the needs of engineering measurement.

Table 4. Measurement results of white wall obstacles at long distances

actual distance /mm	measuring distance /mm	error /mm	Relative error
200	202	2	1%
400	405	5	1.25%
600	603	3	0.5%
800	805	5	0.63%
1000	1008	8	0.8%
1200	1207	7	0.58%
1400	1412	10	0.70%
1600	1590	10	0.67%
1800	1785	15	0.83%
2000	2012	12	0.6%

Table 5. Measurement results of cardboard obstacles at long distances

actual distance /mm	measuring distance /mm	error /mm	Relative error
200	203	3	1.5%
400	406	6	1.5%
600	597	3	0.5%
800	804	4	0.5%
1000	1004	7	0.7%
1200	1206	7	0.58%
1400	1410	10	0.71%
1600	1613	13	0.81%
1800	1775	25	1.38%
2000	1972	28	1.47%

It can be seen from Tables 4 and 5 that in the long-distance measurement mode, with the increase of the measurement distance, the error becomes larger and larger. When the measurement distance reaches 2000mm, the maximum error reaches 28mm, and the relative error is 1.47%; for different obstacles, the error The change is large, especially when it is close to 2000mm, the error is quite different. In summary, if the positioning requirements are high during the measurement, the high-precision mode can be selected, but the measurement distance is small, and the location of the fault point can be measured more accurately; if the positioning requirements are not high, but the need to quickly detect whether there are obstacles, you can The long-distance measurement mode is adopted to provide reference for maintenance personnel.

5. Conclusion

In this paper, the stm32f103 series single chip microcomputer is used as the core to design a cable pipeline fault location device based on dynamic laser scanning, which realizes the function of detecting whether there are obstacles in the pipeline within a certain distance blocking the pipeline.

1) In the high-precision measurement mode, with the increase of the measurement distance, the error does not change much, and the measurement error is basically the same for different obstacles.

2) In the long-distance measurement mode, as the measurement distance increases, the error becomes larger and larger, and different obstacles have a greater impact on the error.

3) For different needs, different measurement modes can be selected to meet the needs of the actual situation.

References

[1] Shan haigang. Research on the Application of Ground Penetrating Radar Technology in Underground Pipeline Detection. Jiangxi Building Materials.2022,(03): 59-60.
[2] Wu yan,Jiao yongqiang,Li yan. Application of 3D Ground Penetrating Radar in Urban Underground Pipeline Detection. Geomatics & Spatial Information Technology.2022,45(01): 212-214.
[3] Chen Yangbiao,Xu Zhihao.Application of 3D Geological Radar in Urban Underground Pipeline Detection. Scientific and TTechnological Innovation.2022(04):189-192.
[4] Wang Wei,Zhao Shaokui. Research status and prospect of pipeline robot [J]. Ordnance Industry Automation,2019,38(12):24-30.
[5] Shandong Fuyue Electric Co., Ltd. a cleaning robot: China, 201610861086.6[P].2016-09-27.
[6] Fan Ji.Research on the Application of Ground Penetrating Radar in Urban Pipeline Survey [J]. Management & Technology of SME.2018(10):
[7] Zhao Hao.Research on fault diagnosis and identification of drainage pipeline based on acoustic wave detection [D]. Kunming University of Science and Technology.2018.
[8] Wang Yuhoong. Research and application of monitoring and early warning system of urban underground comprehensive pipe gallery [D]. Xi'an University of Architecture and Technology.2017.

Design of Lightning Current Monitoring System on Wind Turbines

Guojian Huang[1]*, Jiasi Huang[1], Chunyan Li[1], Yuquan Wu[1], Junjie Zhou[1], Yating Dai[1], Qingtong Li[1]

[1]Guangdong Mechanical & Electrical Polytechnic, Guangzhou 510515, China

*Corresponding author: 3136944@qq.com

ABSTRACT: In recent years, because climate change is a major issue related to human survival and sustainable development, the global climate is warming and glaciers are melting, so the country has been advocating "green environmental protection, low-carbon life". To achieve this goal, China is constantly looking for green and sustainable energy. However, in thunderstorm weather, wind turbines are more likely to be attacked by lightning. Even if we have a set of strict technology for lightning protection of wind turbines, we cannot resist direct lightning strikes. The blade of a wind turbine is the first target of lightning strikes. Therefore, this paper designs a lightning current monitoring system for wind turbines based on Rogowski coils, and designs a current measurement system that can be stably used to monitor the whole process of discharge when wind turbines are struck by lightning, so as to improve the safety of wind turbines. To improve the wind power generation technology due to the economic losses caused by lightning strikes to wind power generation enterprises.

1. Introduction

With the increasing depletion of global fossil energy, the international competition for energy has already started, the earth's energy reserves are becoming increasingly serious, and the development and use of renewable and clean energy has become a strategic pursuit of various countries. In the 1990s, he began to plan a strategic blueprint for clean energy, of which wind turbines are an important part[1~3]. It is clearly stated in the relevant development plan that by the end of the 20th century, China's wind turbine power generation should account for 20% of the world's total[4].

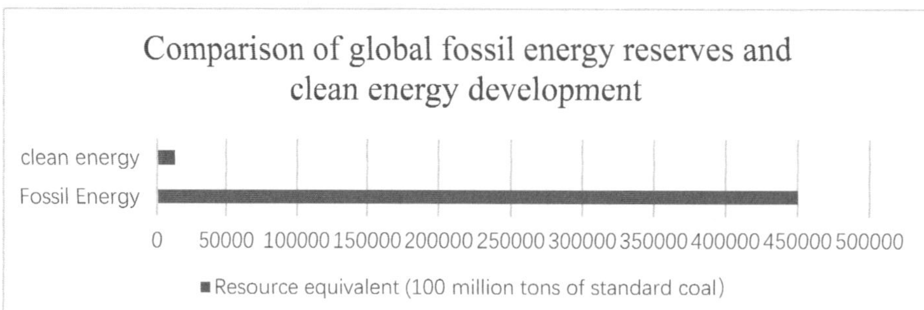

Figure 1 Comparison of global fossil energy reserves and clean energy exploitable capacity

As a new clean energy and renewable energy, wind turbines have made an important contribution to the protection of the environment, and also provided a stable power supply for the economy and national needs[5]. Now it has become one of the very important energy sources. To attract the attention of countries all over the world. According to statistics from the National Renewable Energy Information Management Center, China's approved wind power projects have accumulated a cumulative capacity of more than 40,000 megawatts; the national wind power grid-connected capacity has reached more than 127,000 megawatts[6,7].

Table 1 Expected benefits of independent operation and joint operation

run strategy		Expected return (yuan)
Operate independently	WPP1	128000
	WPP2	76800
	WPP3	51200
	CSP	1500
	sum	257500
joint operation		261500

In recent years, wind power technology has developed rapidly at home and abroad, and the development of technology has brought about the gradual expansion of the scale of its wind farms. A larger-scale power farm means that it can provide higher power generation, but its drawback is that it is The capacity of the single machine in the infield and the height of the tower body have been greatly increased. At present, the position of the blade tip of the existing high-power fan in China can reach up to 220m. This structure makes it easy to become the target of lightning strikes. According to statistics, more than 90% of lightning disasters are caused by induction lightning[8].

2. Design of lightning current monitoring system

2.1 Research status of lightning current monitoring

The most primitive method of measuring lightning current is to use magnetic steel to measure[9]. The principle of the magnetic steel measurement method is to measure the value corresponding to the lightning current according to the change of the magnetic flux during the discharge process of the lightning. The measurement of lightning current in the coastal some countries install the lightning current sensor on the top of the tower and utility pole. This method actually uses a Rogowski coil, as shown in Figure 2. Install Rogowski coils on both sides of the peak end of the pole, and then fix the Rogowski coils on the lightning rod. In this way, the obtained lightning parameters can easily reflect the current lightning activity in the area. The main lightning current measurement device is used to measure the waveform of lightning current. The processing and analysis of the data depend on the analysis of the data terminal[9,10].

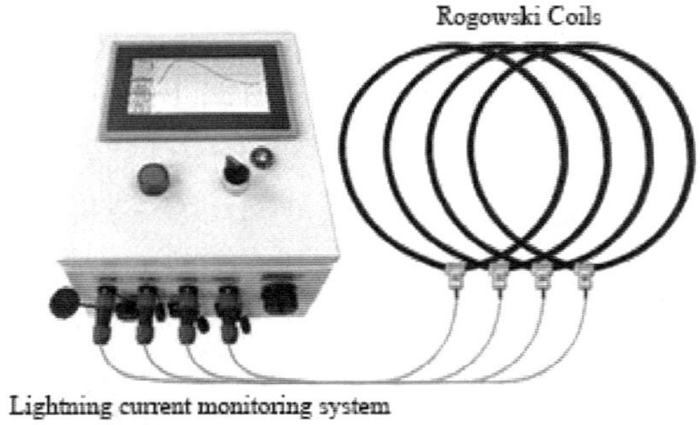

Figure 2 Lightning current monitoring system for large wind turbines

South American countries mainly use different sensing devices. The process is mainly to use the integrated chip of Microcontrollers to collect lightning current, and then use GPRS (General Packet Radio Service) to transmit data to facilitate the real-time observation of monitoring personnel. At present, in terms of equipment and methods for lightning current monitoring and positioning, there are still many places that need to be learned from developed countries.

2.2 The principle of lightning strike wind turbine

The main development principle of lightning strike wind turbine is shown in the development process of lightning strike wind turbine. At the beginning, the thundercloud and lightning generated a downward leader, and under the leading action of the downward leader, a positive charge was induced on the fan. As the descending pilot develops downward, the distance between the descending pilot and the fan decreases. Due to the electric field distortion of the blade tip, the nearby electric field intensity increases rapidly, the air is ionized, and the upward developing upward pilot appears. When the ascending pilot and the descending pilot are confluent and connected in the air, the main discharge process begins, and the fan is struck by lightning. The analysis shows that the lightning current when lightning strikes the wind turbine mainly includes the pre-discharge current in the form of the upstream pilot and the main discharge current in the form of the strike back current, which need to be measured separately.

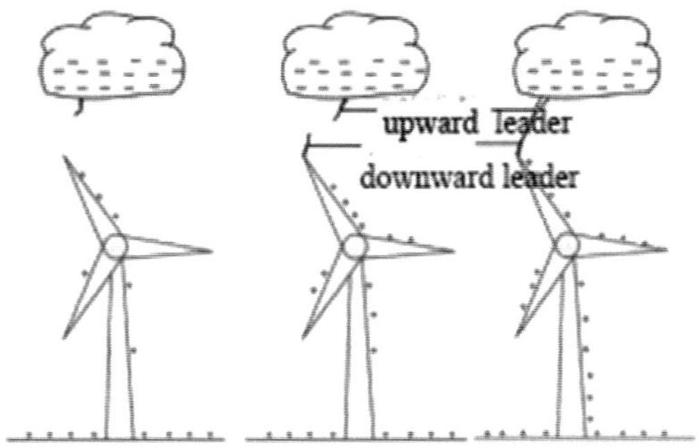

Figure 3 Development process of lightning strike wind turbine

2.3 Lightning current monitoring system design

The lightning current monitoring system is generally composed of a series of devices such as lightning receptors, lightning protection devices (down-conducting devices, side strike lightning protection), lightning current sensors, lightning monitoring boxes, and host computers.

When the wind turbine is hit by a direct lightning strike, the lightning strike hits the lightning receptor, and the lightning receptor leads the lightning current to the lightning current sensor to receive the lightning stroke. The lightning current sensor receives the lightning current signal induced by the lightning protection device and then transmits it to the lightning detection box for further processing. Among them, the lightning current detection unit should be installed near the hub of the wind turbine to avoid too large a phase difference between the detection unit and the lightning current sensor. The lightning current sensor measurement receives the external lightning current signal through the lightning protection device, and the lightning current signal enters the front-end processing of the lightning monitoring box through the cable; then, the lightning current signal is processed by the front-end processor and sent to the controller composed of DSP+CPLD. Using the data to analyze the lightning current waveform and peak current; the output device is used for the team to monitor the field command input and display the measurement data; the modem encodes and demodulates the monitoring data to realize long-distance transmission with the host computer; The 5G signal conducts two-way communication with the lightning monitoring box. Whenever the fan is struck by lightning, the counter built in the lightning monitoring box will be triggered to count, and the number of lightning strikes will be accumulated and counted.

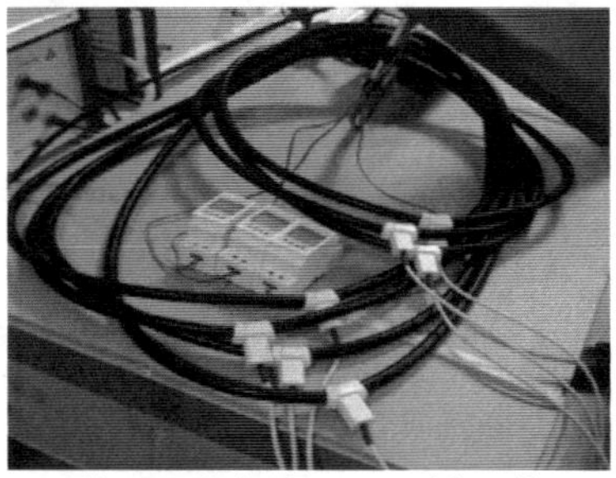

Figure 4 the Rogowski coils

According to the research, the lightning current monitoring system in this paper will use seven Rogowski coils as individual sensing units (FIG 4). Three of them are installed at the down-conductor of the blade at the lower end of the lightning arrester of the fan, and the other four are installed in the front, rear, left, right, and four directions of the tower base. Seven Rogowski coil sensors work together to inductively convert the lightning current flowing through the fan into a small electromotive force, which is used as the input signal for lightning current monitoring.

2.4 Lightning current monitoring system test

The prototype of the wind turbine lightning monitoring system developed in this paper is entrusted to a third-party monitoring agency of the Shanghai Lightning Protection Center Product Testing Center to conduct experimental monitoring. All data are valid.

Figure 5 shows simulated lightning current monitoring performance test (10/350μs), Figure 6 shows simulated lightning current monitoring performance test (8/20μs):

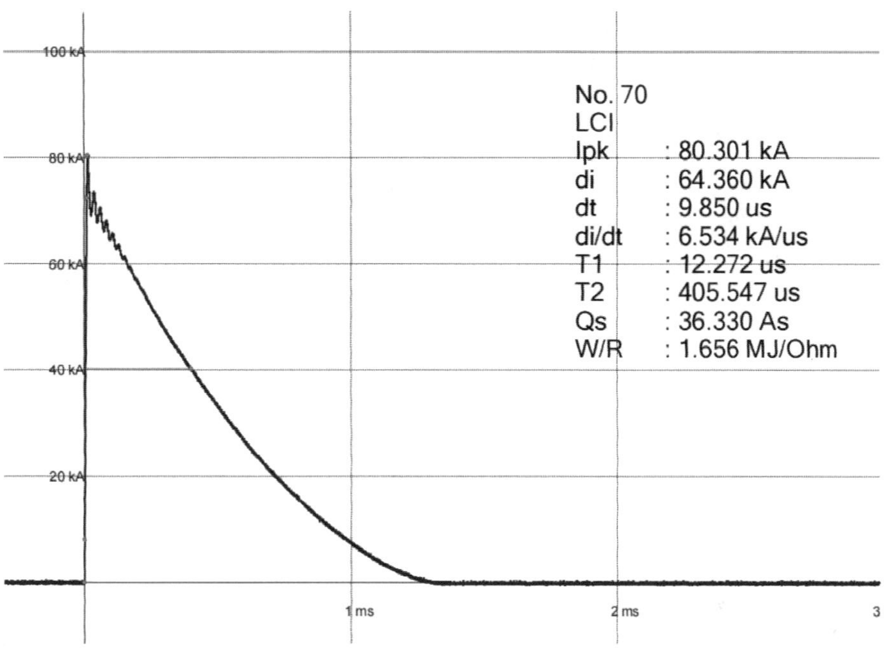

Figure 5 Current shape under the inrush current of the primary protection standard (10/350μs)

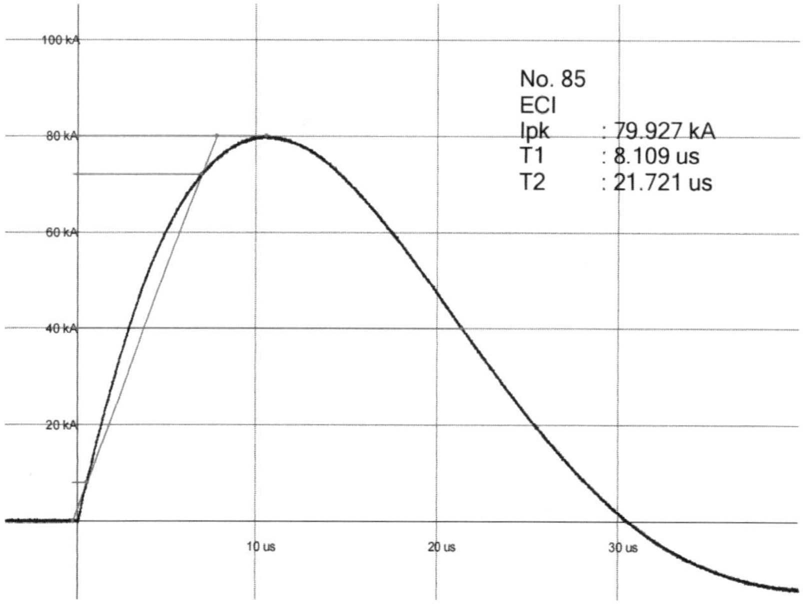

Figure 6 Current waveform under the secondary protection standard (8/20μs) impulse current

Rogowski coil sensor sensing performance simulation:

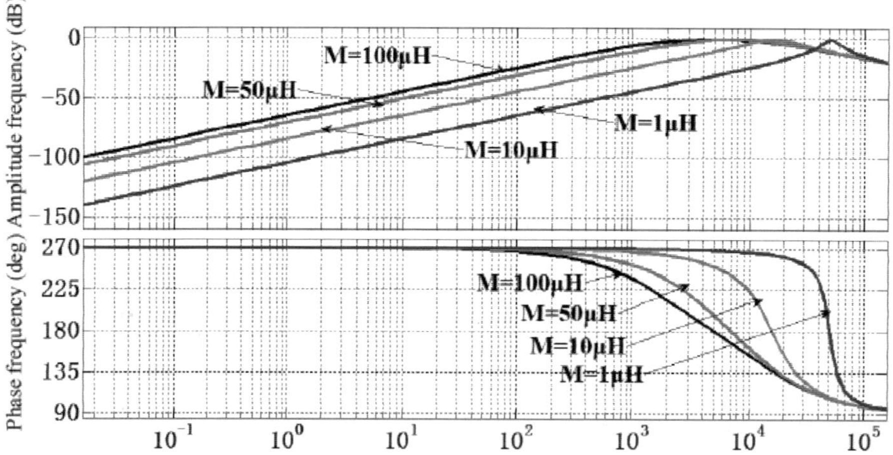

Figure 7 the relationship between the frequency response of the coil and the mutual inductance is implemented

3. The conclusion

Due to the special location of large wind turbines, the probability of being struck by lightning will greatly increase. In actual work, lightning strikes on wind turbines will bring great hidden dangers and damage, and the prices of various components of large wind turbines both are relatively expensive, have low installation convenience, and are relatively complicated to maintain later.

The use of large-scale wind turbines for power generation in China is not long, and the research experience on lightning protection is not sufficient. Therefore, this paper develops a lightning current monitoring system suitable for large-scale wind turbines by analyzing the whole process of lightning strike wind turbine discharge and the propagation path analysis of lightning current hitting wind turbines. The monitoring and analysis of lightning data through this system provides reliable data support for researchers to further study and evaluate accident levels, facilitate the design of lightning protection schemes, and ensure the safe and long-term operation of the entire large-scale wind turbine, thereby promoting wind power generation. The vigorous development of the machine.

Acknowledgment
This work was supported by the Key scientific research platforms and scientific research projects (Science and Technology) of colleges and universities of Guangdong Province (No. 2021ZDZX1108).

References

[1] Souvik Sen, Sourav Ganguly. Opportunities, barriers and issues with renewable energy development - A discussion - ScienceDirect[J]. Renewable and Sustainable Energy Reviews, 2017, 69:1170-1181.
[2] Moriarty P, Honnery D. New Energy Technologies: Microalgae, Photolysis and Airborne Wind Turbines[J]. Sci. 2021; 43(1): 5.
[3] Pan Li, Du Ershun, Wang Jianxiao et al. Research on economic benefits of combined operation of wind power generation and solar thermal power generation[J]. Comprehensive Smart Energy, 2022, 44(01).
[4] Wenjuan Zhang, Qing Meng, Ming Ma et al. Lightning casualties and damages in China from 1997 to 2009[J]. Natural Hazards, 2011. 57(2): 465-476.
[5] Bruce Glushakow, Haiyan Qin. Wind Turbine Lightning Protection of Blades, Electronics and Humans[C], 2019 International Symposium on Lightning Protection (XV SIPDA), 2019: 1-8.
[6] Vin Cent Tai1, Yong Chai Tan1, Nor Faiza Abd Rahman1 et al. A Novel Power Curve Prediction Method for Horizontal-Axis Wind Turbines Using Artificial Neural Networks[J]. Energy Engineering, 2021. 118(3):507-516.
[7] Yoh Y. A new lightning protection system for wind turbines using two ring - shaped electrodes[J]. IEEJ Transactions on Electrical & Electronic Engineering, 2010, 1(3):314-319.
[8] Rachidi F, Rubinstein M, Montanyà J et al. A Review of Current Issues in Lightning Protection of New-Generation Wind-Turbine Blades[J], in IEEE Transactions on Industrial Electronics, 2008, 55(6): 2489-2496.
[9] Fontanes P, Montanyà J, Arcanjo M et al. Experimental investigation of the electrification of wind turbine blades in fair-weather and artificial charge-compensation to mitigate the effects[J]. Journal of Electrostatics, 2022, 115:103669.
[10] Arif Waqas, Qingmin Li, Wanshui Yu et al. A Review on Experimental Study of Wind Turbine Blade Lightning Protection System[C], 2019 IEEE International Conference on Power, Intelligent Computing and Systems (ICPICS), 2019:25-29.

Research on test and evaluation of Li-MnO_2 primary battery under high temperature stress storage

Jin Deng[a], Jinghong Jiang*[a], Zhimao Ming[a]

[a]Guangzhou GRG metrology and test Co., Ltd, Guangzhou, China

* Corresponding author: jiangjh@grgtest.com

ABSTRACT: This paper mainly studies the influence of high temperature conditions on Li-MnO_2 primary battery, and analyzes the influence of high temperature conditions on the performance of Li-MnO_2 primary battery by testing the capacity, voltage and internal resistance of the product. Through impedance test, the different polarization differences between products stored at high temperature and normal temperature were analyzed. The effects of high temperature environment on the internal macro structure and micro material components of the battery were analyzed by means of electronic computed tomography (CT), scanning electron microscope and X-ray energy spectrometer (SEM-EDS) and X-ray photoelectron spectroscopy (XPS). Finally, through various test data, the internal reaction mechanism of the battery at high temperature was analyzed to deepen the research on the performance of Li-MnO_2 primary battery.

1. INTRODUCTION

As a primary battery, Li-MnO_2 battery has been widely used in industrial, medical, military and other fields due to its low cost and stable performance. Li-MnO_2 primary battery are generally made into buckle or cylindrical shape structure [1-3].

In the Li-MnO_2 primary battery, we generally use manganese dioxide (MnO_2) as positive electrode, lithium metal (Li) as the negative electrode, electrolyte is lithium perchlorate ($LiClO_4$), propylene carbonate ($C_6H_6O_3$) and ethylene glycol dimethyl ether ($C_4H_{10}O_2$) mixed solvent, diaphragm is generally polypropylene material $(C_3H_6)_n$. The battery discharge reaction is shown in (1).

$$MnO_2 + x\text{Li} \rightarrow \text{Li}_x MnO_2 \tag{1}$$

High energy density of Li-MnO_2 primary battery can be obtained by adding a large number of active substances in electrode preparation. In order to improve the energy density, the specific surface of the electrode material decreases, which limits the discharge rate performance of the Li-MnO_2 primary battery. In order to pursue excellent battery performance, researchers optimize the material system and manufacturing process of Li-MnO_2 primary battery. Xianling Meng studied the performance differences of lithium fluorocarbon-Manganese dioxide batteries at different temperatures and storage times, and found that with the increase of drying temperature, the gas expansion of batteries was significantly inhibited [4]. Ruirui Liu mainly studied the influence of fluorocarbons on the performance improvement of Li-MnO_2 primary battery [5]. Fen Liu mainly studied the influence of superconducting carbon black, graphene and carbon nanotube conductive agent on the rate performance of Li-MnO_2 primary battery [6].

Due to its low self-discharge rate of disposable batteries, lithium-manganese batteries are used as backup power sources, with a storage life of up to 10 years under certain conditions. Zhonghong Dong studied the performance changes of lithium/thionyl chloride batteries stored for a long period at high temperature, and the results showed that the attenuation trends of accelerated storage at high temperature and storage at room temperature were consistent [7]. However, there are few researches on the performance, internal material mechanism and reaction mechanism of Li-MnO_2 primary battery under high temperature conditions. This paper mainly studies the effect of high temperature storage on the microstructure of electrode materials and the performance of the battery, and analyzes the internal reaction mechanism of the battery at high temperature by means of different gauges.

2. RESULTS AND DISCUSSION

2.1 Electrical performance test of stored samples at high temperature and room temperature

After storage at high temperature (1#, 2#, 3#) and normal temperature (4#, 5#, 6#), the open circuit voltage of the sample is 3.26464V, 3.23479V, 3.2434V, 3.24774V, 3.24708V, 3.2611V, respectively. It can be found that the battery voltage after high temperature storage has no obvious change compared with normal temperature storage condition, and the test results are shown in Fig. 1 below (1#, 5#).

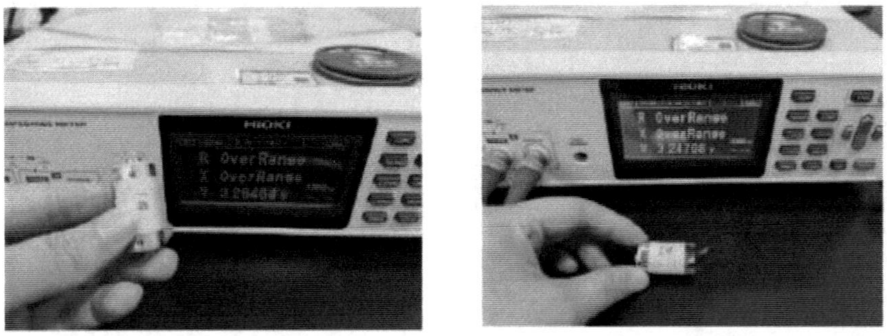

(a) (b)

Figure 1. Battery open-circuit voltage test after high and normal temperature storage. (Figure a is the 1# sample, b is the 5# sample)

Open-circuit voltage test results show that high temperature stress does not cause significant voltage drop of the battery. Then, we conducted electrochemical impedance test on the products stored at high temperature (3#) and room temperature (6#). Fig. 2 shows the impedance spectrum of test samples, and Fig. 3 shows the equivalent circuit diagram constructed by Li-MnO_2 primary battery.

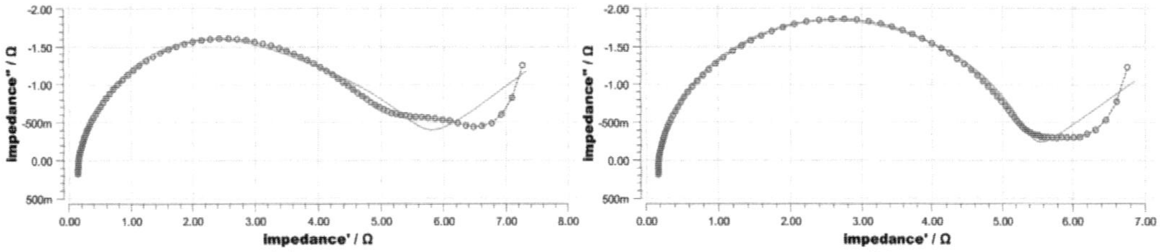

(a) (b)

Figure 2. Impedance spectra of the battery after storage at high and normal temperature. (Figure a is the 3# sample, b is the 6# sample)

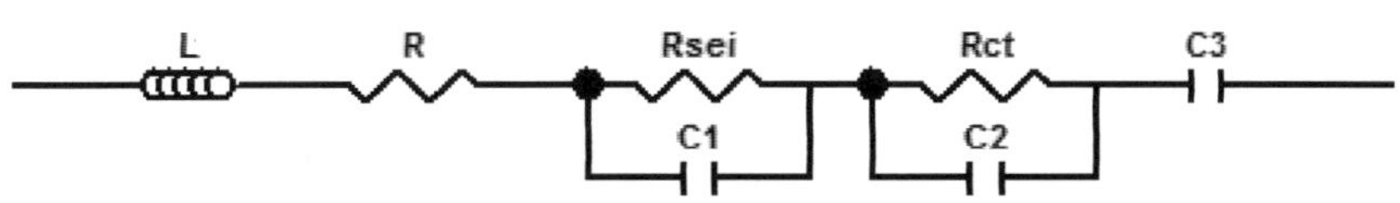

Figure 3. Equivalent circuit diagram.

According to the battery equivalent circuit and impedance test, the battery impedance data was fitted and analyzed, as shown in Fig. 4 and Fig. 5 below. It can be found that the ohmic polarization (R), solid electrolyte impedance (R_{sei}), electrochemical polarization (R_{ct}) and concentration polarization (warburg impedance and ionic diffusion impedance) of the samples stored at high temperature and room temperature have no significant changes. In addition, it can be found that the electrolyte impedance and electrochemical impedance of Li-MnO_2 primary battery are one order of magnitude higher than the ohm resistance of the battery itself, which limits the application of Li-MnO_2 primary battery in high rate scenarios.

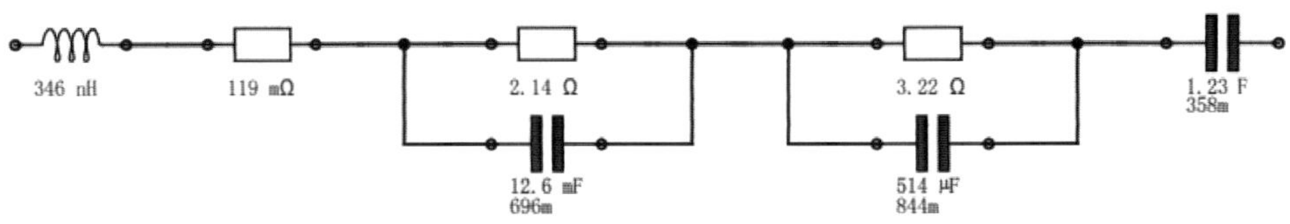

Element	Value	Significance	Error
inductor (L)	346 nH	0.956	2.68%
resistor (R)	119 mΩ	0.802	2.39%
resistor (R)	2.14 Ω	0.342	11.14%
constant-phase-element (V)	12.6 mF	0.166	10.82%
constant-phase-element (exponent)	696 m	0.718	2.00%
resistor (R)	3.22 Ω	0.641	3.00%
constant-phase-element (V)	514 μF	0.579	3.18%
constant-phase-element (exponent)	844 m	4.542	0.44%
constant-phase-element (V)	1.23 F	0.250	5.16%
constant-phase-element (exponent)	358 m	0.270	7.60%

Figure 4. Equivalent circuit diagram of (3#) battery after high temperature storage.

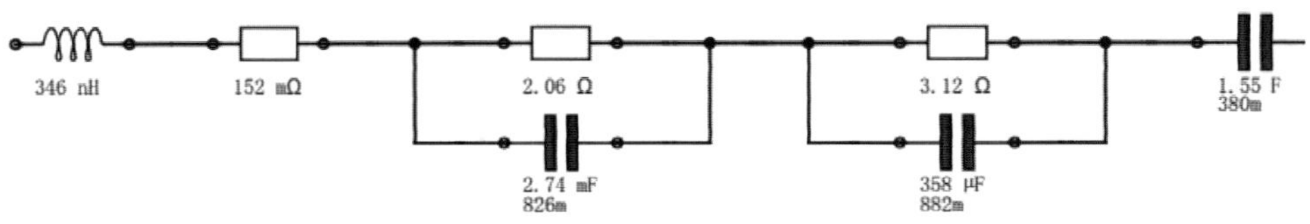

Element	Value	Significance	Error
inductor (L)	346 nH	0.875	2.26%
resistor (R)	152 mΩ	0.887	1.47%
resistor (R)	2.06 Ω	0.363	4.46%
constant-phase-element (V)	2.74 mF	0.194	4.29%
constant-phase-element (exponent)	826 m	1.145	0.62%
resistor (R)	3.12 Ω	0.582	2.14%
constant-phase-element (V)	358 μF	0.552	2.37%
constant-phase-element (exponent)	882 m	4.539	0.38%
constant-phase-element (V)	1.55 F	0.220	13.85%
constant-phase-element (exponent)	380 m	0.153	9.89%

Figure 5. Equivalent circuit diagram of (6#) battery after storage at room temperature.

After the impedance test, the products stored at high temperature (3#) and normal temperature (6#) were subjected to constant discharge, and the discharge curve was shown in Fig. 6. It can be seen that the discharge trend and fluctuation of the two products are consistent in the discharge process, but the discharge platform of the product stored at high temperature is 30 mV lower than the product stored at normal temperature. The change of the discharge platform may be caused by the adverse reaction in the battery under high temperature, leading to the decline of the discharge platform. This is also corresponding to the polarization impedance and electrolyte impedance of the two temperature samples in impedance test. The polarization resistance of the high temperature storage sample is (3.22 Ω) higher than that of the normal temperature storage sample (3.12 Ω). In addition, the electrolyte resistance stored at high temperature was (2.14 Ω) higher than the polarization resistance (2.06 Ω) of the stored sample at normal temperature, indicating that the cathode material had changed under high temperature storage. In order to further understand the internal changes of the battery, we carried out CT test and dismounted physical and chemical analysis of the battery.

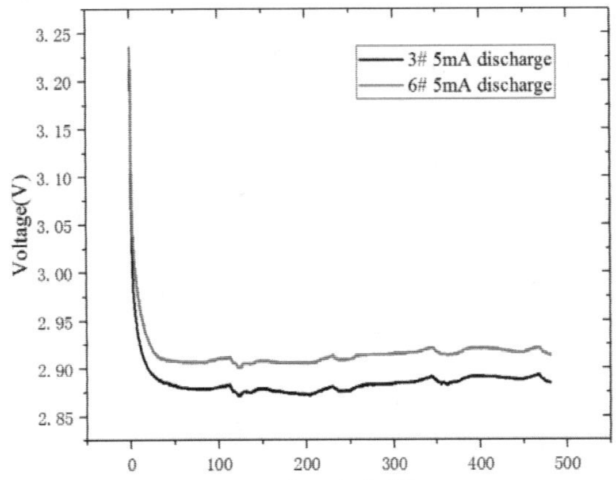

Figure 6. Cell constant discharge curve after high and normal temperature storage.

2.2 CT nondestructive testing of stored samples at high temperature and normal temperature

After storage at high temperature (1#, 2#) and normal temperature (4#, 5#), the battery samples were detected by CT, and the contents of the battery were observed from different sections, as shown in Fig. 7 below.

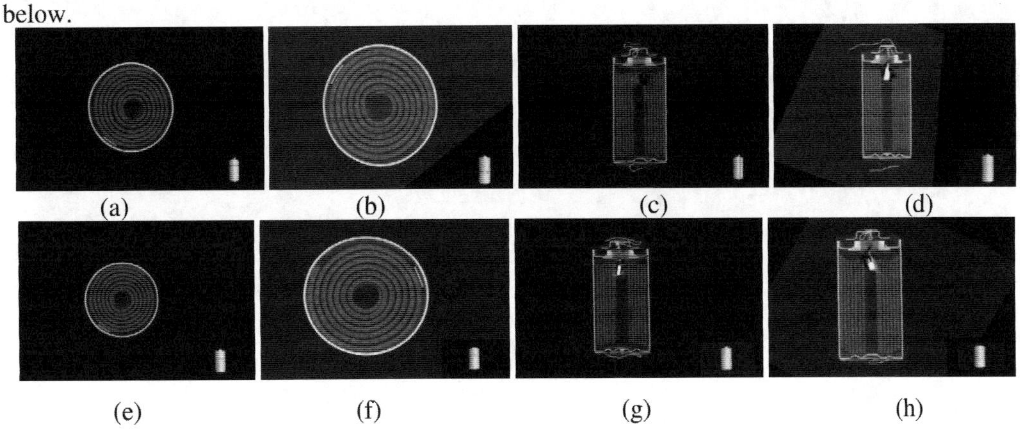

Figure 7. CT images of battery after storage at high and normal temperature. (Figure a is top view of sample 1#, and b is top view of sample 2#, c is the side view of sample 1#, d is the test view of sample 2#, e is the top view of sample 4#, f is the top view of sample 5#, g is the side view of sample 4#, h is the test view of sample 5#)

Through CT detection, it can be seen that the core of the cell in high temperature storage has small chap and powder drop phenomenon, and the cell in normal temperature is smooth.

2.3 SEM-EDS detection of stored samples at high temperature and normal temperature

In order to further understand the change of material structure inside the cell, the high temperature (1#, 2#) and normal temperature (4#, 5#) cells were disassembled, and the cathode materials were characterized by SEM test method, as shown in Fig. 8 below. The surface of the cathode material stored at high temperature is more severely chapped than that stored at room temperature, which is consistent with CT detection.

(a)

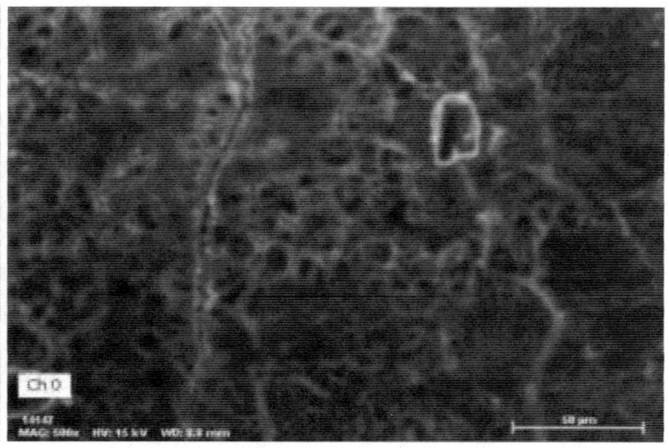

(b)

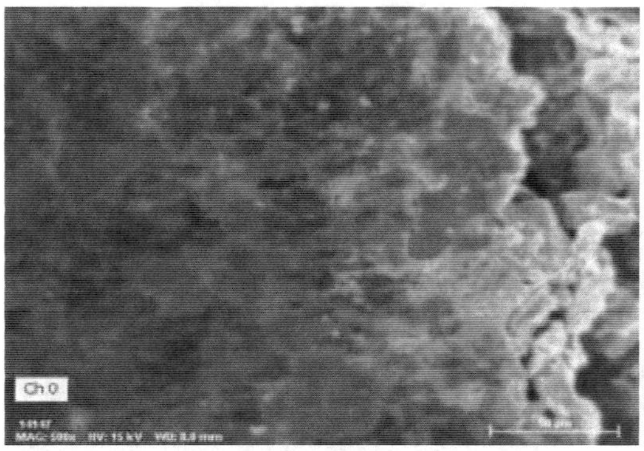

(c)

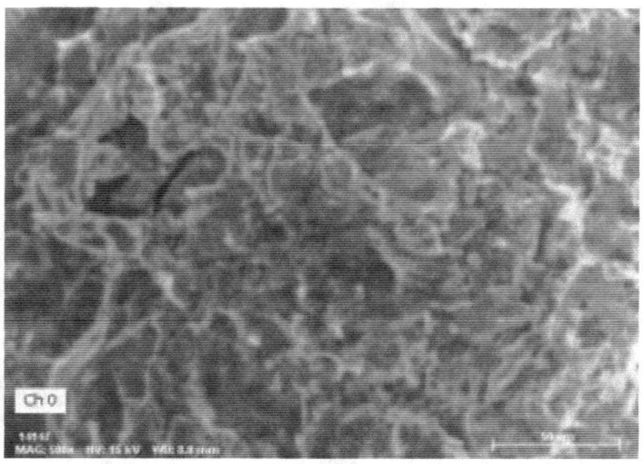

(d)

Figure 8. SEM of cathode material after storage at high temperature and normal temperature. (Figure a is the 1# sample, b is the 2# sample, c is the 4# sample, and d is the 5# sample)

In order to further understand the difference of cathode material surface under high temperature and normal temperature conditions, EDS detection was performed on the samples, as shown in Fig. 9 below.

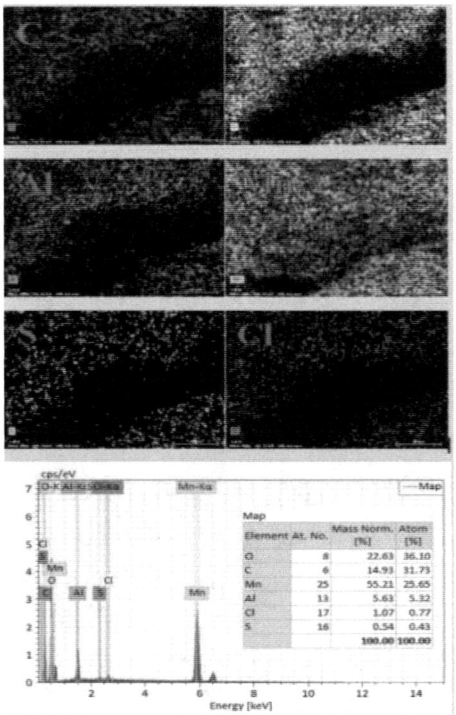

(a)

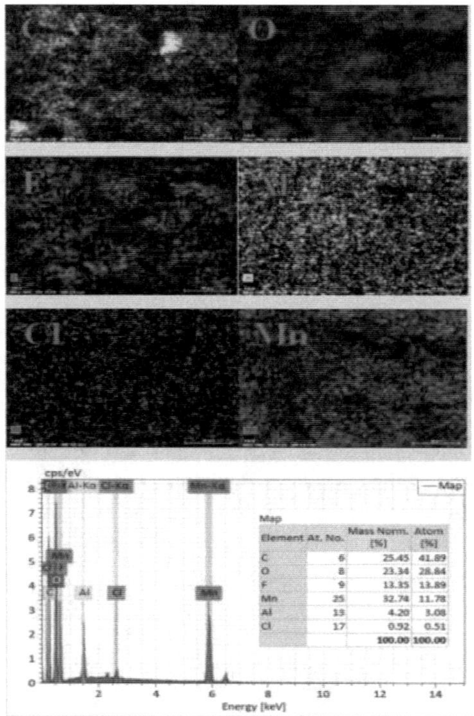

(b)

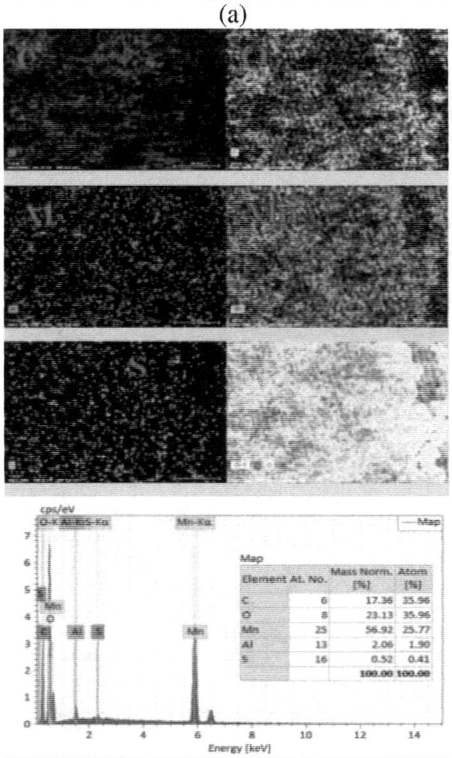

(c)

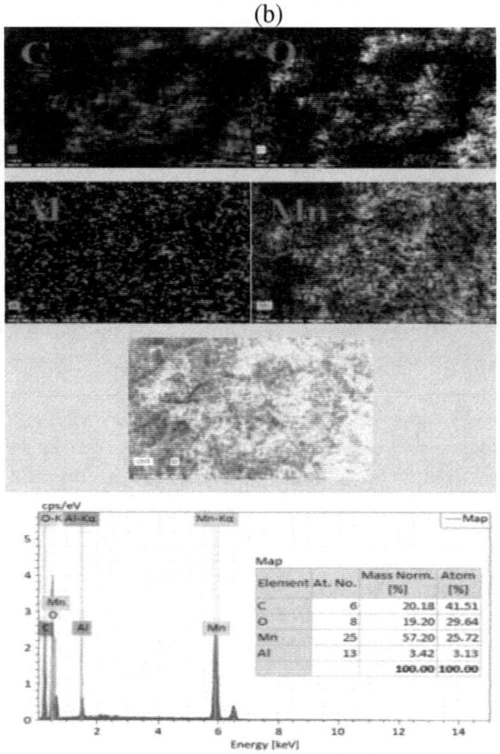

(d)

Figure 9. SEM-EDS diagram of cathode material after storage at high and normal temperature. (Figure a is the 1# sample, b is the 2# sample, c is the 4# sample, and d is the 5# sample)

Through SEM-EDS test characterization, it can be found that Cl and F are introduced into the cathode surface of the high temperature storage cell. The introduction of these two elements may be related to the immersion of electrolyte $LiClO_4$ on the cathode surface and the precipitation of binder polyvinylidene tetrafluoroethylene PVDF from the cathode material under high temperature conditions. This also indicates that under high temperature conditions, side reactions in Li-MnO_2 primary batteries are intensified, and electrolyte salts are more easily contacted with positive electrode materials through infiltration characteristics of electrolyte.

2.4 XPS test of stored samples at high temperature and normal temperature

In order to further understand the surface differences between high temperature (2# cell) and normal temperature (5# cell) lithium-manganese batteries and the surface side reaction mechanism of cathode materials at high temperature, Ar etching technology and XPS analysis technology were used to observe the surface and internal element differences of materials.

Fig. 10 and Fig. 11 show the full spectrum analysis (0nm depth) of the cathode material surface after high temperature storage and the quantitative analysis diagram of elements with different thickness of the cathode material after etching (etching thickness includes 0nm, 50nm, 100nm, 125nm and 150nm). It can be seen that the results of XPS surface elements are in good agreement with those of SEM-EDS analysis. Li element in XPS does not appear in EDS, mainly because EDS can only detect elements above C element.

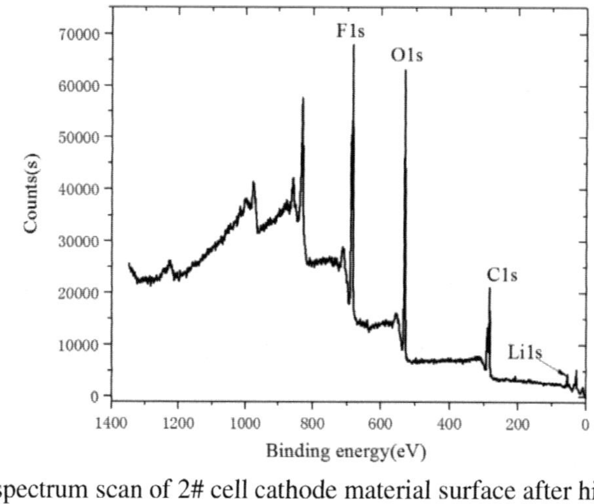

Figure 10. XPS full spectrum scan of 2# cell cathode material surface after high temperature storage.

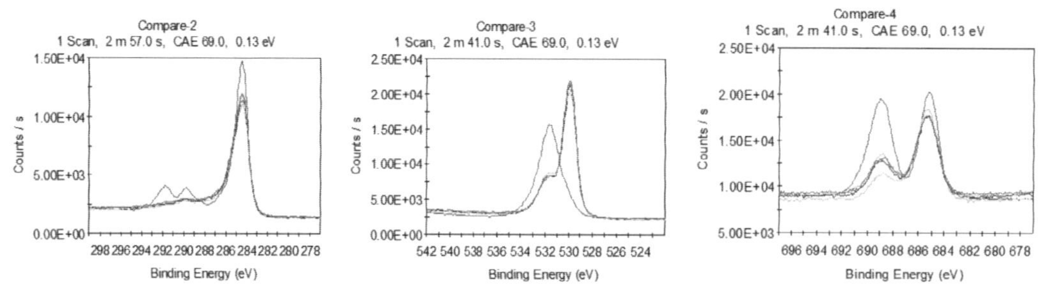

(a) (b) (c)

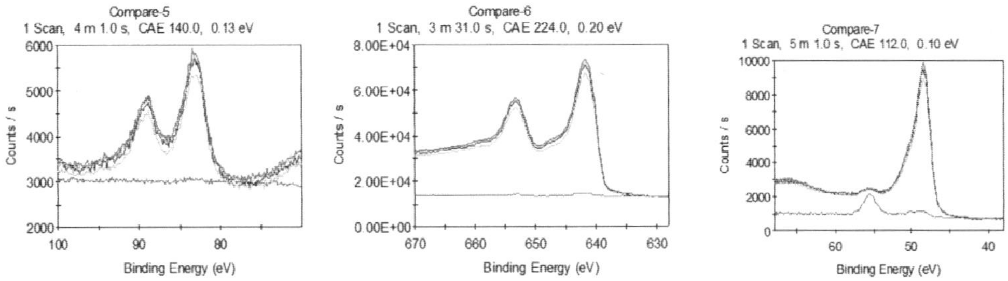

(d) (e) (f)

Figure 11. XPS fine spectrum scanning of 2# cell cathode material with different thickness after high temperature storage. (a、b、c、d、e、f are C1s, O1s, F1s, Mn3s, Mn2p, and Mn3p respectively)

Fig. 12 and Fig. 13 show the full spectrum analysis (0nm depth) of the cathode material surface after storage at normal temperature and the quantitative analysis diagram of elements with different thickness of the cathode material after etching (etching thickness includes 0nm, 50nm, 100nm, 125nm and 150nm).

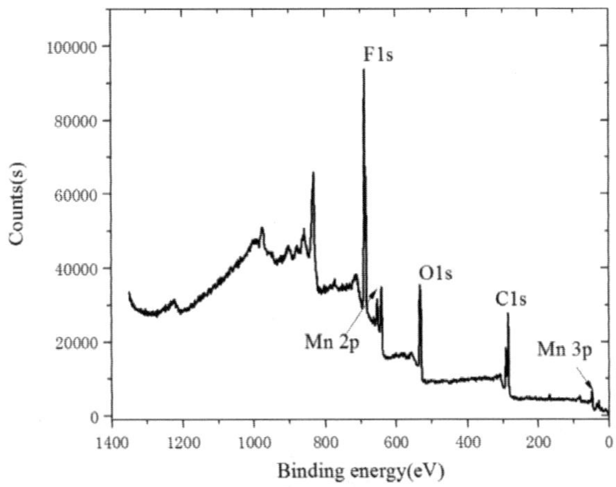

Figure 12. XPS full spectrum scan of 2# cell cathode material surface after nomal temperature storage.

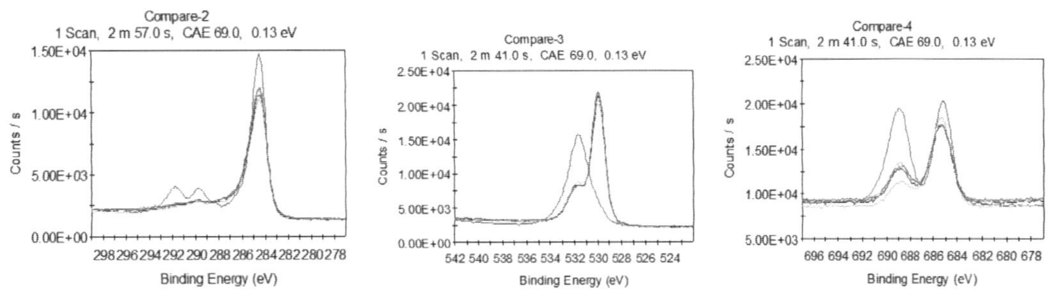

(a) (b) (c)

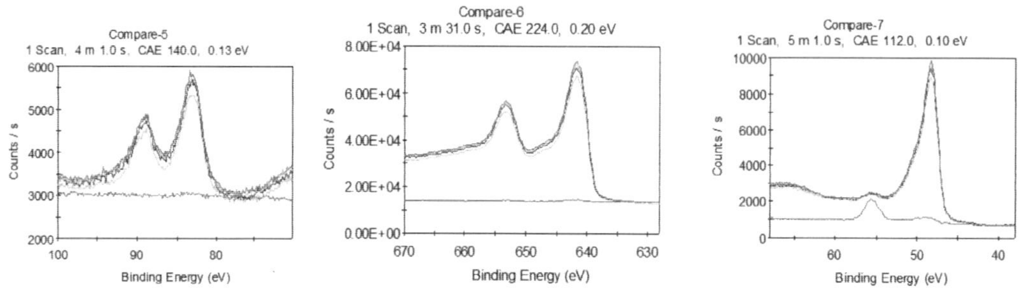

(d) (e) (f)

Figure 13. XPS fine spectrum scanning of 2# cell cathode material with different thickness after normal temperature storage. (a、b、c、d、e、f are C1s, O1s, F1s, Mn3s, Mn2p, and Mn3p respectively)

By comparing the full spectrum of XPS on the surface of positive electrode materials stored at high temperature and normal temperature, we can find that the samples stored at normal temperature have obvious characteristic peaks at 641.5 eV and 653 eV, which correspond to the characteristic peaks of manganese element 2P ($Mn2p_{3/2}$ corresponds to 641.5 eV and $Mn2p_{1/2}$ to 653 eV), as shown in e in Fig. 12 and Fig. 13. It also indicates that manganese exists in +4 valence state, which is consistent with the valence state of MnO_2 [8]. However, no characteristic peak of manganese appeared in the full spectrum of high temperature storage, indicating that other substances deposited on the positive electrode surface after high temperature storage. The result is also consistent with the element content in Fig. 9. The Mn content on the positive electrode surface measured by SEM-EDS in high temperature storage (11.78%) is significantly lower than that in normal temperature storage (25.72%). With the etching of Ar, positive active substance was also detected, as shown in d, e and f in Fig. 11.

In addition, we conducted statistical analysis of element contents at different etching depths, as shown in Fig. 14 below. When etched to about 50 nm, the lithium element stored at normal temperature tends to zero, and the lithium element content stored at high temperature is 10%, while the lithium element content stored at high temperature is still 6% after etched to 150nm. This also proves that under high temperature conditions, side reactions occur on the positive surface, leading to film formation and thickening of the positive surface. The cathode surface deposition and film thickening during high temperature storage also correspond to the increase of R_{sei} in cell EIS under high temperature storage. High temperature storage also results in a lower discharge voltage platform.

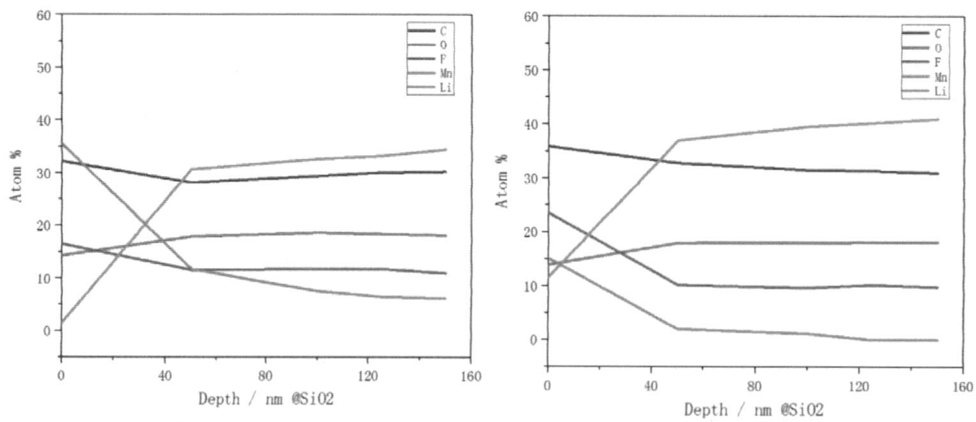

(a) (b)

Figure 14. Plot of element contents in different etched thicknesses. (Figure a is 2# high temperature storage cell cathode material, b is 5# normal temperature storage cell cathode material)

In order to further analyze the internal reaction mechanism, peak fitting was carried out for C1s, O1s, F1s and Li 1S, as shown in Fig. 15 below.

For the analysis of C1s, the peak of 285 eV is attributed to the C-H and C-C bond formed by the decomposition of conductive additive and electrolyte, and the peak of 286.1 eV is attributed to C-O and CH_2 functional group, among which the C-O functional group comes from ROLi, polyether or R-CH_2-OCO_2Li in the positive sediment. The peak of 287.5 eV can be classified as O=C-O functional group, which may come from esters or carboxylate components. The peak of 290.2 eV comes from CO_3, Li_2CO_3 and other substances.The peak of 291.9 eV comes from CF_2 in the binder.

The analysis of O1s mainly contains four peaks, the peak at about 529.7 eV is from Mn-O bond in the cathode material, the peak at about 531.1 eV is from ROLi (lithium alcohol), the peak at about 532.15 eV is from carbonate or carbonate ($ROCO_2Li$), and the peak is from CO_3 /O-C=O group, 533.6 eV comes from the C-O bond (polyether, ester compound).

The XPS spectrum of F1s is mainly divided into two peaks, the peak of 685.4 eV corresponds to LiF in the positive sediment, and the peak of 689 eV mainly comes from C-F in the binder.

The XPS spectrum of Li1s is mainly divided into two peaks. The peak of 55.5 eV corresponds to the Li-containing organics on the positive electrode surface, such as ROLi or R-CH_2-OCO_2Li, and the peak of 56.3 eV mainly comes from LiF in SEI [9-11].

Therefore, we can speculate that the occurrence of adverse reactions in lithium batteries at high temperature plays a promoting role, and the adverse reactions mainly include the following two aspects.

$$2C_4H_6O_3 + 2Li^+ + 2e^- \rightarrow (CH_3CH_2O_2COLi)2\downarrow + C_2H_2\uparrow \quad (2)$$

$$(CH_3CH_2O_2COLi)2\downarrow + H_2O \rightarrow Li_2CO_3\downarrow + CO\uparrow \quad (3)$$

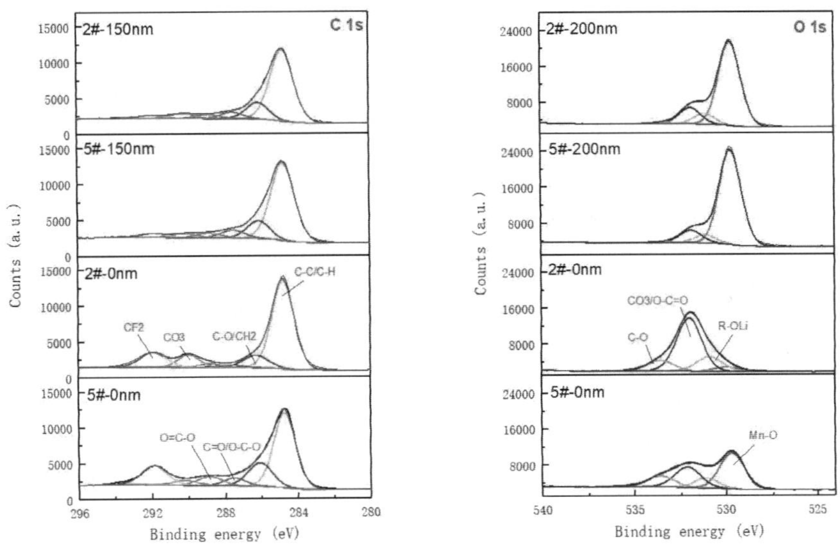

(a) (b)

IOP Publishing
2366 (2022) 012010 doi:10.1088/1742-6596/2366/1/012010

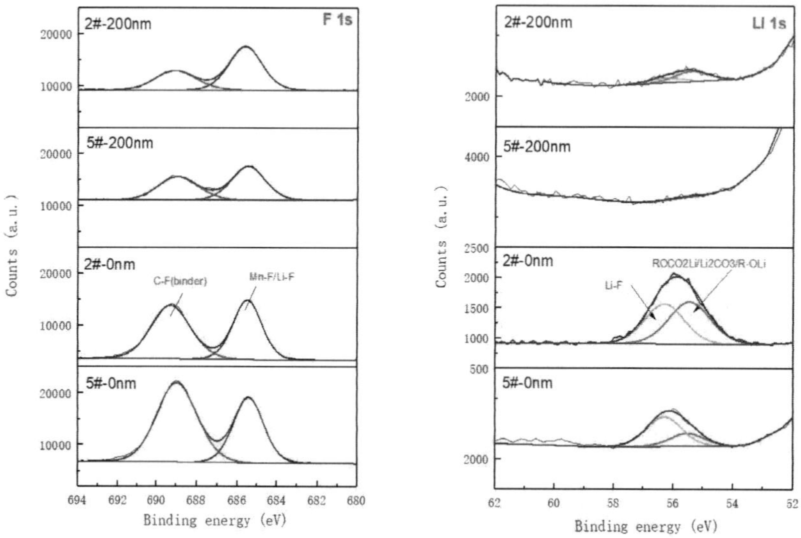

(c) (d)

Figure 15. XPS peak fitting diagram of different elements. (Figure a is C 1s; b is O 1s; c is F 1s; d is Li 1s)

The side reaction was accompanied by acetylene, carbon monoxide and other gases, which was consistent with the apparent ejection of gases during the dismantling process.

3. CONCLUSION

In this paper, by studying the influence of high temperature storage (60℃) on the performance of Li-MnO_2 primary battery, CT, SEM-EDS, XPS and other methods were used to analyze the structure and composition of the internal cathode materials, and the mechanism of adverse reactions on the cathode surface under high temperature was speculated, and the following conclusions were drawn.

High temperature storage conditions will increase the internal impedance of the battery, resulting in the decline of the battery discharge platform, affecting the battery rate performance.

With the increase of storage temperature, the surface chap of battery cathode material deepens, and it is easier to introduce other substances into the cathode material.

Under the condition of high temperature storage, side reaction intensifies, and more side reaction products are deposited on the positive electrode surface, accompanied by gas generation.

The increase of internal resistance in high temperature storage has a great relationship with the deposition of side reaction on the positive electrode surface and the occurrence of gas production under high temperature condition.

4. EXPERIMENTAL SECTION

In this paper, by studying the influence of high temperature storage (60℃) on the performance of Li-MnO_2 primary battery, CT, SEM-EDS, XPS and other methods were used to analyze the structure and composition of the internal cathode materials, and the mechanism of adverse reactions on the cathode surface under high temperature was speculated, and the following conclusions were drawn.

The test sample is cylindrical Li-MnO_2 primary battery, the height of the battery is 33.5 mm, the diameter is 16.5 mm, the rated capacity of the battery is 1200 mAh, the open circuit voltage is 3.26V, the load voltage is 2.75 V, the normal operating temperature of the battery is -20~60℃.

Six fresh batteries were randomly selected and numbered as samples 1#, 2#, 3#, 4#, 5# and 6#.1#, 2#, 3# samples for high temperature storage, 4#, 5#, 6# samples for normal temperature storage. The high temperature storage test was stored at 60 ℃ for 25 days, and the normal temperature storage test was stored at 25 ℃ for 25 days. After the battery is stored at room temperature and high temperature, the performance of the storage battery is tested, and the open-circuit voltage of the 6 samples is tested using an internal resistance tester. 3# and 6# batteries were charged to 2.0 V with 5 mA current by charge-discharge equipment. Electrochemical workstation was used to test the AC impedance of 3# and 6# batteries at a frequency of 10mHz-100kHz.1#, 2#, 4# and 5# samples were tested and analyzed by CT to observe the internal structure of the material. After the test, the 1#, 2#, 4# and 5# samples were disassembled under low humidity condition to observe the appearance state of electrode materials. SEM-EDS analysis was performed on the cathode materials after the disassembled samples 1#, 2#, 4# and 5#, and the microscopic morphology and surface element differences of the cathode materials were observed. XPS test was carried out on the cathode materials after disassembly of 2# and 5# samples, and Ar etching test was used to analyze the element composition on the cathode surface and the variation trend of element composition with thickness.

The main test equipment is shown in Table 1 below.

Table 1. Test equipment information.

Device name	The equipment brand	precision
Battery charge and discharge test system	CT 2001A-5 V 5 mA	±(0.1%RD+0.1%FS)
Electrochemical workstation	ZAHNER	±0.1%
High resolution 3D computed tomography	XTH 225 ST	1 um
Scanning electron microscope	HITACHI	0.8 nm (15 kV)
X-ray photoelectron spectrometer	PHI 5000 VersaProbe III	/
High and low temperature humid heat test chamber	CZ-A-1000G	±0.5 ℃; -70~150 ℃

Acknowledgements

This paper is one of the stage results of the Guangdong Province Key Areas R&D Program Project Traction Battery System Testing and Evaluation Technology Research (2019B090908003).

References

[1] Z. Huihui, Z. Yanlei, and Z. Lijuan, "Research progress in Li-MnO2 primary battery," Battery Bimonthly, vol. 47(03), pp. 188-190, June 2017.

[2] W. Pingan, C. Yongjie, J. Youze, and G. Qing, "Research progress and expectation of Li-MnO2 batteries," Chinese Journal of Power Sources, vol. 36(07), pp. 1058-1061, July 2012.

[3] S. Yundong, H. Lihua, Y. Bing, L. Yong, and M. Hongbin, "Performance comparison of Li-MnO2 primary battery at home and abroad," Chinese Journal of Power Sources, vol. 39(06), pp. 1282-1285, June 2015.

[4] M. Xianling, Y. Zhong, S. Xiaoqian, Y. Fangning, and M. Hao. "Study on influence factors of the storage performance of lithium carbon monofluoride-manganese dioxide battery," Chinese Journal of Power Sources, vol. 45(08), pp. 1012-1015, August 2021.

[5] L. Ruirui, and H. Yuanshuai, "Study on the application of fluorinated graphite in primary lithium manganese battery," Chinese Battery Industry, vol. 22(03), pp. 131-133, June 2018.

[6] L. Fen, L. Liangcheng, F. Shan, and X. Yangming, "High electrical conductivity positive electrode for Li /MnO2 battery," Battery Bimonthly, vol. 49(05), pp. 420-423, October 2019.

[7] D. Zhonghong, G. Feng, D. Qiuyue, and Y. Muting, "Study on storage performance of long-life lithium primary battery for torpedo," Chinese Journal of Power Sources, vol. 44(08), pp. 1161-1164, August 2020.

[8] Z. Wenjing, Y. Xiaoxue, Y. Xuehua, Y. Mingyu, J. Hui, et al., "Tripotassium citrate monohydrate derived carbon nanosheets as a competent assistant to manganese dioxide with remarkable performance in the supercapacitor," Frontiers of Chemical Science and Engineering, vol. 16(3), pp. 420-432, March 2022.
[9] Z. Jing, "A study on the positive electrode of high propermance lithium manganese dioxide cell," Tianjin University, December 2007.
[10] A. F. Stefan, C. Yuhui, P. Zhangquan, M. G. John, J. H. Laurence, et al., "Reactions in the rechargeable lithium–O2 battery with alkyl carbonate electrolytes," vol. 133(20), pp. 8040-8047, May 2011.
[11] L. Qinghao, W. Yi, W. Xuelong, S. Xiaorui, N. Z. Jie, et al., "Investigations on the fundamental process of cathode electrolyte interphase formation and evolution of high-voltage cathodes," ACS applied materials & interfaces, vol. 12(2), pp.2319-2326, January 2020.

A Design of Indoor Air-Quality Monitoring System

Chao Wang[1], Hongli Liu*[1], Jiawen Ji[1], Yangshuang Wu[1]

[1]School of Automation, Wuhan University of Technology, Wuhan, China

*Correspondence: james97@163.com

Abstract: With the deepening of people's understanding of the harm of air pollution, indoor air quality is getting more and more attention. This paper designs an indoor air quality monitoring system based on sensor technology and Internet of things technology, the system function includes air quality detection, real-time data display, server transfer, remote display throught Wechat mini programe and others, the system can detect six important environmental parameters and allows users to view the data on mobile phone.Based on the data collected, this paper also designs an air quality evaluation algorithm.

1. Introduction

In recent years, as people pay more and more attention to environmental pollution, air quality has become one of the focuses of public attention[1]. In 2003, the state issued a national standard on indoor air quality[2], in which a series of air quality parameters and air quality test methods were stipulated. The air quality inspection methods specified in the national standard are chemical or biological methods. These test methods can obtain accurate results but the operation steps are complex, which are not suitable for the requirements of long-term monitoring of air quality in general places.

With the development of Internet of Things (IOT) technology, it has many application in the field of environmental monitoring, such as atmospheric monitoring and water quality detection. Jagriti Saini et al. put forward that IoT has great potential in indoor air quality monitoring, which can ensure lower power consumption, negligible time delay and superior ability to interact with the physical world[3]. In the current common Internet of Things indoor air quality monitoring system, there are often the following problems : firstly, only monitoring the air quality without assessment[4]-[5]; secondly, the evaluation of air quality often focuses on the research of algorithms, which uses public data sets and ignores the acquisition of actual data[6]- [7].

In this paper, the Internet of Things technology is used to propose a scheme for indoor air quality monitoring and evaluation aim at the shortcomings of the existing schemes. This scheme includes two functions : real-time monitoring and quality evaluation. Based on this scheme, a set of air quality monitoring and evaluation system is developed in this paper, and the system has good effect in practical application.

2. Design of Real-time Monitoring System for Environmental Quality

2.1 System design objectives and functions

The purpose of this design is to realize the monitoring and evaluation of indoor air quality. According to the national indoor air quality standards, several typical indicators such as temperature, relative humidity, PM2.5, TVOC and formaldehyde are selected as monitoring parameters.

The environmental quality real-time monitoring system should have the following main functions :

1) accurately detect five indoor air quality parameters
2) upload the data of air parameters to the server
3) store uploaded data by database
4) display of data through a client remotely

Based on the above functional design, the monitoring system adopts the following scheme : using sensors to collect data about environmental parameters, using MCU to process data, using WiFi to upload data to the elastic compute service(ECS), the transmission protocol is MQTT, using MySQL database to storage data, using WeChat mini program to access the server and database to display data remotely. Overall technical programme as shown in figure 1.

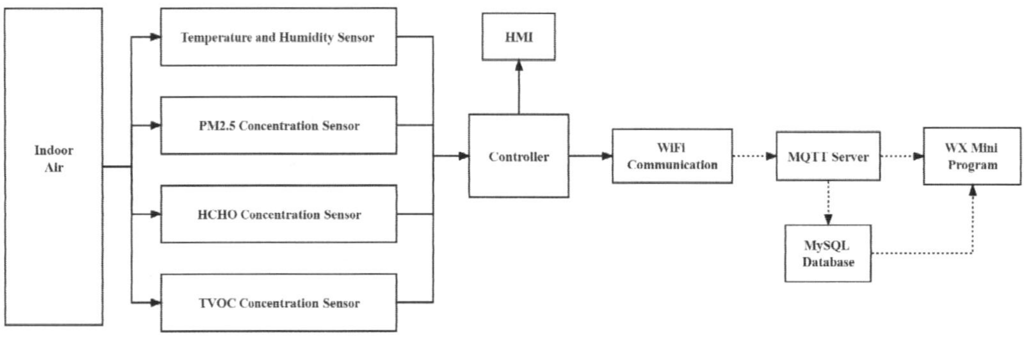

Figure 1 Technical Programme

2.2 Hardware Design of Environmental Monitoring System

An embedded device is designed and fabricated as the hardware platform of the monitoring system. The embedded system can collect, process and transmit data of indoor air quality parameters. The embedded device consists of three parts, controller, sensors and communication module. The controller uses STM32F407ZGT6 as the control chip. The 32 - bit chip is based on the Cortex-M4 core and has rich peripherals, which meets the performance and is economical.

According to the standard and limits of indoor air quality parameters stipulated by national standards, the system selects the following sensors to realize data acquisition function : DHT11 temperature and humidity sensor, FS00202 laser dust sensor, FS00602 TVOC sensor and FS00509 formaldehyde sensor. The sensor selection is shown in table 1.

Table 1 Sensor Selection Table

Parameter	Sensor	Range	Accuracy	Interface
PM2.5	FS00202	0 ~ 1000μg/m³	±10%rd	UART
HCHO	FS00511	0 ~ 1.5mg/m³	±10%rd	UART
TVOC	FS00602	0 ~ 5000μg/m³	±10%rd	UART
Humidity	DHT11	20% ~ 95%	±5%	1-Wire
Temperature	DHT11	0℃ ~ 50℃	±2℃	1-Wire

In this scheme, WiFi is selected as the communication mode of uploading data to the server, and ATK-ESP8266 communication module is adopted. The module has built in TCP / IP protocol stack,Communication module communicates with controller through UART and uploads data to server through WiFi.

This design includes HMI (Human Machine Interface), using 2.8-inch resistive touch screen to display data directly on the device. Based on the above scheme, this paper made a environmental quality monitoring system, as shown in Figure 2.

Figure 2 Picture of Environmental Quality Monitoring System

2.3 Software Design of Environmental Monitoring System

The software design of this scheme includes three parts : embedded device software design, server and database design, and WeChat mini program software design, corresponding to acquisition and upload for data, storage and forwarding for data, and remote display.

The embedded system program design includes the main program design and drivers design. The main program design requires the completion of hardware initialization, air quality data acquisition and air quality assessment. The sensors, HMI screens and communication modules are drived by corresponding drivers.

The purpose of the server is to allow WeChat mini program and MySQL database access server to realize remote display and transfer of data. This scheme uses the Aliyun Internet of Things platform to build the server, and buys the MySQL database from Aliyun to realize the function of the database, so as to avoid server operation and maintenance. The database example is shown in Figure 3.

The development of WeChat mini program is based on the mpvue framework, and the application layer protocol is MQTT. When the WeChat mini program successfully connects to the server and subscribes to the topic, it begins to receive the message. and unpacks the received message in JSON format, so as to obtain the data of the corresponding sensor, and display the data on the mini program page.

id	temperature	humidity	pm25	TVOC	HCHO	time
754	25	54	44	220	59	2022-05-23 14:57:16.000000
755	25	53	44	220	62	2022-05-23 14:58:16.000000
756	26	54	44	235	64	2022-05-23 14:59:16.000000
757	26	54	42	237	66	2022-05-23 15:00:16.000000
758	26	54	44	241	66	2022-05-23 15:01:16.000000
759	26	54	44	242	60	2022-05-23 15:03:13.000000
760	26	54	46	254	66	2022-05-23 15:04:13.000000
761	26	53	42	251	69	2022-05-23 15:05:13.000000
762	26	54	42	259	65	2022-05-23 15:06:13.000000
763	26	53	42	260	63	2022-05-23 15:07:14.000000
764	26	53	44	269	63	2022-05-23 15:08:13.000000

Figure 3 Database Example Diagram

3. Air Quality Assessment Algorithm

At present, the most common method for calculating the comprehensive evaluation index of indoor air quality is to use the air quality index (AQI)[8]. The air quality index of each pollutant is calculated according to the air quality classification standard, and the comprehensive evaluation index is obtained according to appropriate method. the individual index reflected by a single pollutant or parameter and the combined quality index of several pollutants or parameters. According to the national standard of indoor air quality and indoor environmental pollution control standard of civil building engineering[9], this design adopts the air quality classification standard shown in table 2.

Table 2 Air Quality Classification Criteria in This Programme

Index	HCHO (ug/m3)	TVOC (ug/m3)	PM2.5 (ug/m3)
0	0	0	0
50	35	250	35
100	70	450	75
150	105	650	115
200	140	850	150

Individual indoor air quality index (IIAQI) is the quality index calculated by single pollutant, and the calculation method is as follows.

$$P = \frac{P_H - P_L}{BP_H - BP_L}(C_P - BP_L) + P_L \quad (1)$$

In the formula, P is the individual indoor air quality index, C_P is the concentration of single pollutant item in Table 2, BP_H is the high value of the pollutant concentration limit close to C_P, BP_L is the low value of the pollutant concentration limit close to C_P, P_H is the air quality index corresponding to BP_H, and P_L is the air quality index corresponding to BP_L.

Indoor air quality index (IAQI) can be obtained by individual index through a variety of calculation methods. The following introduces several common formulas for calculating the IAQI by IIAQI, where P is the IIAQI, C is the IAQI, and k is the number of calculation indexes, P_{max} is the maximum in P.[10]

1) arithmetic superposition method

$$C = \sum_{i=1}^{k} P_i \quad (2)$$

2) arithmetic average method

$$C = \frac{1}{k}\sum_{i=1}^{k} P_i \quad (3)$$

3) comprehensive evaluation method

$$C = \sqrt{P_{max} \cdot \frac{1}{k} \cdot \sum_{i=1}^{k} P_i} \quad (4)$$

The above several calculation schemes are common methods to calculate the air comprehensive quality index. The arithmetic superposition method reflects the overall pollutant strength, but it is related to the number of pollutants. When the number of pollutants is large, even if the single pollutant slightly exceeds the standard, the comprehensive quality index may be large. The arithmetic average method reflects the average level of pollutants, but it is easy to cover up some serious pollutants. The comprehensive evaluation method takes the geometric average of the maximum and average of the index, which is more applicable than the above two methods. This design uses the comprehensive evaluation method as the calculation method of the comprehensive quality index. Finally, the air quality evaluation grade is determined by air quality comprehensive quality index[11], and the corresponding relationship is shown in table 3.

Table 3 Air Quality Rating

IAQI	Grade
0-50	Good
51-100	Moderate
101-150	Unhealthy
>151	Very Unhealthy

The air quality evaluation results are displayed through the WeChat mini program interface. Figure 4 shows the WeChat mini program interface examples, including the evaluation results and real-time data display.

4. Conclusions
In this paper, a practical air quality monitoring and evaluation system is designed and produced. This system includes data acquisition terminal, server, database and WeChat mini program, which can realize the monitoring of indoor air quality, data upload and storage, quality evaluation and remote real-time display.

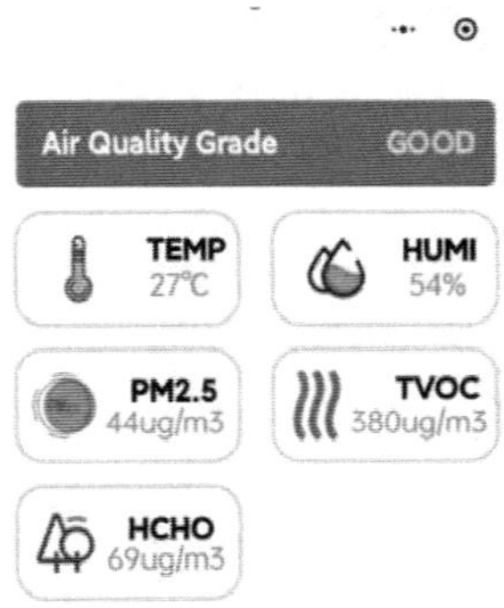

Figure 4 WeChat Mini Program Interface

Combined with the national standard of indoor air quality, this paper provides a suitable calculation method of indoor air quality comprehensive index, and applies this method to the air quality monitoring system. The results show that this method can evaluate the indoor air quality well. Since the calculation method is not complex and occupies less computer memory, it is suitable for real-time comprehensive evaluation of indoor air quality in small air quality monitoring systems such as the Internet of Things[12].

References
[1]Cheek E, Guercio V, Shrubsole C, et al. Portable air purification: Review of impacts on indoor air quality and health[J]. Science of The Total Environment, 2020.
[2] GB/T 18883-2002, "Indoor air quality standard ".
[3] Saini J, Dutta M, Marques G. "A comprehensive review on indoor air quality monitoring systems for enhanced public health" .Sustainable Environment Research, 2020(1).
[4] Yuanyu Liu, Jinwei He. "Indoor air quality monitoring system based on STM32 and Wi-Fi module ". Electronic Technology & Software Engineering, 2019(12).
[5] Ping Li. "Design of indoor air quality monitoring system based on ARM". Southwest Jiaotong University,2010
[6] Kang J, Hwang K I. "A Comprehensive Real-Time Indoor Air-Quality Level Indicator" . Sustainability, 2016(9).
[7] Yu Peng. "Research on portable indoor air Quality Monitor based on STM32 ". Jinan University,2013
[8] Tianjiao Wang. "Design and development of Internet of Things service platform for indoor gas quality monitoring system". Xidian University ,2015(12)
[9] GB 50325-2020, " Standard for indoor environmental pollution control of civil building engineering" .
[10] Yunzhi Zhong."Study on expert system of indoor environment assessment ", 2003(5)
[11] Brainvendra Widi Dionova, M.N. Mohammed, S. Al-Zubaidi, Eddy Yusuf."Environment indoor air quality assessment using fuzzy inference system".ICTE 259 ,2020(4)
[12] Hwang K I, Park S K. Experimental Study of Real-Time Comprehensive Indoor Air Quality . Springer Singapore, 2016.

Design and Application of Seabed Sediment Temperature and Pressure Measurement and Control Software

Shaowei Chen [*a], Qian Zhang[b], Lei Wu[b], Mengyao Pan[a], Yongjun Jiang[b]

[a]Guangdong Polytechnic of Industry and Commerce, Guangzhou City, Guangdong Province, 510510;

[b]School of Electro-mechanical Engineering in Guangdong University of Technology, Guangzhou City, Guangdong Province, 510006

[*] Corresponding author: 1251571695@qq.com

Abstract: Taking the temperature-pressure controllable acoustic measurement system for the seabed sediment as the object, according to the HostLink communication protocol of the Omron CP1H type PLC, FINS and C-mode commands are used to design and realize the control and monitoring of PLC by temperature-pressure monitoring software for seabed sediment. Moreover, SerialPort, Timer, and Chart controls in Visual Studio 2010 platform are used to realize the communication, data acquisition, and real-time curve drawing of the industrial field, developing, and realizing the automatic measurement of seabed sediment based on automatic detection and control of temperature and pressure. In addition, the actual measurement and application verify the reliability and stability of the system design.

1. Introduction

Due to different pressures and temperatures on the seafloor, the acoustics of the surface sediments on the seabed are affected to different degrees [1-8]. The real-time impact of pressure-temperature changes cannot be systematically observed in the seabed in situ acoustic measurements and acoustic telemetry. With the help of acoustic measurement research of pressure-temperature controllable experiments, the change that acoustic characteristics of surface seabed sediments are affected by the environment can be analyzed. Therefore, a temperature-pressure measurement and control system for seabed sediment is developed to simulate the in-situ seabed sediment and any possible changing environment by controlling the pressure and temperature and reproducing the acoustic measurement of the in-situ state of the seabed, which is of research significance to explore the seabed acoustic characteristics.

Chen Shaowei et al. [9] conducted research on the automatic control system of seabed sediment pressure. The system can simulate the seabed pressure environment, but it lacks the control of the ambient temperature. Zhou Shu et al. [10] studied a temperature-pressure controllable system for seabed sediment based on Omron PLC, which can simulate the pressure and temperature of the seabed. However, since the configuration software is used as the host computer, the function expansion is not flexible, and the experimental research operation is inconvenient. Wang Ning et al. [7] developed an automatic measurement program of a sound wave by LabVIEW, but the temperature and pressure can only be adjusted manually. The experimental operation is cumbersome, and the experimental efficiency is relatively low. Building a temperature-pressure controllable system with rich functions, scalability, and easy experimental operation will help improve the efficiency and effect of research on the acoustic characteristics of seabed sediments. Therefore, taking temperature-pressure controllable acoustic measurement and control system for the seabed sediment as the research object, the overall design scheme of the system is first proposed in this paper, analyzing the various components and functions of the system. Then, the communication protocol between the host computer and PLC, and the PLC communication hardware setting process are studied and analyzed, realizing the reliable data transmission between the host computer and the PLC control unit. Moreover, the design methods of key modules such as the communication module, control module, and display module of the host computer are studied and analyzed, and the temperature-pressure measurement and control software for seabed sediments is designed. Finally, the temperature-pressure controllable acoustic measurement system for the seabed sediment is operated based on the software, carrying out the experimental research on the automatic measurement of seabed sediment acoustics to verify the correctness of the system.

2. Overall Design of Temperature- pressure Controllable Acoustic Measurement System for Seabed Sediments

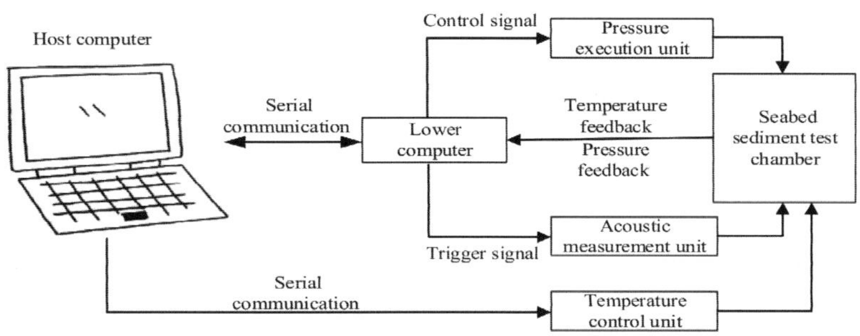

Figure 1. Overall design of temperature-pressure controllable acoustic measurement system for seabed sediments

As shown in Figure 1, the temperature-pressure controllable acoustic measurement system for seabed sediment is mainly composed of a host computer, lower computer, pressure execution unit, acoustic measurement unit, temperature control unit, and seabed sediment test chamber. The host computer is a measurement-control software developed based on the Visual Studio 2010 platform, which mainly includes functions such as temperature-pressure measurement and control of simulated seabed environment, automatic analysis of sound speed, and serial communication settings. The lower computer is a control unit with PLC as the core, which mainly realizes the functions of temperature data acquisition of the test chamber, closed-loop control on the simulated seabed environment pressure of the test chamber, and automatic trigger signal to start the acoustic wave measurement unit for acoustic measurement of seabed sediments. The temperature control unit mainly realizes the control of the test chamber to simulate the seabed environment temperature. The host computer software controls each functional module to work together, and completes the system' s automatic temperature control, pressure control, acoustic measurement, and automatic analysis of experimental data according to the requirements of experimental research. The design pressure range of this system is 0-30 MPa, which can simulate the pressure range of 3000 meters under the sea, and the temperature control range is 0-35 ℃, meeting the requirements of experimental research.

3. Communication Protocol between Host Computer and PLC

The HostLink protocol is a dedicated communication protocol for data communication between the host computer software and Omron PLC. In Visual Studio 2010, the host computer software writes the program to package the data packets according to the HostLink protocol standard and then sends the data to the PLC. After the PLC receives the data packets, it parses the data packets according to the HostLink protocol standard and performs corresponding actions. The HostLink protocol has two different packet formats: C-Mode and FINS. PLC shows different functional characteristics to C-Mode and FINS command execution [11].

3.1 C-Mode Command

In the HostLink communication protocol, the data packets sent by the host computer software to the PLC are called command packets, and sent by the PLC to the host computer software are response packets. The format of the command packet and the response packet is shown in Figures 2 and 3.

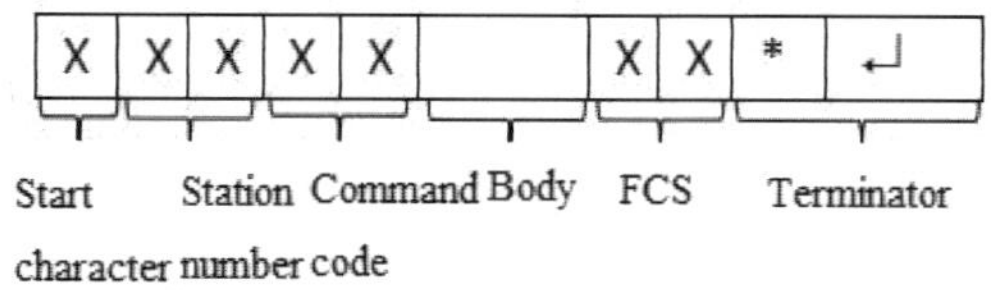

Figure 2. C-Mode command packet

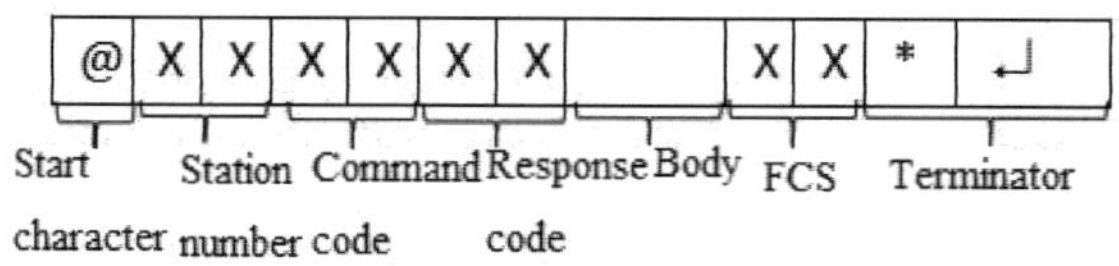

Figure 3. C-Mode response packet

CC-Mode packets start with "@". The identification address of the lower computer PLC is called the station number. There is only one PLC controller in this system, so the station number is set to "00". The length of two bytes is set as the command code, and different command codes such as "RD" and "WD" are used to read different data areas in the PLC. The command content is called the body. The check code FCS is mainly used for frame checking. The end of the C-Mode packet is indicated by "*" and "CR carriage return". Compared with the command packet, the response packet only adds a response code, and different response codes represent different action execution states of the PLC. The C-Mode command has certain limitations. For example, the C-Mode command can only read the PLC memory area as a whole word, which cannot read or write a certain bit of the memory area. The C-Mode command also cannot operate on the *W* register. The C-Mode command can only write in the OMRON PLC programming state, but the read operation can work in the three states of OMRON PLC operation, monitoring, and programming. Therefore, the C-Mode command is suitable for the real-time acquisition of PLC data.

3.2 FINS Command

The FINS command supports the open communication mode. The control operation and data communication can be realized only by writing the corresponding FINS command to the specified address space. Moreover, the FINS command also supports the expansion of address space, which can communicate with PLC in the operation mode of PLC. The temperature-pressure measurement and control software for seabed sediment needs to carry out real-time program control on the PLC. Through studying and analyzing the characteristics of the FINS command, it is found that the FINS command is suitable for real-time communication control applications. The command packet and response packet of FINS are shown in Figure 4 and Figure 5, respectively.

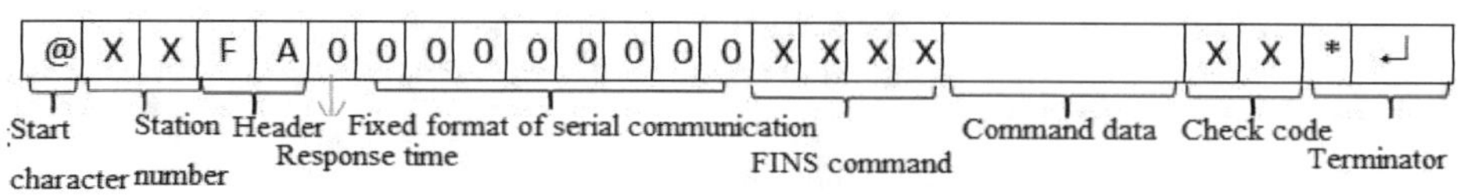

Figure 4. FINS command packet

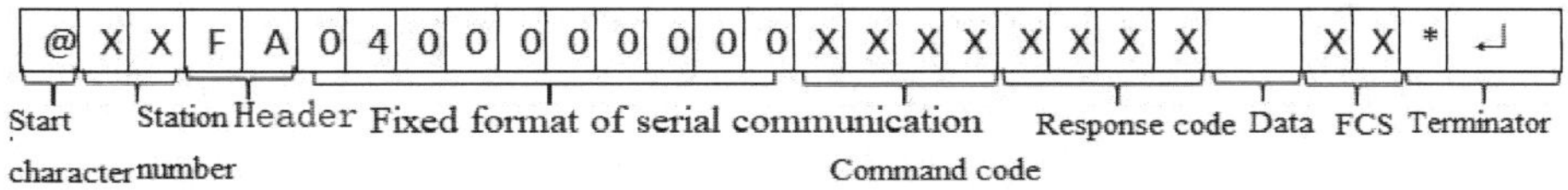

Figure 5. FINS response packet

Both C-Mode and FINS commands belong to the standard data packets of the HostLink protocol. Therefore, the FINS commands also have data packet components such as start symbol, terminator, station number, and FCS check code.

4. PLC Communication Hardware Settings

The work of temperature-pressure measurement and control system for seabed sediment needs to set communication hardware. Disconnecting PLC power supply is disconnected. The dip5 switch of PLC is set to On, and the rs232 serial port is selected. The other areas are set to off. The Cx-Programmer programming software is opened. The network type is configured as SYSMAC WAY, and the baud rate is 9600. The data bit is 7. The parity bit is E. The stop bit is 1 bit. The mode is HostLink. The program file is downloaded to the PLC through the Cx-Programmer programming software to complete the serial port setting.

5. Realization of Temperature-pressure Measurement and Control Software for Seabed Sediment

The upper computer software can be developed and written in high-level languages such as VC, C#, and VB. This host computer software is developed on the VS2010 platform with C# language which has the characteristics of easy learning, a beautiful interface, and a short development cycle. VS2010 provides controls such as serial port communication SerialPort, OnTimer, and Chart, which provides convenience for the development of the host computer. The structure of temperature-pressure measurement and control software is shown in Figure 6. This paper mainly analyzes the reality of the communication module, control module, and data display module.

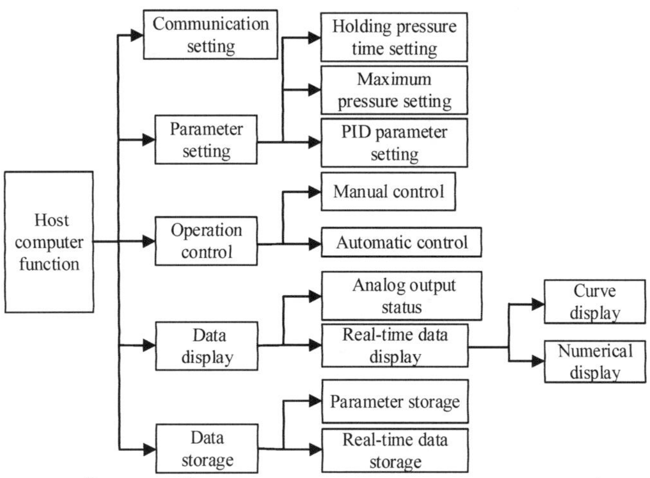

Figure 6. Structure diagram of temperature-pressure measurement and control software

5.1 Design of Communication Module of Host Computer

The serial communication of the host computer is realized based on the SerialPort class, which provides the data read event driver and access to the serial port parameter attributes. The serial port properties are set through parameters such as BaudRate, Parity, DataBits, and StopBits of the SerialPort class. The serial communication settings between the host computer and the PLC must be consistent. The specific procedures are as follows:

```
private void button2_Click(object sender, EventArgs e)
   {    addData_thread.Start();
        if (serialPort1.IsOpen == false)
           { serialPort1.PortName = comboBox1.Text;
             serialPort1.BaudRate = 9600;
             serialPort1.DataBits = 7;
             serialPort1.Parity = Parity.Even;
             serialPort1.StopBits = StopBits.Two;
             serialPort1.Open();
             timer1.Enabled = true ;
             button2.Text = " Close serial port ";}
        else
            { serialPort1.Close();
              timer1.Enabled = false ;
              button2.Text = " Open serial port";}
```

Among them, button 2 clicked to complete the opening and closing of serial communication is a button control. By the statement of "addData _ thread. Start ( ) ;", the thread of real-time data display is opened. By the statement of "timer1.Enabled = true;", the timer is started to provide accurate time nodes for the real-time display of temperature and pressure data.

5.2 Design of the Control Module of the Host Computer

The host computer control module of the temperature-pressure measurement and control software mainly realizes two state working conditions of pressurization and decompression through automatic and manual modes. Taking forward pressurization as an example, the process is shown in Figure 7.

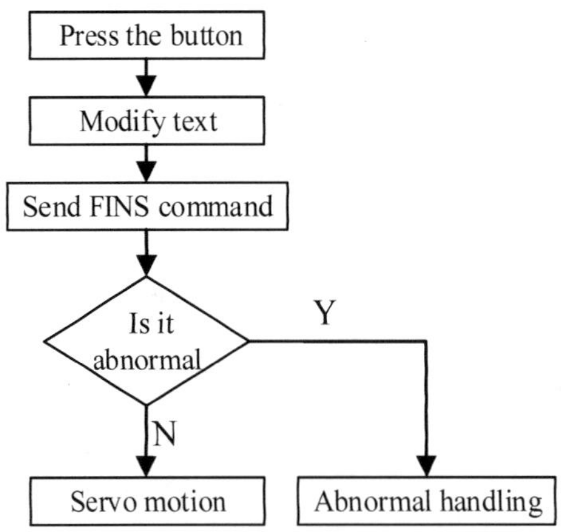

Figure 7. pressurization flow chart

When controlling the forward pressurization, the PLC controller of the lower computer uses the switch of the control W1.00 auxiliary relay to control the advance and stop of the servo motor. When the button3 in the host computer is pressed/bounced, the FINS command will be sent to realize the real-time control of the W1.00 auxiliary relay of the PLC. The control program is as follows:

```
private void button3_Click(object sender, EventArgs e)
  {string Send_Code1 = "";string Send_Code = "";
     Send_Code1 = "@00FA000000000" + "0102" + "31" + "0001" + "00" + "000100" + "76" + "*" +
     (Char)13;
     Send_Code = "@00FA000000000" + "0102" + "31" + "0001" + "00" + "000101" + "77" + "*" +
```

```
(Char)13;
  serialPort1.DiscardInBuffer();
  serialPort1.DiscardOutBuffer();
if (open == true)
  try{ button3.Text = "servo off"; open = !open;
      if (serialPort1.IsOpen == false)
      {MessageBox.Show("Please open the serial port!");}
      else
      {serialPort1.WriteLine(Send_Code);}}
  catch (Exception ex)            {System.Windows.Forms.MessageBox.Show(ex.Message);}
 else
  {try{button3.Text = "Pressurize forward!"; open = !open;
      if (serialPort1.IsOpen == false)
        { MessageBox.Show("Please open the serial port!");}
      else{serialPort1.WriteLine(Send_Code1);}}
   catch(Exception ex)
{System.Windows.Forms.MessageBox.Show(ex.Message);}}
```

The strings Send_Code1 and Send_Code are the command codes for controlling backward pressure reduction and forward pressure respectively. Among them, 77/76 is the FCS check code. The "serialPort1.WriteLine();" is used to send commands to the PLC through the serial port to control the PLC. serialPort1.DiscardInBuffer() is used to clear the data in the receiving buffer area of serial port, and serialPort1.DiscardOutBuffer() is used to clear the data in the sending area of serial port, avoiding interfering with the reading of real-time data. The Try-catch structure is used to catch the exceptions in serial port operation and make prompts [12].

5.3 Design of Data Display Module of Host Computer

The lower computer PLC collects the pressure value of the test chamber in real time through the A/D conversion module and the pressure sensor, and the pressure value is saved in the data register D2. The temperature-pressure measurement and control software of the host computer sends the C-Mode command every 300ms to read the data in D2 through the Timer control. The workflow is shown in Figure 8.

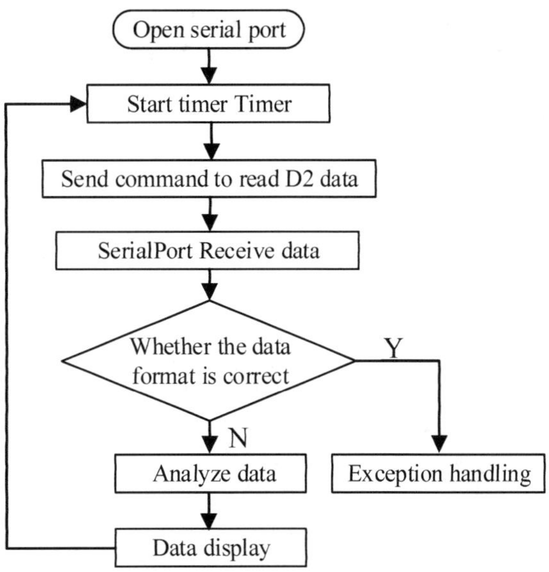

Figure 8. Real-time display flow chart

The program that *Timer* sends C-Mode command to read data in D2 register every 300 ms is as follows :

```
private void timer1_Tick(object sender, EventArgs e)
 {if (serialPort1.IsOpen == true)
     {        string Send_Code = "";
           Send_Code= "@00RD" + "0002" + "0001"+ "55"+ "*" + (Char)13;
             serialPort1.WriteTimeout = 100;
             serialPort1.WriteLine(Send_Code);} }
```

After the serial port is opened, the DataReceived event of the serial port will be automatically triggered when data is uploaded to the host computer. The serial port data receiving procedure is as follows:

```
private void serialPort1_DataReceived(object sender, System.IO.Ports.SerialDataReceivedEventArgs
 e)
 { char[] ReadBuffer = new char[50];
      Double d,data;
      do{ Thread.Sleep(5);}
      while (serialPort1.BytesToRead < 11);
      Thread.Sleep(20);
serialPort1.Read(ReadBuffer, 0,serialPort1.BytesToRead);
string instring = new string(ReadBuffer);
d = Convert.ToInt32(instring.Trim().Substring(7, 4), 16);
     data = (d / 100.0);
     Invoke(new MethodInvoker(delegate()
           {Thread.Sleep(50);
             if (data < 25)
              {data1 = data;
               textBox2.Text = data.ToString();}}));
  }
```

In the above code, the timer1_Tick event is triggered at regular intervals through the Timer control, so that the upper computer can continuously scan and refresh the data collected by the PLC. serialPort1.WriteTimeout() is the data sending timeout setting procedure to ensure the normal sending of real-time commands. When the serialPort1 is opened, the serialPort1.DataReceived () event will be triggered when the data arrives at the host computer, and the response code returned by the PLC will be received. The data will be stored in the data buffer, and the number of bytes of the buffer data will be obtained in real time by using the serialPort1.BytesToRead () command to judge the integrity of the data reception. The function of Thread.Sleep() is to make the thread sleep to ensure the integrity of data reading and analysis. The cached data is stored in the array ReadBuffer by serialPort1.Read (), then the array will be converted to a string, and the character conversion decimal value of the string 7 to 10 bits will be intercepted and stored in the *d* variable. Then, the *d* variable will be divided by 100 to get the real pressure value. textBox2.Text = data.ToString() is to display the pressure value on the interface. Since the data processing is carried out in the auxiliary thread, and the data is displayed on the main thread control, the delegate Invoke(new MethodInvoker(delegate())) of the window handle is used to ensure the transmission of data between the main and auxiliary threads [13]. The pressurization curve of temperature-pressure measurement and control software for the seabed sediment is shown in Figure 9.

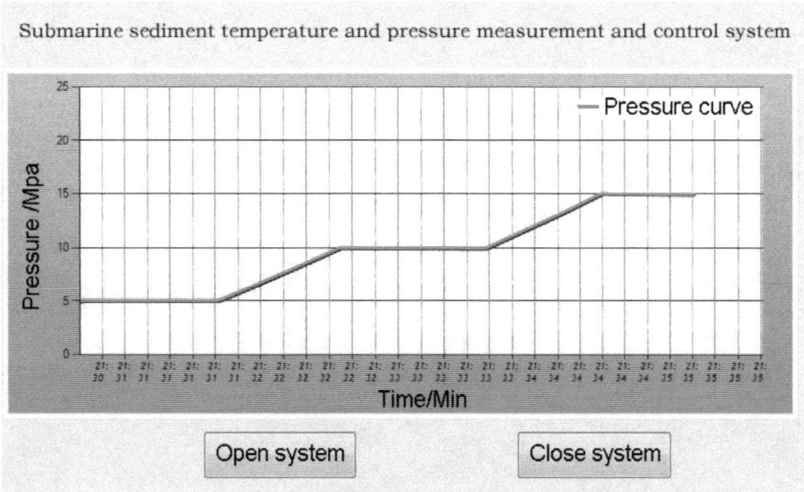

Figure 9. Pressurization curve of temperature- pressure measurement and control software for seabed sediment

6. Temperature-pressure Controllable Acoustic Measurement for Seabed Sediments

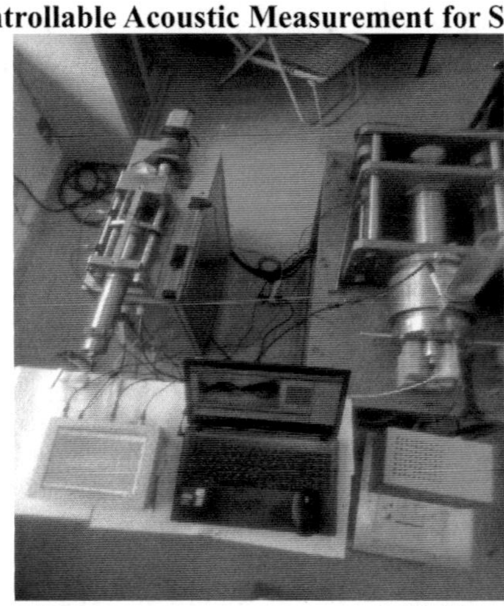

Figure 10. Physical diagram of temperature-pressure controllable acoustic measurement system

The temperature-pressure controllable acoustic measurement system for seabed sediments is shown in Figure 10. The TH-204 multifunctional acoustic wave parameter tester is used, the frequency of the acoustic transducer is 100 kHz, and the temperature control adopts the DHX-010 intelligent constant temperature circulator. Besides, the experimental samples of seabed sediment are taken from the South China Sea, with a length of 286 mm.

In the experiment, parameters such as serial communication, experimental temperature, and experimental pressure were set on the temperature-pressure measurement and control software for seabed sediment. The serial port baud rate is 9,600, the simulated seabed temperature is 17 ℃, and the test chamber pressure is 10 MPa. The experimental temperature value is sent to the intelligent constant temperature circulator through the serial communication of the software platform, and then the constant temperature control is performed on the test chamber. The PLC detects the temperature of the test chamber in real time through the temperature sensor. After reaching the preset temperature value, the PLC automatically starts the closed-loop pressure control of the test chamber. Moreover, the PLC detects the pressure of the test chamber in real time through the pressure sensor. After reaching the preset pressure value, the PLC automatically sends a 0-5 v voltage pulse trigger signal to the multi-function acoustic parameter tester to start the acoustic measurement of seabed sediments. Finally, the acoustic measurement data is imported into the software platform, and the automatic interpretation of the jump point algorithm based on three B-samples is automatically invoked to analyze sound velocity, as shown in Figure 11. The experimental results show that the temperature-pressure measurement and control software for seabed sediment works reliably and achieves the expected experimental goals.

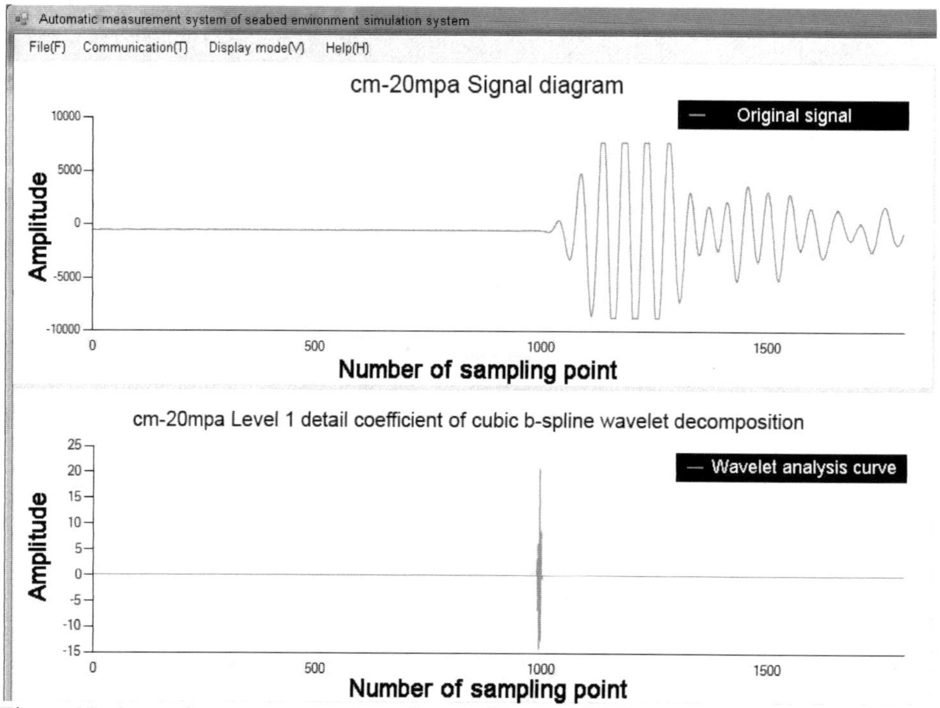

Figure 11. Acoustic measurement waveforms of automatically measured seabed sediments

7. Conclusion

Through the research on the communication protocol of Omron PLC, it is concluded that the C-Mode format of the HostLink protocol is suitable for reading the real-time data of the host computer, and the FINS format is suitable for the control of the real-time working condition of the host computer. On the VS2010 platform, the C# language is used to research and design functional modules such as communication, control, and display, realizing the temperature-pressure measurement and control functions of seabed sediments. Moreover, it provides a more definite acoustic characteristic of seabed sediments for automatic, convenient, and efficient simulation of marine sediment acoustic characteristics measurement and marine acoustic detection under different in-situ environments.

Fund Project

This paper was financially supported by the Youth Innovative Talents Project of General Universities in Guangdong Province (Grant No. 2021KQNCX168) and the National Natural Science Foundation of China (Grant No. 41776043).

About the author

Chen Shaowei (1987-), male, is a postgraduate. His research interests cover intelligent measurement and control of electromechanical hydraulic systems.

References

[1] ZOU Dapeng, YE Guican, LIU Wei, SUN Han, LI Jun and XIAO Tibing. Effect of temperature on the acoustic reflection characteristics of seafloor surface sediments [J]. Oceanic and Coastal Sea Research, 2022 21 (1): 62-68

[2] ZOU Dapeng, LV Hengsheng, KAN Guangming, LIU Wei, XIAO Tibing. Characteristics of environmental factors influencing the sound velocity of seabed surface sediments [J]. Acta Acustica, 2021,42(02): 227-236.

[3] JACKSON D. R., RICHARDSON M. D. High-Frequency Seafloor Acoustics [M]. Translated by Liu Baohua, Kan Guangming, Li Guanbao, etc. High-frequency submarine acoustics [M]. Ocean Press, 2104, Beijing.

[4] Dapeng Zou, Kevin L. Williams, and Eric I. Thorsos. Influence of temperature on the acoustic sound speed and attenuation of sand sediment. IEEE Journal of Oceanic Engineering. 2015, 40(4): 969-980

[5] R. Carbó and A. C. Molero, The effect of temperature on sound wave absorption in a sediment layer, J. Acoust. Soc. Amer., Vol. 108, no. 4, pp. 1545-1547, 2000

[6] WANG Qi, Liu Yanchun, WU Yingzi, et al. Study on the relationship between the acoustic properties of seabed sediments and the change of seawater depth [J]. Applied Acoustics, 2008, 27(3): 217-221.

[7] WANG Ning, WANG Ying, ZHOU Jianping, LV Miaoli. Design of bottom acoustic measurement system based on the virtual instrument [J]. Journal of Zhejiang Sci-Tech University, 2012, 29(02): 214-219.

[8] ZOU Dapeng, WU Baihai, LU Bo, et al. Research on the calibration of laboratory measurements of sound velocity of seabed sediments [J]. Journal of Tropical Oceanography, 2008, 27(1): 27-31.

[9] CHEN Shaowei, YANG Yanming, ZOU Dapeng, et al. Simulation and experimental research on the automatic control system of seabed sediment pressure [J]. Machine Tools and Hydraulics, 2016, 45(7): 34-37.

[10] ZHOU Shu, ZOU Dapeng, XIAO Tibing, ZHANG Qian, CHEN Shaowei. Design and experiment of a test system for sound velocity-pressure characteristics of seabed sediments[J]. Ocean Technology, 2017, 36(05): 55-61.

[11] HUO Gang, SU Qiang, et al. Principles and typical cases of Omron CP1 series PLC [M]. Beijing: China Machine Press, 2016.

[12] DOU Yitong, ZHAO Shiping. Realization of communication between PLC and PC based on c# and HostLink [J]. Embedded Software Application, 2010, 26(2): 104-106.

[13] LI Wei, WANG Jian, et al. C#.NET desktop application project development tutorial [M]. Beijing: Tsinghua University Press, 2016.

Reliability analysis and verification of the rubber support base for battery pack

Limin Fang[1], Linhua Xiong[2], Shuai Liu[2], Jinghong Jiang[3*]

[1]Department of Physics and Information Engineering, Guangdong University of Education, Guangzhou 510303, China

[2]Institute of Reliability and Environmental Engineering, Guangzhou GRG metrology and test Co., Ltd, Guangzhou 510656, China

[3]Guangzhou GRG metrology and test Co., Ltd, Guangzhou China

[*] Corresponding author: jiangjh@grgtest.com

Abstract—In this paper, a design sample of rubber support base of battery pack is analyzed and verified. The status and response of the sample under the actual working scene are simulated and analyzed by using structural finite element simulation. The weak point position and local stress-strain index of the rubber support base are obtained, and the stress variation and fatigue life under two different types of mounting base are compared and analyzed. At the same time, the main failure modes of the rubber support base are verified by the durability test of the real samples and the microscopic characterization analysis of the fracture morphology, which are consistent with the simulation results. On this basis, the design and optimization suggestions of the rubber support base are put forward.

1. INTRODUCTION

The battery pack rubber support base is a support mechanism used for support and vibration reduction to ensure the reliability of the battery pack itself under vibration and impact environment. The support base is mainly composed of a ball stud and a rubber cover ball bowl, in which the internal lubrication protection part can be isolated from the external environment without interfering with the free movement of the ball [1-6]. The common failure mode of the ball stud is abnormal wear of the ball joint and the ball bowl, and most of the reasons for this failure are the invasion of dust and foreign matter after the rubber dust cover damaged. Corrosion and aging are the main factors for the degradation of mechanical properties of rubber. Under the condition of superimposing long-term alternating stress, the rate of rubber damage is further accelerated [7-12]. In the automotive application scenario, if the battery pack is exposed and leaked under the vehicle, the impact of flying stones is also one of the causes of damage, but it is quite accidental. Domestic research on rubber support mainly focuses on structural analysis, material analysis and structural optimization. Lin Demin applied ABAQUS analysis software to perform motion simulation analysis on the dust cover of ball stud assembly, check the motion interference of the dust cover, and improve the structural design [13]. Xu Ying established a nonlinear finite element model of the rubber protective sleeve on the automobile by using ABAQUS, fitted the rubber material parameters based on MooneyRivlin constitutive model, and analyzed the deformation and stress distribution of the rubber protective sleeve under different working conditions [14]. Liu Guangzheng of China North Vehicle Research Institute established a finite element model of the rubber mechanism for the premature failure of the rubber dust cover due to interference problems, and analyzed the stress and strain distribution of the rubber cover during the swinging process of the ball stud [15]. Taking the gap between the deformed rubber cover and the surrounding components as the optimization goal, they solved the interference problem of the dust cover through the optimization design method. Su Zepeng from the School of Transportation and Vehicle Engineering of Shandong University of Technology designed a new type of dust cover to improve the reliability of a certain steering rod dust cover [16]. Kim J H, Lee B Y and others abroad used nonlinear finite element analysis method to evaluate the stress and sealing performance of the ball stud dust cover of the automotive steering system, compared and analyzed the sealing performance of two different types of dust cover [17].

In this paper, the design sample of a ball stud rubber dust cover support base used to support the battery pack is analyzed and verified. The state and response of the ball stud rubber dust cover in the actual working scene are simulated by structural finite element simulation [18-20]. The position and local stress-strain index of the structural weak point of the rubber support base are obtained. Moreover, the stress variation and fatigue life of the rubber support under two different types of mounting bases are compared and analyzed. At the same time, through the durability test of the physical sample and the micro characterization analysis of the sample fracture morphology [21], it is effectively verified that the main failure mode of the rubber support base is consistent with the simulation analysis result, and on this basis, the optimization suggestions for the structural design of the support base are proposed.

2. Simulation Analysis

Fig. 1 shows a newly designed and developed rubber support base for automobile battery pack, in which the dust cover is designed to be installed on two different bases. In the working state, the dust cover will always be in a state of preloading on the top surface, and will be deformed back and forth with the rotation of the ball stud. It is generally believed that during the use of the ball stud rubber dust cover, on the one hand, the rubber is squeezed and pulled for a long time, and on the other hand, it will age itself in the humid heat and radiation environment, and the surface is prone to cracks and penetrations. The mechanical strength simulation analysis, durability test and failure analysis of the dust cover structure are carried out to try to reveal the weak point of the rubber and provide a basis for structural optimization design.

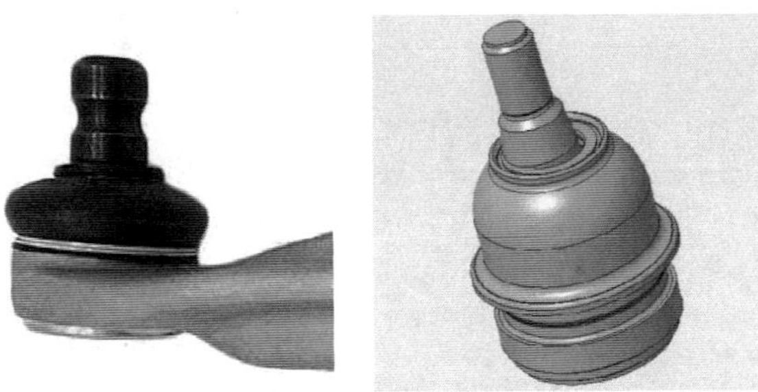

Figure 1. Physical image(left) and geometric model(right) of ball stud and dust cover

The mechanical simulation analysis of the rubber support base is carried out to obtain the weak point, stress and stress distribution of the structure in the service mechanical environment, and the stress index of the weak point of the rubber dust cover under the two bases is compared and analyzed.

As shown in Fig. 2, the bottom of the rubber support base is installed on two different bases, and the top is wrapped around the ball joint shaft in an interference fit state. During installation, there is a pre-pressure on the top to press down the whole top by 7.5mm. During operation, the ball joint shaft with the dust cover rotates around the center of the ball joint by 25° at most. The rotation action reciprocates several times during the entire product service period. The base is divided into model 1 and model 2, where the model 2 has grooves on the edge of the base.

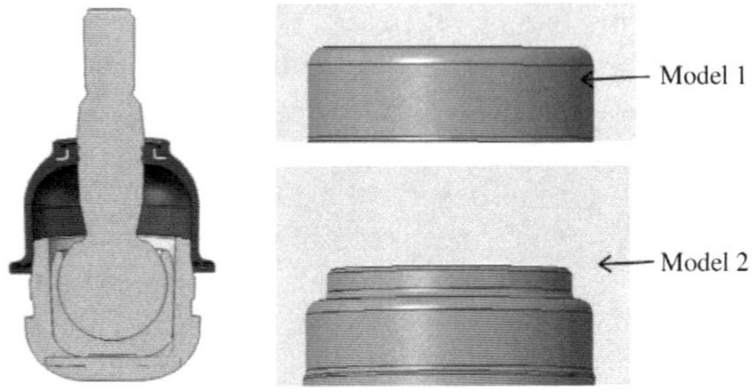

Figure 2. Sectional view of ball stud and dust cover and two types of base

The whole machine is mainly composed of ball stud and rubber dust cover. The ball stud is made of steel and the rubber dust cover material is C5121KL. The mechanical properties parameters are shown in Table 1.

TABLE 1 MATERIAL PROPERTIES PARAMETER TABLE

Material	Young's modulus of elasticity(GPa)	Density(g/cm^3)
C5121KL	0.00205	1.35
Steel	210	7.85

Material	Poisson's ratio	Yield Strength (MPa)	Tensile Strength (MPa)
C5121KL	0.49	/	14.6
Steel	0.3	355	600

In order to reduce the amount of calculation and ensure the accuracy of the results, the model of the non-concerned areas is simplified, and the materials are given to the corresponding components. The simplified model and coordinate system are shown in Fig. 2 (left). The geometric model is meshed, and the number of finite element meshes of the simplified model is 10828. The mesh model of the rubber is shown in Fig. 2 (right), and the mesh type is hexahedron.

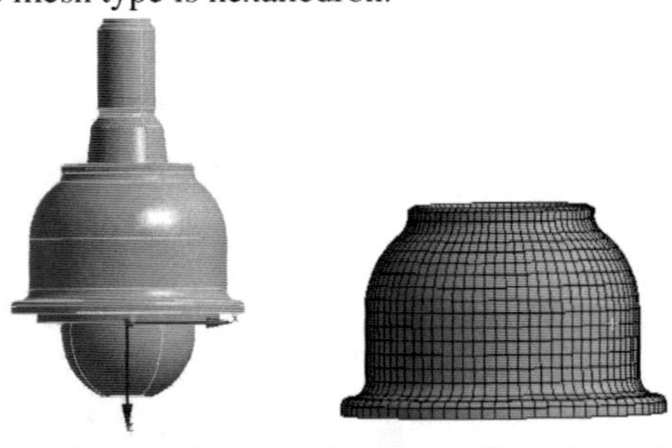

Figure 3. Finite element model of ball stud and dust cover

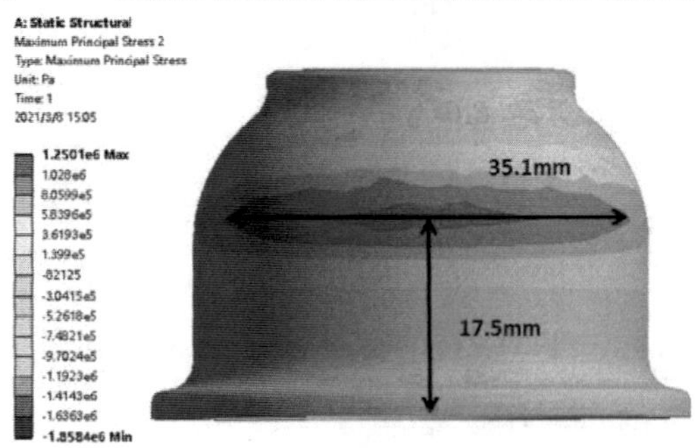

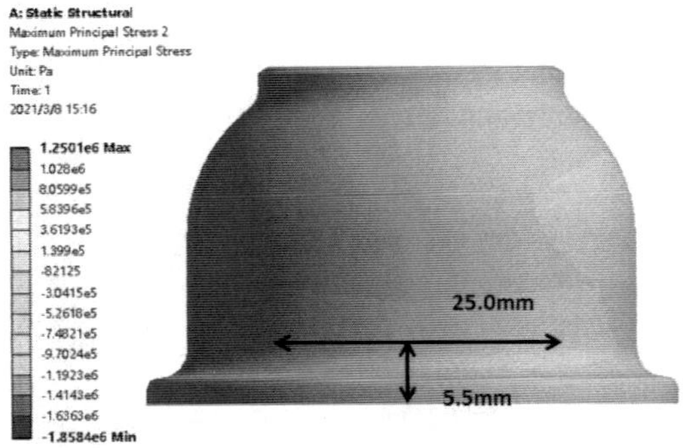

Figure 4. Simulation results corresponding to the base of model 1

In order to analyze the stress condition of the dust cover ball stud in the rotation and rotation state, the bottom is fixed, and the following working conditions are applied to the upper part of the ball stud: the rotation angle around the X axis is 25°, and the rubber cover head is pressed down by 7.5mm.

When the ball joint type is model 1, the overall stress on the ball joint stud of the dust cover is shown in Fig. 4 under the condition that the ball joint rotates 25° around the X axis and the dust cover head is pressed down by 7.5mm. There are two stress concentration areas on the outside of the dust cover. One area has the maximum tensile principal stress value of about 1.25Mpa (lower than the rubber tensile limit of 14.6Mpa), and the maximum tensile principal stress in the other area is about 0.52Mpa.

When the ball joint type is model 2, the overall stress on the ball joint stud of the dust cover is shown in Fig. 5 under the condition that the ball joint rotates 25° around the X axis and the dust cover head is pressed down by 7.5mm. There are two stress concentration areas on the outside of the dust cover. One area has the maximum tensile principal stress value of about 1.27Mpa (lower than the rubber tensile limit of 14.6Mpa), and the maximum tensile principal stress in the other area is about 0.59Mpa. According to the deformation diagram, the cover is extruded in the area with a height of about 17.5mm-17.8mm, and the outer surface is greatly stretched. The cover is greatly stretched in the area with a height of about 5.2mm-5.5mm. The principal stresses in the above two areas are large and belong to weak areas.

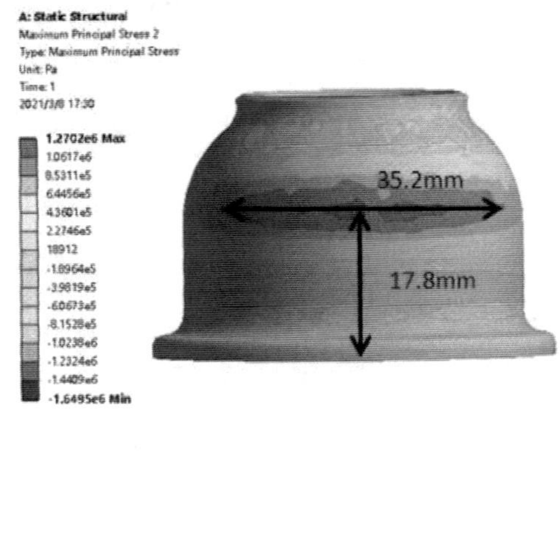

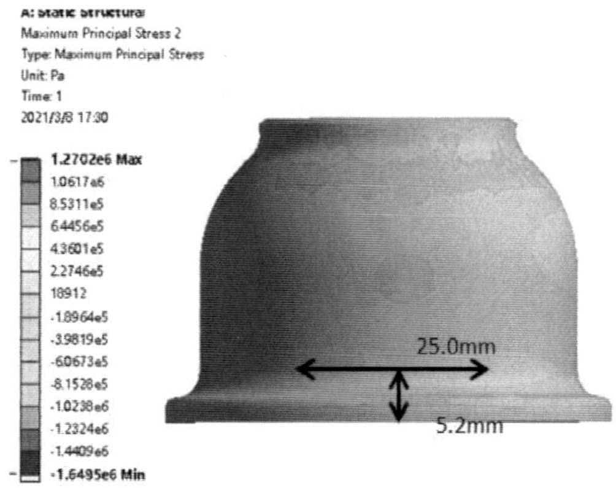

Figure 5. Simulation results corresponding to the base of model 2

3. TEST VERIFICATION

In order to verify the deformation state and durability index of the dust cover, a durability test was designed and carried out. In the test, the electric servo actuator provides the reciprocating rotation of the ball stud. The test frequency is 2Hz, and the number of samples is 6. Each base corresponds to 3 samples respectively. Carry out several reciprocating movements until the rubber cracks, and then carry out the crack analysis.

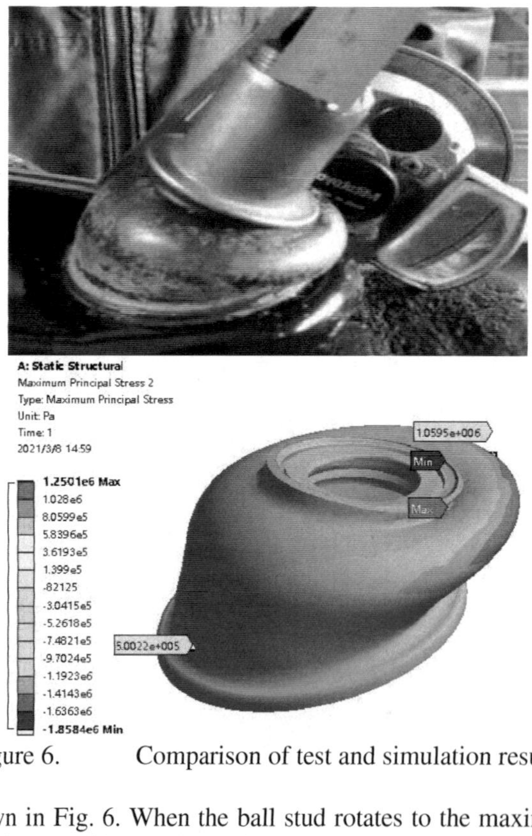

Figure 6. Comparison of test and simulation results

The test photo is shown in Fig. 6. When the ball stud rotates to the maximum position, the overall deformation state is consistent with the simulation results. The actual measurement shows that the maximum deformation and the deformation at key positions are consistent with the simulation results. The crack appears in one of the stress concentration areas, that is, about 5mm-6mm from the bottom surface, and the crack does not occur in the theoretical maximum stress area. The crack of the sample corresponding to the mounting base of model 2 is lower, and the crack almost occurs at the edge of the base support in contact with the rubber cover. The rubber surface at this position is in tension and tearing state. Through visual observation of the sample, there are slight friction marks between the inner surface of the rubber and the base.

The test results are shown in Table 2. The service life of the dust cover corresponding to the base of model 1 is significantly longer than that of model 1, indicating that the design of base model 1 can effectively reduce the stress concentration of the rubber cover and is relatively more conducive to extending the service life of the rubber cover.

TABLE 2 TEST VERIFICATION RESULTS

Sample Number	Base Type	Reciprocating Cycles	Crack Location
01	1	15623	About 6mm from the bottom
02	1	18831	About 6mm from the bottom
03	1	17902	About 6mm from the bottom
04	2	13351	About 5.5mm from the bottom
05	2	12980	About 5.5mm from the bottom
06	2	10943	About 5.5mm from the bottom

4. CRACK GROWTH ANALYSIS

Through the crack detection and crack growth analysis of the cracked specimen, the whole process of crack generation and propagation to crack penetration is judged and reproduced, so as to realize the tracking and positioning of the main failure source.

The common cracks of this type of rubber support base are shown in Fig. 7. The six failure samples all show that the cracks are roughly distributed in a straight line, parallel to the bottom, and the crack length is between 8mm-12mm. The cracked part of the failed sample was sampled and prepared, as shown in Fig. 8, and its macro and micro morphology were observed under SEM. Fig. 8(left) shows the macro morphology of the fracture, and Fig. 9-13 show the micro morphology of different regions in Fig. 8(right). From the macroscopic observation of the section morphology, it can be seen that the section has obvious fracture direction, the fracture trend morphology is from inside to outside, and the fracture source has obvious wear concave surface. From the microscopic observation of the section morphology, it can be seen that the microscopic morphology of the five detection areas is mainly a large number of step morphology. The fracture mode belongs to brittle fracture.

In conclusion, the stress concentration caused by surface wear is the main reason for the fracture of the dust cover. When the rubber is not subject to surface wear, the alternating stress generated by the ball stud operation does not exceed the material limit, and there is no risk of damage to the rubber cover. In the middle and late period of service, after the mechanical properties of rubber materials decline significantly due to wear, environmental corrosion, material aging and other reasons, the alternating stress finally exceeds the allowable limit of the material and failure fracture occurs.

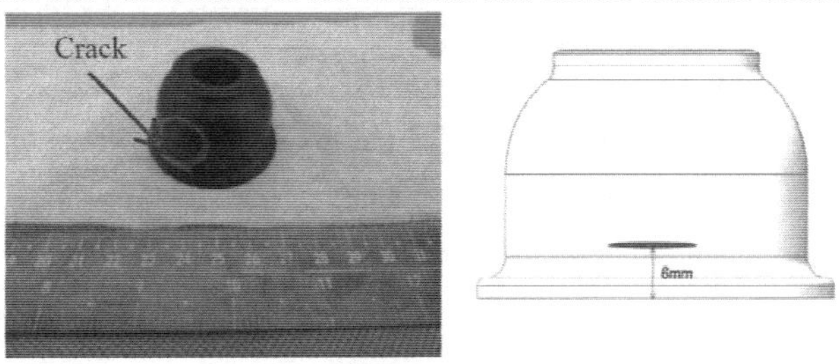

Figure 7. Schematic diagram of the rubber cracking position after the dust cover test

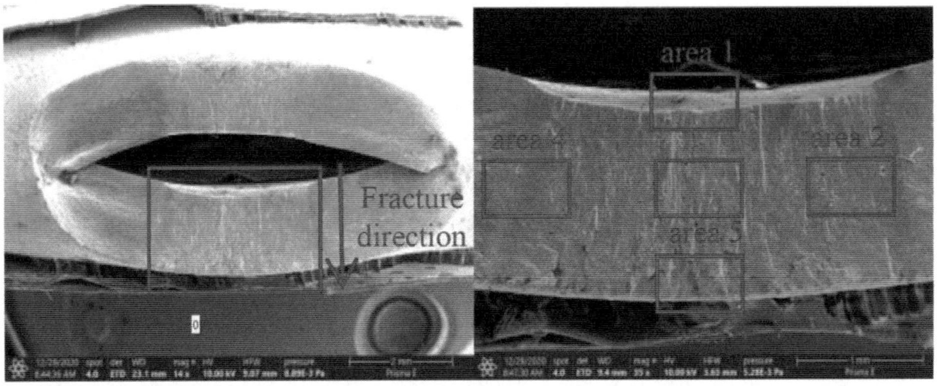

Figure 8. Macroscopic overall appearance of fracture

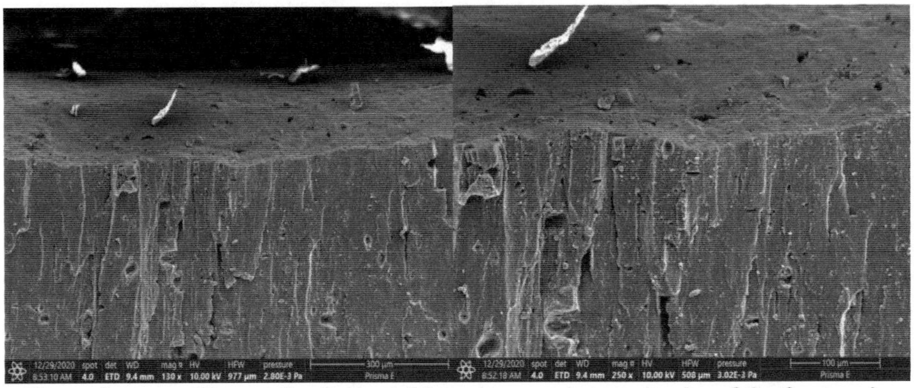

Figure 9. Observation image of the microscopic topography of the fracture in inspection area 1 (130×~250×)

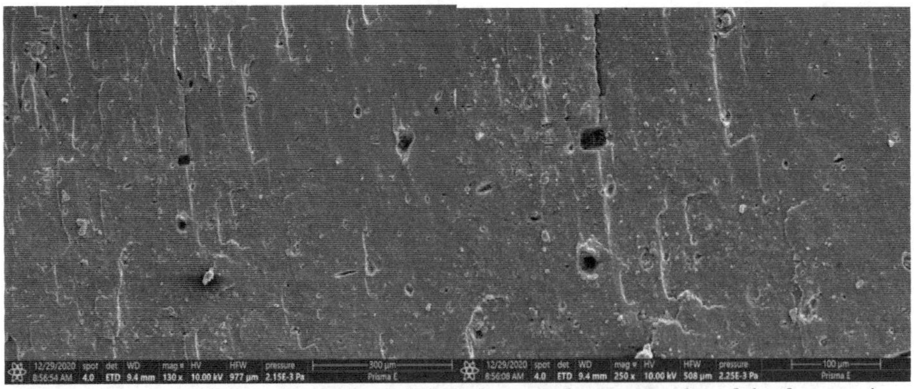

Figure 10. Observation image of the microscopic topography of the fracture in inspection area 2 (130×~250×)

Figure 11. Observation image of the microscopic topography of the fracture in inspection area 3 (130×~250×)

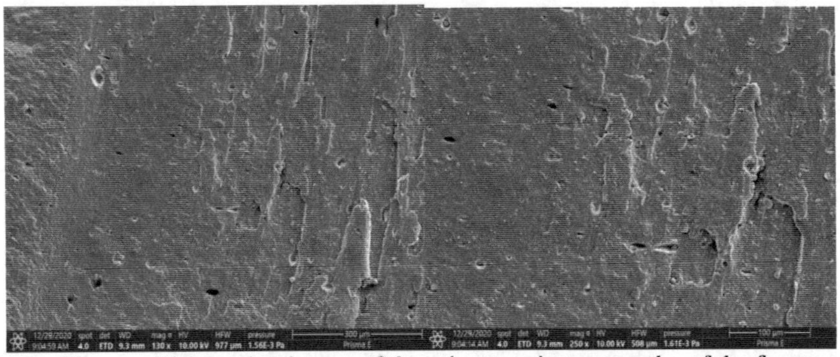

Figure 12. Observation image of the microscopic topography of the fracture in inspection area 4 (130×~250×)

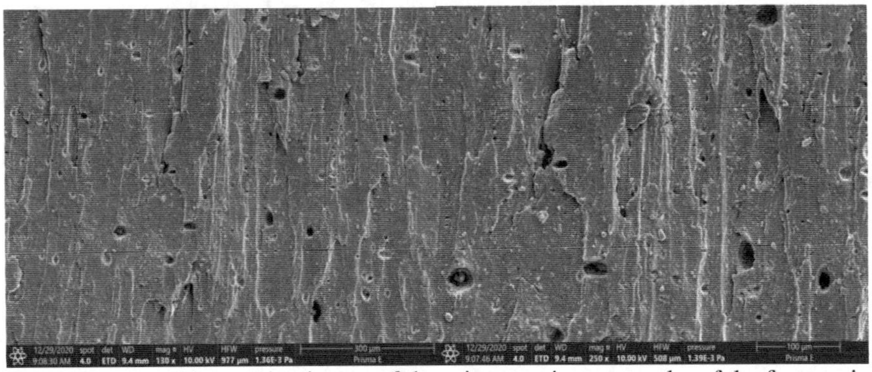

Figure 13. Observation image of the microscopic topography of the fracture in inspection area 5 (130×~250×)

5. Optimization suggestions

According to the analysis results, in order to improve the reliability and service life of the rubber support base of the battery pack, the following structural design optimization suggestions are put forward:

- Use the base of model 1,and appropriately increase the smoothness of the contact edge between the base and the rubber to reduce the friction effect.
- Increase the rubber thickness below the waist of the dust cover to reduce the stress concentration factor in the weak area.

- Appropriately increase the cross-sectional diameter of the waist of the rubber cover to reduce the "tension" effect caused by the rotation of the ball joint stud.

6. Conclusion

In this paper, the reliability of a newly designed rubber support base for battery pack is evaluated by means of structural finite element simulation, physical test verification and micro characterization of crack morphology.

The simulation results show that under normal installation and working conditions, there are two stress concentration areas in the rubber support base of battery pack, but the deformation and stress in the stress concentration areas do not exceed the allowable limit of the material design. Compared with the maximum stress, the calculation result of ball stud based model 1 is lower than that of model 2.

The test results show that during the reciprocating rotation of the ball stud, the rubber support base of the battery pack and the edge of the base are rubbed and worn, and the fracture occurs at the worn part, which is also one of the stress concentration areas. The life of the dust cover corresponding to the base of model 1 is significantly longer than that of model 2.

Acknowledgements

This paper is one of the stage results of the Guangdong Province Key Areas R&D Program Project Traction Battery System Testing and Evaluation Technology Research (2019B090908003).

References

[1] T. Sukumar, B. R. R. Bapu, and B. D. Prasad, “Numerical analysis and experimental evaluation of rubber boot design,” Engineering Failure Analysis, vol.118, pp. 104867, December 2020.

[2] A. Tarafdar, M. J. Oh, N. P. Quynh, and J. H. Kwon, “Profiling and potential cancer risk assessment on children exposed to PAHs in playground dust/soil: a comparative study on poured rubber surfaced and classical soil playgrounds in Seoul,” Environmental geochemistry and health, vol. 42(6), pp. 1691-1704, June 2020.

[3] B. Parrott, P. C. Zanini, A. Shehri, K. Kotsovos, and I. Gereige, “Automated, robotic dry-cleaning of solar panels in Thuwal, Saudi Arabia using a silicone rubber brush,” Solar energy, vol. 171, pp. 526-533, September 2018.

[4] C. W. Doud, and T. W. Phillips, “Responses of red flour beetle adults, Tribolium castaneum (Coleoptera: Tenebrionidae), and other stored product beetles to different pheromone trap designs,” Insects, vol. 11(11), pp. 733, November 2020.

[5] R. Juniah, D. Susetyo, and H. Rahmi, “Technical review of land usage of former limestone mine for rubber plantation in PT Semen BaturajaTbk for sustainable mining environment,” Journal of Physics: Conference Series, IOP Publishing, vol. 1338(1), pp. 012024, November 2019.

[6] L. L. Halle, A. Palmqvist, K. Kampmann, and F. R. Khan, “Ecotoxicology of micronized tire rubber: Past, present and future considerations,” Science of the Total Environment, vol. 706, pp. 135694, March 2020.

[7] K. Yamashita, S. Miyabe, and T. Yamashita, “High quality custom EPDM rubber bellows dust cover,” Surgical Infections, vol. 20(8), pp. 665, 2019.

[8] A. O. Al-Sulttani, A. Ahsan, A. N. Hanoon, A. Rahman, N. N. N. Daud, et al., “Hourly yield prediction of a double-slope solar still hybrid with rubber scrapers in low-latitude areas based on the particle swarm optimization technique,” Applied Energy, vol. 203, pp. 280-303, October 2017.

[9] B. Aloufi, K. Behdinan, and J. Zu, “Vibro-acoustic model of an active aircraft cabin window,” Journal of Sound and Vibration, vol. 398, pp. 1-27, June 2017.

[10] B. Guo, E. Y. T. Chen, W. Javed, and B. Figgis, “Characterization of an electrodynamic dust shield device for PV panel soiling mitigation,” Energy Sustainability, American Society of Mechanical Engineers, vol. 57595, pp. V001T08A001, November 2017.

[11] M. Sienkiewicz, H. Janik, K. Borzędowska-Labuda, and J. Kucińska-Lipka, "Environmentally friendly polymer-rubber composites obtained from waste tyres: A review," Journal of cleaner production, vol. 147, pp. 560-571, March 2017.
[12] T. V. Prokof'eva, A. V. Kiryushin, V. A. Shishkov, and F. A. Ivannikov, "The importance of dust material in urban soil formation: the experience on study of two young Technosols on dust depositions," Journal of Soils and Sediments, vol. 17(2), pp. 515-524, February 2017.
[13] L. Demin, "CAE simulation analysis of ball stud assembly dust cover based on Abaqus," Journal of Guilin University of Aerospace Technology, vol. 18(02), pp. 151-153, June 2013.
[14] X. Ying, "Application of ABAQUS rubber material in dust cap simulation analysis," Automobile Parts, vol. 06, pp. 71-74, June 2014.
[15] L. Guangzheng, G. Jianghua, S. Rui, and Z. Song, "Improved design of ball joint boot," Vehicle & Power Technology, vol. 01, pp. 55-58+64, March 2017.
[16] S. Zepeng, J. Xuejian, Y. Jianqiang, and Y. Lei, "Design simulation and performance research of a steering outer pull rod dust cover," Journal of Shandong University of Technology (Natural Science Edition), vol. 34(06), pp. 20-24, July 2020.
[17] J. H. Kim, and B. Y. Lee, "Nonlinear Stress Analysis of Dust Covers for Ball Joint of Automotive Steering System," Transactions of the Korean Society of Mechanical Engineers A, vol. 37(10), pp. 1297-1303, July 2013.
[18] Z. Jianmin, C. Chao, T. Wenbing, L. Yi, and H. Yanbin, "Composite layer defect detection method based on infrared heat wave technology, finite element and SVM," Chinese Journal of Scientific Instrument, vol. 41(03), pp. 29-38, March 2020.
[19] L. Xueping, and L. Xing, "Finite element simulation analysis of PMNT transducer," Foreign Electronic Measurement Technology, vol. 39(08), pp. 54-57, August 2020.
[20] Z. Ying, X. Hongyan, Z. Xu, Z. Qi, and L. Jun, "Simulation analysis of soil moisture sensor probe," Journal of Electronic Measurement and Instrumentation, vol. 35(04), pp. 55-65, April 2021.
[21] L. Longteng, J. Bo, H. Jiaxing, Z. Yulin, and L. Zhengang, "Degradation law analysis of QFP package interconnection structure based on PCMD health index," Journal of Electronic Measurement and Instrumentation, vol. 33(12), pp. 100-108, December 2019.

Online Quartz Crystal Microbalance Analyzing Device of Mass Concentration of Formaldehyde Gas Based On FPGA

Gangxin Chen[1,2], Jian Qiu*[1,2], Li Peng[1,2], Peng Han[1,2], Kaiqing Luo[1,2], Dongmei Liu[1,2]

[1] School of Physics and Telecommunication Engineering, South China Normal University, Guangdong, Guangzhou 510006, China

[2] Qingyuan Institute of Science and Technology Innovation, South China Normal University, Guangdong, Qingyuan 511517, China

* Corresponding author: qiuj@scnu.edu.cn

Abstract. As one of the main indoor gas pollutants, formaldehyde gas (HCHO) seriously threatens human health and ranks second among toxic chemicals in China. In order to monitor the concentration of formaldehyde in the environment in real time, an online Quartz Crystal Microbalance (QCM) analyzing device based on FPGA is developed in this paper. This device is composed of QCM sensor module and FPGA signal processor. Firstly, graphene oxide film is prepared on QCM chip to adsorb formaldehyde, which causes the change of resonance frequency. Secondly, to monitor the quality of adsorbed formaldehyde through frequency change, FPGA signal processor is mainly integrated with Direct Digital Synthesis module, Analogue-to-Digital Conversion module, Low-Pass Filter module and Frequency Domain Analysis module. These modules have the characteristics of high speed, low power consumption and parallel processing, which provide the best solution for the online implementation of the analyzing device. Finally, the standard gas experiment shows that the device is feasible and reliable for the analysis of formaldehyde gas mass concentration, with a resolution of about 10 counts per 1% concentration.

1. Introduction

As a basic chemical material, formaldehyde is widely used in the manufacture of paint, linoleum, varnish, interior decoration, wood, plastic and other building materials and household products [1]. The formaldehyde content of decoration materials and furniture in newly decorated houses is high, which will release a large amount of formaldehyde. Decorative materials and furniture can continue to release formaldehyde for 3 to 15 years [2]. As is known, formaldehyde is a harmful air pollutant. Exposure to formaldehyde for a long time will cause discomfort symptoms, such as burning, swelling, tears, pharyngeal edema, pharyngitis, cough, etc. In serious cases, it will cause bronchospasm and bronchial asthma [3-5]. Formaldehyde seriously endangers human health. Nowadays, indoor formaldehyde pollution is serious in China. The average concentrations in 2003~2018 exceeded the national standard limit of 0.1 mg/m^3, and the over-standard rates were more than 50% in most years [15]. Therefore, a device that can monitor the mass concentration of environmental formaldehyde online is needed.

Various analytical methods for detecting formaldehyde have been reported in the literature [6-10]. However, all these methods including spectrophotometry, liquid chromatography, gas chromatography, electrochemical detection and conductance analysis photometry have their own shortcomings and limitations. For example, the sensitivity is not high enough, the experimental equipment is too large, the operation is complex and online monitoring is not possible. Therefore, it is necessary to develop a simple, convenient and fast device that can monitor the mass concentration of formaldehyde online. The use of QCM for the detection of gaseous compound had been reported [11-14]. Inspired by this, QCM sensor can be used as a sensor module in the device in this paper. The working principle of QCM sensor is that the quartz crystal coated with sensing material adsorbs the detected target material to increase the quality of the crystal, thereby causing the resonant frequency of the crystal to change [18]. By monitoring the frequency change of piezoelectric QCM, the mass of the adsorption target material can be calculated. The resonant frequency of quartz crystal is a high-frequency signal and its frequency is usually 5MHz or 10MHz, and some even reach 30MHz and 50MHz. Therefore, it is necessary to develop a signal processor that can process high-frequency signals with low delay to meet the requirements of online monitoring. FPGA has the characteristics of high speed and parallel processing which meets the above requirements, so the signal processor in this paper is developed by FPGA. FPGA can realize all the functions of signal processing in this paper so it realizes the miniaturization of the device at the same time.

2. Gas Sensing System

2.1 QCM Theory

QCM is a general-purpose mass sensor that uses the piezoelectric effect of quartz crystal to transform mass change into resonant frequency change. According to Sauerbrey equation, the relationship between resonant frequency change and mass change is as follows

$$\Delta f = -\frac{2f_0^{\ 2}}{A\sqrt{\rho_q \mu_q}}\Delta m \tag{1}$$

Where Δf is frequency shift, Δm is the mass change, f_0 is the oscillation frequency of quartz, and A is the electrode area, ρ_q is the density of quartz, μ_q is the shear mode of quartz [16]. In this paper, AT-cut quartz crystal (resonant frequency 6Mhz) with formaldehyde sensitive layer on one side is used. Graphene oxide is used as the formaldehyde sensitive layer in this paper which has good adsorption capacity and selectivity for formaldehyde [17]. As shown in the figure 1, when the QCM modified by graphene oxide is exposed to formaldehyde, the graphene oxide film selectively adsorbs formaldehyde so the quartz crystal quality increases and the resonance frequency changes.

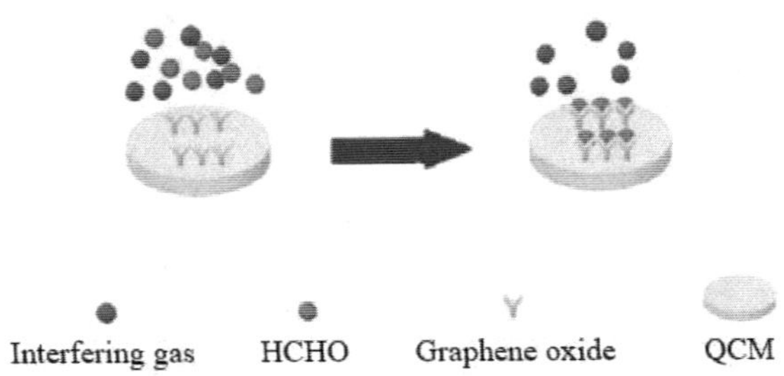

Figure 1. Sensing process of QCM adsorbing HCHO

2.2 Hardware Components

The experimental device in this paper is shown in figure 2, including the gas circuit device composed of air pump, QCM gas chamber, flowmeter and solution storage gas chamber, and the sensor device composed of oscillation circuit, FPGA and PC.

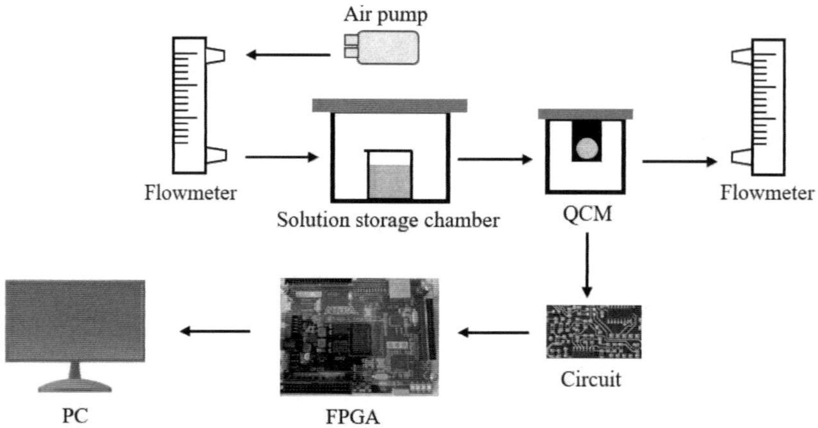

Figure 2. Schematic diagram of experimental device

The gas pipe in the gas circuit device is a transparent PU pipe, which is hard to deform and leak. The flow rate of the air pump is about 2L/min, and the flowmeter at the gas inlet and outlet is used to check the air tightness of the device. QCM chamber and solution storage chamber are designed as cube box, using acrylic material. The raised platform on four sides of the upper part of the box is used to place silicone pad. In order to ensure that the top cover will not deform after the top cover and the box are screwed, the top cover material is metal but not acrylic. The QCM chamber has a QCM probe inside and the QCM chip is stored in the probe. The QCM probe is connected to the oscillation circuit with SMA RF line which can stable transmission of high-frequency signals. Pierce oscillation circuit is used as the oscillation circuit [18] which is shown in figure 3. The 5V power supply of the oscillation circuit is provided by the FPGA board, and its signal output is connected to the FPGA board, which sends the processed data to the PC for display.

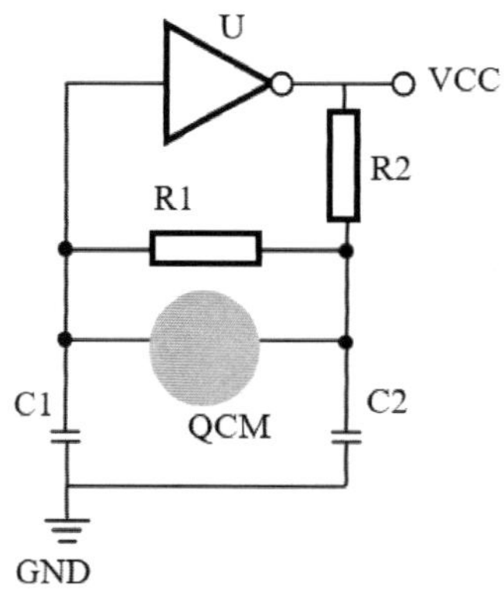

Figure 3. Pierce oscillation circuit

2.3 Detection Device based on FPGA

The signal in QCM detection device is high-frequency signal, and a series of data processing requires a lot of circuit resources. At the same time, online monitoring also requires low device delay. FPGA can perform signal processing in parallel, which has the characteristics of high speed and low power consumption. Therefore, it provides the best solution for high-speed signal processing. The FPGA developed in this paper includes the following modules: DDS module, AD conversion module, multiplier module, low-pass filter module, FFT module. EP4CE30 FPGA board is shown in figure 4. All the above functions are integrated on this board, realizing the miniaturization of the device.

Figure 4. Picture of EP4CE30 FPGA board

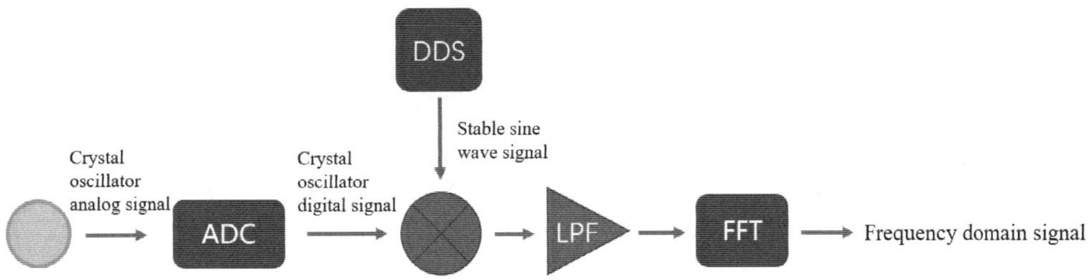

Figure 5. FPGA signal processing schematic diagram

FPGA signal processing schematic diagram is shown in figure 5. First, the analog signal of crystal oscillator is converted into digital signal of crystal oscillator through AD conversion. Secondly, the crystal oscillator digital signal and the stable sine wave signal generated by DDS get the mixing signal through the multiplier. Thirdly, the frequency difference signal between the crystal oscillator signal and DDS sine wave signal can be obtained through the low-pass filter. Finally, the difference frequency signal is processed by FFT to get its frequency domain signal.

(1) ADC Module

Because the analog signal cannot be processed in FPGA, the analog signal must be converted into digital signal by AD conversion first. This paper selects AD9226 chip. This chip is a single chip, 12 bit, 65 MSPs analog-to-digital converter (ADC), which is powered by a single power supply and has an on-chip high-performance sampling. Just provide the clock signal and the chip can directly collect data, which is a very simple operation.

(2) DDS Module

In order to get the frequency change directly, a constant sine wave signal is needed which can perform differential frequency processing with crystal oscillator signal. The quartz chip used in this paper has an oscillation frequency of 6M, so DDS is used to generate a sine wave digital signal with a frequency of 6M. DDS module is mainly composed of three parts: phase accumulator, sine function table (waveform memory) and table lookup module. The phase accumulator is composed of an adder and a phase register. Every time a clock comes, the adder adds the frequency control word to the phase data output by the accumulation register, and then inputs it to the address of the sinusoidal query table. The sine query table contains the digital amplitude information of a periodic sine wave, and each address corresponds to a phase point in the range of 0-2π in the sine wave. The look-up table maps the input address phase information into a digital signal of sine wave amplitude. Due to the limited resources of FPGA development board, the signal generated by DDS is an 8-bits signal.

(3) Multiplier Module

The 12-bit crystal oscillator signal converted by AD and the 8-bit sine wave signal generated by DDS are passed through the multiplier module to obtain a 20-bit mixing signal.

(4) Low-pass Filter Module.

In this paper, FIR filter is used as the low-pass filter, and the fully parallel distributed algorithm structure is adopted. The upper limit frequency of the filter is set to 200kHz, and the variable frequency signal Δf can be obtained by passing the mixing signal through the low-pass filter.

(5) FFT Module

In order to get the frequency domain signal of the signal, FFT processing is also needed. 2048-point FFT operation is used in this module. The Streaming FFT is used as I/O Data Flow. The streaming FFT allows continuous processing of input data, and outputs a continuous complex data stream without the need to halt the data flow in or out of the FFT IP core. The streaming FFT generates a design with a quad output FFT engine and the minimum number of parallel FFT engines for the required throughput. Quad-output FFT engine can minimize transform time. Quad-output refers to the throughput of the internal FFT butterfly processor. The engine implementation computes all four radix-4 butterfly complex outputs in a single clock cycle. Schematic diagram of the Quad-Output FFT Engine is shown in figure 6.

The FFT reads complex data samples x[k,m] from internal memory in parallel and reorders by switch (SW). Next, the radix-4 butterfly processor processes the ordered samples to form the complex outputs G[k,m]. Because of the inherent mathematics of the radix-4 DIF decomposition, only three complex multipliers perform the three non-trivial twiddle-factor multiplications on the outputs of the butterfly processor. To discern the maximum dynamic range of the samples, the block-floating point units (BFPU) evaluate the four outputs in parallel. The FFT discards the appropriate LSBs and rounds and reorders the complex values before writing them back to internal memory.

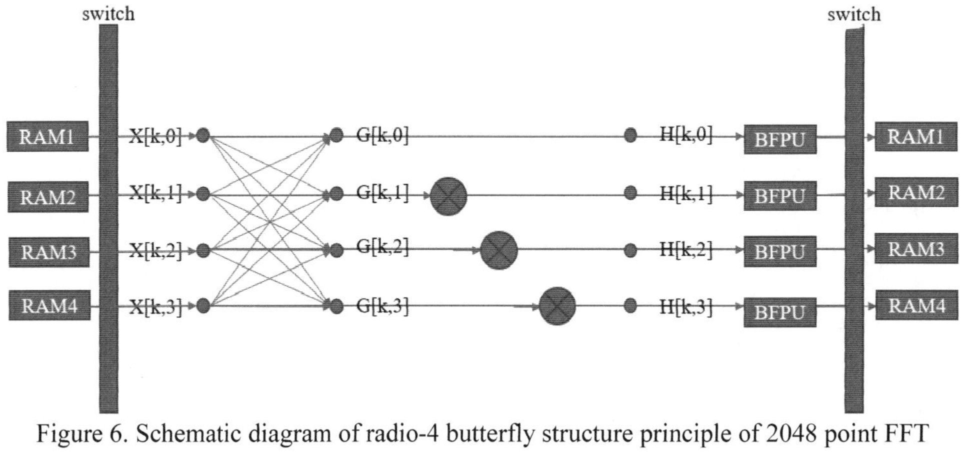

Figure 6. Schematic diagram of radio-4 butterfly structure principle of 2048 point FFT

3. Experiment

3.1 Materials

Graphene oxide solution, 37% formaldehyde solution, 6 MHz AT-cut quartz crystal, electric blast box, ultrasonic cleaning machine, airbrush, pure water, beaker.

3.2 Preparation Process

3.2.1 Cleaning of Experimental Devices

(1) Clean the beaker, fill the beaker with pure water, put the beaker into the ultrasonic cleaner, and vibrate for 5 minutes. After cleaning, put it into the electric blast box for drying.

(2) Clean QCM, put quartz crystal into beaker, add pure water, and vibrate in ultrasonic cleaner for 5 minutes. After cleaning, put it into the electric blast box for drying.

(3) Clean the airbrush, fill the airbrush storage chamber with pure water, and place it in the ultrasonic

cleaning machine for 5 minutes. Fill the airbrush storage chamber with pure water, connect the air pump to spray until the pure water in the chamber is empty. Repeat This step for five times.

3.2.2 Preparation of Thin Films
In order to control the spraying area of the wafer during the spraying process and prevent spraying to the area outside the electrode, a spraying auxiliary mold is designed by using 3D printing technology. Before coating, it is necessary to build an experimental platform for coating. The schematic diagram of the platform is showed in figure 7. The cross clamp is clamped on the iron frame platform to fix the micro clamp. The micro clamp is used to fix the airbrush so that the airbrush is perpendicular to the horizontal plane.

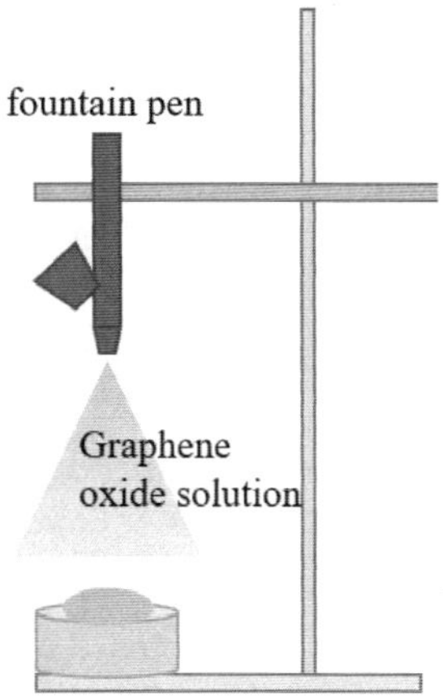

Figure 7. The schematic diagram of the coating platform

The specific steps of coating are as follows:

(1) Put the graphene oxide solution into the ultrasonic cleaner and oscillate for 5 minutes. After oscillation, pour a certain amount of solution into the airbrush storage chamber, and then put the airbrush into the ultrasonic cleaner and oscillate for 5 minutes to make the solution disperse evenly.

(2) Adjust the position of the airbrush so that the height of the nozzle from the quartz wafer is 15cm, and keep it unchanged.

(3) After spraying, put QCM into electric blower and dry it at 60 ℃.

3.3 Experimental Procedure
The experimental process is as follows:

(1) Clean the gas circuit of the gas chamber. Clean the gas circuit of the gas chamber with nitrogen for 1 min to remove the impurity gas in the device.

(2) Turn on the air pump to ventilate for 10 mins, observe and record the QCM no-load frequency.

(3) Put 20ml formaldehyde solution into a cleaned beaker and then put the beaker into the solution storage chamber. Then observe and record the frequency change until the frequency is stable and last take out the solution.

(4) Inject nitrogen, which can purge the formaldehyde adsorbed on the chip and remove the residual gas in the device.

4. Results and Discussion

4.1 System Sensitivity
The experiment was carried out with 37% solution, and the frequency response curve is shown in the

figure 8.

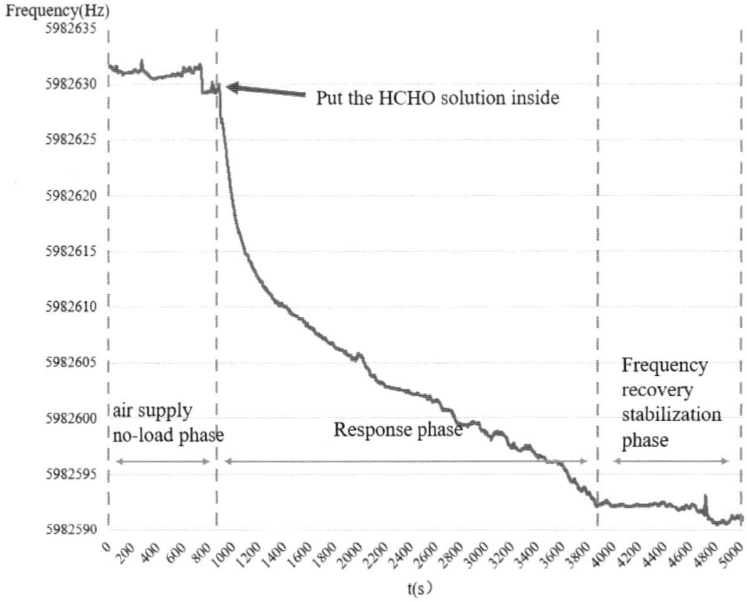

Figure 8. Experimental frequency response curve of 37% formaldehyde solution

As shown in the figure 8, the complete process of the on-line measurement experiment of 37% formaldehyde solution.

Process analysis:

(1) No-load phase: Pump air in for 15 minutes without putting the solution inside. Because the film material on QCM is graphene oxide which has little adsorption to other substances in the air. So there is almost no adsorbed substance in this phase and the resonant frequency changes little with a variation range of ± 2Hz.

(2) Response phase: fifteen minutes later, put the beaker containing 37% formaldehyde solution into the air chamber. The formaldehyde solution is highly volatile. The incoming air will bring the volatile formaldehyde to the QCM chamber. The QCM chip begins to adsorb formaldehyde, the mass increases and the resonant frequency begins to decline rapidly.

(3) Frequency recovery stabilization phase: when the adsorption of formaldehyde by graphene oxide film reaches saturation, the mass will not continue to increase and the frequency tends to stabilize. After reaching the lowest resonant frequency in about 4000 seconds, the resonant frequency gradually stabilized.

4.2 System Reusability and Repeatability

The adsorption of formaldehyde on graphene oxide film is physical adsorption. Purging the quartz wafer with nitrogen can blow away the formaldehyde adsorbed on the film to achieve the purpose of desorption. Therefore, the device is also reusable.

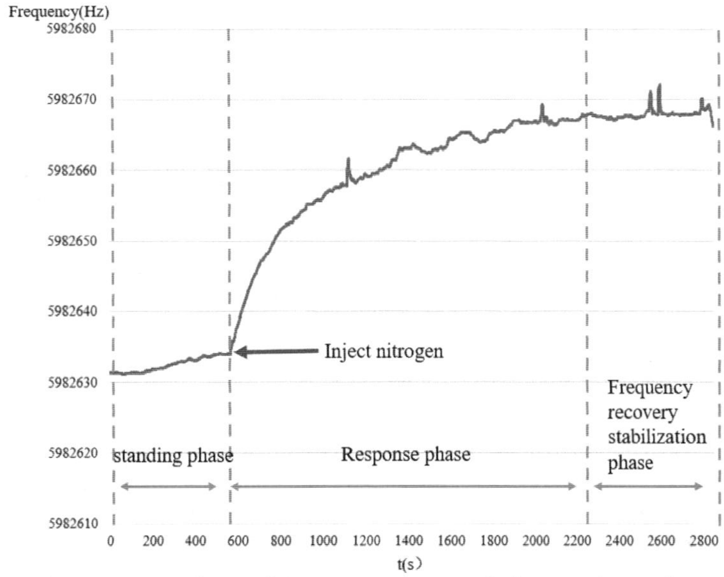

Figure 9. Experimental frequency change of nitrogen desorption

As shown in the figure 9, the device stood still for ten minutes. After ten minutes, start to inject nitrogen to purge the QCM chip that has adsorbed formaldehyde. At this time, the frequency showed a rapid upward trend, indicating that formaldehyde separated from the QCM chip and the chip quality decreased. After a period of time, formaldehyde desorption was completed and the frequency gradually stabilized. The final frequency change is 35.2Hz, which is not so different from the 39Hz frequency change after formaldehyde adsorption above.

In order to verify the reliability of the device, repeat the experiment with 37% formaldehyde solution for several times, and the experimental results are shown in Table 1.

Table 1. the experiment with 37% formaldehyde solution

37% solution Experiment	Frequency change（/0.1Hz）
1	392.9
2	389.2
3	332.8
4	332.5
5	330.2
Average	355.5
STD	32.4

The experiments were repeated for 4 times, and the frequency change obtained by adding the previous experimental data were 392.9, 389.2, 332.8, 332.5 and 330.2. Their average is 355.5 and standard deviation is 32.4. The results show that the device has good repeatability and its resolution of about 10 counts per 1% concentration.

5. Conclusion

This paper develops an on-line quartz crystal microbalance analyzing device of mass concentration of formaldehyde gas based on FPGA. Using FPGA as the signal processor has the characteristics of high speed, low power consumption and parallel processing, which solves the problems of on-line monitoring and high-frequency signal processing. DDS module generates an 8-bit sine wave and the frequency is stable at 6MHz. The sampling rate of AD module reaches 65MHz and the 6MHz crystal oscillator signal can be accurately restored by sampling and conversion. The FFT module adopts 2048-point processing and streaming data input which can continuously carry out FFT processing to obtain frequency domain signals. QCM sensor uses graphene oxide film which has good adsorption and selectivity for formaldehyde. The results with 37% formaldehyde solution experiment show that the device has good performance for formaldehyde measurement and its resolution is about 10 counts per 1% concentration.

Acknowledgments
This paper was supported by: Funding. National Natural Science Foundation of China (61975058); Natural Science Foundation of Guangdong Province (2019A1515011401); the Science and Technology Program of Guangzhou (2019050001).

References

[1] Hu, W., Chen, S., Liu, L., Ding, B., & Wang, H. (2011). Formaldehyde sensors based on nanofibrous polyethyleneimine/bacterial cellulose membranes coated quartz crystal microbalance. *Sensors and Actuators B: Chemical*, *157*(2), 554-559.
[2] Zhai, L., Zhao, J., Xu, B., Deng, Y., & Xu, Z. (2013). Influence of indoor formaldehyde pollution on respiratory system health in the urban area of Shenyang, China. *African health sciences*, *13*(1), 137-143.
[3] Rumchev, K. B., Spickett, J. T., Bulsara, M. K., Phillips, M. R., & Stick, S. M. (2002). Domestic exposure to formaldehyde significantly increases the risk of asthma in young children. *European Respiratory Journal*, *20*(2), 403-408.
[4] Beane Freeman, L. E., Blair, A., Lubin, J. H., Stewart, P. A., Hayes, R. B., Hoover, R. N., & Hauptmann, M. (2009). Mortality from lymphohematopoietic malignancies among workers in formaldehyde industries: the National Cancer Institute Cohort. *JNCI: Journal of the National Cancer Institute*, *101*(10), 751-761.
[5] Kerns, W. D., Pavkov, K. L., Donofrio, D. J., Gralla, E. J., & Swenberg, J. A. (1983). Carcinogenicity of formaldehyde in rats and mice after long-term inhalation exposure. Cancer research, 43(9), 4382-4392.
[6] Chen, H. W., Zheng, J. A., Ming, L., Song, Q. Y., Liu, C., Yu, A. M., & Jin, Q. H. (2004). Methods and instruments for formaldehyde monitoring. *Chinese Journal of Analytical Chemistry*, *32*(7), 969-972.
[7] Ding, B., Wang, M., Yu, J., & Sun, G. (2009). Gas sensors based on electrospun nanofibers. *Sensors*, *9*(3), 1609-1624.
[8] Kim, S., & Kim, H. J. (2005). Comparison of standard methods and gas chromatography method in determination of formaldehyde emission from MDF bonded with formaldehyde-based resins. *Bioresource Technology*, *96*(13), 1457-1464.
[9] Shimomura, T., Itoh, T., Sumiya, T., Mizukami, F., & Ono, M. (2008). Electrochemical biosensor for the detection of formaldehyde based on enzyme immobilization in mesoporous silica materials. *Sensors and Actuators B: Chemical*, *135*(1), 268-275.
[10] Dzyadevych, S. V., Arkhypova, V. N., Korpan, Y. I., Anna, V., Soldatkin, A. P., Jaffrezic-Renault, N., & Martelet, C. (2001). Conductometric formaldehyde sensitive biosensor with specifically adapted analytical characteristics. *Analytica chimica acta*, *445*(1), 47-55.
[11] Matsuguchi, M., Kadowaki, Y., & Tanaka, M. (2005). A QCM-based NO2 gas detector using morpholine-functional cross-linked copolymer coatings. *Sensors and Actuators B: Chemical*, *108*(1-2), 572-575.
[12] Vilaseca, M., Yagüe, C., Coronas, J., & Santamaria, J. (2006). Development of QCM sensors modified by AlPO4-18 films. *Sensors and Actuators B: Chemical*, *117*(1), 143-150.
[13] Sasaki, I., Tsuchiya, H., Nishioka, M., Sadakata, M., & Okubo, T. (2002). Gas sensing with zeolite-coated quartz crystal microbalances—principal component analysis approach. *Sensors and Actuators B: Chemical*, *86*(1), 26-33.

[14] Mirmohseni, A., & Oladegaragoze, A. (2003). Construction of a sensor for determination of ammonia and aliphatic amines using polyvinylpyrrolidone coated quartz crystal microbalance. *Sensors and Actuators B: Chemical*, *89*(1-2), 164-172.
[15] Qiong, W., & Kang, X. (2021). Indoor Formaldehyde Concentrations and the Influencing Factors in Urban China. *Progress in Chemistry*, *34*(3), 743.
[16] Aydemir, F., & Ebeoğlu, M. A. (2017). A QCM sensor array-based electronic tongue with the optimized oscillator circuit using FPGA. *IEEE Transactions on Instrumentation and Measurement*, *67*(2), 431-438.
[17] Yang, M., & He, J. (2016). Graphene oxide as quartz crystal microbalance sensing layers for detection of formaldehyde. *Sensors and Actuators B: Chemical*, *228*, 486-490.
[18] Yang, M., & He, J. (2018). A copper–manganese composite oxide as QCM sensing layers for detection of formaldehyde gas. *RSC advances*, *8*(1), 22-27.

Low temperature characteristics of FBG with free arc shape for low temperature monitoring

Huanlang Lu[1], Shuai Zhang[1], Yunshuai Yao[2], Yukun Yan[2], Zecong Lin[1] and Pingyu Zhu[1,3]

[1]School of Mechanical and Electrical Engineering Guangzhou University Guangzhou, China

[2]Zhuzhou Times New Material Technology Co. LtD, Zhuzhou, China

[3]Author to whom any correspondence should be addressed: pingyu.zhu@gzhu.edu.cn

Abstract. It is important to monitor the iced condition for operating wind turbine blades. However, most sensors are difficult to be used in this quite cold and lightweight environment. Although the fiber bragg grating (FBG) is a good candidate for the case. With the change of the temperature, the thermal expansion coefficient and thermo-optical coeffecient of FBG material will affect the accuracy of characteristics of FBG sensor. Aiming at the application of fiber optic sensor in low temperature environment and ultrathin requirement, the low temperature characteristics of FBG sensor with free arc shape package has been studied. The low temperature calibration experiments of two packaged FBG samples are tested from -35℃ ~20℃ in a constant temperature freezer with alcohol tank. The effects of low temperature characteristics of two FBGs packaged with free straight and arc shape tubes are compared. The results show that arc shape can effectively reduce the repeatability error and improve its accuracy due to providing possible low temperature contracting of the optical fibers. Flat arc shape makes it possible to package ultrathin FBG temperature sensor.

1. Introduction

FBG sensor utilizes the photosensitivity of germanium-doped fiber, and uses ultraviolet exposure and femtosecond laser to fabricate a sensor with a core periodic change grating called bragg grating. FBG sensors have been greatly developed due to their advantages such as anti-electromagnetic interference, corrosion resistance, low transmission loss, small size and easy embedding structure. Compared with temperature sensors such as semiconductor diodes, thermocouples and metal resistances, FBG sensors have obvious advantages in the field of wind turbine blade icing monitoring.

FBG is sensitive to both temperature and strain, which makes it complicated to measure each variable independently. When FBG is used for temperature measurement, it is necessary to use packaging and other methods to deal with the influence of strain on wavelength drift [1]. In order to ensure the uniformity and accuracy of the sensor values, the newly developed sensor must be calibrated and tested before it can be applied. The calibration results directly affect the sensor's measurement of physical quantities such as temperature and strain [2]. An Yang et al. proposed and discussed a new method for temperature calibration of fiber bragg grating (FBG) sensors based on BP neural network. The processing and analysis of the experimental data show that the method fits the complex relationship between the center wavelength and temperature well [3]. Xudong Ding et al. studied the structure of an all-metallized package FBG sensor, which can improve the linearity and temperature sensitivity of the sensor in low temperature environment [4]. Bowen Li et al. analyzed the changes of thermo - optical coefficient and thermal expansion coefficient of FBG sensor in low temperature environment, and verified that FBG sensor still had good response and repeatability in low temperature environment [5]. Mingyao Liu et al. found that the fiber grating sensor packaged in a brass tube can improve the temperature sensitivity and linearity of the sensor in low temperature environment [6]. Zhaohui Lu et al. used the pre-stretching process to package the fiber grating sensor, and verified the feasibility of the process to solve the chirp phenomenon at low temperature [7]. Jianhua Zhou et al. studied the low temperature sensing characteristics of fiber gratings and gave a reasonable explanation for the formation mechanism of the low-temperature chirping phenomenon of fiber gratings [8].

In the application of FBG sensor in temperature measurement, relevant researches mainly focus on the ambient temperature above room temperature, and there are few researches on the low temperature sensing characteristics below 0℃. For the problems of chirp phenomenon and certain repeatability error of temperature measurement results in the process of conventional FBG low temperature monitoring work, this paper studies the low temperature characteristics of the FBG sensor with arc tension package in the gate area, and it is verified by low temperature calibration experiments. The low temperature calibration experiment proves that this method can effectively reduce the repeatability error of measurement results and improve the accuracy of temperature measurement in low temperature environment of FBG sensor. Compared with the traditional tube-type and metal-type package solutions, the arc-shaped package structure can achieve a lighter and thinner plane package effect, which has important guiding significance for the application of fiber bragg grating sensors in low temperature environment.

2. Experimental principle

The reflection wavelength of FBG sensor has a linear relationship with physical quantities such as strain, temperature, and pressure. When the temperature around the grating changes, the grating period or refractive index of the fiber core will change, resulting in the grating Bragg signal wavelength shift. By monitoring the wavelength shift, the temperature to be measured can be obtained. Generally speaking, the reflected wavelength satisfies the Bragg condition:

$$\lambda_B = 2\eta_{eff}\Lambda \tag{1}$$

Where, λ_B is the reflection wavelength; η_{eff} is the effective refractive index of fiber core. Λ is the grating period. When the broadband light source is incident into the FBG, the periodic structure of the grating will reflect back a reflected light with a wavelength of λ_B, and the wavelength of the reflected light is related to the refractive index modulation period of the grating and the effective refractive index of the fiber core. The refractive index modulation period of the grating and the effective refractive index of the fiber core change with the change of ambient temperature [6]. According to grating theory, the reflection center wavelength of FBG changes as follows:

$$\frac{\Delta\lambda_B}{\lambda_B} = (\alpha + \xi)\Delta T + (1 - \rho_e)\varepsilon \tag{2}$$

Where $\Delta\lambda_B$ is the central wavelength drift, α is thermal expansion coefficient of optical fiber; ξ is the optical fiber thermo-optic coefficient, ρ_e is the optical elastic coefficient of the optical fiber, ΔT is the temperature change, ε is the axial strain of the grating.

Let $K_T = (\alpha + \xi)\lambda_B$, $K_\varepsilon = (1 - \rho_e)\lambda_B$, then equation (2) can be expressed as:

$$\Delta\lambda_B = K_T\Delta T + K_\varepsilon\varepsilon \tag{3}$$

Where, K_T and K_ε represent temperature sensitivity and strain sensitivity respectively. When FBG is in a free state free from strain action, the change of the sensor's central wavelength is only affected by temperature change. Thus, Equation (3) can be simplified as:

$$\Delta\lambda_B = K_T\Delta T \tag{4}$$

It can be seen from the above that the temperature change of the sensor can be detected by detecting the change of FBG wavelength. Due to the different writing processes of fiber and grating used by FBG sensors, the temperature sensitivity of sensors will be different, and the ambient temperature will also affect the thermal-optical coefficient and thermal expansion coefficient of fiber. At low temperature, the temperature sensitivity K_T of the sensor will decrease with the decrease of temperature.

3. Experiment

The low temperature calibration system of FBG sensor is shown in figure 1 below. The system consists of FBG interrogator SI155, FBG sensor, thermocouple sensor, computer and constant temperature freezer. The constant temperature freezer has a working temperature range of -50℃~20℃. The thermocouple sensor of the calibration system gives a standard input temperature value to the fiber bragg grating sensor, and the center wavelength of the FBG is read by the interrogator and output through the demodulation system.

When the temperature of the refrigerating constant temperature tank is uniform, insert the FBG sensor into the cooling water tank, as shown in figure 1. Keep the FBG temperature consistent with the tank temperature for a period of time. Record the output temperature of thermocouple sensor and wavelength data of FBG sensor after the temperature indicator of constant temperature freezer and wavelength value of demodulation instrument are stable.

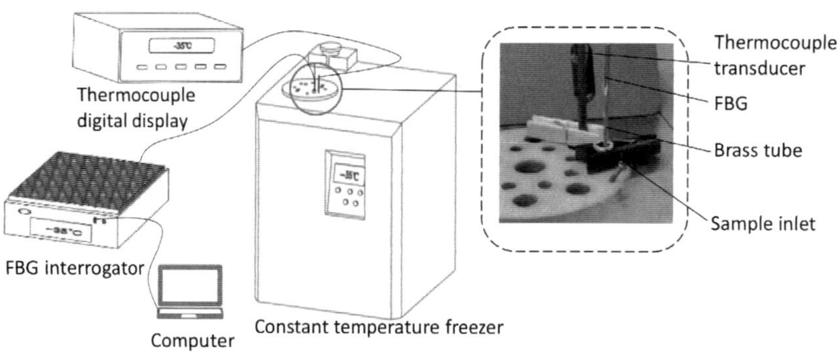

Figure 1. FBG low temperature calibration test system diagram.

In order to ensure that the sensors are in a free state and not affected by strain during the experiment, it is necessary to encapsulate the gate area of the two FBG sensors with a PI tube with a length of 50 mm before the experiment. Both ends of the sensor behind the casing are fixed by UV glue. FBG1 is packaged with a conventional flat and straight sleeve, and the arc shape can compensate for fiber shrinkage at low temperature. Therefore, FBG2 needs to be flat micro-bent sleeve on a special arc-shaped mold. The use of this mold can ensure uniform dispensing and prevent virtual bonding. The schematic diagram of the packaging mold is shown in figure 2. The encapsulation method can also hot-press a layer of glass fiber epoxy resin prepreg on the surface of the mold to encapsulate the FBG plane. The flat arc shape makes it possible to encapsulate the thin fiber grating temperature sensor. Since the UV glue has a certain elasticity, the optical fibers at both ends of the PI tube will rebound after bonding, and in order to prevent the chirp phenomenon in the sensor measurement during the low-temperature calibration process, the grid area of the optical fiber sensor needs to be pre-tensioned. After that, the two ends of the optical fiber are tensioned and bonded to the hollow brass tube to ensure that the optical fiber can be kept in a tensioned state. The packaged sensor is shown in figure 3.

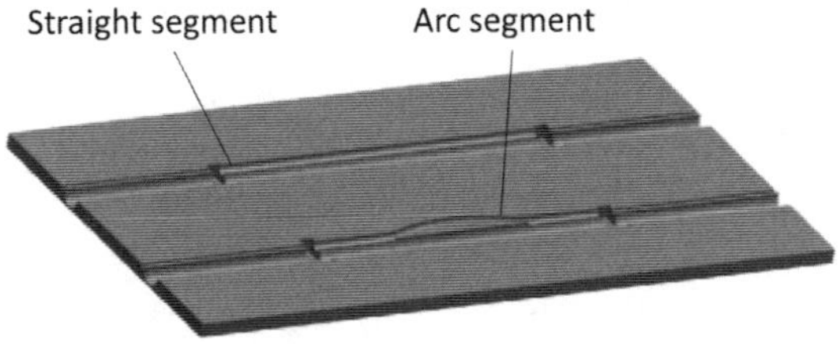

Figure 2. Schematic diagram of packaging mold.

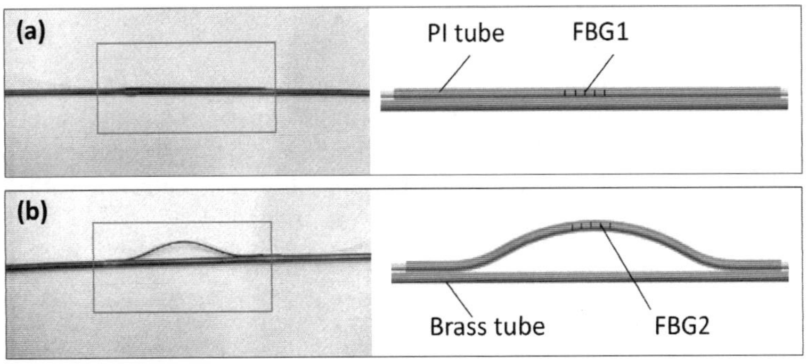

Figure 3. (a) FBG1 structure. (b) FBG2 structure.

The inside of the constant temperature freezer is circulated through water to ensure uniform internal temperature. In order to further ensure that the sensor grid area will not be strained due to the circulating flow of water, the FBG sensor string is placed in a hollow stainless steel tube, as shown in figure 1. Transformer oil with large thermal conductivity should be added to the hollow stainless steel tube to ensure uniform temperature inside the stainless steel round tube.

4. Experimental result

In this experiment, two processes of heating and cooling are carried out between -35 °C and 20 °C to calibrate the two sensors of FBG1 and FBG2. Five identical temperature points were taken in the heating process and the cooling process, and the distance between adjacent temperature points was 11 °C.

In the process of temperature measurement, there is still a small range of drift when the sensor wavelength value reaches a relatively stable state. In order to ensure the accuracy of the results, each temperature measurement point needs to record four sets of central wavelength values, and take the average value of the four sets of data. Experimental data are shown in table 1 and table 2.

Table 1. FBG1 temperature and cooling data.

FBG1 heating process		FBG1 cooling process	
Central wavelength/nm	Temperature /℃	Central wavelength/nm	Temperature /℃
1562.5937	-34.9794	1564.37225	20.0306
1562.9556	-23.9979	1563.9968	9.0493
1563.2867	-13.0200	1563.6341	-1.8020
1563.6934	-1.9842	1563.277	-12.7739
1564.0873	9.1630	1562.9257	-23.9397
1564.4885	20.1488	1562.5937	-34.9794

Table 2. FBG2 temperature and cooling data.

FBG2 heating process		FBG2 cooling process	
Central wavelength/nm	Temperature /℃	Central wavelength/nm	Temperature /℃
1565.5583	-34.9781	1567.3000	20.1400
1565.9471	-23.9984	1566.9212	9.0515
1566.2528	-13.0190	1566.5211	-1.8017
1566.5348	-1.9864	1566.2495	-12.7711
1566.9348	9.1675	1565.9431	-23.9377
1567.3096	20.1502	1565.5583	-34.9781

Through the input (temperature)/output (average value of central wavelength) relationship of the two FBG sensors, the least square method is used to fit the experimental data, and the fitting results are shown in figure 4 and figure 5. The abscissa is the average central wavelength of the reflected spectrum of the fiber bragg grating, and the ordinate is the calibration temperature point.

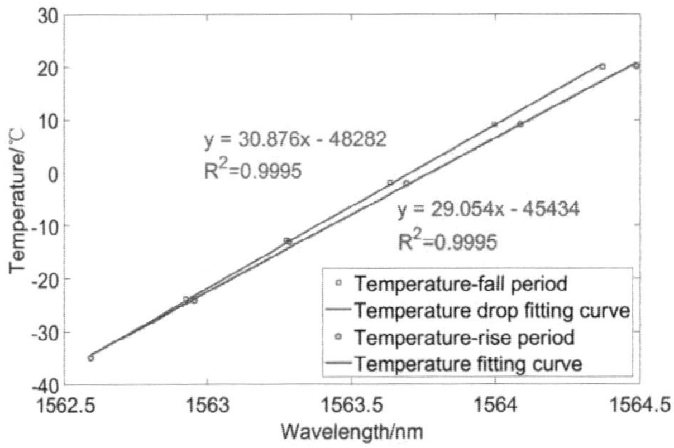

Figure 4. FBG1 temperature calibration scatter plot and curve fitting.

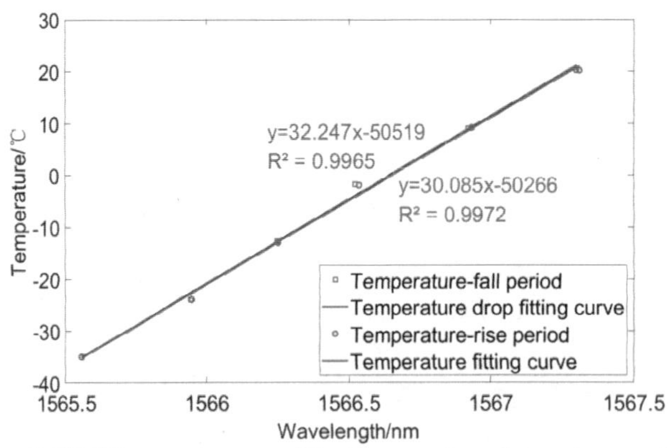

Figure 5. FBG2 temperature calibration scatter plot and curve fitting.

It can be seen from the above temperature-wavelength fitting curves in figure 4 and figure 5 that during the heating and cooling process, the wavelength drift of the two FBGs after being packaged by the sleeve has a good linear relationship with the temperature, and the linearity of the two sets of sensors heating and cooling fitting curves are as high as 0.9965 or more. Comparing the repeatability of the heating and cooling curves of the two groups of sensors, the degree of inconsistency of the fitting curves between the heating formation and the cooling stroke of FBG1 with a flat package is significantly greater than that of FBG2 with a circular arc package. At low temperature, the difference of thermal expansion coefficient of fiber sensor will cause inconsistent strain, resulting in micro-bending loss [8]. Although a PI tube was used to encapsulate the FBG1, the uncoordinated thermal expansion strain of the coating layer and the fiber caused by low temperature will still affect the wavelength change of the grating. Compared with FBG1, FBG2 adopts the arc casing package structure, which can compensate the shrinkage and elongation of sensor contracting under the influence of temperature to a certain extent， and ensure that the fiber is in a relaxed stress state at all times, thereby reducing the deformation response caused by temperature changes. Therefore, FBG2 can show better repeatability and more stable low temperature monitoring effect in the process of rising and cooling.

5. Conclusion

In view of the chirp phenomenon and the repeatability error of temperature measurement results in the process of FBG cryogenic monitoring, the low temperature characteristics of FBG sensors with arc tensioning package in gate area for low temperature monitoring were studied, and the low temperature sensing characteristics of two groups of FBG. The packaged FBGs samples are calibrated from -35℃ ~20℃ in a constant temperature freezer. The fitting curve of FBG2 calibration scatter has better linearity than FBG1. It shows that the arc shape is an effective method to produce enough length for low temperature contracting of the optical fibers. Therefore, it can improve its measurement accuracy well.

Acknowledgments

The project was supported by Fund of Zhuzhou Times New Material Technology Co. LtD (Grant XCF XCFDJS-2022-00004495) and the Natural Science Foundation of Guangdong Province, China (Grant No. 2021A1515010698).

References

[1] Hao J, Cai Z, Gong Y, Ng J H, Varghese P and Takahashi S (2008) A simple passive arc-shape temperature-independent load sensor using a pair of fiber bragg gratings *J. International Journal of Optomechatronics*. **2**(1): 16-31.

[2] Zhang C, Wang L and Jiang D (2006) Research on calibration method of FBG demodulation system C. ISTAI'2006 Proceedings of the First International Symposium on Test Automation & Instrumentation. Vol.**3**: 1448-1452.

[3] An Y, Wang X, Qu Z, Liao T and Nan Z (2018) Fiber bragg grating temperature calibration based on BP neural network *J. Optik*. **172**: 753-759

[4] Ding X, Zhang Y, Xia J, Zhang W and Zhu L (2017) Study on ultralow temperature characteristics of metallic packaging FBG sensor *J. Laser & Inerared*. **47**(06): 773-777.

[5] Li B, Liu Y, Song X, Fu H, Jia Z and Gao H (2020) Research on low temperature sensing characteristics of fiber bragg grating *J .Piezoelectirics & Acoustooptics*. **42**(06): 787-790.

[6] Liu M, Zhang W and Song H (2022) Study on temperatuer sensing characteristics of FBG in low temperatuer environment *J. Semiconductor optoelectronics*. **43**(02): 327-331.

[7] Lu Z and Hu Y (2013) Analysis on optical fiber sensor temperature chiped phenomenon *J. Cryogenics & superconductivity*. **41**(09): 85-88.

[8] Zhou J, Zhang D and Fu R (2010) Abnormality of fiber bragg grating spectrum at cryogenic temperatures *J. Journal of WUT (information & management engineering)*. **32**(05): 733-737.

Structural Deformation Reconstruction of Polymer Distributed Optical Fiber Sensing Tape Based on Ko Theory

Jianming Cheng[1,a], Huanlang Lu[1,b], Fumin Xie[1,c], Pingyu Zhu[1,d]*

[1]School of Mechanical and Electrical Engineering, Guangzhou University, Guangzhou, China

[a]2112107001@e.gzhu.edu.cn, [b]2112107073@e.gzhu.edu.cn, [c]2112107089@e.gzhu.edu.cn, [d]*pingyu.zhu@gzhu.edu.cn

Abstract—The progress of material technology makes the structural deformation of composites more complex and more difficult to monitor. The distributed optical fiber sensor has a wide monitoring range and can continuously obtain the strain data of the structure in space for structural deformation reconstruction. In this paper, the optical frequency domain reflectometer (OFDR) and distributed optical fiber sensor are used as measurement tools to design the deformation reconstruction experiment of a cantilever beam with constant cross section. Based on the Ko theory, the strain signals collected by the distributed optical fiber sensors are converted into displacement signals, and the deformation reconstruction of the cantilever beam is carried out, and then compared and analyzed with the actual measured values. The experiment verifies the effectiveness of using distributed fiber optic sensors for cantilever beam bending deformation based on Ko theory, which provides a favorable reference for structural health monitoring.

1. Introduction

Plastic deformation of engineering structure or equipment has certain harmfulness, and there are many kinds of deformation monitoring methods, for example, using satellite positioning to monitor the ground subsidence and bridge deflection, using total station to monitor the deflection of large structures, and using digital photography to monitor the landslide mass [1]. Optical fiber sensor has the advantages of small size, light weight, strong corrosion resistance, strong anti-electromagnetic interference ability and long service life. In recent years, with the continuous development of optical fiber sensing technology, optical fiber has shown great potential in shape measurement, especially in medical equipment such as endoscope, robot surgical instruments, key structures in aerospace field, long-distance pipeline shape detection and other aspects with significant application advantages [2].

Yi et al [3] studied the deformation reconstruction of airfoil structures using quasi-distributed fiber grating to measure strain, which helps obtaining the curvature information of the corresponding points of the airfoil, then interpolating the curvature to achieve airfoil reconstruction, and developing an experimental model reconstruction visualization platform with OpenGL. Wang et al [4] used B-spline interpolation method to study the dynamics of a flexible rectangular thin plate. The B-spline interpolation method was used to discretize the deformation field of the flexible thin plate, and the kinetic equations of the flexible thin plate with a wide range of motion were derived based on Lagrange's equations. The simulation results of B-spline interpolation method, hypothetical modal method and finite element method are compared and analyzed through dynamics simulation, and it is verified that the B-spline interpolation method has excellent performance in dealing with the large deformation problem of flexible thin plate in terms of computational accuracy. However, the algorithms above have limitations, such as the quasi-distributed fiber grating sensor can only measure the strain at the fiber grating, and the number of gratings is limited, while the B-spline interpolation method focuses on the analysis of theoretical methods, and the corresponding algorithm has high complexity and complex solution process, and cannot directly calculate the deformation based on the strain, so it is not suitable for structural deformation reconstruction based on distributed strain.

Ko et al [5] proposed the Ko displacement theory, which can estimate the deformation value of a structure very accurately and in a simple and easy-to-operate way, by simply arranging strain sensors on the surface of the structure to be measured and collecting the strain information on the surface of the structure, the deformation value of the structure can be estimated based on these strains. However, the application of this theory is currently limited to the field of aircraft. The advantages of simple arrangement, low cost and high spatial resolution of distributed fiber optic sensors [6] can well meet the core requirements of Ko displacement theory.

Xia et al [7] chose distributed fiber optic sensors as a measurement tool and designed an experiment on deformation estimation of equal strength beam based on Ko theory. The experiments verified the validity of Ko theory for deformation estimation on equal strength beam, but the applicability in large-scale structural health monitoring was not mentioned.

Considering that the strength of large beam structure is different in practical engineering, in this paper, the cantilever beam with constant cross section is selected to design the experiment. The polymer-encapsulated distributed optical fiber sensor is attached to a cantilever beam with a uniform cross section for deformation experiment, the deformation values of the cantilever beam is calculated and reconstructed through Ko theory, and the actual measured values are compared and analyzed, so as to explore the effectiveness of the polymer-encapsulated distributed optical fiber sensor for reconstructing the bending deformation of the cantilever beam.

2. Experimental Equipment and Methods

2.1. Deformation Reconstruction of Cantilever Beam

The Ko theory is base on that assumption of traditional material mechanics and the geometrical and physical relationship of beam in ben deformation, and proposes that the beam is divided into n parts with the length of Δl, the strain $\varepsilon(x)$ is in a piecewise linear relationship along the x-axis direction, thus, the problem of quadratic integration of structural deformation is transformed into a repeated summation problem. Therefore, the deformation information of the structure can be demodulated only with acquiring the surface strain signal of the cantilever beam in the bending state by using the distributed optical fiber sensor.

When $x_{i-1} \le x \le x_i$, the local bending curvature $tan\theta(x)$ in the x-direction can be expressed as:

$$\begin{aligned} tan\theta(x) &= \int_{x_i}^{x} \frac{d^2y}{dx^2} dx + tan\theta_{x_{i-1}} \\ &= \int_{x_{i-1}}^{x} \frac{\varepsilon(x)}{c} dx + tan\theta_{i-1}; \quad x_{i-1} \le x \le x_i \end{aligned} \tag{1}$$

Where x is the length coordinate of the structure, y is the displacement on the vertical axis at the measuring point, $\varepsilon(x)$ is the strain at x, $tan\theta_{x_{i-1}}$ is the slope of the structural plane at x_{i-1}, and c is the distance between the beam surface and the neutral plane at x.

Since that strain $\varepsilon(x)$ is in a piecewise linear relationship along the x-axis direction, equation (1) can be converted into a repeated summation problem:

$$tan\theta_n = \frac{\Delta l}{2c}\left(\varepsilon_0 + 2\sum_{j=1}^{n-1} \varepsilon_j + \varepsilon_n\right) + tan\theta_0 \tag{2}$$

The cantilever beam satisfies the boundary condition that one end is fixed and the other end is free, so fixed end satisfies $tan\theta_0 = 0$, equation (2) can be express as:

$$tan\theta_n = \frac{\Delta l}{2c}\left(\varepsilon_0 + 2\sum_{j=1}^{n-1} \varepsilon_j + \varepsilon_n\right) \tag{3}$$

Integrate equation (1) again, the deformation information of every point on that structural plane along the x direction can be obtain in the region of $x_{i-1} \le x \le x_i$:

$$y(x) = \int_{x_{i-1}}^{x} tan\theta(x)dx + y_{x_{i-1}}$$
$$= \int_{x_{i-1}}^{x}\int_{x_{i-1}}^{x} \frac{\varepsilon(x)}{c}dxdx + \int_{x_{i-1}}^{x} tan\theta_{i-1}dx + y_{x_{i-1}};$$
$$x_{i-1} \le x \le x_i \quad (4)$$

Equation (4) is converted into a repeated summation problem to obtain:

$$y_n = \frac{(\Delta l)^2}{6c}(2\varepsilon_0 + 3\sum_{i=1}^{n-1}\varepsilon_i + \varepsilon_n) + \Delta l\sum_{i=1}^{n-1} tan\theta_i \quad (5)$$

According to equation (3), the last term of equation (5) above can be expressed as:

$$\sum_{i=1}^{n-1} tan\theta_i = \frac{\Delta l}{2c}\{(n-1)\varepsilon_0 + \sum_{i=1}^{n-1}[2(n-i)-1]\varepsilon_i\} \quad (6)$$

Substituting equation (6) into equation (5) yields:

$$y_n = \frac{(\Delta l)^2}{6c}[(3n-1)\varepsilon_0 + 6\sum_{i=1}^{n-1}(n-i)\varepsilon_i + \varepsilon_n] \quad (7)$$

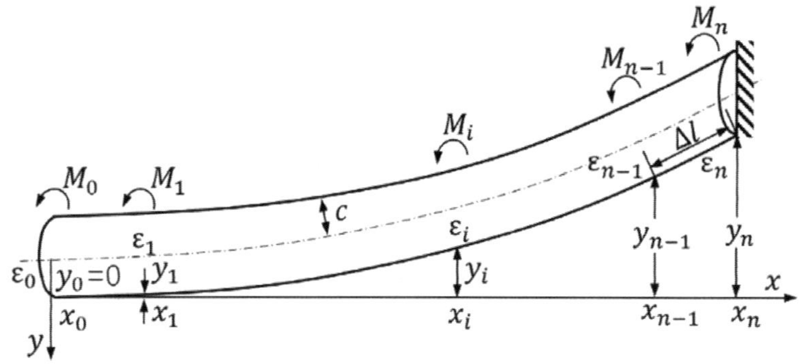

Figure 1. The strain $\varepsilon(x)$ follow the piece-linear relationship along the axial coordinate x

2.2. Experimental Setup and Procedures

Attaching the polymer-packaged distributed optical fiber sensor on the cantilever beam with the uniform section to complete the layout of the distributed optical fiber sensor on the cantilever beam. The actual layout is shown in Figure 2.

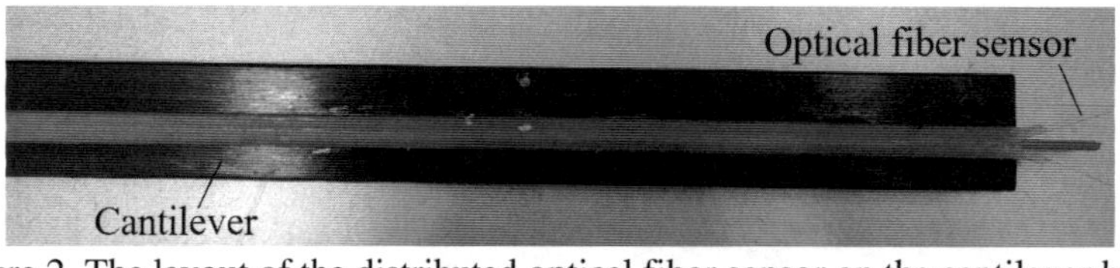

Figure 2. The layout of the distributed optical fiber sensor on the cantilever beam

The joint of the distributed optical fiber sensors were connected to the LUNA ODiSI6100 demodulator to complete the construction of the deformation reconstruction system of the uniform cross-section cantilever beam. The whole system can be seen in Figure 3.

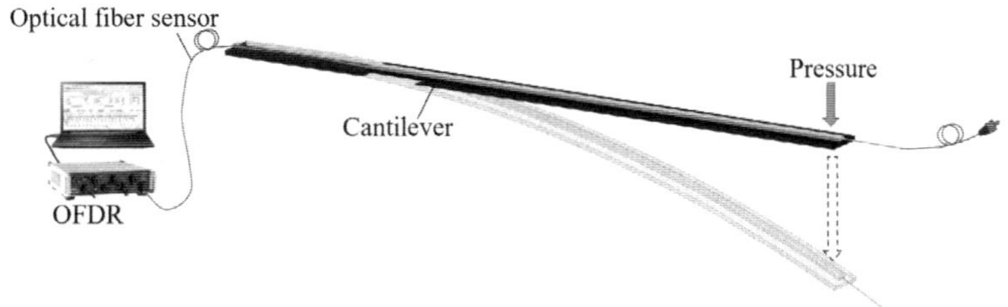

Figure 3. The construction of the deformation reconstruction system of the uniform cross-section cantilever beam

The cantilever beam deformation experiment is divided into three steps as follow:

1) Calibration of the start and end points of the polymer encapsulated distributed fiber optic sensor to determine the sensing segment.

2) Apply a slowly increasing pressure in the downward direction to the free end of the equal-section cantilever beam to bend the cantilever beam, and then reduce the pressure to slowly restore the cantilever beam from the bent state to its natural state as in Figure 3, and repeat the above steps. Strain signal acquisition was maintained throughout the experiment.

3) With the strain signal obtained from the distributed optical fiber acquisition in step 2, the program is written based on Ko theory to convert the strain signal into displacement signal to realize the deformation reconstruction process of the cantilever beam.

3. Results and Analysis of Experiments

3.1. The Result of the Strain Conversion to Displacement

Figure 4 shows the process of two deformation experiments of a cantilever beam with a constant cross section. The curve in the figure describes the result of converting the maximum strain value of the distributed optical fiber sensor at the end measuring point of the uniform section cantilever beam into the displacement in the direction of the vertical axis. It can be seen that, as time goes on, in the first pressure stage, the end of the measuring point displacement value along the negative direction of the vertical axis increases from 0 to 140mm or so, and then in the first pressure relief stage, the displacement value along the negative direction of the vertical axis decay to 0. In the second pressure stage, the displacement value of the end measure point increases from 0 to about 140mm along the negative direction of the longitudinal axis, and then in the second pressure release stage, the displacement value decays to 0 along the negative direction of the longitudinal axis. The process above meets that operation of the actual experiment, namely, the free end of the cantilever beam with the uniform cross section is slowly bent under downward and increasing pressure in a natural state, then the pressure is reduced, the cantilever beam slowly recovers from the bend state to the natural state, and the experiment is repeated. The reconstruction process is in good agreement with the experimental process.

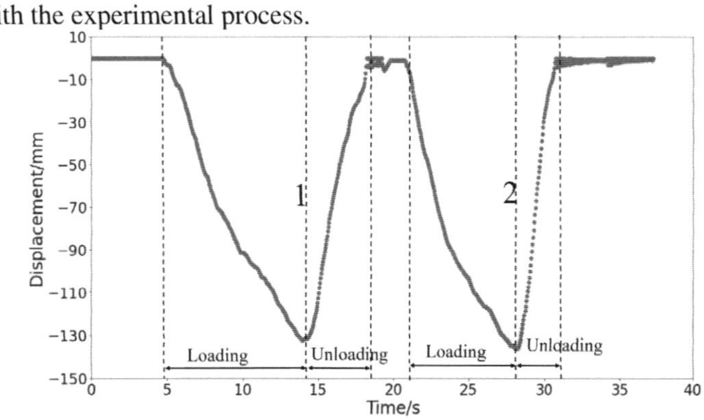

Figure 4. Displacement of the distributed optical fiber sensor at the end measuring point

3.2.Error Analysis

In two deformation experiments, the absolute error and the relative error are calculated by using the result of strain displacement in Figure 4 and the actual measured values. It can be seen from the figure that the absolute errors of the two experiments are 16.81 mm and 17.79 mm respectively, and the relative errors are 11.267% and 11.524% respectively. Within the allowable error range, it can be considered that the sensing and reconstruction of the deformation of the cantilever beam can be completed by using the polymer-encapsulated distributed optical fiber sensor based on the Ko theory.

TABLE I. ERROR ANALYSIS OF THE EXPERIMENT

Experiment	Absolute error	Relative error
1	16.81mm	11.267%
2	17.79mm	11.524%

4. CONCLUSION

The strain values of a cantilever beam with constant cross-section under bending are obtained through distributed optical fiber sensors. Based on Ko theory, the transformation formula from bending strain to displacement of constant cross-section cantilever beam is derived. Using this formula, the deformation of the cantilever beam is reconstructed by using the strain data from the distributed optical fiber sensors on its surface. The errors between the experimental results and the actual measured values are less than 12%. It shows the Ko theory algorithm realizes the reconstruction of the bending deformation of the cantilever beam with equal cross-section within the range of engineering allowable error range, and provides an effective tool for the monitoring of the composites structure deformation.

ACKNOWLEDGMENT

The project was supported by the Natural Science Foundation of Guangdong Province, China (Grant No. 2021A1515010698).

REFERENCES

[1] Huaguang Wang. Research on the Application of GPS Technique in Xipan Expressway Landslide Monitoring[D].Southwest Jiaotong University,2007.(In Chinese)

[2] Shiyuan Zhao, Jiwen Cui, Mengmeng Chen. Review on Optical Fiber Shape Sensing Technology[J]. Optics and Precision Engineering,2020,28(01):10-29. (In Chinese)

[3] Jincong Yi. Shape Perception and Active Monitoring for Smart Structure Using FBG Sensor Array[D].Shanghai University,2014. (In Chinese)

[4] Yong Wang, Dingguo Zhang, Jihua Fan, Liang Li. B-spline Interpolation Method for the Dynamic Analysis of Flexible Rectangular Plates [J]. Journal of Vibration Engineering,2019,32(05). (In Chinese)

[5] Ko W L , Richards W L , Tran V T . Displacement Theories for In-Flight Deformed Shape Predictions of Aerospace Structures. 2007.

[6] Bao X , Chen L . Recent Progress in Distributed Fiber Optic Sensors[J]. Sensors, 2012, 12(12):8601-8639.

[7] Li Xia, Dapeng Wang. Study of Deformation Estimation Combining DFDR and Ko Displacement Theory[J]. Optical Communication Technology, 2021. (In Chinese)

IOP Publishing
2366 (2022) 012017 doi:10.1088/1742-6596/2366/1/012017

Stress analysis of gantry crane cylinder structure

Aihua Jiang[1,a], Xinhua Wang[1], Shan He[1], Min Chen[1]

1.Guangzhou Academy of Special Equipment Inspection & Testing, Guangzhou, China

[a]aiwa9104@126.com

Abstract. By building three finite element analysis models for its four-bar metal structure under different amplitudes (the boom span is 13.5m, 22.5m and 34.5m), the relationship between stress, amplitude and load at key parts is established by linear interpolation method; At the same time, the relationship between the measured strain data of key parts and the load and amplitude is linearly fitted. The analysis and comparison between the finite element calculation data and the measured data are well verified. In order to provide a feasible analysis method for solving the working performance of four-bar linkage structure of portal crane. At the same time, it provides a more convenient method for the compilation of typical load, which is the important basic data for the calculation and analysis of residual fatigue life of later structures.

1. Introduction

The four-bar portal crane is a general-purpose portal crane. It is an important and representative type of intermittent action rail operated lifting machinery with short, repeated, periodic cycle, frequent starting and braking and large impact load. It is widely used in port terminal loading and unloading, shipyards and steel companies[1,2]. As the main metal structure of the portal crane, the four-bar linkage bears the self weight and external load of the equipment, and constitutes the necessary working system and movement space to complete various functions of the crane. The working performance and technical status of its metal structure directly affect the safety and reliability of the whole machine. Therefore, how to detect and evaluate the overall working performance of the metal structure of the four-bar linkage of the old portal crane has become an important issue to be solved in the industry[3,4].

In order to fully understand the stress distribution of the four-bar linkage structure, identify the more dangerous positions of the structure, and help determine the strain test positions; Analyze the influence of self weight on structural stress[5,6]. In order to simplify the analysis model, the finite element analysis in this paper is mainly carried out for the four-bar structure model. The analysis models are established for three different boom spans of 13.5m, 22.5m and 34.5m. In the modeling process, some geometric details were simplified. For example, a large number of welds and their details in the crane were not reflected in the model. The nodes and cell numbers of each component in the three models are consistent. As shown in Figure 1.

Fig. 1 Finite element model of four-bar linkage structure.

2.Analysis and Discussions

The model is mainly composed of shell63 plate shell element and beam188 beam element. The crane boom structure, elephant nose beam structure, large tie rod structure, A-frame structure, luffing mechanism and turntable structure are all welded from steel plates, and the structural characteristics conform to the shell element shell63, so this element is selected. Beam element beam188 is adopted for balance beam structure, small tie rod structure and luffing tie rod.

The arrangement of stress and strain measuring points mainly considers the load key control points and large stress points of the four-bar linkage of the gate seat, and also considers the possibility and operability of the actual patch. The strain measuring points are all unidirectional strain gauges arranged along the main direction of stress.

In this test, four key parts of the elephant nose bridge (the center of the left and right upper flange plates of the fourth hole) and the large pull rod (the center of the left and right lower flange plates of the rear end) were selected for testing and finite element calculation.As shown in Table 1.

Table 1.Description of specific information of measuring points

Channel No.	Channel Title	Unit	Parts	Sensor location description	Collector
1	strain_A4-1	uE	elephant nose bridge	Center of left upper wing plate	TML
2	strain_A4-3	uE	elephant nose bridge	Center of right upper wing plate	TML
3	strain_B2-2	uE	Large tie rod	Center of left lower wing plate	TML
4	strain_B2-4	uE	Large tie rod	Center of right lower wing plate	TML

The test condition is the actual simulation condition, which includes the change of boom span. In order to obtain the strain value of each measuring point under different boom spans (boom spans are 13.6m, 18m, 22m, 23.6m and 32.5m respectively), we intercepted the measured signals. From these intercepted signal segments, the mean value of strain can be calculated and used as the strain value of the current boom span position.

The relationship between load and strain should be basically linear, which has been verified. This figure compares the strain data of all channels 8 with the test boom span of 22 ~ 24m. It can be seen that the data of the two tests are very consistent, and all the data show a linear trend of change.

The relationship between the boom span and strain when the load is constant. When the load is about 60t, all the test data of channel 8 are included. It can also be seen that the strain basically changes linearly with the increase of the boom span. Furthermore, the linear equations between the strain of 8 measuring points of the channel and the boom span under three different loads are given, which are obtained from the linear fitting of the test points. Similar linear relationship can be obtained for other typical measuring points. As mentioned above, there should also be a linear relationship between strain and load, and when the load is zero, the strain value (regardless of self weight) should also be zero. Based on this, the data were subjected to load orthogonalization and linear regression.

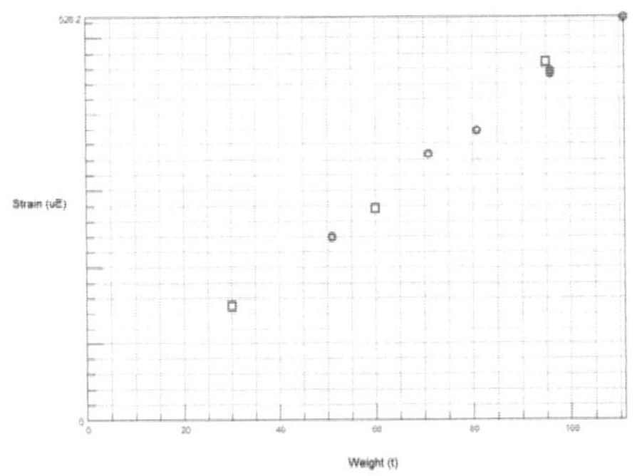

Fig.2 Load strain relationship when boom span is basically unchanged (22-24m)

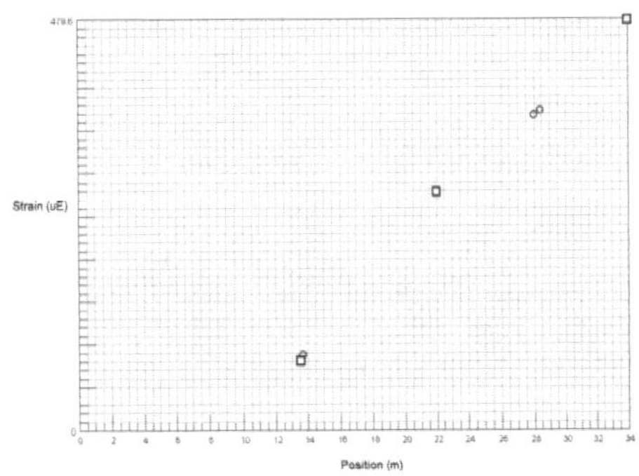

Fig. 3 The relationship between boom span and strain when the load is basically unchanged (about 60t)

Since there are many combinations of boom positions and lifting weights in the test conditions, we use the strain linear interpolation method to calculate the strain value of a certain boom and lifting weight from the finite element results of three different unit lifting weights. The interpolation formula is as follows:

$$\varepsilon(L) = W\left[\alpha(L)\varepsilon_{L=LA} + \beta(L)\varepsilon_{L=LB} + \gamma(L)\varepsilon_{L=LC}\right]$$

Including $\alpha = 1$； $\beta = 0$； $\gamma = 0$

When $L_A \le L < L_B$： $\alpha(L) = (L_B - L)/(L_B - L_A)$； $\beta(L) = (L - L_A)/(L_B - L_A)$； $\gamma = 0$

When $L_B \le L < L_C$： $\alpha = 0$； $\beta(L) = (L_C - L)/(L_C - L_B)$； $\gamma(L) = (L - L_B)/(L_C - L_B)$

When $L \ge L_C$： $\alpha = 0$； $\beta = 0$； $\gamma = 1$

In the above formula, W and L are the crane and boom span, and are respectively the finite element strain results of three different boom positions (13.5m), (22.5m) and (34.5m) under 1t lifting load;, And are interpolation coefficients of these three positions respectively.

Extract the strain value of a strain measuring point from the finite element results, and use the virtual strain gauge function to determine the type and location of the strain gauge in the finite element model.When the load is kept at about 60t, the difference between the measured strain and the analyzed strain of A4-3 at different boom positions is compared. It can be found that the two agree well when the span is 22.5m, the measured value is lower than the analyzed value when the span is 13.5m, and the measured value is about 25% higher than the analyzed value when the span is 34m. By comparing the measured and analyzed strains, it can be found that the two agree well in the middle span range, but when the span is large, the measured strain is larger than the analyzed strain, but the error is generally not more than 50%. Based on the orthogonalization strain, the measured and analyzed strains are compared comprehensively. The results show that the measured values have greater dispersion, but are basically in the upper and lower parts of the analyzed values, and they tend to be basically consistent with the changes of the boom span.

3. Conclusion

Through the analysis of three finite element models with the boom span positions of 13.5m, 22.5m and 34.5m respectively in the four-bar linkage structure, the calculated strain values of the key positions of the nose bridge are compared with the measured strain values, and the results are in good agreement, indicating that the finite element analysis has better analysis accuracy. An interpolation method is proposed to calculate the stress and strain of the four-bar linkage structure under any load from the static analysis results of a finite number of boom spans. In order to provide a feasible analysis method for solving the working performance of four-bar linkage structure of portal crane.

Acknowledgement

Guangzhou Administration for Market Regulation 2021kj24/2021kj25

References

[1]Chen Qingrong, Cheng Wenming,Gao Ling chong,Du Run. Dynamic Response of a Gantry Crane's Beam Subjected to a Two-Axle Moving Trolley[J]. Mathematical Problems in Engineering,2020,(2020).

[2]Qihui Peng,Wenming Cheng,Hongyu Jia,Peng Guo. Fragility Analysis of Gantry Crane Subjected to Near-Field Ground Motions[J]. Applied Sciences,2020,10(12).

[3]P. A. Ospina-Henao,Framsol López-Suspes. Dynamic analysis and control PID path of a model type gantry crane[J]. Journal of Physics: Conference Series,2017,850(1).

[4]Chun Xin Mu,Ming Quan Shi,Zhen Feng Han,Qi Min Li. Fuzzy-LQR Based Anti-Swing Control of Gantry Crane[J]. Advanced Materials Research,2014,3470(1030-1032).

[5]Da Peng Zhang,Wen Ming Cheng,Kun Cai. Study on Prediction Deflection of at the Girder Midspan of Large Tonnage Gantry Crane Rigid Legs[J]. Applied Mechanics and Materials,2014,3436(628-628).

[6]Jie Dong,Wen Ming Cheng,Yang Zhi Ren,Yu Pu Wang. Research on the Structural Performance of Large-Tonnage Gantry Crane[J]. Advanced Materials Research,2013,2731(823-823).

Research on impact localization method based on passive tomography of piezoelectric sensors

Weixi Yang*[a], Xin Cao[a], Hongxiao Chao[a]

[a]Northwest Institute of Mechanical and Electrical Engineering, Xianyang Shannxi, 712099

[*] Corresponding author: 1013189110@qq.com

ABSTRACT: Impact monitoring of plate structures has always been a hot spot in structural health monitoring research. In order to improve the accuracy of impact monitoring, this paper proposes a beam-focusing impact localization algorithm based on passive tomography using a sensor arrangement method in a ring array. The impact damage monitoring of large-sized structures in different directions is performed by each linear array of the ring array and the impact damage localization distribution maps of the corresponding linear arrays are generated. The localization accuracy of the linear arrays in different directions is obtained experimentally to assign the weights of each linear array to achieve image reconstruction of the impact damage. The experimental results show that the method can achieve a more accurate visualization of the impact location, and the reconstructed impact laminar image can well reflect the impact location distribution information. It can be used for the localization of impact damage and extended monitoring.

1. INTRODUCTION

Due to the high randomness and hazard of impact, it has been one of the hot spots as well as difficult areas in structural health monitoring. Impact damage will have a direct impact on the service life of the plate, mechanical properties, etc., and then affect the actual work efficiency of the components. When the defects continue to accumulate and expand to a certain scale, the service life of the components will be greatly shortened, and may even cause serious safety accidents[1,2]. Tomography technology has a very wide range of applications in medical, biological and industrial fields[3]. Its basic principle is to use a certain sensor structure to extract the relevant information of the measured object in a hierarchical manner, and then use a certain algorithm to reproduce the required information, and the superposition of multiple layers of information can reproduce the three-dimensional information of the object[4]. In recent years, Lamb wave detection technology in the field of structural health inspection has been combined with Lamb wave laminar imaging method, which has shown great application prospects in crack detection and defect monitoring[5]. Compared with the traditional Lamb wave detection and monitoring methods, this technique can intuitively reconstruct the location and size of the damage, showing a better research prospect for damage monitoring and evaluation. Since the shock signal is a non-smooth random signal with transient, sudden and random characteristics, and has more frequencies and modes, which poses difficulties in reconstruction of shock damage. Therefore, there are fewer results in the existing studies of laminar imaging methods for imaging monitoring and evaluation of structural impacts. In this paper, based on the current state of the technology, the traditional beam focusing and positioning method is improved, and the impact source is observed from different angles by multiple one-dimensional linear arrays to achieve a layered collection of impact source information from different angles. And because each array sensor is not the same in terms of monitoring angle, timeliness and accuracy, the monitoring ability of each direction for different orientation of the impact damage is not the same, and the accuracy of the impact source positioning is fitted by experimental analysis of each angle to achieve image reconstruction of the impact source.

2. LAMINAR IMAGING METHOD BASED ON BEAM FOCUS LOCALIZATION ALGORITHM

2.1 Basic principle of acoustic emission detection

The shock signal can be monitored based on acoustic emission technology. When the structural member under test is stressed, tiny deformations will occur inside the structure, and these tiny deformations will accumulate in the structure and form defects, thereby generating stress waves in the material, resulting in Acoustic emission[6]. The basic principle of acoustic emission technology is to collect the acoustic emission signal of the structure in real time by arranging high-sensitivity piezoelectric sensors on the surface of the structure, and realize the damage monitoring through the analysis of the acoustic emission signal[7].

2.2 Linear array beam focus imaging algorithm

In an isotropic plate, the impact signal propagates outward in concentric circles with the impact source as the center, and the impact signal has different wave velocities depending on the plate material[8]. The schematic diagram of impact source localization is shown in Figure 1: Assume that the distance between different sensors and the impact source is Ri, and the impact signal wave intersects each sensor on the way.

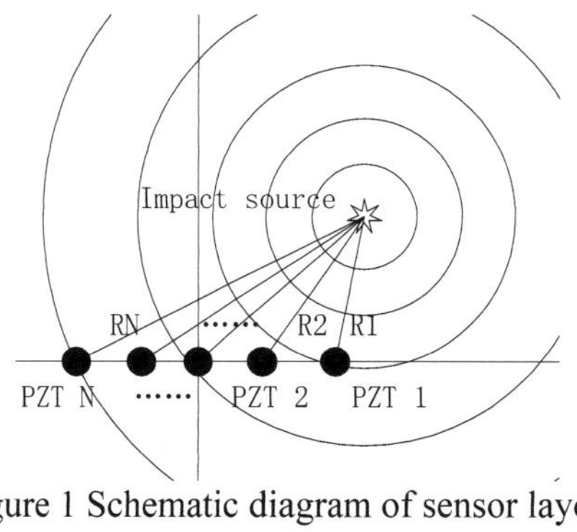

Figure 1 Schematic diagram of sensor layout

If the coordinate position of the impact point is set as (X, Y), and the placement of the piezoelectric sensor is (a_i, b_i).Then the length of R_i in the figure can be expressed as.

$$R_i^2 = (X - a_i)^2 + (Y - b_i)^2, i = 1, 2, 3 \cdots \quad (1)$$

The distance difference δ_{ij} between the two sensors and the impact point in the figure is,

$$|R_i - R_j| = \delta_{ij}, i \neq j \quad (2)$$

If the speed of the wave is constant at v, the delay between the two sensors receiving the shock signal is ΔT_{ij}, there is a formula

$$\delta_{ij} = v \cdot \Delta T_{ij} \quad (3)$$

Since the equation is an indeterminate equation, at least three piezoelectric sensors are required to achieve accurate positioning of the CEP on a two-dimensional plane. By combining the above formulas, a set of hyperbolas can be obtained. It can be seen from the figure that the solution of a single set of indeterminate equations is any point on the corresponding hyperbola, so two sets of hyperbolas are required, and the intersection of the two sets of hyperbolas is the location of the actual calculated impact source.

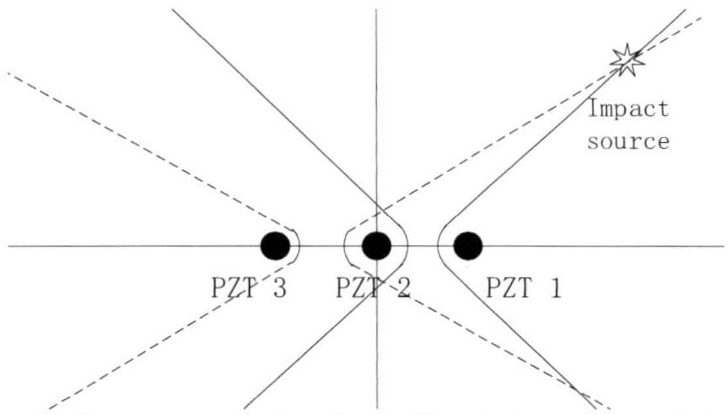

Figure 2 Schematic diagram of the location of the intersection point of the hyperbola

The monitoring range of linear array sensor is 0°-180°, which can not be monitored in all directions, and can only monitor the sector area, that is, the best positioning effect in the direction perpendicular to the linear array, while the positioning effect near 0° and 180° is poor or even invalid. At the same time, there is mirror interference in the imaging of linear arrays, i.e., false imaging results will occur in the symmetric position of the array. Although arranging the linear array sensors at the edge of the system can avoid the interference of mirror localization, however, this is achieved with the loss of half of the localization range. In order to solve the problem of beam focusing in impact localization, a circular array sensor arrangement is used in this paper, as shown in Figure 3.

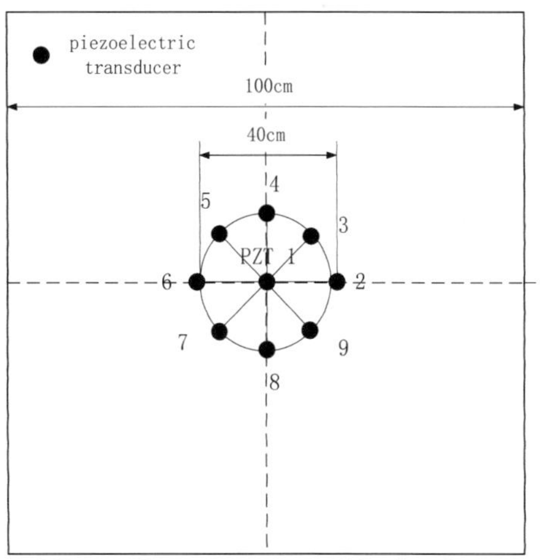

Figure 3 Beam focus sensor positioning impact map

The ring array sensor uses a combination of multiple linear arrays, and the different angles of the arrays can be realized from different angles of the impact signal acquisition, to obtain a multi-level impact damage description.

2.3 Laminar imaging localization method based on beam focusing algorithm
The shock localization method proposed in this paper is mainly divided into four steps: First, wavelet transform is performed on the obtained shock signal to obtain its dominant frequency envelope. Secondly, the time delay of the two signals is obtained by cross-correlating the wavelet transformed signals. Again, the localization distribution of each array is derived separately. Finally, the positioning

distribution maps of different arrays are reconstructed to obtain the final impact positioning result map. The algorithm flow chart is shown in Figure 4.

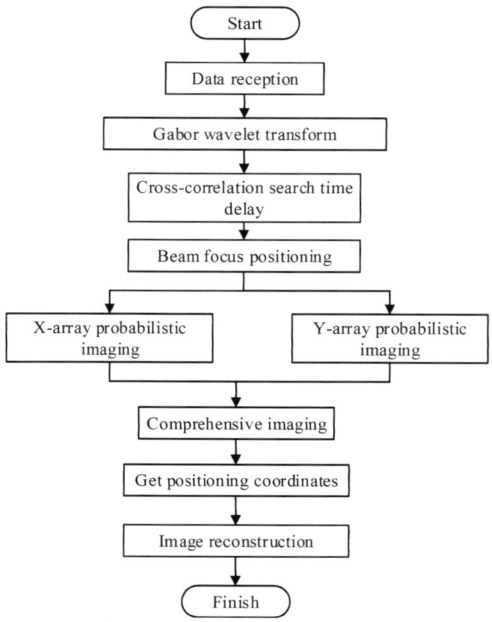

Figure 4 Algorithm flowchart

2.3.1 Linear array beam focused imaging method

When performing impact location based on the method proposed in this paper, the measurement result of wave velocity is very important to the location accuracy. The error sources of the wave velocity measurement results are mainly the error introduced by the sensor itself, the error introduced by the signal processing and the error caused by the temperature during the experiment. The conventional localization algorithm uses multiple measurements of wave speed to obtain the average value for damage localization, and the localization results are deterministic results determined by the quality of wave speed sampling, and its accuracy fluctuates with wave speed measurement. In this paper, we adopt a probability distribution-based approach to wave velocity measurement, so as to obtain a probability distribution map for impact localization and achieve a more scientific and reasonable probabilistic description of impact localization.

Using three adjacent sensors in the array to receive shock signals at different distances, multiple shocks are performed at each distance point, and the resulting wave velocity v is calculated as

$$v = \frac{s_i - s_j}{t_i - t_j}, i \neq j \tag{4}$$

In the formula, i and j are the corresponding numbers of the piezoelectric sensors respectively; s is the distance between the impact point position and the placement position of the sensor, and t is the arrival time of the signal received by the corresponding sensor.

The shock signal received by the piezoelectric sensor is a non-stationary random signal with many frequencies and modes, so the signal processing method is required to have the ability of time-frequency analysis. The time-frequency information of the signal can be obtained through wavelet transform. First, the main frequency range of the signal is obtained through Fourier transformation, and then the Gabor wavelet is used to analyze the time-variable window, filter out irrelevant frequency information, and extract the required main frequency signal. time domain information. The wavelet transform expression is.

$$\Psi_g(t) = \frac{1}{\sqrt[4]{\pi}} \sqrt{\frac{\omega_0}{\gamma}} \exp\left[-\frac{(\omega_0 / \gamma)^2}{2} t^2 + \mathrm{i}\omega_0 t \right] \tag{5}$$

Its Fourier transform is,

$$\hat{\Psi}_g(\omega) = \frac{\sqrt{2\pi}}{\sqrt[4]{\pi}} \sqrt{\frac{\gamma}{\omega_0}} \exp\left[-\frac{(\gamma / \omega_0 \gamma)^2}{2} (\omega - \omega_0)^2 \right] \tag{6}$$

The signal obtained from the wavelet transform is normalized and then subjected to the inter-correlation operation to obtain the signal, $m_i'' = (\hat{R}_{x_1,x_2}(\hat{\tau}), \hat{R}_{x_1,x_3}(\hat{\tau}), \ldots\ldots, \hat{R}_{x_1,x_5}(\hat{\tau}))$. Where:

$$\hat{R}_{x_1,x_2}(\hat{\tau}) = \int_{-\infty}^{+\infty} P_{x_1,x_2}(f) \mathrm{e}^{2\pi i f \hat{\tau}} \mathrm{d}f \tag{7}$$

Take the peak value $\hat{\tau}$ of the two inter-correlation signals $\hat{R}_{x_1,x_2}(\hat{\tau})$, the resulting $\hat{\tau}$ is the difference between the wave arrival times t of the data received by the two sensors from the inter-correlation calculations, and record as ΔT_{ij}. The required wave speed v can be obtained by bringing it into equation (6).

The information of N sets of wave velocities is obtained through N trials, and the statistical analysis can have the probability distribution of wave velocities conforming to the normal distribution $V \sim N(\bullet)$, is denoted as: $V(\bullet)$.When the time delay is ΔT_{1i} and ΔT1j and the wave speed is V(•), the x- and y-coordinate probability distributions of the shock source (denoted as $Zx(\bullet)$ and $Zy(\bullet)$, respectively) can be found in Eqs. (4) and (5) as shown in Eq. (8).

$$\begin{cases} Z_x(\bullet) = -\dfrac{\left(L^2 - \Delta T_{1i} \times \Delta T_{1j} \times (V(\bullet))^2\right) \times \left(\Delta T_{1i} + \Delta T_{1j}\right)}{2L \times \left(\Delta T_{1i} - \Delta T_{1j}\right)} \\ Z_y(\bullet) = \pm \dfrac{\sqrt{\left(k^2 - (\Delta T_{1i})^2\right) \times \left(k^2 - (\Delta T_{1j})^2\right) \times \left(2k - (V(\bullet)) \times \Delta T_{1i} - (V(\bullet)) \times \Delta T_{1j}\right) \times \left((V(\bullet)) \times \Delta T_{1i} + (V(\bullet)) \times \Delta T_{1j}\right)}}{2L \times (V(\bullet)) \times \left(\Delta T_{1i} - \Delta T_{1j}\right)} \end{cases} \tag{8}$$

The one-dimensional distribution functions $Zx(\bullet)$ and $Zy(\bullet)$ of the x-axis and y-axis of the shock source can be obtained by equation (10), and the two-dimensional distribution functions Z_{xy}(x=xi|y=yi) of the shock source can be obtained by the joint distribution function in equation (11), which is written as $Z_{xy}(\bullet)$. Where,

$$Z_{xy}\left(x = x_i \mid y = y_i\right) = \frac{Z_x\left(x = x_i\right) \cap Z_y\left(y = y_i\right)}{Z_y\left(y = y_i\right)} \tag{9}$$

$Z_{xy}(\bullet)$is the two-dimensional probability distribution function of the impact source calculated by the array, and the probability imaging map of the array can be obtained by the composition of this function.

2.3.2 Impact source image reconstruction method

In order to verify the positioning accuracy of the method proposed in this paper in the range of 0° to 180°, as shown in Figure 5: multiple impact experiments were conducted at 10° intervals on a circle with radius R from the center sensor to obtain the sensitivity of the linear array sensor to different angular impact sources.

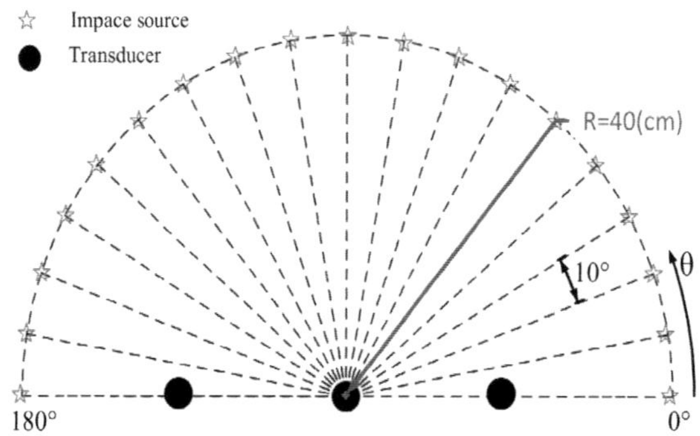

Figure 5 Distribution of impact points

Multiple impact positioning experiments were conducted for different angles shown in Figure 4, and the relative errors between the positioning results and the actual impact source were counted. The smaller the error, the higher the positioning accuracy, i.e., the accuracy of each point is inversely proportional to the average error of the point positioning results, where the relationship between the relative error δ_θ of positioning at different angles and the positioning accuracy $\Phi(\theta)$ is shown in Equation 10.

$$\Phi(\theta)=\frac{1}{\delta_\theta} \tag{10}$$

where δ_θ is the relative error corresponding to the angle θ.

Statistical analysis of the relative errors obtained from the impact experiments shows that the positioning errors δ_θ for different angles θ in the range of 0° to 180° correspond to the accuracy $\Phi(\theta)$ in accordance with the radial basis function, which can be fitted by the Gaussian kernel function in the radial basis function, and the error between the positioning accuracy and the angle can be as shown in Equation (11).

$$\Phi(\theta)=\begin{cases} A_1 \cdot e^{\left(\frac{-(\theta-B)^2}{C_1}\right)}+A_2 \cdot e^{\left(\frac{-(\theta-B)^2}{C_2}\right)}, & \theta \subseteq 0°-180° \\ A_1 \cdot e^{\left(\frac{-((360°-\theta)-B)^2}{C_1}\right)}+A_2 \cdot e^{\left(\frac{-((360°-\theta)-B)^2}{C_2}\right)}, & \theta \subseteq 180°-360° \end{cases} \tag{11}$$

where A1 is 0.37, A2 is 0.9, B is an angular offset of 90°, C1 is 16.84, and C2 is 57.72.

The impact source has different angles with respect to different linear arrays, and the linear arrays have different resolution for different angles of impact, thus achieving a hierarchical damage information acquisition. The impact information is acquired from different angles, and the reconstruction of the impact image can be achieved by giving different weights to each array through the positioning accuracy of different angles.

3. EXPERIMENTAL STUDY

3.1 Impact experimental system

The proposed impact localization algorithm is experimentally studied in this paper. The composition of the experimental system is shown in Fig. 6. An aluminum plate is used, and all four sides are fixed with screws.

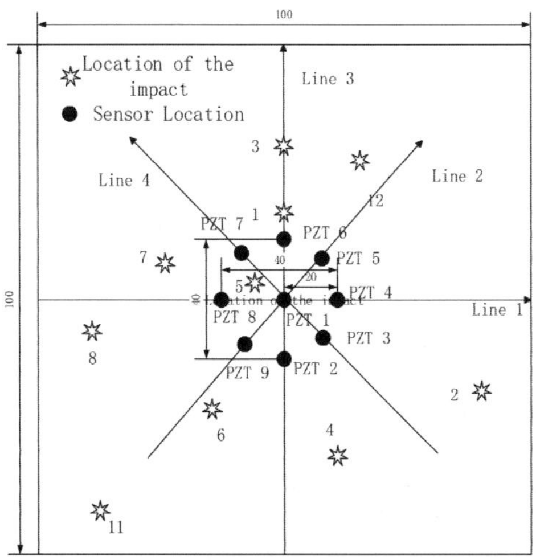

Figure 6 Sensor paste schematic

3.2 Impact signal analysis

Due to the isotropic nature of the plate used in the experiment, it is assumed that the propagation speed of the wave is constant. Within the range of 25-45 cm from the sensor, a point is selected every 5 cm to calibrate the wave speed, as shown in Figure 7. The calculation result of the wave propagation velocity obtained from the experiment conforms to the normal distribution V~N (1591.8, 41.74), so it is used as the standard wave velocity in the positioning algorithm for positioning calculation.

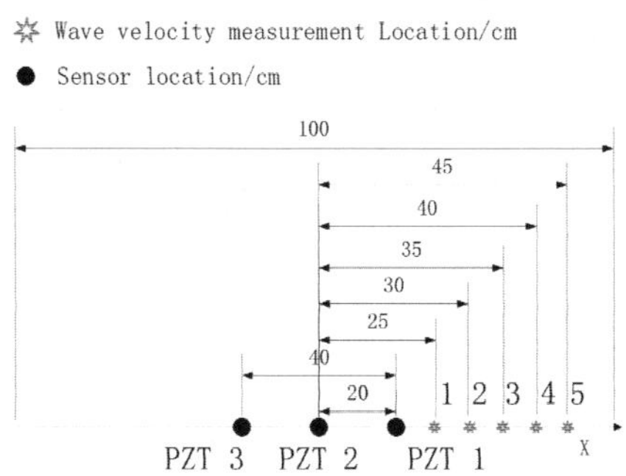

Figure 7 Schematic diagram of wave speed check point

The lead break experiment was performed at the impact position (20,30) as shown in Figure 5, the signal curve received by the sensor is shown in Figure 8.

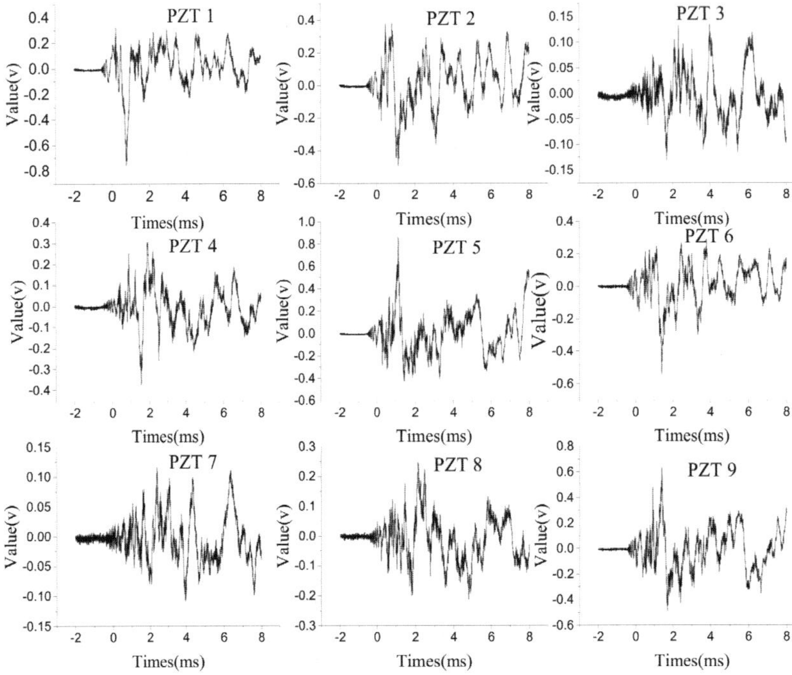

Figure 8 Original signal waveform diagram

In this paper, the time delay parameters of the signal are extracted by using the method based on generalized mutual correlation, and the filtering extracts the signal in the interval where the main energy is located, and it can be found that the main energy distribution of the acoustic emission signal is in 0-80 kHz, and the energy is most concentrated between 20-40 kHz, so this paper takes 30 kHz as the central frequency of the signal for the signal using Gabor wavelet for The time eigenvalues are extracted using Gabor wavelet analysis with variable time windows. The time delay of the envelope of the Gabor wavelet transformed signal is obtained by using the intercorrelation algorithm and its time delay is shown in the following table.

Table 1 Time delay

Sensor serial number	2-1	3-1	4-1	5-1	6-1	7-1	8-1	9-1
Time delay (ms)	1.1	-0.875	0.81	0.38	0.9	0.41	1.04	-0.83

3.3 Analysis of positioning results

Carry out the experimental research on the location of the impact point with the coordinates (20, 30), and bring the probability distribution function and time delay of the wave propagation speed into the formula (8), and then the corresponding probability distribution function of the abscissa and the ordinate can be obtained. . The probability distribution map of the impact point estimated by the algorithm can be obtained by combining the probability distribution functions of the horizontal and vertical coordinates. After reconstructing it as shown in Figure 9, the location of the peak of the probability map is the estimated location of the impact source.

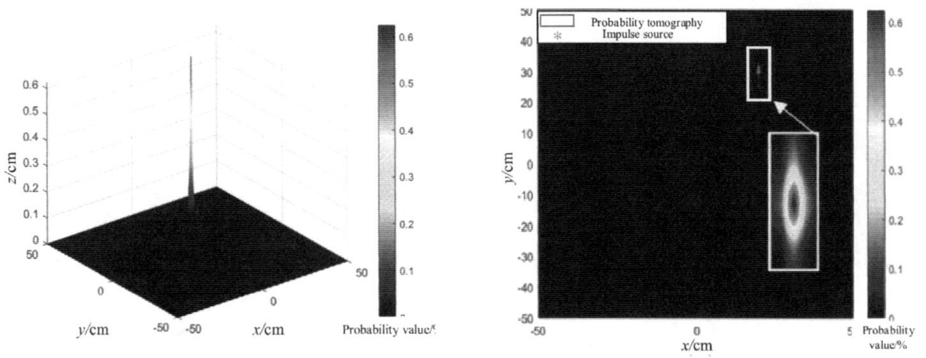

(a) Three-dimensional distribution map (b) Two-dimensional floor plans

Figure 9 Impact Source Reconfiguration

The coordinates of the peak point (20.05,30.35) in the probability plot are the coordinates of the requested impact source, which has an absolute error of 0.353 cm; the relative error D=1%.

4. CONCLUSION

In this paper, a laminar imaging method based on beam focusing algorithm is proposed for impact damage monitoring. The method adopts a ring-shaped sensor arrangement method to collect impact source information from multiple directions through the mutual collaboration of multiple linear arrays, and reconstructs the impact source by experimentally fitting the localization accuracy of the linear arrays for different angles of the impact source. The experimental study of impact monitoring for aluminum plates proves that the method has high impact localization accuracy and good reliability.

References

[1] Shrestha P, Kim J-H, Park Y, et al. Impact localization on composite structure using FBG sensors and novel impact localization technique based on error outliers[J]. Composite Structures, 2016, 142: 263-271.

[2] Shrestha P, Park Y, Kim C-G. Low velocity impact localization on composite wing structure using error outlier based algorithm and FBG sensors[J]. Composites Part B: Engineering, 2017, 116: 298-312.

[3] Yang Jianghua,Lin Fengxian,Duan jiping, et al.The application of ultrasonic CT in bridge T-team quality testing[J].Journal of Yunnan University (Natural Sciences Edition), 2012, 34(S2): 248-251.

[4] Su Chenhui,Jiang Mingshun,Liang Jianying, et al.Lamb wave tomography damage imaging of carbon fiber reinforced polymer composite structures in strong noise environment[J].Acta Materiae Compositae Sinica, 2020, 37(04): 886-895.

[5] Ramadas C, Padiyar J, Balasubramaniam K, et al. Lamb wave based ultrasonic imaging of interface delamination in a composite T-joint[J]. NDT & E International, 2011, 44(6): 523-530.

[6] Shrestha P, Park Y, Kwon H, et al. Error outlier with weighted Median Absolute Deviation threshold algorithm and FBG sensor based impact localization on composite wing structure[J]. Composite Structures, 2017, 180: 412-419.

[7] ZHONG Yongteng,YUAN Shenfang,QIU Lei. Omni-directional impact localization method on composite structure using plum blossom array[J]. Acta Materiae Compositae Sinica | Acta Mater Compos Sin, 2014, 31(05): 1369-1374.

[8] Shahsavari V, Chouinard L, Bastien J. Wavelet-based analysis of mode shapes for statistical detection and localization of damage in beams using likelihood ratio test[J]. Engineering Structures, 2017, 132: 494-507.

Research on Integrated Control Strategy of Reactive Power and Voltage for Distributed Generation

Xirong Guo*[a]

[a] Chifeng Industrial Vocational and Technical College,024000, No. 22, Balin North Street, Xincheng District, Chifeng City, Inner Mongolia, China; 0476 5890921

* Corresponding author: 1264057134@qq.com

ABSTRACT: The access of distributed photovoltaic power sources will change the power flow distribution in the distribution network. When the line load is light and the photovoltaic access capacity is large, the distribution network will become a multi-source network, which will cause the reverse flow of the line, increase the line voltage, and affect the power quality of the distribution network. Aiming at this problem, this paper studies the voltage control strategy after photovoltaic grid-connected, and proposes a four-level reactive voltage control strategy to solve the voltage problem when photovoltaic large-scale grid-connected. Finally, the method is verified by an example.

1. Introduction

In power systems, maintaining voltage stability is critical to improving the quality of power supply. Since the medium voltage distribution system press close to the user, the quality of voltage of the busbar of the 10kV substation plays a vital role. Usually, in the comprehensive automatic control of substation voltage and reactive power, the dual-parameter control of reactive power on the low-voltage side and high-voltage side of the substation is realized by combining the on-load voltage regulating transformer and the capacitor bank. As distributed photovoltaic power is gradually connected to the distribution network, its randomness and volatility make the control strategy easy to lead to the inability to solve the security problems of the distribution network in time, which greatly reduces the service life of the equipment. In view of this problem[1-2]. This paper presents a four-level reactive power and voltage control strategy based on the nine-zone diagram and photovoltaic inverter. Numerical examples are used to conduct control simulations to verify the effectiveness of the proposed control strategy.

2. Research and application of reactive power and voltage comprehensive control strategy

2.1 Traditional voltage and reactive power control method

Electrical equipment in a power system is designed and fabricated to rated voltage. The general control methods are mainly to adjust the capacity of the reactive power compensation device and change the transformer tap. The control principle of voltage and reactive power in a typical substation is mainly based on the nine-zone diagram control strategy, and through control means to ensure that the voltage and power factor of the substation meet the requirements of grid operation. The nine-zone diagram control strategy is simple and foolproof to implement, so it is diffusely used in substation voltage and reactive power control strategy[3].

Commonly used nine-area diagram for comprehensive command of voltage and reactive power in substations. According to the reactive power Q and voltage U during the operation of the substation, the coordinate plane is divided into 9 regions, as shown in the figure. When the voltage and reactive power is in the abnormal area, it can run to the normal area through the corresponding control strategy. The following is the adjustment method: (0 area is the normal area, no action)

Area 1: Adjust the tap to reduce the voltage. If the tap is adjusted to the upper limit, the voltage still does not meet the requirements, and the capacitor is cut.

Area 2: First adjust the tap to reduce the voltage, after the voltage is qualified, if the reactive power exceeds the limit, then add the capacitor. (At the moment, the voltage is high, and the voltage will be further increased by casting the capacitor first.)

Area 3: cast capacitor, if there is no capacitor to cast, the reactive power is still higher than the upper limit, adjust the tap to reduce the voltage.

Area 4: First cast the capacitor, after the reactive power is qualified, if the voltage is out of bounds, then adjust the tap to boost the voltage. (At this time, there is a lack of reactive power. If you adjust the tap to boost the voltage, the reactive power will be further lacking.)

Area 5: Adjust the tap to boost the voltage. If the tap is adjusted to the lower limit, the voltage still does not meet the requirements, and the capacitor is switched.

Area 6: First adjust the tap to boost the voltage. After the voltage is qualified, if the reactive power exceeds the limit, then cut off the capacitor. (When the voltage is low at this time, cutting the capacitor first will further reduce the voltage.)

Area 7: Cut the capacitor. If there is no capacitor to cut, and the reactive power is still lower than the lower limit, adjust the tap to boost the voltage.

Area 8: Cut the capacitor first, after the reactive power is qualified, if the voltage exceeds the limit, then adjust the tap to reduce the voltage. (At this time, the reactive power is abundant. If the tap is adjusted to reduce the pressure, the reactive power will be further excessive.)

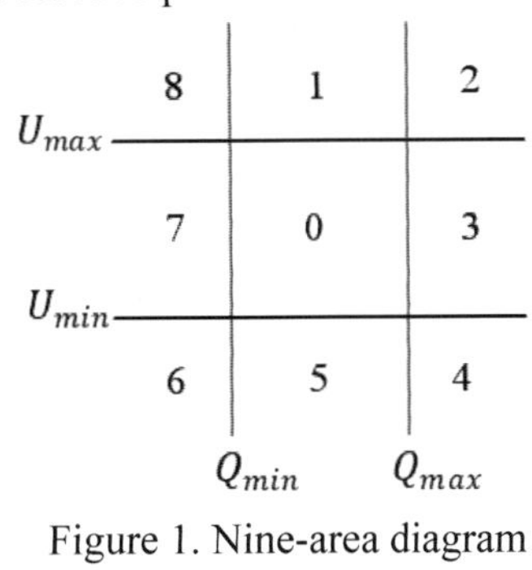

Figure 1. Nine-area diagram

2.2 Photovoltaic voltage reactive power control method

1.PV inverter voltage regulation

Photovoltaic mainly uses the reactive power regulation ability of the inverter to regulate the voltage. The value of the power factor of the traditional distributed power control method is connected with the active power of the photovoltaic power supply. When the voltage level is low, the photovoltaic power factor increases with the enlargement of the photovoltaic active power output, reducing the reactive power output value to avoid overcompensation; when the level of voltage is high, the photovoltaic power will increase with the output of photovoltaic active power. When the voltage exceeds the limit, the power factor of photovoltaic inverter runs at the minimum value, and the voltage is adjusted to the maximum capacity; when the voltage is better, the photovoltaic inverter uses the unit power Factorial operation, no voltage regulation[4-5].

2. Photovoltaic active output regulation

For the purpose of ensuring the user's demand for power quality and the secure operation of the power grid, the voltage of the distribution network can be regulated by reducing the active output of photovoltaics, so that the distribution network will not exceed the limit and increase the photovoltaic power generation system access capability. In order to improve the utilization rate of clean energy, the majorization goal is to minimize the sum of the active power reduction of each photovoltaic power station under the condition of qualified voltage. Inequality constraints include active and reactive output constraints of photovoltaic power plants, and the voltage meets the voltage requirements of the distribution grid[6].

$$\begin{cases} P_{pvi\min} \le P_{pvi} \le P_{pvi\max} \\ Q_{pvi\min} \le Q_{pvi} \le Q_{pvi\max} \\ U_{\min} \le U \le U_{\max} \end{cases} \tag{1}$$

In the formula: $P_{pvi\max}, P_{pvi\min}, Q_{pvi\max}, Q_{pvi\min}$ are upper and lower bounds of the active and reactive power output of the photovoltaic power station i, respectively, and $U_{\min}, U_{\max}$ are the least value and peak value of qualified voltage respectively.

Since photovoltaic active power output regulation is a multi-constrained nonlinear optimization problem, this paper chooses particle swarm optimization algorithm to solve it.

Particle Swarm optimization (PSO) is an evolutionary computing method, which first proposed by Dr Eberhart and Dr Kennedy in 1995. In the particle swarm optimization (PSO) algorithm, each solution of optimization problem is a bird in the search space. We call them "particles". All particles have a fitness value determined by the optimization function, and each particle also has a velocity that determines their flight direction and distance. The particle then follows the current optimal particle to search in the solution space The calculation process of particle swarm is shown in the figure below:

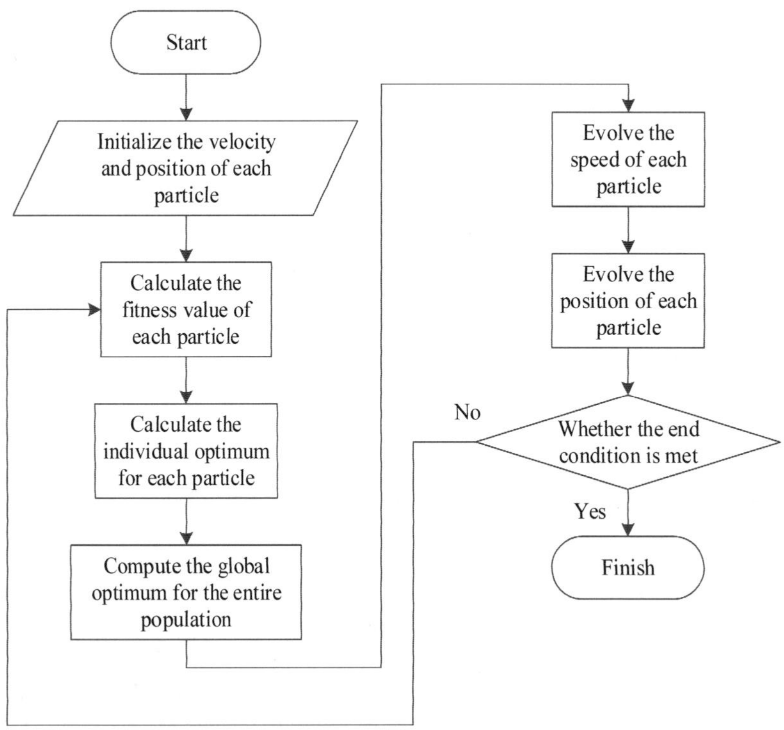

Figure 2. Calculation process of particle swarm optimization

The use of photovoltaic active power output regulation is generally only applicable to the case that the peak period of photovoltaic output accounts for a small part of the effective photovoltaic time of the whole year, so that more photovoltaic power can be obtained in other periods. If you are in the situation of using photovoltaic active power output to adjust the voltage for a long time, it is recommended to transform the power distribution equipment, install reactive power compensation or energy storage devices and other methods to improve the utilization rate of clean energy.

3. Reactive power and voltage integrated control strategy considering large-scale distributed photovoltaic

The access of distributed power sources has changed the operation mode of the original distribution network, and it is difficult to meet the grid voltage operation requirements only by relying on traditional voltage regulation methods. In order to adapt to DG access with high penetration rate and make brimful and rational use of network voltage regulation resources, it is necessary to classify voltage regulation methods such as distributed photovoltaic power supply, reactive power output, and transformer taps in the network according to their characteristics, and use them in the comprehensive voltage. In the control, it is controlled in stages to realize multi-level reactive power and voltage control including distributed photovoltaic power distribution network, so as to improve the voltage regulation effect.

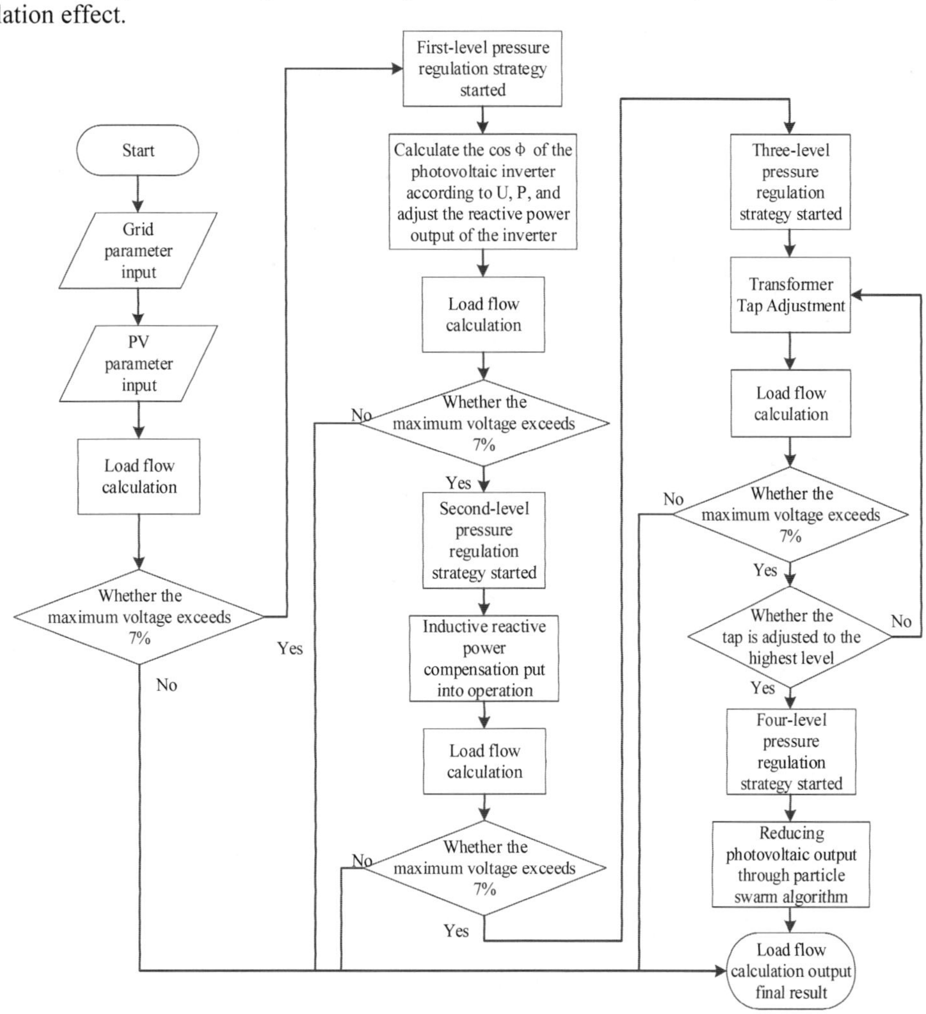

Figure 3. Four-level voltage regulation strategy considering distributed power supply

4. Simulation analysis

The simulation is carried out for a certain land area. The installed capacity of distributed photovoltaics in this area is small. Generally, new 0.4kV and 10kV lines are considered to be connected. At present, due to the low original load rate, the access of distributed photovoltaics has a great impact on the voltage of the distribution network in the region, and the voltage regulation strategy needs to take into account the actual situation of the distribution network.

The pressure regulation strategies at all levels are described as follows:

Primary voltage regulation: There are many small distributed photovoltaics, so the potential of primary voltage regulation should be fully exploited, photovoltaic inverters should be used for voltage regulation as much as possible, and control strategies should be used to achieve voltage regulation tasks.

Secondary voltage regulation: Considering that the current medium and low voltage lines and photovoltaics along the line are not equipped with inductive reactive power compensation devices, secondary voltage regulation measures cannot be adopted. However, the secondary pressure regulation should be used as the main pressure regulation method in the future, which will be analyzed and explained later.

Three-stage voltage regulation: The voltage regulation capability of photovoltaic inverters is very limited. Coordinating and controlling the tap position of the transformer to regulate the system voltage is the most important voltage regulation method at present.

Four-stage voltage regulation: reducing the output or even cutting the machine is the last resort, and it is also the most effective pressure regulation method. In order to ensure the absorption of clean energy as much as possible, this study takes the minimum reduction of the active power of each photovoltaic power station as the optimization goal, and uses the particle swarm algorithm to calculate..

Taking the Lihe Line as an example, the current line load rate is relatively low, and one photovoltaic power station has been put into operation, and four photovoltaic power plants are planned to be put into operation, with a total capacity of 4086kW.

The details of the Lihe line photovoltaic power station are shown in the table below:

Table 1. Photovoltaic power station information

Number	PV power station	Installed capacity (kW)	Installation location
1#	Xiaozhuang Village Power Station	175.77	113 Miaozuishan branch line of Lihe line 13# pole Majiashanzhuang distribution transformer
2#	Lidian Lidian First Village Power Station	1227.68	113 The 22# pole T-connection of the main line of Lihe line newly added a photovoltaic box transformer in Lidian
3#	Lidian Lidian Second Village Power Station	1300.96	113 Lihe line main line 22# pole T connection newly added Lidian second photovoltaic box transformer
4#	Changhe Shengyi Village Power Station	1261	113 The 166# pole T-connection of the main line of Lihe Line adds a new Shengyi photovoltaic box transformer
5#	Changhe Sunchuan Village Power Station	120.96	113 The 45# pole T-connection of the Kangliang branch line of the Lihe line added the second distribution transformer of Sun Chuan

In the future, with large-scale photovoltaic access, the voltage will first rise and then fall with the increase of power supply distance. The voltage deviation of the node with the largest line voltage deviation will reach +9.68%, and the problem of excessive voltage after photovoltaic access is more serious.

The photovoltaic inverter cannot provide reactive power, and the total installed photovoltaic capacity is 7822kW.

Adopt one-level and three-level voltage regulation measures, when the transformer tap gear is adjusted to 35+2*2.5%kV, the maximum voltage deviation of the line is +7.62%, which still does not meet the requirements of the regulations. At this time, the fourth-level voltage control is activated. Particle swarm optimization is used to key off photovoltaics.After iterative calculation, the minimum photovoltaic output of 335kW needs to be cut off at the Changhe Shengyi Village Power Station. At this time, the maximum voltage deviation is +6.99%, which meets the voltage requirement of ±7%. The following figure shows the calculation process of the particle swarm algorithm.

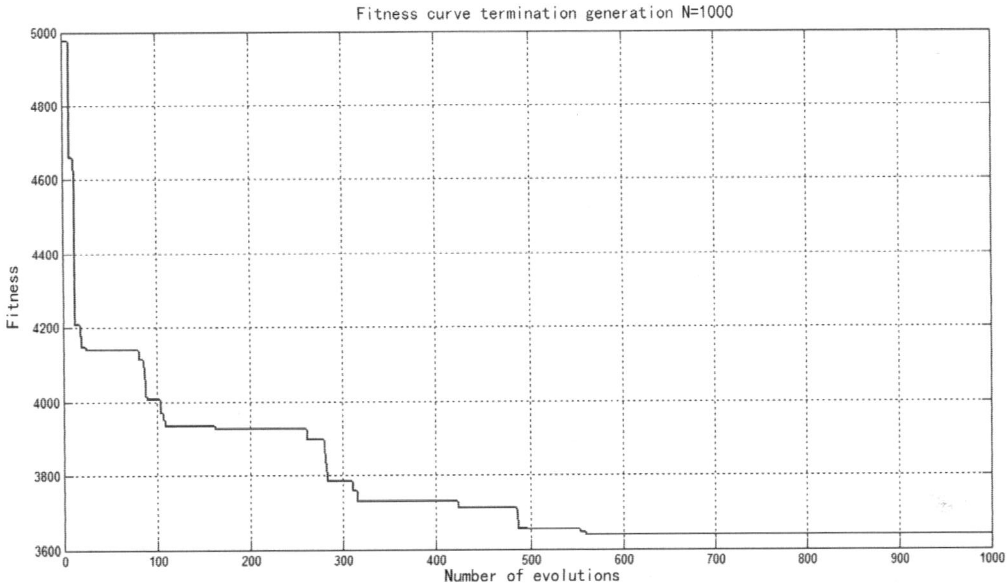

Figure 4. Calculation process of particle swarm algorithm

	Node	Belongs	Voltage level	Amplitude	Deviation	Phase	Active compen..	Reactive comps..	Active contri..
1	301#	--	10	11.8872	18.87	13.3	0	0	2.9452
2	297#	--	10	11.807	18.07	12.9	0	0	0
3	204#	--	10	11.807	18.07	12.9	0	0	0
4	203#	--	10	11.8067	18.07	12.9	0	0	0
5	198#	--	10	11.8057	18.06	12.9	0	0	0
6	217#	--	10	11.8057	18.06	12.9	0	0	0
7	216#	--	10	11.8057	18.06	12.9	0	0	0
8	225#	--	10	11.8052	18.05	12.9	0	0	0
9	224#	--	10	11.8052	18.05	12.9	0	0	0
10	219#	--	10	11.805	18.05	12.9	0	0	0
11	218#	--	10	11.805	18.05	12.9	0	0	0
12	221#	--	10	11.8048	18.05	12.9	0	0	0
13	220#	--	10	11.8048	18.05	12.9	0	0	0

Figure 5. Simulation results before voltage regulation

	Node	Belongs	Voltage level	Amplitude	Deviation ▲	Phase	Active compens..	Reactive comps..	Active contrib..
1	300#	--	0.4	0.4776	19.41	29.9	0	0	0.284
2	263#	--	0.4	0.4603	15.08	23.5	0	0	0.3842
3	301#	--	10	10.7218	7.22	13.8	0	0	2.187
4	297#	--	10	10.6603	6.6	13.4	0	0	0
5	204#	--	10	10.6603	6.6	13.4	0	0	0
6	203#	--	10	10.6601	6.6	13.4	0	0	0
7	198#	--	10	10.6591	6.59	13.4	0	0	0
8	217#	--	10	10.659	6.59	13.4	0	0	0
9	216#	--	10	10.659	6.59	13.4	0	0	0
10	225#	--	10	10.6585	6.59	13.4	0	0	0
11	224#	--	10	10.6585	6.59	13.4	0	0	0
12	219#	--	10	10.6584	6.58	13.4	0	0	0
13	218#	--	10	10.6584	6.58	13.4	0	0	0

Figure 6. Simulation results after voltage regulation:

Through the calculation and analysis of different examples, it can be seen that the current and future voltage stabilization requirements can be satisfied by using reactive power compensation and transformer tap regulation under the condition of sufficient reactive power compensation; when the reactive power compensation is insufficient, the transformer tap adjustment is relatively frequent. In large-scale access, it is even necessary to limit the photovoltaic output to maintain the voltage level. It is needful to astrict the capacity of the line photovoltaic access or reasonably transform the line in advance.

5. Conclusion

By combining traditional voltage regulation technology and photovoltaic voltage regulation technology, a four-level reactive power and voltage control strategy is formed. Through the corresponding simulation calculation verifies the accuracy of the control strategy and algorithm. In the future, in-depth analysis of new technology applications such as energy storage can be carried out to reduce or even avoid the occurrence of abandoned light.

References

[1] PEI Xiaojuan, WANG Qian, JIN Yingbo. "The influence of distributed generation on the voltage of distribution network, "Journal of Electric Power,2010,25 (6) :480-483.

[2] WANG Qi, LI Bin, DENG Xiaobo,et al. "Distribution network voltage control based on coordinated optimization of PV and air-conditioning, "International Journal of Photoenergy,2018,2018 (2) :1-7.

[3] XU Zhicheng, ZHAO Bo, DING Ming,et al. "Photovoltaic hosting capacity evaluation of distribution networks and inverter parameters optimization based on node voltage sensitivity, "Proceedings of the CSEE,2016,36 (6) :1578-1587.

[4] ZHOUU Lin, REN Wei, LIAO Bo,et al. "Reactive power and voltage control for grid-connected PV power plants, "Transactions of China Electrotechnical Society, 2015,30 (20) : 168-175.

[5] LIU Shuang, ZHANG Jianzhou, WANG Hanlin,et al. "Two stage reactive power and voltage coordinated control strategy for photovoltaic power station considering multiple reactive power sources, "Automation of Electric Power Systems,2017,41 (11) :120-125.

[6] WANG Hongkun, WANG Shouxiang, PAN Zhixin,et al. "Optimized dispatching method for flexibility improvement of distribution network with high-penetration distributed generation, "Automation of Electric Power Systems,2018,42 (15) :86-93.

Detection method of water channel defects in engine block based on deep learning

Yaohua Deng[1], WeiJie Li[1], ShengYu Lin[1]*, Qiwen Lu[1]*, Xiali Liu[1]

[1]School of Electromechanical Engineering,Guangdong University of Technology, 510006

Correspondence should be addressed to ShengYu Lin, Qiwen Lu

Lin: lsy18679930915@126.com

Abstract. Aiming at the low efficiency of manual detection in the detection of water channel defects of engine cylinder block and the poor generalization ability of traditional machine vision in manual design features, an improved water channel defect detection model of cylinder block based on fast RCNN network is proposed. Restnet50 is selected as the feature extraction network of waterway defects, and feature pyramid network (FPN) is introduced to improve the detection ability of small defects; The anchor box is optimized by k-means++ clustering algorithm to improve the positioning of the target box. Experiments show that the map of the improved network in the engine cylinder block waterway defect data set reaches 88.74%, the accuracy is increased by 4% compared with the original fast RCNN, and the recall rate is increased by 1.67%. The recognition and detection effect in real samples is good, and it can be effectively used for the engine cylinder block waterway defect detection.

1. Introduction

As a part of the automobile engine cooling system, the water channel of the engine block is used to cool the engine with coolant. Under modern production conditions, affected by the design of the process system, casting defects such as pores, sand holes and cracks will appear in the water channel, causing serious safety hazards to the engine. Due to the narrow and long surface shape and deep bottom of the water channel, the internal natural light irradiation is seriously insufficient, which makes the internal surface difficult to observe and the existing minor defects difficult to be detected. The traditional manual lighting visual inspection method is greatly affected by subjectivity, and the detection efficiency is low, which can not meet the growing demand for high-quality production. The defect detection method using traditional machine vision is easy to be affected by environment, light and production process, so it is difficult to design feature extraction method. It needs to consider the specific detection environment and the characteristics of cylinder block waterway defects. Small changes in factors may affect the feature extraction results, so it is not suitable for cylinder block waterway defect detection scene.

With the continuous development of machine vision technology, defect detection technology based on deep learning is relatively mature in theory and practical application. It has been successfully applied in defect detection of wood [1], glass [2], strip steel [3], battery [4], chip [5], semiconductor [6] and other products, improving the efficiency and accuracy of defect detection. At present, the research of using deep learning technology to detect the water channel defects of engine cylinder block is rare, and there is a problem that the size of some kinds of defects in the water channel is small, which makes it difficult to detect. Based on the above reasons, this paper introduces the visual detection technology based on deep learning in the detection of the surface defects of the water channel of the engine block, and proposes an improved faster RCNN model for the detection of the water channel defects of the engine block. In this paper, resnet50 deep convolution neural network is used to extract defect features, improve the classification ability of the detection network, and feature pyramid network is introduced to fuse the features of different convolution layers to enhance the feature strength of small defects in the water channel of the engine block; Using k-means++ clustering method to generate an anchor frame suitable for the size of waterway defects, improve the accuracy and efficiency of defect detection results. Finally, through the defect detection experiment on the established waterway defect data set, the feasibility and effectiveness of this model are verified.

2. Improvement and design of fast RCNN algorithm

2.1 Network structure of fast RCNN algorithm

Fast rcnn[7] is one of the mainstream target detection algorithms for deep learning at present. The network structure is shown in Figure 1, which mainly includes convolutional neural network (CNN), regional suggestion network (RPN), ROI pooling, full connected layers (FC) and classification and location regression network. The target detection steps of the algorithm are divided into feature extraction, candidate region generation, classification and location regression. These steps are integrated into a network to design an efficient and unified end-to-end defect detection model, which effectively improves the detection efficiency.

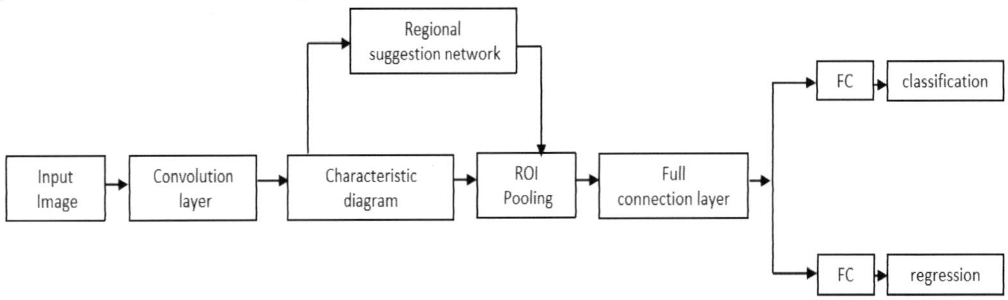

Figure 1 network structure of fast r-cnn algorithm

2.2 Improvement of feature extraction network

The feature extraction ability of convolutional neural network affects the result of defect detection. In the original fast RCNN, vgg16 is used as the feature extraction network. The number of layers of vgg16 network is relatively shallow, so it is difficult to ensure the full extraction of defect features to a certain extent. Deeper networks can extract defect features more fully, but with the increase of network depth, degradation will occur, and restnet[8] network uses residual structure to solve this problem well. In order to better and more accurately extract the features of cylinder block waterway surface defects, restnet50 is used as the feature extraction network in this paper. The first five groups of convolution structure of restnet50 network are shown in Table 1.

Table 1 Top 5 volume layer structure of restnet 50 network

Convolution layer	Conv1	Conv2_x	Conv3_x	Conv4_x	Conv5_x
Output size	112×112	56×56	28×28	14×14	7×7
structure		3×3 maxpool, stride2			

7×7, 64, stride 2	$\begin{bmatrix}1\times1,64\\3\times3,64\\1\times1,256\end{bmatrix}\times3$	$\begin{bmatrix}1\times1,128\\3\times3,128\\1\times1,512\end{bmatrix}\times4$	$\begin{bmatrix}1\times1,256\\3\times3,256\\1\times1,1024\end{bmatrix}\times6$	$\begin{bmatrix}1\times1,512\\3\times3,512\\1\times1,2048\end{bmatrix}\times3$

CNN network shallow feature map has high resolution and rich location information, which is suitable for small target detection, while deep feature map has low resolution and rich semantics, which is suitable for medium and large target detection [9]. The water channel of the engine block has tiny shrinkage holes, pores and other defects. The defect detection based on the top-level characteristics of convolutional neural network can not completely reflect the information of small defects, which reduces the recognition ability of the network to small defects. To solve this problem, this paper constructs a feature pyramid network based on restnet50, and improves the detection accuracy of small targets by integrating the characteristics of different levels of restnet50 network. The characteristic pyramid network structure based on restnet50 is shown in Figure 2.

The feature pyramid FPN structure is composed of three parts: bottom-up, horizontal connection and top-down. Through this structural design, the high-level features of low-resolution, high semantic information and the low-level features of high-resolution, low semantic information are horizontally connected from top to bottom, and a feature map with richer semantics is obtained. In the feature pyramid network based on restnet50, Select conv2 of restnet50_ x~conv5_ The last layer of the X stage outputs the feature map $C2$~ $C5$ convolution operation is performed to obtain a 256 dimensional feature map, Then, the interpolation algorithm is used to double up sample from the top level, and the up sampled feature map is fused with the feature map after the convolution operation of $C2$, $C3$, $C4$, and $C5$.

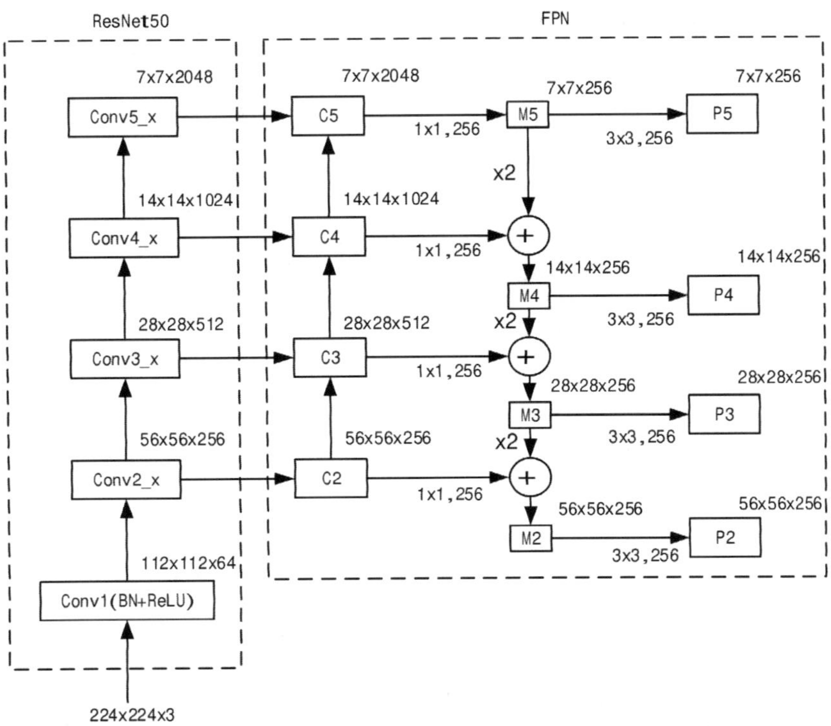

Figure 2 characteristic pyramid network structure based on restnet50

Finally, the convolution check is used to convolute each fused feature map in order to eliminate the aliasing effect of up sampling, and finally get a multi-level feature map with richer semantics $\{P2, P3, P4, P5\}$, These feature maps will be shared by regional candidate networks and detection classification regression networks. When candidate regions are generated, they need to be mapped back to the feature map. For candidate regions of different sizes, the mapping relationship between them is

$$k = k_0 + \log_2\left(\frac{\sqrt{w \times h}}{224}\right) \quad (1)$$

In the above formula, 224 is the standard input of Imagenet, k_0 is the reference value, is set to 5, represents the output of P5 layer, and w, h is the length and width of ROI area.

2.3 Improved anchor box generation scheme based on k-means++

The default candidate box in fast RCNN algorithm is designed for the detection target in Pascal VOC data set, and cannot be adaptively changed according to the size of defects in the actual input image. For cylinder water jacket defect detection, the difference between the default candidate box size and the water jacket defect size is too large, and a lot of regression work needs to be done when generating the regional candidate box, resulting in slow convergence of the network model and affecting the detection efficiency. In order to improve the detection performance of cylinder block waterway surface defects and improve the anchor frame generation scheme, this paper adopts the k-means++[10] clustering method, takes the IOU distance as the k-means++ distance measurement formula, sets the number of clustering centers K, and takes the clustering results as the width and height of the anchor frame.

IOU is used to evaluate the similarity between two target frames. It is the ratio of the area of the intersection area of the two target frames to the area of the merging part. Since clustering only cares about the width and height of the frames and does not care about their positions, it is assumed that the center points of the two frames coincide when calculating the IOU. Then the IOU of dimension boxes a and B in the watercourse defect data set is shown in equation (2), in which dimension boxes a and B and the intersection and merge ratio are shown in Figure 3.

$$\text{IOU}(A, B) = \frac{\min(w_a, w_b)\Box\min(h_a, h_b)}{w_a h_a + w_b h_b - \min(w_a, w_b)\Box\min(h_a, h_b)} \quad (2)$$

In the above formula, (w_a, h_a) it represents the width and height of dimension box a, (w_b, h_b) it represents the width and height of dimension Box B, $\min(w_a, w_b)$ means taking the smaller value of w_a, w_b , $\min(h_a, h_b)$ means taking the smaller value of h_a, h_b .

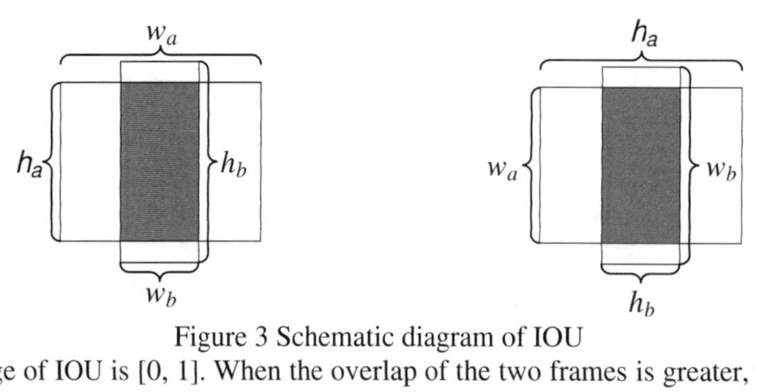

Figure 3 Schematic diagram of IOU

The value range of IOU is [0, 1]. When the overlap of the two frames is greater, the larger the IOU value is, and the more similar the shape and position of the two frames are. In order to make the measurement value negatively correlated with the similarity, that is, the smaller the measurement value is, the greater the similarity is, equation (3) is used as the distance measurement formula

$$d_{iou} = 1 - IOU \quad (3)$$

The specific steps of clustering callout boxes using kmeans++ algorithm are as follows:

(1) Extract the width and height data of all defect callout boxes in the waterway surface defect training set, and randomly select a callout box as the width and height of the clustering center c_1;

(2) Calculate the distance d_{iou} between each dimension box and the existing cluster center c_1, and calculate the probability that each dimension box is recognized as the next cluster center, and select the dimension box with the greatest probability as the next cluster center;

(3) Repeat step (2) until K clustering centers are selected;

(4) Calculate the distance d_{iou} from each dimension box to each cluster center, and assign it to the class corresponding to the cluster center with the smallest distance d_{iou};

(5) For each class, the cluster center of each class is recalculated according to the centroids of all samples of the class;

(6) Repeat steps (4) and (5) until the calculation result is the same as the previous one, and finally output the clustering result.

3. Experiment and result analysis

3.1 experimental data

A total of 469 pieces of image data of cylinder water channel were collected in the experiment, including hole defects (pores and sand holes) and crack defects. In order to improve the robustness of the model and achieve better network training effect, image augmentation methods such as flipping, rotating, scaling and clipping are used to expand the data set, which is divided into training set and test set according to 4:1. Some data images are shown in Figure 4.

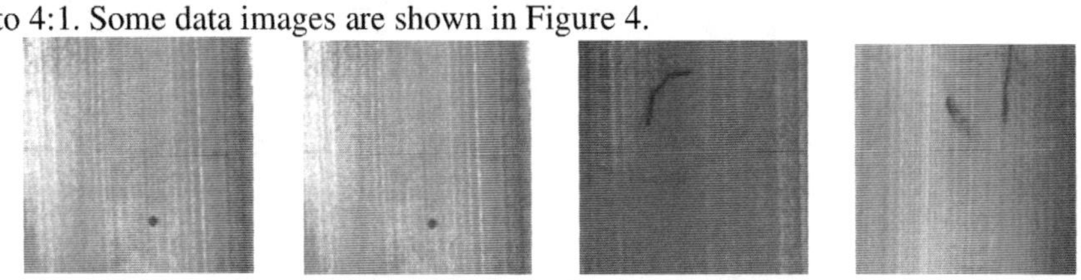

(a) Hole defect (b) Hole defect (c) Crack defect (d) Crack defect

Figure 4 partial defect sample image

3.2 Experimental environment

The built model is trained and tested on PC. the hardware configuration of the experiment: the CPU is Intel (R) core (TM) i5-9300h (memory 16g), the GPU is NVIDIA GTX 1080ti (video memory 11g), and the system is Ubuntu, tensorflow framework.

This paper selects seven widely used indicators: accuracy, precision, recall, score (-score), AP value, mean average precision (MPA) and FPs to evaluate the performance of the model. The accuracy rate is the proportion of samples with correct prediction results in the total samples; Accuracy refers to the proportion of actual positive samples in the number of predicted positive samples; The recall rate is the probability of predicted positive samples in the actual positive samples; Scores combined with accuracy and recall can comprehensively evaluate the performance of the model; AP represents the average precision of a category in the classification; MPa is the average of the average precision of all categories; FPS indicates the number of detection frames per second.

3.3 experimental result

3.3.1 Comparative experiment of convolutional neural network

In order to compare the defect feature extraction ability of different convolutional neural networks, zfnet, vgg16 and resnet50 networks are used to extract the characteristics of the water channel surface of the engine block, and the recognition accuracy is taken as the evaluation standard. The recognition results of the three networks on various defects in the data set are shown in Table 2. Through the experimental results, it can be seen that the recognition accuracy of resnet50 network is higher than that of the other two models, which shows that resnet50 is indeed better than the other two networks in the extraction of defect feature information.

Table 2 recognition rate of different convolution networks in waterway defect data set

Convolutional neural network model	Hole	crackle
ZFNet	87.31%	85.76%
VGG16	89.85%	86.17%
ResNet50	90.57%	88.49%

3.3.2 Anchor frame generation scheme experiment

Aiming at the problem that the default candidate box in fast RCNN algorithm is not suitable for cylinder block waterway defect detection, a k-means++ clustering method is proposed to improve it. Because the feature golden network designed in this paper generates four feature maps, it needs to set a scale on each feature map, so the number of clustering centers is 4. The width and height data clustering results of the waterway surface defect callout box are shown in Figure 5.It can be seen from Figure 5 that the final coordinates of the four cluster centers are (12, 12), (40, 19), (19, 23), (24, 64). After analyzing the above data and combining the anchor frame design of the regional candidate network, set the anchor frame scale as (12×12, 19×19, 40×40, 64×64) and the length width ratio as (1:2, 1:1, 2:1) respectively on the $\{P2, P3, P4, P5\}$ feature map.

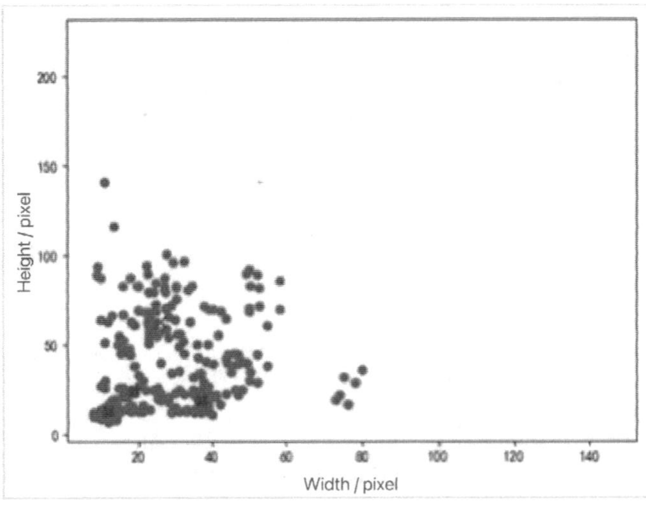

Figure 5 clustering results of k-means++ algorithm

3.3.3 Testing experiment of water channel defect of engine block

The loss function curve of the improved network model training in this paper is shown in Figure 6. It can be found that with the increase of iteration times, the loss value of training continues to decline, indicating that the improved network model is suitable for the detection of water channel surface defects in the engine cylinder block.

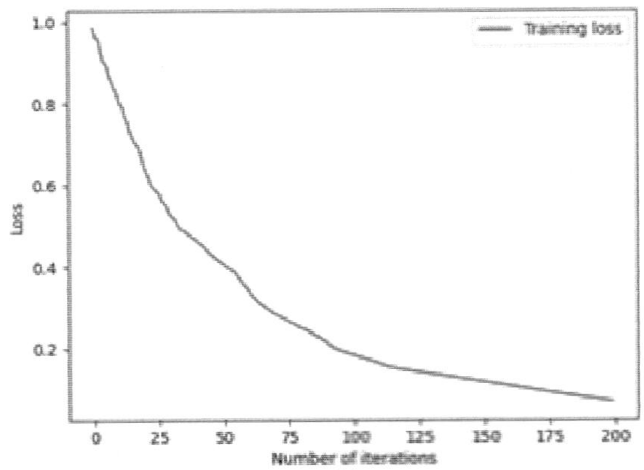

Figure 6 line chart of training loss

The detection and recognition effect of the improved network model on the waterway surface defect data set is shown in Figure 7. It can be seen from the figure that the model in this paper can correctly identify and locate water channel defects, and classify them with high confidence. For the small target defects in figure (a), the confidence also reaches 84.7%, indicating that the model can better learn the information of such small defects.

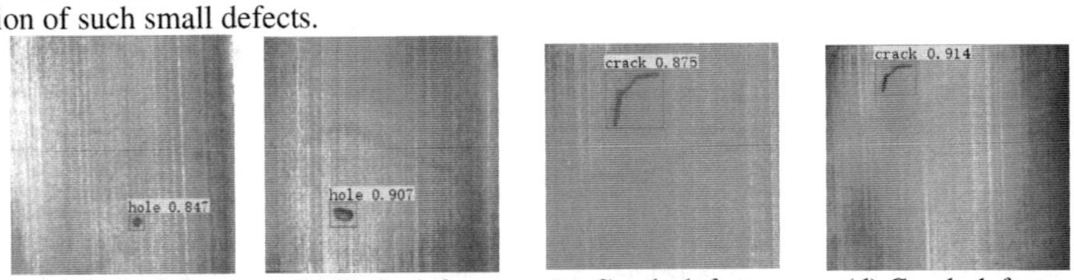

(a) Hole defect (b) Hole defect (c) Crack defect (d) Crack defect

Figure 7 test results of waterway data set

In order to further evaluate the improved defect detection model, the improved network model is compared with the original fast RCNN and yolov3. The experiment selects 100 images to be tested, and compares and analyzes them from AP, map, FPS, accuracy, recall and six model evaluation indicators, so as to evaluate the performance of the model more intuitively and comprehensively. Table 3 shows the quantitative comparison results of AP, map and FPS values of the three detection models. Table 4 shows the comparison results of accuracy, recall and scores of the three models.

From the data comparison in the above table, it can be seen that the AP value, map value, accuracy, recall rate and score of the two types of defects detected on the image to be tested by the defect detection model proposed in this paper are higher than those of the original fast RCNN and yolov3 network model, and the detection speed is faster than that of the original fast RCNN. Compared with the yolov3 model, although the detection time of the model in this paper is relatively long, the detection time is far less than the time for the micro camera to collect an image. The real-time detection can be realized in the whole process of cylinder water channel detection, and the overall performance is good.

Table 3 comparison results of AP, map and FPS values

network model	hole	crackle	mAP	FPS
Faster-RCNN	89.45%	84.85%	87.15%	4.56
YOLOv3	85.72%	82.17%	83.95%	28.20
This paper improves the model	92.21%	85.26%	88.74%	8.03

Table 4 comparison results of accuracy rate, recall rate and F1 score

network model	Accuracy	recall	F_1
Faster-RCNN	77.68%	80.47%	0.791
YOLOv3	77.25%	70.72%	0.738
improves the model	81.68%	82.14%	0.819

4. Conclusion

Aiming at the problem that it is difficult to design feature extraction in traditional machine vision in engine cylinder block waterway defect detection, this paper proposes an improved cylinder block waterway defect detection method based on fast RCNN network. Through experiments, three different convolutional neural networks are compared and resnet50 is used for feature extraction; Using k-means++ algorithm and IOU distance as the distance measurement formula, the anchor frame generation scheme in RPN network is improved to improve the defect detection performance of the waterway surface of the lift cylinder. Compared with the original fast RCNN and yolov3, the improved algorithm significantly improves the accuracy of waterway defect detection and recognition.

Funding statement

The present work is financially supported by National Natural Science Foundation of China (Grant No.: 52175457) and the Key-Area Research and Development Program of Guangdong Province (Grant 2019B010154001).

References

[1] Hu Zhongkang, Liu Ying, Zhou Xiaolin, et al Research on solid wood plate defect and texture recognition based on depth confidence network [j] Computer application research, 2019, 36 (12): 3889-3892

[2] Jin Y, Zhang D, Li M, et al. A fuzzy support vector machine-enhanced convolutional neural network for recognition of glass defects[J]. International Journal of Fuzzy Systems, 2019, 21(6): 1870-1881.

[3] Hao Yaru Research on visual recognition method of strip steel surface defects [d] Chongqing Jiaotong University, 2018

[4] Xu Chaofan Algorithm design and implementation of industrial appearance defect detection platform based on stereo light source [d] Harbin Institute of technology, 2019

[5] Bian Jingshuai Research on automatic detection method for chip surface defects [d] Hefei University of technology, 2019

[6] Yu J, Zheng X, Liu J. Stacked convolutional sparse denoising auto-encoder for identification of defect patterns in semiconductor wafer map[J]. Computers in Industry, 2019, 109: 121-133.

[7] Ren S, He K, Girshick R, et al. Faster r-cnn: Towards real-time object detection with region proposal networks[J]. Advances in neural information processing systems, 2015, 28.

[8] He K, Zhang X, Ren S, et al. Identity mappings in deep residual networks[C]//European conference on computer vision. Springer, Cham, 2016: 630-645.

[9] Zhao Jingbo, Du baoshuai Development of small target detection technology based on deep learning [j] Electro optic and control

[10] Yang junchuang, Zhao Chao A review of K-means clustering algorithm [j] Computer engineering and application, 2019, 55 (23): 7-14

Reliability Evaluation Method of Pseudo-failure Life Distribution for Industrial Robot Servo System Based on The Generalized Lambda Distribution

Shaomin Tang[1], Guixiong Liu[1*], Xiaobing Li[2]

[1]School of Mechanical and Automotive Engineering, South China University of Technology, Guangzhou 510640, China;

[2]CEPREI, Guangzhou 510610, China

* Corresponding author: megxliu@scut.edu.cn

ABSTRACT: A reliability evaluation method of pseudo-failure life distribution for Industrial Robot Servo System (IRSS) based on The Generalized Lambda Distribution (GLD) was proposed in the paper. The IRSS pseudo-failure life GLD distribution was established. Quantile estimation method was applied to estimate the GLD parameters. The proposed method had no need of presetting specific distribution morphology, and thus improve the accuracy of IRSS pseudo-failure reliability modeling. In the application experiment, the IRSS pseudo-failure life distribution reliability evaluation method based on GLD was used to obtain the IRSS failure life, the reliability evaluation result was consistent with the design failure life, which indicated the validity of the proposed method.

1. introduction

Industrial Robot Servo System (IRSS) is the execution unit used to complete the specific trajectory motion of Industrial robots, and is the core component of Industrial robots. The reliability of IRSS is directly related to the normal service life of Industrial robots [1]. IRSS are typical products with long life and high reliability. Its reliability evaluation is faced with the problem of insufficient reliability test data due to the lack of test samples and limited test time. The reliability evaluation method based on performance degradation can better study the reliability evaluation of long-life and high-reliability products [2-3], and its reliability evaluation modeling method includes PFA (Physics of Failure Analysis)-based method and PSA (Performance degradation data Statistics Analysis)-based method. PFA-based method assumes that IRSS failure is caused by the degradation of a single key part (such as bearing, motor rotor, stator insulation, etc.), and the model established by the method is hard to meet the system-level reliability research and analysis under the joint action of multiple failure mechanisms [4-5]. Whereas, the PSA-method establishes the degradation model by statistical analysis method, can better analyze the IRSS degradation process [6-8]. However, the PSA-method should be based on a reasonable distribution assumption.

Address to the issue, the paper proposed a reliability evaluation method of pseudo-failure life distribution for IRSS based on The Generalized Lambda Distribution GLD. The idea is that making use of the characteristics of various distribution forms of GLD to modeling the IRSS pseudo-failure life distribution, so as to improve the accuracy of pseudo-failure reliability modeling of IRSS without presetting specific distribution forms.

2. the gld-based pseudo-failure life distribution reliability evaluation methodology

Figure 1 is the flow chart of IRSS pseudo-failure life distribution reliability evaluation method, including: i. Set pseudo-failure level, and obtain the IRSS pseudo-failure life data set based on the IRSS performance degradation data; ii. Expand IRSS pseudo-failure life data set used the Bootstrap method; iii. Establish the IRSS pseudo failure life distribution based on the GLD distribution method; vi. Extrapolate IRSS life distribution under failure threshold to complete reliability assessment.

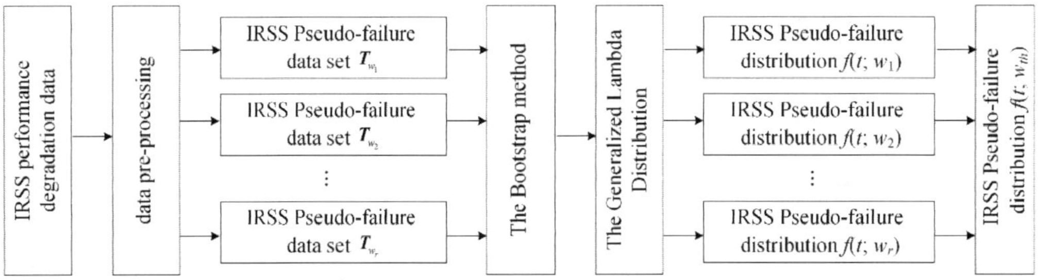

Figure 1. The flow chart of IRSS pseudo-failure life distribution reliability evaluation method

2.1 GLD-based method for IRSS pseudo-failure life distribution estimation

Suppose that the position and scale parameters of generalized λ distribution are λ_1 and λ_2, and the shape parameters are λ_3 and λ_4, Table 1 shows the relationship between typical probability distribution and the generalized λ distribution parameters. It can be concluded that the generalized λ distribution can approximate the morphological characteristics of a variety of distributions, without the prior distribution preset, which is very suitable for IRSS pseudo-failure life distribution fitting.

Table 1. The relationship between typical probability distribution and the generalized λ distribution parameters

Typical probability distribution	Distribution expression	Distribution parameters	The generalized λ distribution parameters
Normal distribution	$f(x)=\frac{1}{\sqrt{2\pi}\sigma}\cdot e^{-(x-\mu)^2/2\sigma^2}$	μ=0； σ=1	λ_1=0; λ_2=0.1975 λ_3=0.1349; λ_4=0.1349
Logarithmic normal distribution	$f(x)=\frac{1}{x\sqrt{2\pi}\sigma}\cdot e^{-(\ln x-\mu)^2/2\sigma^2}$	μ=0； σ=1/3	λ_1=0.8451; λ_2=0.1085 λ_3=0.0102; λ_4=0.0342
Exponential distribution	$f(x)=\mu e^{-\mu x}$	μ=1	λ_1=0.0100; λ_2=-10^{-3} λ_3=-10^{-6}; λ_4=-10^{-3}
Gamma distribution	$f(x)=\frac{x^{\alpha-1}e^{-x/\beta}}{\Gamma(\alpha)\cdot\beta^{\alpha}}$	α=5； β=3	λ_1=10.761; λ_2=0.0145 λ_3=0.0252; λ_4=0.0939
Weibull distribution	$f(x)=\alpha\beta x^{\beta\text{-}1}\exp(-x^{-\alpha x^{\beta}})$	α=5； β=1	λ_1=0.9935; λ_2=1.0491 λ_3=0.2121; λ_4=0.1061

Suppose $T_{w_i}^{j*}$ from the IRSS pseudo-failure extended data set $\boldsymbol{T}_{w_i}^{*}$ obey the generalized λ distribution, $Q(u)$ is the quantile function of the generalized λ distribution, where $j\in[1,\ N_{w_i}]$, and N_{w_i} is the number of samples of $\boldsymbol{T}_{w_i}^{*}$, $u\in[0,1]$. Thus $T_{w_i}^{j*}$ = Q(u) generates a random variable:

$$Q(u)=\lambda_1+\frac{1}{\lambda_2}\cdot[u^{\lambda_3}-(1-u)^{\lambda_4}],\ 0\le u\le 1 \tag{1}$$

According to the inverse function derivation rule, the probability density function of IRSS pseudo-failure life distribution can be obtained [9]:

$$f(l)=\frac{\lambda_2}{[\lambda_3 u^{\lambda_3-1}+\lambda_4(1-u)^{\lambda_4-1}]}\quad l=Q(u) \tag{2}$$

2.2 IRSS pseudo-failure life distribution GLD parameter solution

The relationship between the first four moments of the generalized λ distribution and parameters (λ_1, λ_2, λ_3, λ_4) is more complex than that between the quantile and parameters (λ_1, λ_2, λ_3, λ_4). Therefore, the Quantile-based GLD parameter estimation method is applied [10]. Based on the IRSS pseudo-failure extended data set $T^*_{w_i}$, estimate the median $\hat{\rho}_1$, 10-quartile difference $\hat{\rho}_2$, 10-quartile ratio $\hat{\rho}_3$ and 10-quartile difference ratio $\hat{\rho}_4$ as the following:

$$\hat{\rho}_1=\pi_{0.5}\;;\quad \hat{\rho}_2=\pi_{0.9}-\pi_{0.1}\;;\quad \hat{\rho}_3=\frac{\pi_{0.5}-\pi_{0.1}}{\pi_{0.9}-\pi_{0.5}}\;;\quad \hat{\rho}_4=\frac{\pi_{0.75}-\pi_{0.25}}{\pi_{0.9}-\pi_{0.1}} \tag{3}$$

where, π_p is the p-quantile of $T^*_{w_i}$.

The median ρ_1, 10-quartile difference ρ_2, 10-quartile ratio ρ_3 and 10-quartile difference ratio ρ_4 of the GLD can be obtain based on the GLD quantile function $Q(u)$ as the following:

$$\begin{aligned}
&\rho_1=Q(0.5)=\lambda_1+\frac{1}{\lambda_2}\cdot(0.5^{\lambda_3}-0.5^{\lambda_4})\;;\\
&\rho_2=Q(0.9)-Q(0.1)=\frac{1}{\lambda_2}\cdot[(0.9^{\lambda_3}-0.9^{\lambda_4})-(0.1^{\lambda_3}-0.1^{\lambda_4})]\;;\\
&\rho_3=\frac{Q(0.5)-Q(0.1)}{Q(0.9)-Q(0.5)}=\frac{[(0.5^{\lambda_3}-0.5^{\lambda_4})-(0.1^{\lambda_3}-0.1^{\lambda_4})]}{[(0.9^{\lambda_3}-0.9^{\lambda_4})-(0.5^{\lambda_3}-0.5^{\lambda_4})]}\;;\\
&\rho_4=\frac{Q(0.75)-Q(0.25)}{Q(0.9)-Q(0.1)}=\frac{[(0.75^{\lambda_3}-0.75^{\lambda_4})-(0.25^{\lambda_3}-0.25^{\lambda_4})]}{[(0.9^{\lambda_3}-0.9^{\lambda_4})-(0.1^{\lambda_3}-0.1^{\lambda_4})]}\,.
\end{aligned} \tag{4}$$

Establish the corresponding relationship between $\{\hat{\rho}_1, \hat{\rho}_2, \hat{\rho}_3, \hat{\rho}_4\}$ and $\{\rho_1, \rho_2, \rho_3, \rho_4\}$:

$$\rho_k=\hat{\rho}_k\,,\ k=1, 2, 3, 4 \tag{5}$$

It is difficult to directly solve λ_1, λ_2, λ_3, and λ_4. The initial values of λ_3 and λ_4 can be obtained by referring to the table in [9], and then applied Gauss-Newton method iteratively solve λ3, λ4 based on the function as below:

$$\rho_3=\hat{\rho}_3\,,\quad \rho_4=\hat{\rho}_4 \tag{6}$$

The objective function is as below:

$$\min[(\rho_3-\hat{\rho}_3)^2+(\rho_4-\hat{\rho}_4)^2]<\varepsilon \tag{7}$$

where, ε is the iteration stop condition parameter.

Substitute λ_3 and λ_4 into equation $\rho_2=\hat{\rho}_2$、$\rho_1=\hat{\rho}_1$, then λ_1, λ_2 solved. Substitute λ_1, λ_2, λ_3, and λ_4 into equation (1) and (2), the IRSS pseudo-failure life distribution obtained.

3. application of the proposed method in the reliability evaluation of industrial robot servo system

Table 2 shows the IRSS pseudo-failure life data based on the performance degradation data of a certain type IRSS. Suppose T_{w_i} is the IRSS pseudo-failure life data set under pseudo-failure level w_i ($i\in[1, 7]$), apply the Bootstrap method expand T_{w_1}, T_{w_2}, T_{w_3}, T_{w_4}, T_{w_5}, T_{w_6}, T_{w_7}. The repeated sampling times is 100, IRSS pseudo-failure life expanded data set $T^*_{w_1}$, $T^*_{w_2}$, $T^*_{w_3}$, $T^*_{w_4}$, $T^*_{w_5}$, $T^*_{w_6}$, $T^*_{w_7}$ obtained.

Table 2 IRSS pseudo-failure life data (unit: h)

IRSS No.	Pseudo-failure level						
	w_1	w_2	w_3	w_4	w_5	w_6	w_7
1	104.08	232.42	405.09	542.07	703.75	852.99	987.99

2	108.08	266.73	410.62	558.09	707.83	841.19	991.83
3	101.33	247.65	411.56	550.13	698.85	865.74	985.56
4	89.74	244.16	435.11	570.63	721.53	837.22	989.39
5	116.09	261.53	416.84	551.88	693.67	848.89	987.72

Apply the GLD method to obtain the IRSS pseudo-failure life distribution based on the IRSS pseudo-failure life expanded data set. Table 3 shows the GLD parameters of each pseudo-failure level w_i ($i \in [1, 7]$).

Table 3 The GLD parameters of each pseudo-failure level

GLD parameters	Pseudo-failure level						
	w_1	w_2	w_3	w_4	w_5	w_6	w_7
λ_1	103.6518	259.1295	422.5463	557.4715	691.5579	840.7216	985.9855
λ_2	0.0391	0.0247	0.0122	0.0210	0.0236	0.0248	0.1087
λ_3	0.2966	0.4433	0.3417	0.1770	0.0060	0.2372	0.0361
λ_4	0.2474	0.1421	0.0799	0.1525	0.5111	0.5904	0.5945

Extrapolate the GLD parameters under IRSS failure threshold w_{th}, they are λ_1=11237.2016, λ_2=0.0363, λ_3=0.2197, and λ_4=0.3311. Substitute λ_1, λ_2, λ_3, and λ_4 into equation (1) and (2), IRSS failure life distribution quantile function $Q(u)$ ($u \in [0, 1]$) and probability density function $f(t; w_{th})$ are obtained.

$$Q(u) = 11237.2016 + \frac{1}{0.0363} \cdot [u^{0.2197} - (1-u)^{0.3311}];$$

$$f(t; w_{th}) = \frac{0.0363}{\left[0.2197u^{0.2197-1} + 0.3311(1-u)^{0.3311-1}\right]} \quad t = Q(u);$$

Figure 2 (a) and (b) shows the IRSS failure life distribution probability density function diagram and failure distribution diagram. IRSS failure life $T_{failure}$, characteristic life $T(e^{-1})$ and median life $T(0.5)$ are 11264.75h, 11238.96h and 11242.32h respectively, which are close to the IRSS design failure life 10000h. It shows the validity of the proposed pseudo-failure life distribution reliability estimation method based on GLD.

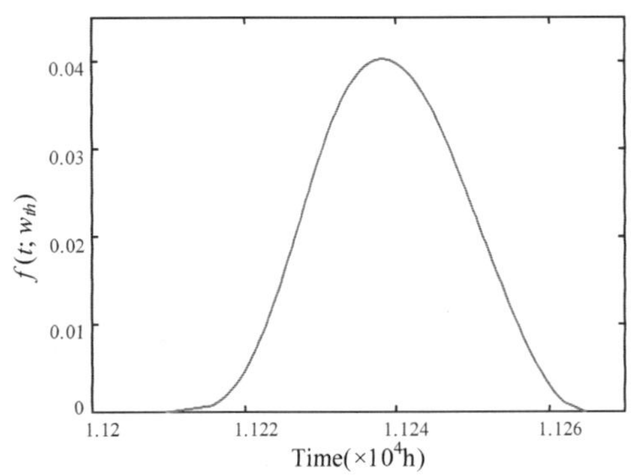

(a) IRSS failure life distribution probability density function diagram

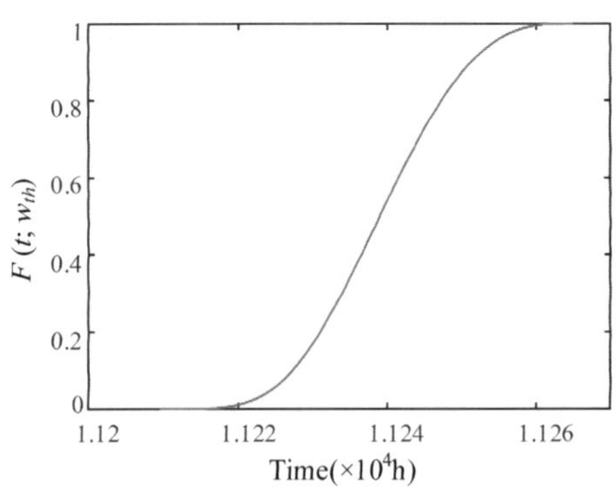

(b) IRSS failure distribution diagram

Figure 2. IRSS reliability evaluation result

4. conclusion

A reliability evaluation method of pseudo-failure life distribution for Industrial Robot Servo System (IRSS) based on The Generalized Lambda Distribution (GLD) was proposed in the paper addressed to the issue of insufficient reliability test data due to the lack of test samples and limited test time. The proposed method has no need to preset the life distribution, overcome the problem of the lack of prior distributions for the IRSS life distribution. The reliability evaluation of IRSS performance degradation is carried out by experiments, and the reliability evaluation method of IRSS pseudo-failure life distribution based on GLD series method proposed in this paper is applied. The reliability evaluation result is: $T_{failure}$= 11264.75 h, $T(e^{-1})$= 11238.96 h, $T(0.5)$= 11242.32 h, which are closed to the IRSS designed failure life, indicating the validity of the proposed pseudo-failure life distribution reliability estimation method based on GLD.

Acknowledgment

This work is supported by Guangdong High-end Equipment Manufacturing Plan Project (Grant no. 2017B0909 14003).

References

[1] CAI B P, SHENG C Y, GAO C T, et al. Artificial Intelligence Enhanced Reliability Assessment Methodology with Small Samples [J]. IEEE Transactions on Neural Networks and Learning Systems, 2021: 13.
[2] LIU G X, HE J L, YU R B. Research status and penetration of reliability evaluating methods on PV modules[J]. Modern Manufacturing Engineering, 2014(12) : 123-126.
[3] LIU D T, ZHOU J B, PENG Y. Data-driven prognostics and remaining useful life estimation for lithium ion battery: A review [J]. Instrumentation, 2014, 1(1) : 59-70.
[4] Behzad M, Arghan H A, Bastami A R, et al. Prognostics of rolling element bearings with the combination of Paris law and reliability method [C]. Prognostics and System Health Management Conference, 2017.
[5] Sikanen E, Nerg J, Heikkinen J E, et al. Fatigue life calculation procedure for the rotor of an embedded magnet traction motor taking into account thermomechanical loads [J]. Mechanical Systems and Signal Processing, 2018, 111: 36-46.
[6] Banerjee A, Gupta S K, Putcha C. Degradation Data-Driven Analysis for Estimation of the Remaining Useful Life of a Motor [J]. ASCE-ASME Journal of Risk and Uncertainty in Engineering Systems Part a-Civil Engineering, 2021, 7(2).1
[7] Kelley J, Hagan M. New Fault Diagnosis Procedure and Demonstration on Hydraulic Servo-Motor for Single Faults [J]. IEEE-ASME Transactions on Mechatronics, 2020, 25(3): 1499-509.
[8] Gebraeel N Z, Lawley M A. A neural network degradation model for computing and updating residual life distributions [J]. IEEE Transactions on Automation Science and Engineering, 2008, 5(1): 154-63.
[9] Karian Z A, Dudewicz E J. Fitting Statistical Distributions: The Generalized Lambda Distribution and Generalized Bootstrap Method [M]. USA: CRC Press, 2000.
[10] Lin Honghua. Measurement Error Analysis and Some Key Points of Data Processing (Part 3) -- The Recommended Application of Stochastic Distribution Method [J]. Automation and information engineering, 2020,41(03):1-6+16.

Research on Time Sidelobe Analysis on Pulse Compression Signal

Bocheng Wang[a], Xiaolong Chen*[a], Yue Li[a], Qinshan Zhou[b], Yifan Li[a]

[a]Xi'an Key Laboratory of Intelligent Instrument and Packaging Test, Xidian University, Shaanxi, 710071, China

[b]China Electronics Technology Instruments Co., Ltd, Qingdao ,266555, China

[*]Corresponding author: xlchen@mail.xidian.edu.cn

Abstract. As a measuring instrument, the signal analyzer is widely used in communication, microwave, aviation and other fields. It can not only measure the quality of the input signal, but also quickly demodulate the measured signal to obtain the information contained in the signal. In this paper, a method of pulse compression time sidelobe analysis applied to the signal analyzer is proposed to analyze and test the pulse measurement signal. First, the correlation between the pulse measurement signal and the pulse reference signal is processed by pulse compression. Secondly, the sidelobe suppression analysis is carried out by using the window function method for the reference signal. Finally, according to the pulse compression correlation amplitude results, the main side lobes are separated, and the time side lobes are further analyzed and the corresponding parameters are calculated. The simulation results show that the root mean square error of parameter calculation is less than 5%, which meets the standard of signal analyzer for pulse compression time sidelobe analysis.

1. Introduction

Signal analyzer as a multifunctional measuring instrument can complete the signal vector analysis, not only from the time domain, frequency domain to test the signal, but also can demodulate the digital modulation signal measurement, widely used in wireless communication testing [1]. For the radar pulse compression signal, a high range sidelobe will be generated when the receiver performs matching processing on the echo signal, which will cause interference to the echo signal or the sidelobe of multiple target signals will form false targets [2-3], thus making it impossible to accurately test the target signal.

In order to enable the vector signal analyzer to test and analyze radar echo information more accurately and reduce the influence of range sidelobe on signal detection, early scholars proposed adaptive pulse compression technology [4-5] based on minimum mean square error to extract target information through iterative processing. The disadvantage of this method is that it requires a large amount of calculation. Some scholars have proposed the method of iterative least squares [6-7] to measure the pulse compression signal. In recent years, some scholars have also used floating point genetic algorithm [8] to generate filter coefficients. The above method involves a lot of iteration and search process and is difficult to implement in vector signal analyzer.

This paper is mainly divided into five parts. The first part introduces the significance of pulse compression time sidelobe analysis. The second part describes the relative processing of the pulse compression reference signal and the measurement signal, and uses the method of adding a window function to the reference signal to suppress the side lobes. The third part is to perform time side lobe analysis to obtain the main lobe and side lobe positions of the single-pulse correlation amplitude. The fourth part verifies the algorithm by MATLAB simulation and calculating the corresponding time sidelobe parameters. The fifth part summarizes the full text.

2. Pulse Analysis Process

This paper is mainly to analyze and test the time side lobes of the pulse compression signal, and it is necessary to perform pulse compression correlation processing on the measurement signal first. Pulse compression is achieved by measuring the correlation of the pulsed signal with a reference paulsed signal, which is usually the transmit signal. The specific operation process is shown in Figure 1:

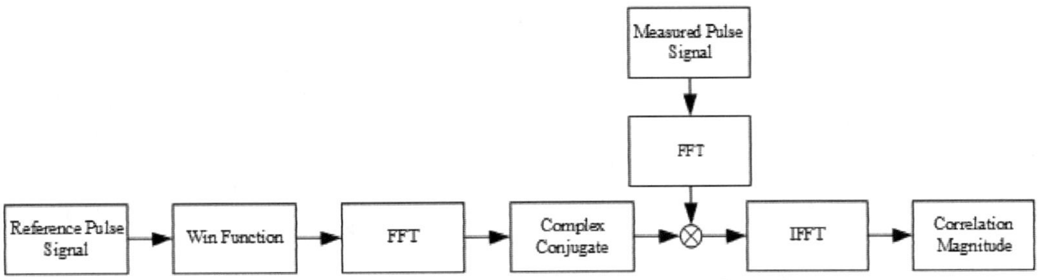

Figure 1. Pulse compression related processing flow chart.

2.1. Pulse Compression related processing

In order to reduce the computation in the process of convolution, the Inverse Fast Fourier Transform (IFFT) of frequency domain multiplication is used to obtain the convolution result in time domain. Firstly, the reference signal and measurement signal are transformed into frequency domain by the Fast Fourier Transform (FFT). The complex conjugate operation is then performed on the frequency domain data of the reference signal and the result is multiplied by the frequency domain data of the measured signal. Then the IFFT is used to convert the frequency domain signal into the time domain signal, and the calculated result is the relevant IQ data. Finally, the square of IQ signal amplitude is calculated by the relevant IQ data, and the relevant amplitude results of pulse compression are obtained. The correlation between the measurement signal and the reference signal is expressed by the following formula:

$$P_{corr}(n) = \left| \sum_{k=1}^{N} IQ_{meas}(t_{k+n}) \cdot (IQ_{ref}(t_k))^* \right|^2 \tag{1}$$

where n is the sample offset of the signal sampling point after the measurement signal correlation calculation. For the measurement signal $IQ_{meas}(t_n)$ sampling point n is $1 \le n \le M$, where M is the total sampling points of the measurement signal. The k value range of the reference signal $IQ_{ref}(t_k)$ is $1 \le k \le N$, where N is the total sampling points of the reference signal.

The above correlation operation is equivalent to the linear time-invariant filtering operation, and the impulse response of the filter is the time-reversal and complex conjugate result of the reference signal. As shown in formula (2):

$$h(t) = R(-t)^* \tag{2}$$

The above correlation process is also equivalent to the convolution process of the measurement signal and the matched filter, as shown in equation (3):

$$y_{corr}(t) = h(t) \otimes IQ_{meas}(t) \tag{3}$$

2.2. Analysis of Sidelobe Suppression

The algorithm can be used to determine the correlation amplitude of the pulse compression signal and extract the correlation amplitude of a single pulse. Each pulse will have a peak position, in which the index position of the largest peak is called the main lobe, and the index position of the other peak is called the side lobe. The main lobe usually carries the echo information of pulse compression signal. In order to accurately obtain the signal, it is necessary to suppress the sidelobe [9]. In this paper, the sidelobe suppression is performed by adding window function before input reference pulse signal. The window functions that can be added are rectangular window, Hamming window, Hanning window, Blackman window and Bartlett window. Figure 2 shows the spectral characteristics of several window functions:

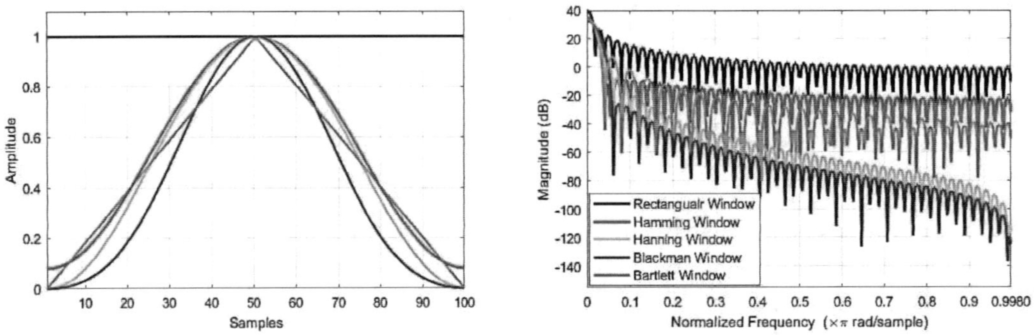

Figure 2. Spectral characteristics of window function.

3. Pulse Compression Time Sidelobe Analysis Process

The single-pulse correlation amplitude result is obtained by the above algorithm, and the time side lobe parameters are calculated according to the main side lobe characteristics. The specific operation process is shown in Figure 3:

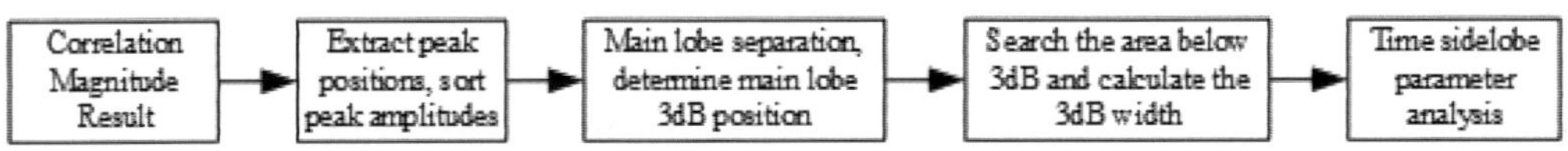

Figure 3. The pulse compression time sidelobe analysis process.

Firstly, according to the relevant amplitude results calculated by the above algorithm, the position of peak point n_{peak} is determined according to Formula (4). The maximum peak point is the peak position of the main lobe $n_{peak,m}$, and the secondary peak point is the maximum peak position of the sidelobe $n_{peak,s}$.

$$p(n_{peak}) \times diff(p(n_{peak}+1)) \leq 0 \tag{4}$$

After determining the peak position of the main lobe, it is necessary to sort the peak amplitude and separate the main side lobe. The next step is to determine the 3dB width of the main lobe as the left and right end points, so as to calculate the 3dB width of the main lobe. According to Formula (5), all the points below 3dB of the main lobe are determined, and the points less than $n_{peak,m}$ are the left end of the region below 3dB. Greater than $n_{peak,m}$ is the right end of the region below 3dB. The last point of the left part is the left endpoint of 3dB width $n_{3dB,l}$, and the right part is the right endpoint of 3dB width $n_{3dB,r}$. The 3dB width can be calculated according to Formula (6).

$$p(n_{3dB,under}) \leq p(n_{peak,m}) - 3 \tag{5}$$

$$W_{3dB,m} = (n_{3dB,r} - n_{3dB,l}) \times ts \tag{6}$$

4. Simulation Test

The input signal is a linear frequency modulation pulse signal, a total of ten groups of pulses. The above algorithm process is used to perform pulse compression processing on input signals, and then the relevant amplitude results of pulse measurement signals are obtained, as shown in Figure 4:

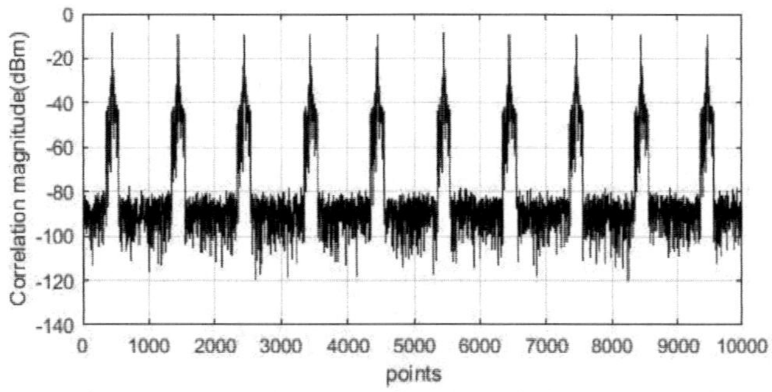

Figure 4. Pulse Compression Correlation Amplitude

According to all the pulse correlation amplitude results of the measured signal, the single pulse correlation amplitude results are extracted, as shown in Figure 5:

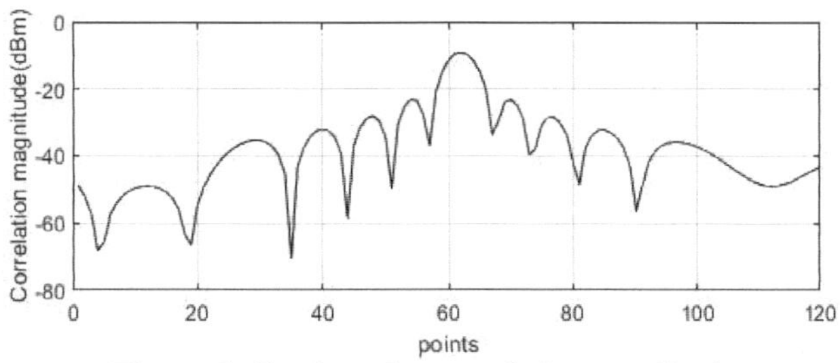

Figure 5. Single-pulse correlation amplitude.

The maximum peak of the correlation amplitude results in Figure 5 is called the main lobe, and the other peak positions are called side lobes. Use the window function listed above to add window function to the pulse pressure reference signal, and analyze the sidelobe suppression of the relevant amplitude results as shown in Figure 6:

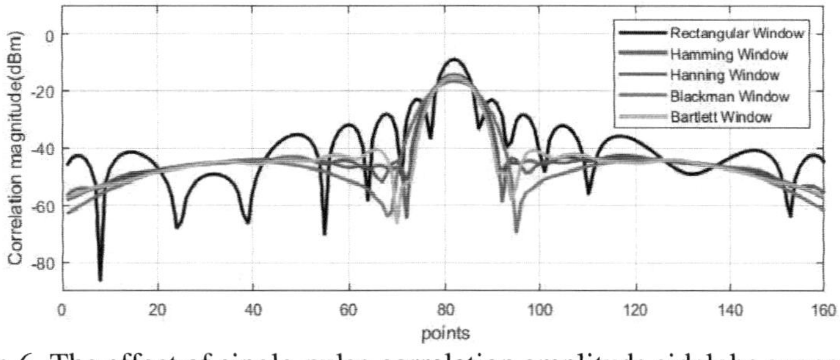

Figure 6. The effect of single-pulse correlation amplitude sidelobe suppression.

It can be seen from the relevant amplitude results that the side-lobe suppression effect is the same as the theoretical analysis. The main lobe width of rectangular window is the smallest, while the side-lobe suppression effect of Blackman window is the best, but the width of the main lobe is widened. The time sidelobe analysis algorithm was used to index the peak waveform and separate the main sidelobe. As shown in Figure 7, red represents the peak value of the main lobe, and the index position is the maximum index position of the peak value of the correlation amplitude of a single pulse. Blue indicates the maximum sidelobe peak, and the index position is the sub-maximum value of the index position of the peak correlation amplitude of a single pulse.

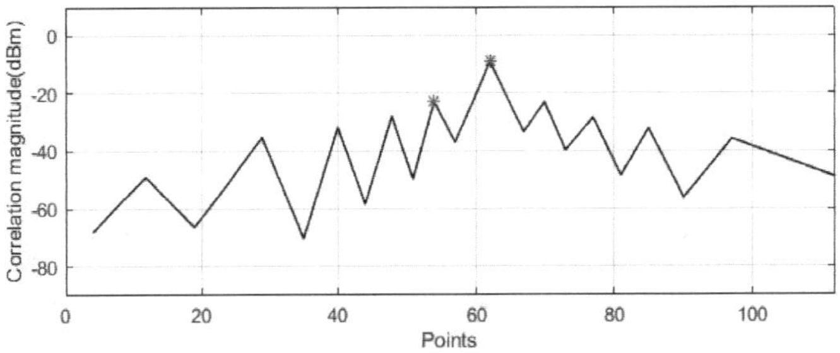

Figure 7. Main side lobe peak index.

After the main lobe peak index position is determined, the main side lobe separation of the correlation magnitude is achieved. According to the above algorithm, the 3dB width of the main lobe is calculated, and the result is shown in Figure 8:

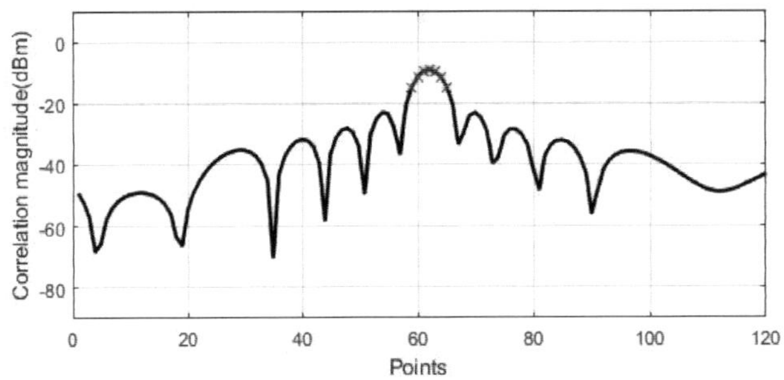

Figure 8. Main lobe 3dB width

The red asterisks in Figure 8 represent the sampling points of the correlation amplitude in the 3dB region of the main lobe, and the left and right endpoints of the 3dB region are used to calculate through equation (6). Using the R&S VSE pulse analysis function, an experimental test platform is built, and the above simulated pulse signal is input to test the time sidelobe parameters. Taking the VSE result as the true value of the signal parameter, the relative error analysis of the parameter result obtained from the simulation is carried out. The error results of the time sidelobe parameters are shown in Table 1:

Table 1. Measurement results and error analysis.

Parameters	Relative error (%)	Simulation Results	R&S Result
Peak Sidelobe Leve	0.361	-13.9380 (dB)	-13.8879 (dB)
Integrated Sidelobe Level	3.356	-9.0115 (dB)	-8.7189 (dB)
Main lobe 3dB Width	0.141	6.0e-08 (s)	5.9916e-08 (s)
Sidelobe Delay	0.458	-8.0e-08 (s)	-7.9635e-08 (s)
Compression Ratio	0.875	16.6667 (1)	16.5221(1)
Main lobe Power (Integrated)	0.677	-14.2493 (dBm)	-14.1535 (dBm)
Main lobe Power (Average)	0.906	-34.2493 (dBm)	-33.9418 (dBm)
Peak Correlation	2.678	0.0030 (1)	0.00292 (1)

5. Conclusion
In order to realize the analysis and test of the pulse compression signal by the signal analyzer, the Fast Fourier Transform method is used to carry out the correlation operation in the pulse compression processing part, so as to reduce the number of iterations and algorithm complexity of the convolution process in time domain. For the time sidelobe analysis of pulse compression signal, the sidelobe suppression is carried out by window function method suitable for signal analyzer to reduce the influence of sidelobe effect on measurement parameters. It can be seen from the results that the main lobe width of the rectangular window is the smallest, and the side lobe suppression effect of the Blackman window is the best, but the main lobe width is broadened. Finally, the main sidelobe is separated, and the error of the result is less than 5% by measuring the time parameter of pulse compression, which realizes the time sidelobe analysis function of the pulse compression signal, and has certain engineering application value.

Acknowledgments
This work has been financially supported by the National Natural Science Foundation of China with Grant No. 61727804

References
[1] Li, H and Ye, WH. 2015. Study of Digital Modulation Signal Error Vector Magnitude Based on Vector Signal Analyser. In: *2015 8th International Symposium on Computational Intelligence and Design (ISCID)*, Hangzhou, China, pp. 100-103.
[2] M. J. Lindenfeld. (2021) Mismatched Filters for Incoherent Pulse Compression in Laser Radar. In IEEE Transactions on Aerospace and Electronic Systems, vol. 57, no. 2, pp. 1252-1260.
[3] Dai J, Hao XH, Li P, Li Z, Yan XP. (2019). Antijamming Design and Analysis of a Novel Pulse Compression Radar Signal Based on Radar Identity and Chaotic Encryption. In *IEEE Access*, vol. 8, pp. 5873-5884.
[4] S. D. Blunt and K. Gerlach. 2004. Joint adaptive pulse compression to enable multistatic radar. In: 2004 International Waveform Diversity & Design Conference, Edinburgh, UK, pp. 1-5.
[5] S. D. Blunt and T. Higgins. (2010) Dimensionality Reduction Techniques for Efficient Adaptive Pulse Compression. In IEEE Transactions on Aerospace and Electronic Systems, vol. 46, no. 1, pp. 349-362.
[6] A. J. Zejak, S. Simić and I. S. Simić. 2011. Doppler optimized iterative reweighed lest square mismatch pulse compression algorithm realization on the FPGA. In: 2011 10th International Conference on Telecommunication in Modern Satellite Cable and Broadcasting Services (TELSIKS), Nis, Serbia, pp. 567-570.
[7] Cheng RC. 2010. Range profile specific optimal waveforms for minimum mean square error estimation. In: 2010 IEEE Radar Conference, Arlington, VA, USA, pp. 1042-1047.
[8] Zhao H, Ye WB, Yu YJ. 2013. Sparse FIR filter design based on Genetic Algorithm. In: *2013 IEEE International Symposium on Circuits and Systems (ISCAS)*, Beijing, China, pp. 97-100.
[9] Zhang W, Wang J and Liu X. 2020. Adaptive Sidelobe Suppression of SAR Images with Arbitrary Doppler Centroids and Bandwidths. In: *IGARSS 2020 - 2020 IEEE International Geoscience and Remote Sensing Symposium*, Waikoloa, HI, USA, pp. 909-912.

Fault injection and diagnosis of the centrifugal fan

Jing Tang[a, b], Guangkuo Guo[a], Ji Wang[a], Wei Xu*[c]

[a]The Fifth Electronics Research Institute of Ministry of Industry and Information Technology, Guangzhou 511370, China

[b]Guangdong Provincial Key Laboratory of Electronic Information Products Reliability Technology, Guangzhou 511370, China

[c]No. 719 Research Institute of China Shipbuilding Industry Corporation, Wuhan 430205, China

* Corresponding author: 46658173@qq.com

ABSTRACT: Centrifugal fan is widely used in industry, vehicle and ship application to provide air cycle power. To detect the fan fault at early stage, this paper explores the fault feature in the vibration signal and fault diagnosis method through the fault injection test, where impeller unbalance, rotor unbalance, and bearing fault are involved. Firstly, the fault mechanism is introduced; then, the fault injection methods of the three faults are introduced to obtain the faulty fans; finally, health and fault tests are performed, where the vibration sensors are employed and distributed on the fan in the axial and radial direction. Moreover, the acquired vibration data is analyzed by using time-domain and frequency-domain methods, and further the fault features between health and fault are compared and discussed. The analysis results indicate that the three faults can be detected through the vibration intensity and magnitude comparison at their individual characteristic frequencies in the spectrum.

1. Introduction

Fan is a fluid machine that converts the mechanical energy of the prime mover to the pressure energy and kinetic energy, which is a typical rotating machine [1]. Fans are widely used in various applications, such as industry, electric power, and agriculture, which are of great significance to the economy development [2]. The condition monitoring, analysis and fault diagnosis of rotating machines are popular in recent years. Fans usually work in harsh environment, leading to its faults. When the fan fault occurs, the machine and even the factory will be shut down, resulting in economic losses and even serious consequences such as personal safety damage and environmental pollution [3]. Therefore, it is very important to develop the condition monitoring and fault diagnosis of the fan to ensure the operation reliability of the fan.

Generally the work environment of the fan is very harsh, which will lead to fan faults. The failures of the fan can be divided into three types: motor-related faults, the impeller-related faults, the others [4]. The main causes for the failure include unbalance, shaft bending, eccentricity, rotating stall and surge, friction and looseness, rotor cracks, seal instability and rolling bearing failures [4].

Fault diagnosis methods of the rotating machine are developed in recent years, where the fault diagnosis method based on the vibration signal analysis is regarded as the most useful method for the mechanical faults [5]. The vibration based method usually extracts the fault features in the vibration signal by using the time domain [6], [7] and frequency domain method [8], and then diagnoses the fault through the threshold comparison. Vibration intensity and kurtosis are the representative fault features in the time domain, and the vibration magnitudes at the fault frequencies in the spectrum are the characteristics in the frequency domain.

This paper researches on the fault characteristics of the impeller unbalance, rotor unbalance and bearing fault of the fan in the vibration signal, and then the fan faults are detected through the threshold comparison. The following sections are organized as: Section 2 introduces the basic structure and the failure mechanism of the three faults; Section 3 introduces the fault injection methods for the three faults by considering the fault severity and location; In section 4, the time domain and frequency domain analysis methods are adopted to extract fault features from the sampled data, and then the results of health and fault under different work condition are compared and discussed to obtain the sensitive measurement points and fault features of the fan, which can be used for the fan fault diagnosis.

2. Fault mechanism for the centrifugal fan

2.1 Basic structure and fundamental

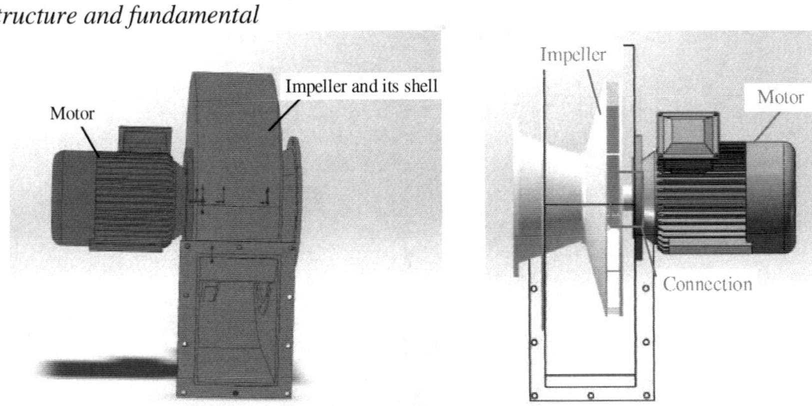

Figure 1. Fan structure.

The fan is mainly composed of motor, impeller and shell components, as shown in Figure 1, where the motor connects to the impeller directly through the shaft. When the motor is rotating, the connected impeller rotates. Then the impeller will produce centrifugal force to make air gather in the impeller case, where the air speed is slow and pressure is high. As a result, the air will be exhausted to the outlet pipe. Consequently, the negative pressure in the impeller pushes the outside new air to the impeller. Hence, the air flows in the pipe under the work of the centrifugal fan.

2.2 Impeller unbalance fault

Centrifugal fan impeller unbalance fault is mainly caused by the impeller wear. When the fan is working, the airflow mixed with dust enters the fan, and then the dust particles rub against the surface of the impeller. Furthermore, the particles collide with the impeller surface, resulting in the impeller to wear, whose degree depends on the impeller material and the hardness of the dust particles. Consequently, the erosion of dust particles on the surface of the impeller leads to impeller weight or volume loss, resulting in impeller unbalance fault.

2.3 Rotor unbalance fault

Rotor unbalance fault is one of the common faults in the centrifugal fan. It is usually caused by the factors of uneven mass distribution, errors in processing and assembly, corrosion and impurities. When the rotor is unbalanced, the vibration frequency of the fault is the rotational speed frequency, and the vibration amplitude increases with the increase of the rotational speed regardless of the load, and the vibration amplitude in the radial direction is larger than the amplitude in the axial direction.

2.4 Bearing fault
Bearing fault is a frequently fault in centrifugal fan. Rolling bearings are the key components that support the rotational of the motor. The inner and outer rings operate relative to each other and bear the combined load of axial and radial force at the same time. Long-term operation will lead to bearing damage. Wear, pitting, crack and surface peeling are the main failure modes for the bearing. When the bearing fault occurs, it will increase the vibration amplitude and noise of the bearing signal during its operation.

3. Fault injection considering severity and location

3.1 Impeller unbalance fault
To simulate the impeller unbalance fault, a 4.5g rubber block is stuck on the surface of the impeller at the outer edge, as shown in Figure 2. The dynamic data of the impeller before and after the fault injection is given in Table 1.

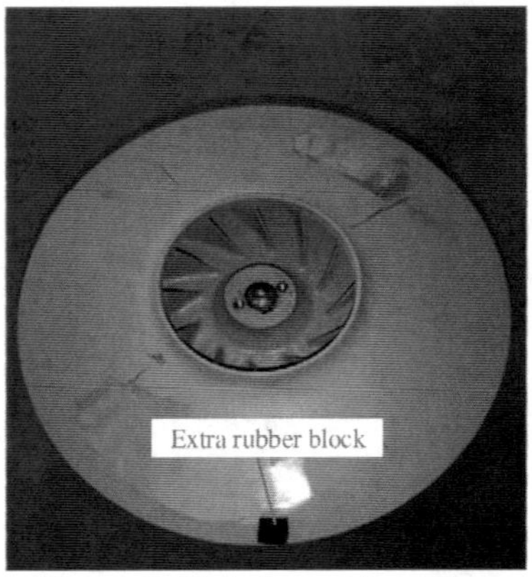

Figure 2. Unbalanced impeller.

Table 1. Dynamic data of the impeller before and after the fault injection.

Component	Health impeller		Impeller with fault injection	
	residual unbalance(I side)	residual unbalance(II side)	residual unbalance(I side)	residual unbalance(II side)
Impeller	8.56 g.mm	8.29 g.mm	16.9 g.mm	8.29 g.mm

It can be seen from Table 1 that the residual unbalance between I and II side has significant difference, and thus the impeller is unbalanced.

3.2 Rotor unbalance fault
To simulate the rotor unbalance fault, eight gaskets are riveted on two balance columns of the rotor, where the diameter of the gasket is 5 mm and total weight of the gaskets is 11 g. The dynamic data of the rotor before and after the fault injection is given in Table 2.

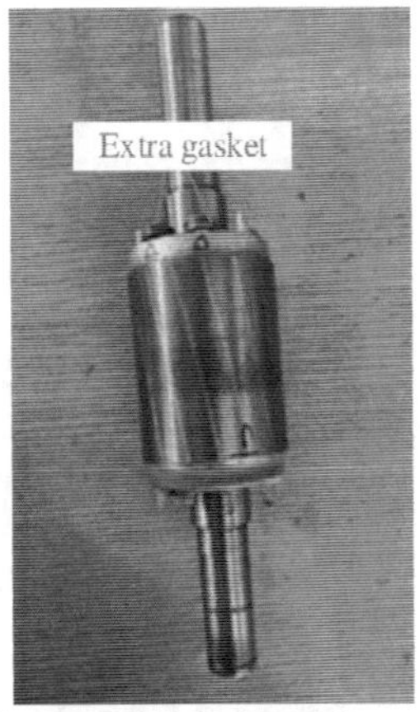

Figure 3. Unbalanced rotor.

Table 2. Dynamic data of the rotor before and after the fault injection.

Component	Health rotor		Rotor with fault injection	
	residual unbalance(I side)	residual unbalance(II side)	residual unbalance(I side)	residual unbalance(II side)
Rotor	6.5 g.mm	6.6 g.mm	12.8 g.mm	6.6 g.mm

According to Table 2, the residual unbalance of the I side is 12.8 g.mm whereas the residual unbalance of the II side is 6.6 g.mm, where the residual unbalance between I and II side has significant difference. This is because extra gaskets are added on one end of the rotor, leading to rotor unbalance. Hence the rotor is unbalanced.

3.3 Bearing fault

To simulate the bearing fault, a crack of 2 mm in width and 1.5 mm in depth is cut in the outer race of the bearing, as shown in Figure 4. When we roll the bearing slowly, it will feel slightly stuck. Therefore, the bearing has fault.

Figure 4. Bearing with outer race fault.

4. Experiments

4.1 Test bench

Figure 5. Test bench.

The test bench consists of test hanger, wind pipe and fan, where the fan is installed and fixed on the test hanger and the wind pipe is used to add load for the fan. There is an air door in the air inlet pipe, and thus the fan load can be adjusted by control the area of the door. Parameters of the fan involved in this paper are given in Table 3.

Table 3. Fan parameters.

Volume flow	Total pressure	Power	Speed	Noise
1000 m^3/h	2000 pa	1.5 kW	2840-2890 r/min	≤75 dB(A)

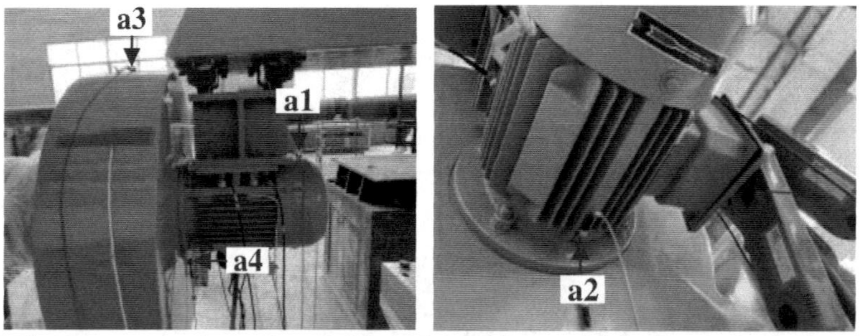

Figure 6. Vibration measuring points, where a1 and a2 are radial vibration acceleration sensor distributed at the motor; a3 and a4 are radial and axial vibration acceleration sensor distributed at the impeller respectively.

The vibration acceleration sensors are distributed in radial and axial direction on the fan, as shown in Figure 6. To demonstrate the fan fault under different work condition, the volume flow and pressure under different condition shown in Table 4 are performed in the experiment. The sampled frequency of the vibration signal is 10 kHz and sampled length is 10s, where it samples six groups of data for one work condition, as shown in Table 4.

Table 4. Acquired data groups under different work condition.

Work condition		Sampled data groups
Volume flow	Total pressure	
654 m^3/h	2380 pa	6 groups, from data 1 to data 6

1094 m^3/h	2123 pa	6 groups, from data 7 to data 12
1550 m^3/h	1643 pa	6 groups, from data 13 to data 18

The fan test can be performed according to the following steps:

a. Install the fan on the test rig and deploy the vibration sensors on the fan according to Figure 6.

b. Adjust the door area of the air inlet pipe to make the fan work in the condition shown in Table 4.

c. Acquire and store the vibration data when the fan works in steady state, where the sample frequency is 10 kHz and length is 10 s.

d. Repeat step b to c until vibration data under all work condition is obtained.

4.2 Data analysis result for the impeller unbalance fault

The time domain method is used to analyze the acquired vibration data. The features of health and impeller fault are observed and compared, and it finds that the intensity feature of impeller fault is significantly larger than the health. The intensity of the 18 groups of data is calculated and described as Figure 7, where figure 7 (a) is for the vibration a1, (b) is for the vibration a2, (c) is for the vibration a3, and (d) is for the vibration a4. It can be seen from the figure that the intensity of the impeller fault is larger than that of the health in all the work condition, where the increases of vibration a1, a2, and a3 are more obvious than a4.

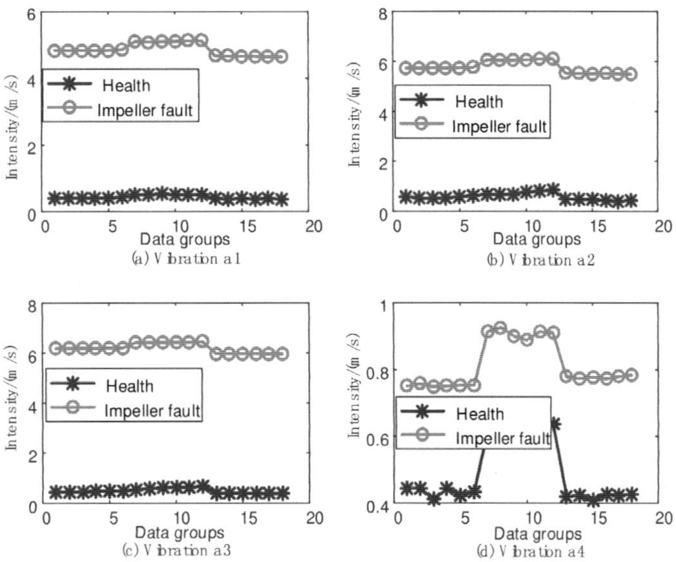

Figure 7. Vibration intensity of the health and impeller fault.

Furthermore, the Fast Fourier Transform (FFT) is used to analyze the vibration signal. By comparing the vibration spectrum of health and impeller unbalance fault, it can observe that the impeller fault has larger magnitude than the health at the rotational frequency. Hence, the magnitudes of all sampled data for health and impeller unbalance fault at the rotational frequency are calculated and shown in Figure 8. It can be seen from the figure that the all sampled data of the impeller fault has larger magnitudes than the health at the rotational frequency, and the results of vibration a1, a2, and a3 are more obvious.

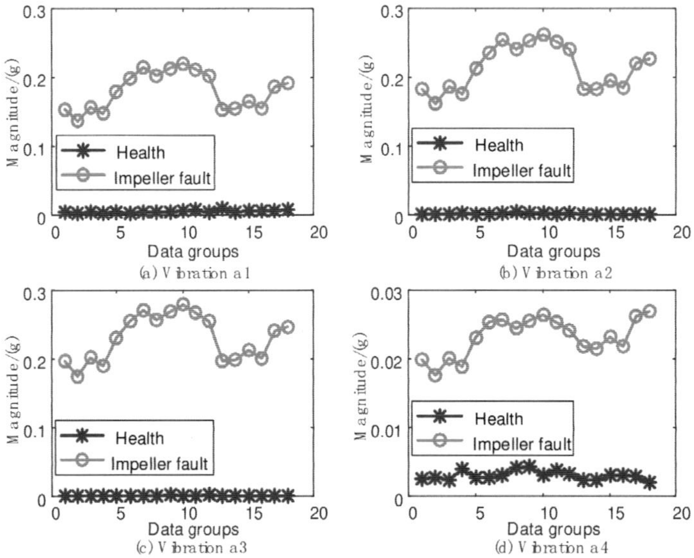

Figure 8. Vibration magnitude at the rotational frequency of the health and impeller fault.

4.3 Data analysis result of the rotor unbalance fault

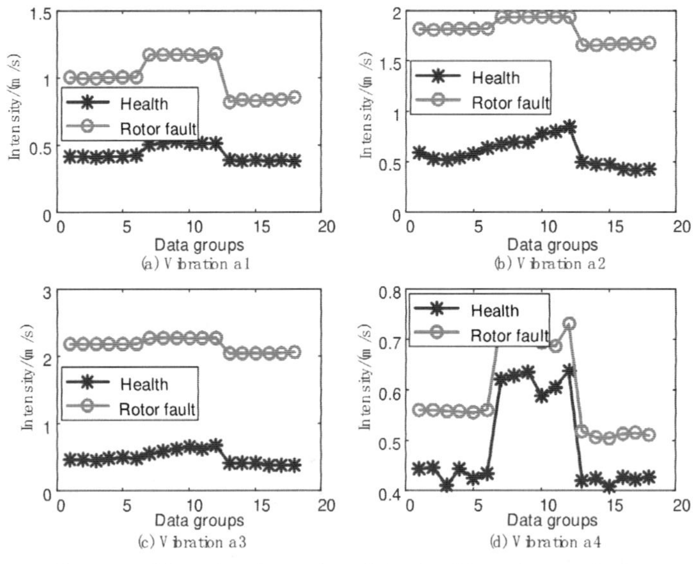

Figure 9. Vibration intensity of the health and rotor fault.

The intensity features of health and impeller fault are calculated and compared, and it finds that the intensity feature of rotor unbalance fault is significantly larger than the health. The intensity of the 18 groups of data is calculated and obtained as shown Figure 9. It can be seen from the figure that the intensity of the rotor fault is larger than that of the health in all the work condition, where the vibration a1, a2, and a3 are more obvious than a4.

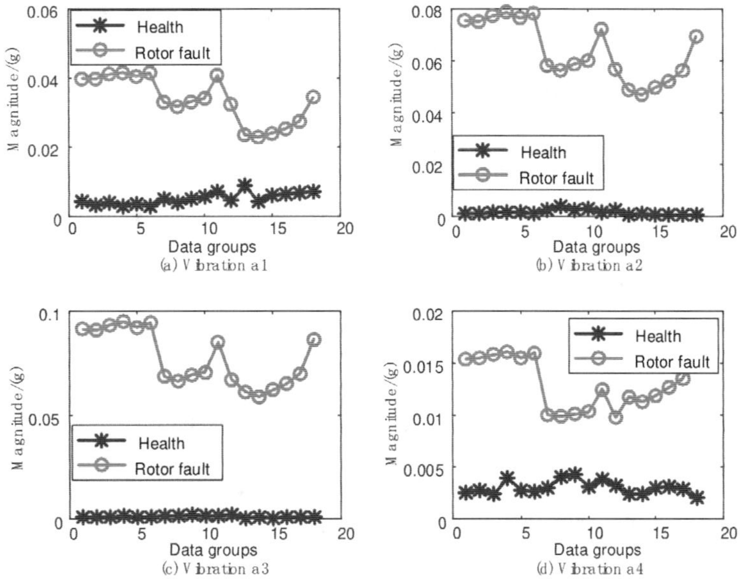

Figure 10. Vibration magnitude at the rotational frequency of the health and rotor fault.

By comparing the vibration spectrum of health and rotor unbalance fault, it can observe that the rotor fault has larger magnitude than the health at the rotational frequency. Hence, the magnitudes of all sampled data for health and rotor unbalance fault at the rotational frequency are calculated and shown in Figure 10. It can be seen from the figure that the all sampled data of the rotor fault has larger magnitudes than the health at the rotational frequency, and the results of vibration a1, a2, and a3 are more obvious.

4.4 Data analysis result of bearing fault

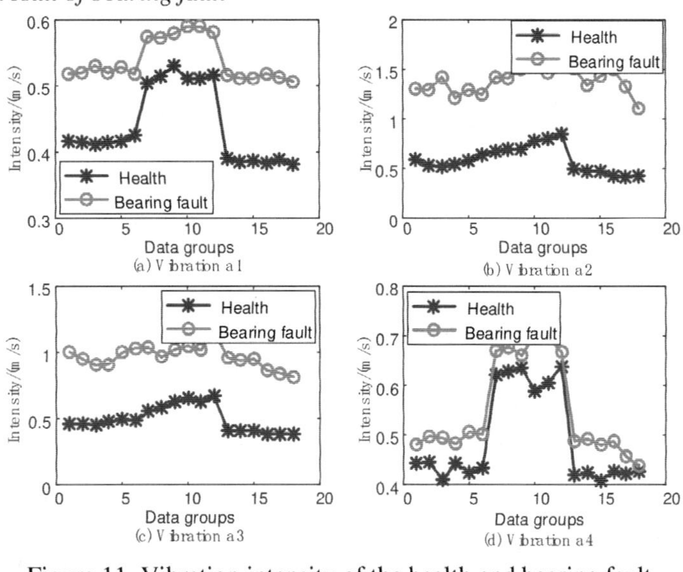

Figure 11. Vibration intensity of the health and bearing fault.

It can be seen from Figure 11 that the intensity of the bearing fault is larger than the health, and the vibration a2 and a3 are obvious in all vibration signal due to the distributed locations of these two vibration sensors are close to the bearing.

Table 5. Bearing parameters.

Diameter of outer ring D	Diameter of inner ring d	Width B	The number of the balls Z	ball diameter da
52mm	25mm	15mm	9	7.94mm

The bearing parameters are shown in Table 5, and then the bearing outer-race fault frequency can be calculated through the following formula

$$\text{Bearing fault frequency (Hz)} = \frac{Z}{2}\times\left(1-\frac{da}{Da}\cos\alpha\right)\times\frac{n}{60}$$

where Da is the pitch diameter and Da = (D+d)/2; α is the bearing contact angle; n is the rotational speed (r/min). According to the formula, the fault frequency is determined by the actual speed n. When the fan works in no load condition, the calculated fault frequency is about 178Hz. Hence the vibration spectra are analyzed and the magnitudes nearby 178Hz are extracted. The extracted magnitudes at the bearing fault frequency for all sampled data are shown in Figure 12. It can be seen from the figure that the magnitude of vibration a2 of the bearing fault is larger than the health, whereas this phenomenon is not obvious for the other vibration signal. This is because the measuring point a2 is close to the bearing.

According to the time domain and frequency domain analysis results, it can be known that the extracted features of measuring point a2 are larger than the health. Therefore, a2 is the sensitive measuring point of the bearing fault.

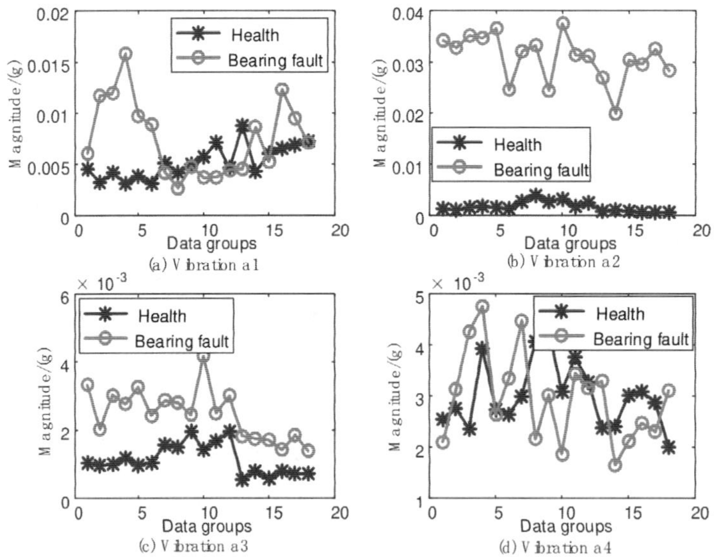

Figure 12. Vibration magnitude at the bearing outer fault frequency of the health and bearing fault.

5. Conclusion

Impeller unbalance fault, rotor unbalance fault, and bearing fault are investigated in this paper. The faults are injected in the components of the fan by considering fault severity and location. Experiments of health and fault are performed, where the vibration sensors are distributed at different location and direction on the fan. The acquired vibration data is analyzed by using the time domain and frequency domain method, and then the features of the health and fault are compared. The results can be summarized as

a. When the impeller unbalance fault occurs, the vibration intensity will increase and the magnitude of the vibration spectrum at the rotational frequency is larger than the health, where the vibration a1, a2, and a3 are valuable for the impeller unbalance fault diagnosis.

b. Similar to the impeller unbalance fault, the rotor unbalance fault will lead to the intensity increasing and the rotational frequency is the fault characteristic frequency in the vibration spectrum. The features of vibration a1, a2, and a3 are obvious, which can be used for the rotor unbalance fault diagnosis.

c. The intensity of the bearing fault is larger than that of the health, and the magnitude of the calculated frequency of the bearing fault becomes larger when bearing fault occurs. Both vibration a2 and a3 are the sensitive measuring points of the bearing fault, which are valuable for the fault detection.

Acknowledgement

This project is supported by the Science and Technology Program of Guangzhou (Grant No. 202201010298), the Key Research and Development Program of Shaanxi Province (Grant No. 2021GY-335), Ministry of industry and information technology project (Grant No. TC210804R-1).

References

[1] Yulin Wu, Qingguang Chen, Shuhong Liu. Fans and Compressors [M]. Beijing: Tsinghua University Press 2005. (in chinese)

[2] Lijun Duan, Liqing Miao. A New Method Exploration for Fan Fault Diagnosis [J]. Journal of science and technology. 2011(2): 101-102. (in chinese)

[3] Guangyang Wang, Yilian Zhou. Survey on Faults Diagnosis of Fan Vibration [J]. Journal of Anhui University of Technology. 2006. 23(1): 64-68. (in Chinese)

[4] He Ma. Research of Fault Diagnosis of Fans on Neural Network [D]. University of Science and Technology Liaoning, 2014. (in Chinese)

[5] Verbruggen TW. Wind turbine operation & maintenance based on condition monitoring WT-O [J]. Final report, ECN-C-03-047, April 2003.

[6] Bing Li, Rui Han, Yigang He, Xiaoyi Zhang, Jinbo Hou. Applications of the Improved Random Forest Algorithm in Fault Diagnosis of Motor Bearings [J]. Proceedings of the CSEE, 2020, 40(04): 1310-1319+1422. (in Chinese)

[7] Zhang Y, Gong Z W, Zheng Y, et al. Joint study on health state assessment and degradation trend prediction of industrial equipment. Sci Sin Tech, 2022, 52: 180–197. (in Chinese)

[8] Yupeng Sun, Xiaoming Xue, Yu Xu. Research on the vibration characteristics and rapid diagnosis method for marine motor fault [J]. China Equipment Engineering, 2022(01): 161-162. (in Chinese)

Time Difference Correction and Location Method for USB Asynchronous Microphone Arrays

Zhisheng Wang, Hanze Yu, Junning Cui

Center of Ultra-precision Optoelectronic Instrument engineering, Harbin Institute of Technology, Key Lab of Ultra-precision Intelligent Instrumentation (Harbin Institute of technology), Ministry of Industry and Information Technology, Harbin, 150080, China

*cuijunning@hit.edu.cn

Abstract: To address the problem of asynchronous delay caused by combining at least two array platforms in 2D/3D location, in this paper, we analyze the impact of time difference between asynchronous devices on time-delay estimation from the experimental phenomenon. Then, we propose a USB asynchronous correction algorithm which takes the known propagation delay as the benchmark, the time difference of asynchronous equipment is calculated by offseting the time delays on the left and right sides.The experimental results show it can correct the time difference for USB asynchronous microphone arrays, which achieve a great location accuracy.

1 Introduction

Acoustic location has been widely applied in many fields, including modern military reconnaissance, mobile communication, and so on.[1] Algorithms for sound source localization can be divided into three main categories: Time difference of arrival (TDOA) method [1-12]; high-resolution spatial spectral estimation based method[11]; and beamforming based method[12]. TDOA-based method is easy to implement in real-time in principle[4-6]. As a result, this paper is based on TDOA to carry out high-precision real-time localisation research.

The microphone array is usually used to determine the sound source location using the relative position relationship of microphone array elements to achieve high-precision 2D or 3D location, but the time difference between multiple asynchronous microphone arrays makes it impossible to estimate the exact estimated time delay, which affects localisation. Some methods for addressing asynchronous device time differences include using a distributed system node layout[18] and incorporating clock compensation into the cost function optimization process[14-15]. These methods can effectively reduce the impact of asynchronous device time differences on location accuracy, but they require a large amount of a priori data as support to build prediction models, and they typically use complex and computationally intensive algorithms. To that end, we employs a pre-designed independent array as the research platform for high-precision real-time sound source localisation, and it develops a USB asynchronous microphone sound source localisation algorithm for this platform to address the time difference problem between asynchronous devices.

2 Method

2.1 Hardware platform construction

We uses a pre-designed array as the base platform for high-precision real-time sound source localisation research, and then conducts localisation research on this platform. Regarding the array platform, it mainly adopts the general and convenient STM32F4 as the main control unit[1,10] and acquires two digital microphone signals by means of a peripheral SAI interface. The platform performs synchronous and omnidirectional acoustic wave signal acquisition on two channels at a high sampling rate of 96 kHz. Considering that the realization of two/three dimensional sound source localisation requires at least three simultaneous signal acquisitions, and in order to be able to carry out precise localisation research based on this platform, this paper investigates the realization of sound source localisation based on TDOA[1-12] by combining two array platforms, i.e., using four digital MEMS microphones to form an array model as the base platform. Based on TDOA, the paper investigates the accurate two-dimensional sound source localisation. The schematic diagram of the asynchronous array model developed is shown in Fig. 1.

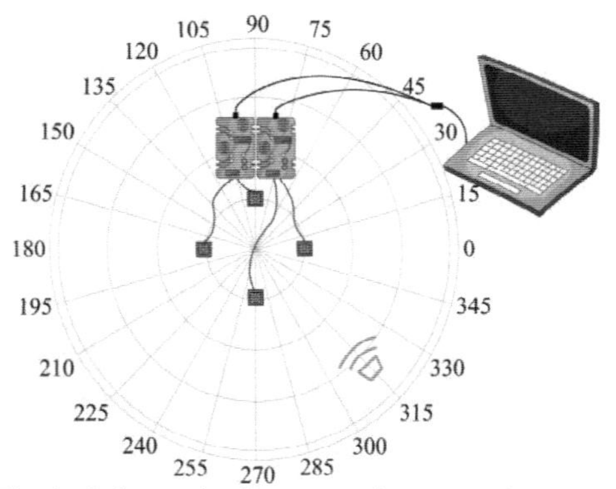

Fig 1. Schematic diagram of rectangular array

2.2 Time delay estimation and time difference problem between asynchronous devices

TDOA-based location is a method that uses the time difference between the signal and each array element for location. The time-delay estimation is the key to accurate localisation. The GCC-PHAT algorithm calculates the time delay between the two signals by directly calculating the cross-correlation coefficient of the two signals [1-12]. In practice, the correlation function is calculated using the inverse Fourier transform of the reciprocal power spectral density.

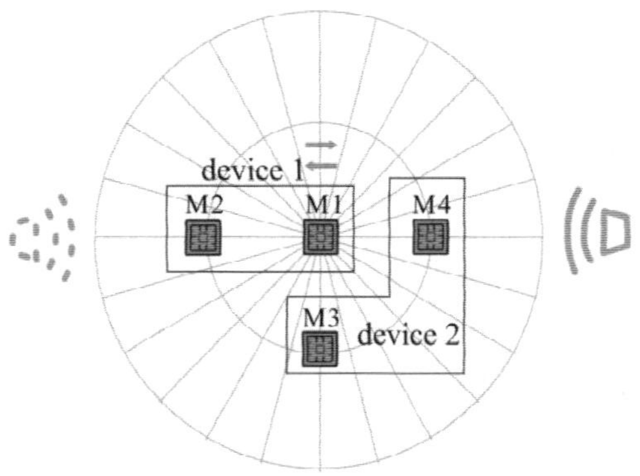

Fig 2. T-Array Schematic

As shown in Fig. 2, microphones M1-M4 form a T-array model, where M1 and M2 are connected to the two signal channels of device 1, M3 and M4 are connected to the two signal channels of device 2, and the sound source is placed in the positive direction of the X-axis, with M1 as the reference, and M2 and M4 are in the same line as M1, with the same relative distance.

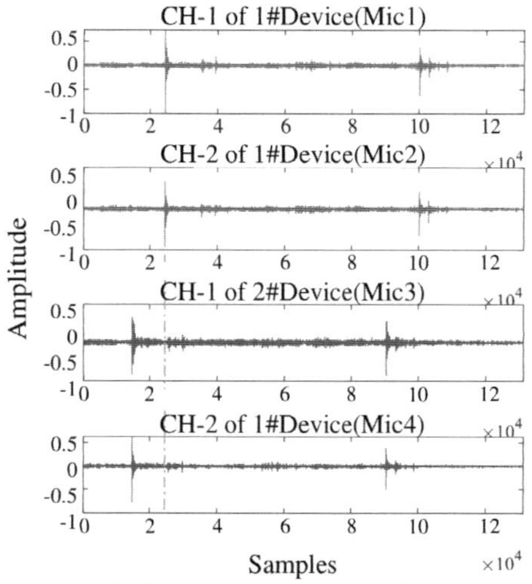

Fig 3. Four-way signal diagram

The four signals acquired by the array are shown in Fig 3. The time delay between two microphones M1 relative to M2 and M3 relative to M4 of the same device is found to be small. Because sampling delay exists, the time delay between two microphones M1 relative to M3 and M1 relative to M2 of different devices is greater. This is due to the time difference between the two devices, so the time difference between the acoustic signals received by the two microphones in different devices, in addition to the propagation time delay generated by the distance between the two microphones, there is also a time difference between the asynchronous devices.

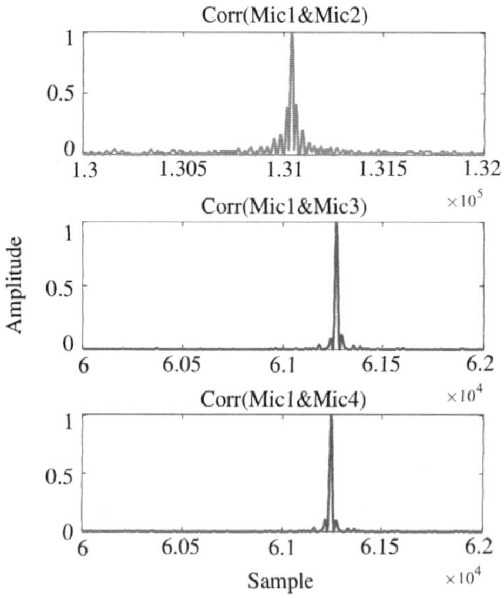

Fig 4. Cross-correlation curves

Using the microphone M1 data as a reference, the four signals are directly subjected to the mutual correlation operation, and three sets of mutual correlation sequences are obtained, as shown in Fig. 5. The delay points obtained from two microphones of different devices performing the mutual correlation operation have large deviations, resulting in incorrect time delay estimation and a significant impact on position estimation.

3 USB asynchronous microphone sound source localisation algorithm

3.1 Methodological framework

In this paper, while studying the implementation of a high precision real-time sound source localisation algorithm on the asynchronous USB microphone array platform, we propose a time difference correction algorithm between asynchronous USB microphone arrays for the time difference problem between asynchronous devices[3], followed by position estimation by obtaining accurate time delay values to achieve high precision real-time sound source localisation.

Fig. 5 shows the main technical routine of the sound location, which is divided into two parts: time delay and location estimation.

The recording data is first stored in a FIFO and pre-processed by band-pass filtering before the endpoint identification of the acoustic signal is performed using a combination of short-time over-zero rate and short-time energy[4] to determine whether it is a valid signal. The inter-correlation sequence between microphones is then calculated using the GCC-PHAT algorithm. The USB asynchronous correction algorithm is designed to compensate for the time difference between asynchronous devices in the total time delay, to ensure the accuracy of asynchronous array time-delay estimation. After correction, the accurate time-delay sequence is obtained by peak detection and outlier rejection of the resulting correlation sequence[5], which is fed into the spherical interpolation algorithm[14] , to obtain coordinates.

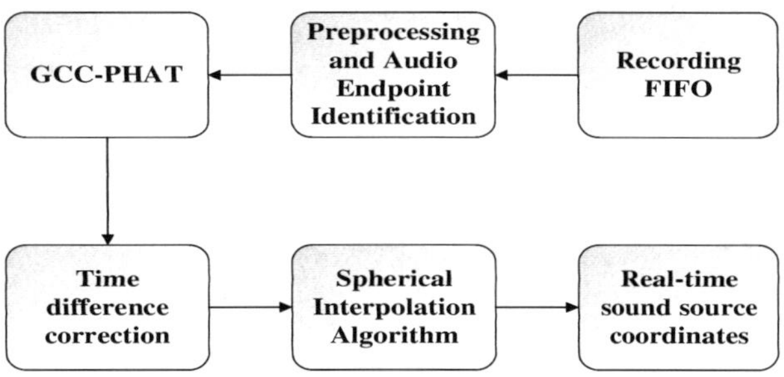

Fig 5. Location framework.

3.2 USB asynchronous correction algorithm

As mentioned in above Fig.2, taking the T-array model as an example, it is necessary to calculate the value of this time-delay and compensate for it in the total time delay. Let the TDOA between M1 and M4 be t_0 , and the sampling delay between device 1 and device 2 be t_1 . When the sound source is located on the left side of the line, the measured delay point delay is $\tau = t_0 + t_1$. When the sound source is located on the right side, the measured delay is $\tau' = -t_0 + t_1$, and the delay value of the sound source is located on the left and right side respectively, subtracted from $\tau_0 = \tau - \tau' = 2t_0$, which is two times the propagation delay of two microphones of the same device, by setting the initial distance between the two microphones, the propagation delay value is specified within the spacing range. When t_0 is within the spacing range, it means that the system has obtained the correct stable delay. At this time, the system completes the sound source left and right side correction, and then calculates $(t + t')/2$ to get the time difference between the asynchronous devices and compensates for the total time delay asynchronously.

The GCC-PHAT method introduced in Section 2.2 is executed to obtain the mutual correlation sequence. Before correction, the microphone array must continuously sample the acoustic signal to ensure that the propagation delay of M1 and M2 measured in the buffer is always within the spacing range, ensuring the reliability of the calculated asynchronous device time difference. Table 1 shows the basic USB asynchronous correction algorithm.

Table 1 USB asynchronous correction algorithm

1: Input: effective acoustic signal: Echo i, Echo j
2: Output: time difference between asynchronous devices middlelag
3: li ←max(gcc(Echo i, Echo j)) //gcc() is a correlation operation, max() is find the index of the peak.
4: lag i ← li - T //extract delay points, T is the half length of the correlation sequence.
5: if (-T1>=mode(lag i))&&(mode(lag i)>=-T2) //left side time-delay is within the spacing, T1 and T2 are values determined by known spacing
6: left_lag t←mode(lag t) // Obtain different delays when the sound is on the left
7: else if (T2>=mode(lag i))&&(mode(lag i)>=T1) //right side time-delay is within the spacing
8: right_lag t←mode(lag t) // Obtain different device time-delays when the sound is on the right side
9: if (left_lag t~=0)&&(right_lag t~=0)
10: deltlag t ←left_lag t-right_lag t //complete left and right side correction, left and right subtracted to get propagation time delay
11: if (deltlag t >= 2T2)&&(deltlag t <= 2T1) //different devices TDOA within the spacing
12: middlelag t← (right_lag t + left_lag t)/2 //get the time difference between asynchronous devices
13: if (rlag t<=T1) // time delay value is within the spacing range
14: Return middlelag t

3.3 Location estimation algorithm
As shown in Fig.8, the spherical interpolation method is an estimation algorithm based on the least square (LS) error function, which is suitable for near-field locaiton[8,10]. Its error criterion is the error function of the distance from the source to each microphone[8].

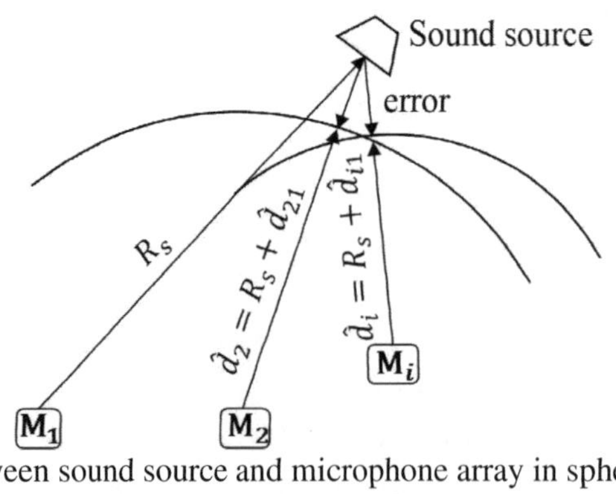

Fig.6 Position between sound source and microphone array in spherical interpolation.

In Fig.6, the $\hat{d}_i = R_s + \hat{d}_{i1}$ denotes the estimated distance from the sound source to the *ith* array elementM_i,R_s denotes the distance value from the sound source to the reference array elementM_1, and$\hat{d}_{i1}$ is the distance difference converted from the time-delay estimate, i.e., the distance difference between the soundM_1 and sourceM_i, respectively.

Using d_i to denote the true value of the distance of the sound source s to reach M_i, we satisfy $d_i^2 = \|\mathbf{r}_i - \mathbf{r}_s\|^2 = R_i^2 - 2\mathbf{r}_i^T\mathbf{r}_s + R_s^2$, where $\mathbf{r}_i$ and $\mathbf{r}_s$ denote the vectors of M_i and the sound source in the coordinate system, respectively, R_i and R_s denote the distances of M_i and the sound source relative to the coordinate origin, respectively. In summary, the spherical LS error function can be described as

$$\varepsilon_{sp,i} = \frac{1}{2}\left(\hat{d}_i^2 - d_i^2\right)$$
$$= \mathbf{r}_i^T\mathbf{r}_s + \hat{d}_{i1}R_s - \frac{1}{2}\left(R_i^2 - \hat{d}_{i1}\right) \quad (1)$$

Some errors can be expressed in matrix form as follows:

$$\boldsymbol{\varepsilon}_{sp} = \mathbf{A}\boldsymbol{\theta} - \mathbf{b} \quad (2)$$

where $\mathbf{A} = \left[\mathbf{S}|\hat{\mathbf{d}}\right] = \left[\mathbf{x}, \mathbf{y}, \hat{\mathbf{d}}\right]$, and $\mathbf{S} = [\mathbf{x}, \mathbf{y}]$ denotes the matrix of $M_2 - M_i$ coordinates, $\hat{\mathbf{d}} = \left[\hat{d}_{21}, \hat{d}_{31}, ..., \hat{d}_{i1}\right]^T$; $\boldsymbol{\theta}$ denotes the vector of the source to be estimated $\boldsymbol{\theta} = [x_s, y_s, r_s]^T$. $\boldsymbol{b} = 0.5[R_2^2 - \hat{d}_{21}^2,\ R_3^2 - \hat{d}_{31}^2,\ R_i^2 - \hat{d}_{i1}^2]^T$. $\mathbf{b} = 0.5\left[R_2^2 - \hat{d}_{21}^2,\ R_3^2 - \hat{d}_{31}^2,\ R_i^2 - \hat{d}_{i1}^2\right]^T$

The corresponding cost function of the spherical LS error can be expressed as:

$$J_h = \boldsymbol{\varepsilon}_h^T\boldsymbol{\varepsilon}_h = \sum_{i=1}^{N}\left[\frac{1}{2}\left(\hat{d}_i^2 - d_i^2\right)\right]^2$$
$$= [\mathbf{A}\boldsymbol{\theta} - \mathbf{b}]^T[\mathbf{A}\boldsymbol{\theta} - \mathbf{b}] \quad (3)$$

Given that R_s is known, the LS algorithm can be used to find the coordinates of the sound source $\mathbf{r}_s$ as:

$$\mathbf{r}_s = \mathbf{S}^{\dagger}(\mathbf{b} - R_s\mathbf{d}) \quad (4)$$

where $\mathbf{S}^{\dagger} = (\mathbf{S}^T\mathbf{S})^{-1}\mathbf{S}^T$ denotes the matrix $\mathbf{S}$ of the pseudo-inverse matrix.

The spherical interpolation method solves for the sound source coordinates by treating $\hat{R}_s$ and $\hat{\mathbf{r}}_s$ are considered as two independent and unconstrained unknown variables, and there is no constraint relationship[13]. Substituting Eq. (5) into $\mathbf{A}\boldsymbol{\theta} = \mathbf{b}$ gives

$$R_s\mathbf{P}_s\hat{\mathbf{d}} = \mathbf{P}_s\mathbf{b} \quad (5)$$

where $\mathbf{P}_s = \mathbf{I}_{N\times N} - \mathbf{S}\mathbf{S}^{\dagger}$ is the projection matrix; $\mathbf{I}_{N\times N}$ is the $N \times N$ unit matrix of size, $\hat{R}_s$ can be solved as follows

$$\hat{R}_s = \frac{\hat{\mathbf{d}}\mathbf{P}_s\mathbf{b}}{\hat{\mathbf{d}}^T\mathbf{P}_s\hat{\mathbf{d}}} \quad (6)$$

Substituting the above equation into equation (5), we obtain the estimated coordinates of the sound source $\hat{\mathbf{r}}_s$ as

$$\hat{\mathbf{r}}_s = \mathbf{S}^{\dagger}\left[\mathbf{I}_{N\times N} - \frac{\hat{\mathbf{d}}\hat{\mathbf{d}}^T\mathbf{P}_s}{\hat{\mathbf{d}}^T\mathbf{P}_s\hat{\mathbf{d}}}\right]\mathbf{b} \quad (7)$$

4 Experimental verification

As shown in Fig.7, to verify the accuracy of the USB asynchronous correction algorithm on the improvement of location, we design the experiment under an interior design experiment of 6.5 × 5 × 3.5 m^3. The interior floor is covered with drawn Cartesian coordinate drawings, and two arrays have four microphones. The distance between the microphones was set to 12 cm.

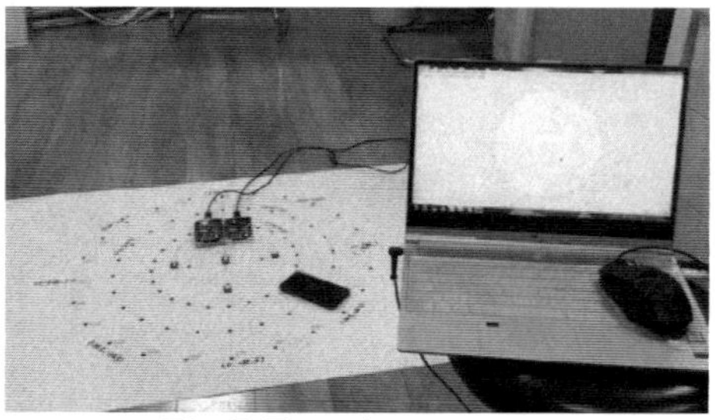

Fig 7. Experiment setup

Table 2 Distance estimation results for different sound source directions

Order number	Sound source distance/cm	Estimated distance/cm	Relative Error
1	8.65	8.20	5.22%
2	8.65	8.40	2.80%
3	17.3	17.28	0.13%
4	17.3	17.34	0.22%
5	25.95	25.68	1.04%
6	25.95	26.40	1.74%
7	34.60	35.26	1.90%
8	34.60	34.32	0.08%

A mobile phone was used to simulate the sound source, and signals were collected at 8.65 cm, 17.3 cm, 25.95 cm, and 34.6 cm from the origin of the coordinate system, and sound signals in each direction from 30° to 330°were collected at an interval of 15° when the distance was kept constant, without the acquisition of the 165°, 180°, and 195° directions. 18 acquisition points in each group, 3 rounds of complete experiments, acquisition of fixed points for time-delay estimation, and calculation of the absolute distance. Table 2 displays the relative error results for 4 groups of measured distance versus actual distance. Table 2 shows this system real-time accurate estimation of the sound source position in different distances and directions. By averaging multiple groups of relative errors, the average relative error is 1.64%.

5 Conclusion

Due to the time difference between the asynchronous devices, the asynchronous USB microphone arrays cannot be used to location accurately. In this paper, we propose a USB asynchronous correction algorithm to ensure the measurement accuracy. The experimental results show that this algorithm can achieve a great accuracy when using USB microphone arrays to location. However, we find that when the system operates continuously for a long time, some deviations will occur. In the future, we wound consider to design a robust long-term correction method.

References

[1] Li Yanshui, Shen Yi, Pan Xinyu, Sun Yunfei, Xia Wenhao, Guo Xiaofei. & Zhou Cen. (2021). Research on microphone array-based sound source localisation system. Internet of Things Technology (07), 26-28.

[2] Liu, Y.-Q., Wu, T.-M. & Li, C.-J. (2020). Construction of sound source localisation model based on microphone array. *Sino-Arabian Science and Technology Forum (in English and Chinese)* (08), 123-125.

[3] Kim, C.-W., Kang, M.-S., Sui, Q.-M., Sai, Y.-Z., Lu, S.-Z., Cao, Y.-Q.. & Jia Lei. (2013). Acoustic emission localisation technique based on generalized mutual correlation time delay estimation algorithm. Journal of Sensing Technology (11), 1513-1518.

[4] Qu, Shunbiao, Yu, Hua, Lu, Zhumao, Shen, Chong, Han, Yu & Wang, Chen-Guang. (2021). An improved generalized mutual correlation time delay estimation method for sound source

localisation. *Navigation Positioning and Timing* (06), 118-124.
[5] Guo, Longxia. (2021). A two-dimensional sound source localisation system based on mutual correlation algorithm. *China Science and Technology Information*(12),98-99.
[6] Yu WJ, He L, Cui LL, Li RB, Wu Qian. Research progress of time delay estimation methods in acoustic source localisation[C]//.Proceedings of the 2016 Annual Academic Conference on Acoustic Technology. ,2016:236-245.
[7] Yu Z, Fu X, Wang J, Hu ZL & Liu Y. (2011). Research on target tracking system based on acoustic wireless sensor network. Journal of the University of Electronic Science and Technology (04), 568-572.
[8] Liu Lulu. Research on sound source localisation technology based on microphone array [D]. Xi'an University of Electronic Science and Technology,2019.
[9] SMITH J O, ABEL J S. Closed-form least-squares source location estimation from range-difference measurements [J]. IEEE Transactions on Acoustics, Speech, and Signal Processing, 1987, 35 (12): 1661-1669.
[10] Wu X. P., Gu Z. H., Shu H. B., et al. A linear least squares method for precise localisation of sound source targets[J]. Journal of Acoustics, 2016(1):87-93.
[11] Takeda R ,Komatani K. Performance comparison of MUSIC-based sound localisation methods on small humanoid under low SNR conditions[C]// IEEE-RAS International Conference on Humanoid Robots. IEEE, 2015.
[12] Yardibi T, Li J, Stoica P , et al. Sparse representations and sphere decoding for array signal processing[J]. Digital Signal Processing, 2012.
[13]Yan Qingli,Chen Jianfeng. Distributed sound source localisation algorithm for sound velocity correction in wind farm environment[J]. Journal of Acoustics,2017,42(04):421-426.
[14] Zhu, S., & Ding, Z. (2010). Joint synchronization and localisation using toas: a linearization based wls solution. IEEE Journal on Selected Areas in Communications, 28(7), 1017-1025.
[15] Canclini, A., Antonacci, F. , Sarti, A. , & Tubaro, S. . (2013). Acoustic source localisation with distributed asynchronous microphone networks. IEEE Transactions on Audio Speech & Language Processing, 21(2).

Research on key technology of ultrasonic temperature measuring instrument in high-temperature solid structure

Yue Cui[a], Jing Xiang[a], Dong Wei*[a], Youan Shi[b], Yanxia Du [a], Yewei Gui[b]

[a] State Key Laboratory of Aerodynamics, China Aerodynamics Research and development Center, Mianyang, China 621000;

[b] Computational Aerodynamics Institute, China Aerodynamics Research and development Center Mianyang, China 621000;

[*] Corresponding author: ncu2016nwpu2020@163.com

Abstract: Aiming at the key technical requirements of ultrasonic measurement of internal temperature of high temperature solid structure, firstly, the mechanical structure of ultrasonic transducer is miniaturized and high-temperature resistant design is carried out, so that it can withstand the high temperature of 800℃. Meanwhile, the multi-waveform, multi-channel and multi-frequency excitation detection is achieved, and the integration of structure and function is realized. In addition, the high-speed data acquisition module adopts a 1GSPS sampling rate acquisition card combined with the phase-based acoustic time integral zero sub-measurement method to achieve the nanosecond acoustic time measurement accuracy required by ultrasonic temperature measurement. Finally, the reconstructed internal temperature field of steel structure is carried out by using the developed electromagnetic ultrasonic temperature measuring instrument and compared with the temperature measuring result of thermocouple, which verifies the reliability of high temperature resistant electromagnetic ultrasonic probe and high-speed acquisition module. The research shows that the integration of ultrasonic temperature measurement technology and the design and development of instruments will provide effective analysis methods, research means and testing tools for exploring the internal temperature parameters of structures.

1 Introduction

The boundary conditions of the temperature field inside the structure contain the information of the surface, so compared with the surface temperature measurement, the internal temperature measurement can more comprehensively reflect the thermal safety state of the structure [1-3], which has a strong application demand in aerospace, machinery manufacturing, petrochemical and other industrial fields. At present, the internal temperature measurement of the structure mainly adopts the contact method [4], which has certain limitations, such as punching structure internal temperature sensor installation can cause local temperature within a structure or stress concentration. However, non-destructive measurement methods such as spectroscopy or radiation thermometry based on non-contact measurement are mostly aimed at the temperature measurement of the surface of the structure and the fluid, and cannot effectively measure the temperature inside the solid structure [5-8].

Ultrasonic inspection plays a vital role in the field of non-destructive testing, can realize non-contact and online measurement of internal temperature of structures. In recent years, great progress has been made in the reconstruction of internal temperature field of solid structures [9-12]. For example, based on the ultrasonic transit time and the conjugate gradient method in parameter identification, Shi [13]

IOP Publishing
2366 (2022) 012025 doi:10.1088/1742-6596/2366/1/012025

developed a high-resolution reconstruction method for the transient temperature field inside the structure. In addition, aiming at the problem of non-contact measurement of the internal temperature field of castings and forgings under severe high temperature conditions, Shi Wenze[14] proposed an electromagnetic ultrasonic oblique incident shear wave reconstruction method based on the bending effect and the triangle forward expansion method. In general, when the thickness of the structure is on the order of centimeters, in order to ensure a temperature resolution of 1°C, the nanosecond acoustic time measurement accuracy required for ultrasonic temperature measurement puts forward higher requirements for the design of the relevant hardware and system of the instrument [15]. At the same time, in the face of the demand for online temperature measurement inside the structure in some narrow spaces or in a high temperature environment, the miniaturized design of the mechanical structure and the high-temperature resistance performance of the ultrasonic transducer have also become the key to the accuracy of the temperature measurement results. As for the design and development of ultrasonic temperature measurement system and instrument, Cui [16] designed an ultrasonic temperature measurement system based on TDC7200, which has a resolution of nanosecond in transit time and can accurately measure the temperature field in gas space. Wang [17] designed a portable ultrasonic guided wave thermometer based on the guided wave pole temperature measurement method, and realized the real-time detection of the temperature of 300℃~1800℃. In addition, it is worth pointing out that most of the research on high-temperature resistance and structural design optimization of ultrasonic probes focus on ultrasonic defect detection or thickness measurement [18,19], and there are still few researches on ultrasonic probe design for internal temperature measurement of solid structures in high-temperature environment. The above literature analysis shows that there is still a lack of relevant key technologies and instruments for the internal temperature measurement of high temperature solid structures.

To improve the ultrasonic probe and heat-resistant properties under high temperature and solid structure internal ultrasonic measuring temperature required nanoseconds sound acquisition accuracy, this paper designed developed high temperature resistant, miniaturization of electromagnetic ultrasonic probe, and through the study of the selection of high-speed data acquisition module and other related hardware design has realized the nanosecond ultrasonic acoustic data collection. Finally, the reconstruction experiment of internal temperature field of solid structure is carried out to further verify the reliability of high-temperature electromagnetic ultrasonic probe and high-speed acquisition module, which provides key technical support for the research and development of ultrasonic temperature measuring instrument of solid structure and non-destructive testing in high temperature environment.

2 Design of Miniaturized Electromagnetic Ultrasonic Transducer with High-temperature Resistance

Electromagnetic Acoustic Transducer (EMAT) is the key equipment for transmitting and receiving ultrasonic signals and temperature measurement. Considering the actual measurement and application requirements of EMAT in high-temperature engineering fields such as aerospace and petrochemicals, therefore, it needs to be designed for high-temperature resistance to meet the needs of calibration experiments and detection in harsh heating environment. In addition, when detecting in a narrow space, the ultrasonic probe needs to be miniaturized.

2.1 Design of high-temperature resistance

The ultrasonic probe itself is composed of coil, shell, wear layer, permanent magnet and bonding material. In order to make it could withstand the high temperature of 800℃, it is necessary to choose the material of each part of the probe.

The coil is made of high temperature resistant enameled wire, and the high temperature adhesive resistant to 1300℃ is selected to seal the coil, so as to ensure that the coil can withstand high temperature above 800℃ for a long time. The shell is made of high temperature alloy material that can withstand high temperature above 800℃. In order to ensure that the electromagnetic ultrasonic coil can normally excite eddy currents and alternating magnetic fields, the wear layer of the coil is made of non-metallic and non-conductive materials, and it is required to have high wear resistance. The permanent magnet is made of samarium cobalt material with high magnetic energy product, not easy to demagnetize, and temperature resistance up to 350 °C.

When conducting ultrasonic detection, no couplant is required between the probe and the specimen. When the probe is placed on the high-temperature workpiece, the heat will be transmitted slowly from the workpiece to the permanent magnet through the shell. Therefore, to improve the high-temperature tolerance time of the probe, it is necessary to consider slowing the heat transfer from the heat transfer pathway. Therefore, in terms of mechanical structure design, the permanent magnet should not be close to the probe shell as much as possible, and the permanent magnet and shell are connected by clamping column support as shown in Fig. 1. Since the air thermal resistance is much higher than the metal shell, heat conduction to the permanent magnet can be greatly slowed down.

At the same time, a movable support structure is designed at the bottom of the probe, and an external thread is set on the outer side of the outer casing near the detection end. The connection between the probe and the support structure is realized by threaded connection, and the probe can be moved in the axial direction through the knob movable support structure, so as to ensure that the probe and the specimen have a gap of 1~2mm to slow down the heat transfer. In addition, an approximate U-shaped heat dissipation grooves are set at the free end of the movable support structure, and an air gap will be formed between the probe and the specimen through the heat dissipation groove. The existence of the air gap slows down the heat conduction to the permanent magnet, thus improving the temperature tolerance limit of the ultrasonic probe.

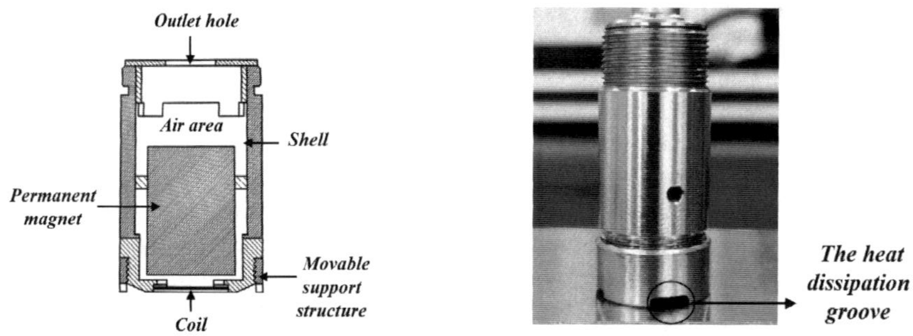

(a) Cutaway view (b) Physical picture

Fig. 1. High temperature resistant and miniaturized electromagnetic ultrasonic probe

Fig. 2 shows the simulation results of the two-dimensional temperature distribution inside the ultrasonic probe at 300s before and after optimization under the transient heat transfer condition when the heating boundary is 500 °C, and the temperature variations process at a position 5 mm from the heating end on the probe centerline. It can be seen that the existence of air gaps and heat dissipation grooves can effectively slow down the heat transfer to the permanent magnet, which provides certain technical support for ultrasonic nondestructive testing in high temperature environment.

IOP Publishing
2366 (2022) 012025 doi:10.1088/1742-6596/2366/1/012025

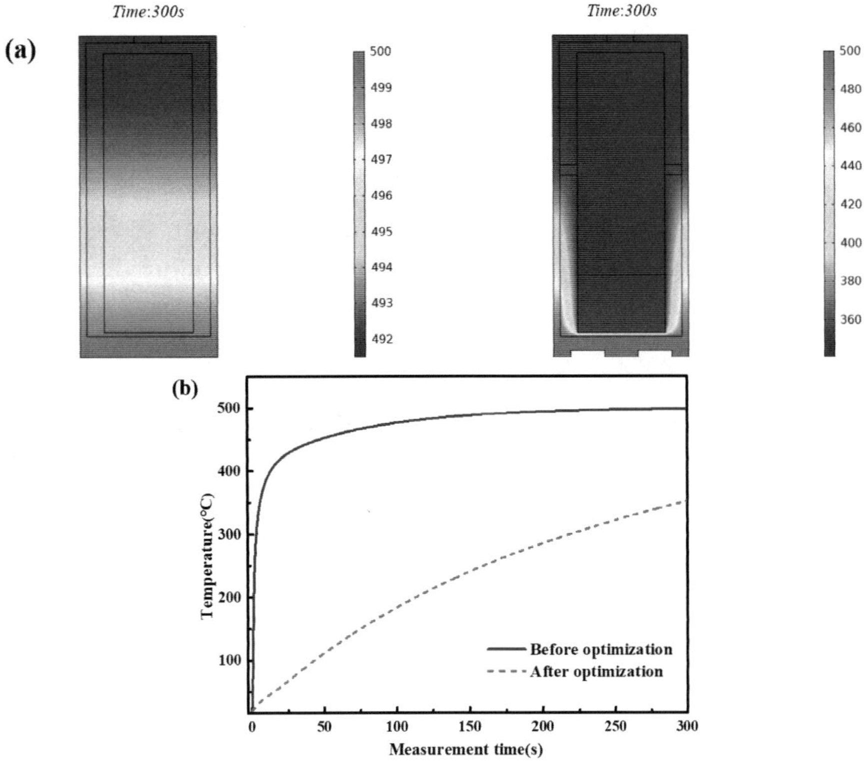

Fig. 2. (a): Simulation results of temperature distribution at 300s; (b): Internal temperature variations before and after optimization design

In order to test and verify the temperature resistance of the electromagnetic ultrasonic probe designed in this work, put a 10mm thick steel cylindrical specimen into a high temperature furnace and heat it to above 800°C, then take it out and quickly put the probe on it to test it for 1 minute. The echo signal is clear and the probe surface is not damaged.

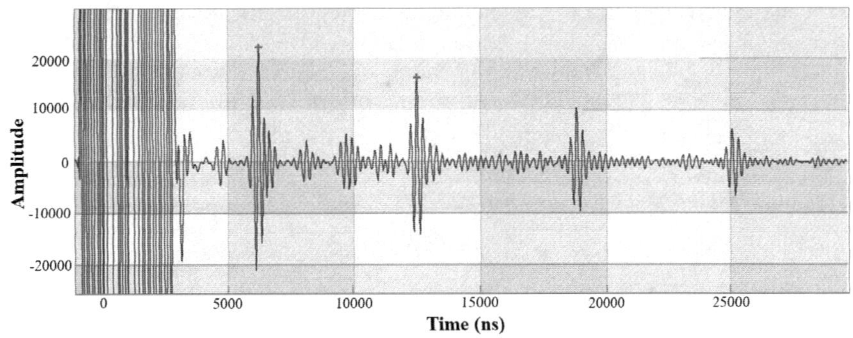

Fig. 3. Ultrasonic detection echo signal at 800°C

2.2 Design of mechanical structure miniaturization

Of all the components of an electromagnetic ultrasonic probe, the permanent magnet and the shell are considered to be the most likely to be miniaturized, On the premise of ensuring that the ultrasonic probe can achieve multi-waveform, multi-frequency excitation and its working efficiency, the thickness of the shell and the size of the permanent magnet are further optimized to obtain the smallest overall size of the electromagnetic ultrasonic probe. Fig.4 shows the actual picture of two existing probes in the field of ultrasonic temperature measurement and the miniaturized electromagnetic ultrasonic probe designed in this work, Table 1 makes a simple comparison of the relevant performance indicators of the three electromagnetic ultrasonic probes. It can be seen that the miniaturized and high-temperature resistant electromagnetic ultrasonic probs designed and developed in this work have achieved the integration of its structure and function.

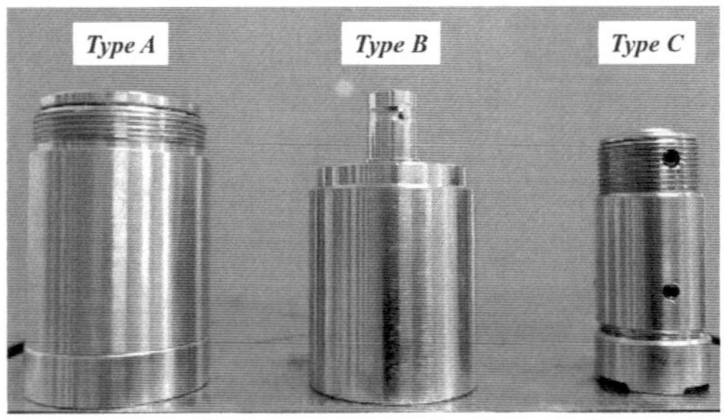

Fig. 4. Different types of electromagnetic ultrasonic temperature probes

Table 1. Comparison of different types of electromagnetic ultrasonic temperature probes

Type	Diameter (mm)	Mass (g)	Upper temperature resistance (℃)
A	33	290	600
B	31	212	350
C	20	78	800

3 Design of High-speed Acquisition Module

Acoustic time is the carrier of perceiving the temperature distribution inside the structure. If the thickness of the structure is on the order of centimeters, in order to ensure a temperature resolution of 1 °C, the measurement of acoustic time should meet the accuracy requirements of nanoseconds. Therefore, during ultrasonic detection, the acquisition accuracy of acoustic time data is particularly important for the accuracy of temperature measurement results.

3.1 Realization of measurement accuracy of nanosecond acoustic time

In order to achieve the acoustic time measurement accuracy of 1ns, an analog-to-digital conversion sampling frequency of at least 1GSPS is required, so the ADQ12DC acquisition card with 12-bit and 1GSPS sampling rate is selected. As shown in Fig.5, two acquisition cards with 2 channels and 4 channels are used to form a 6-channel acquisition. In addition, from the perspective of phase measurement, combined with the characteristics of the direct counting method and the quantization error conversion method, the phase-based acoustic time-to-zero-division method [20] is used to reduce the quantization error, so as to realize nanosecond acoustic time measurement.

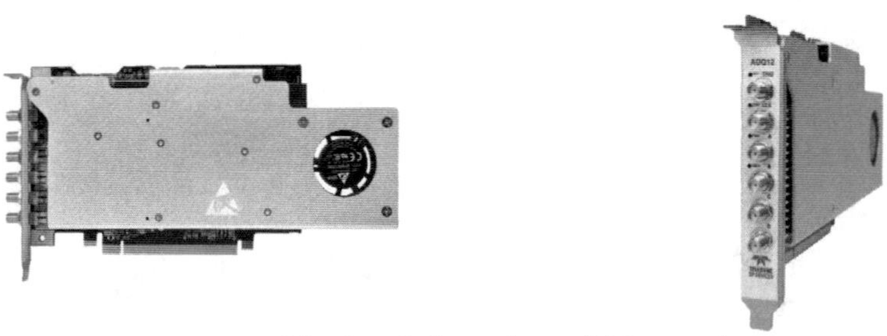

Fig. 5. 6-channel acquisition card

Fig.6 is the functional block diagram of high-speed data acquisition module. The ADC chip performs real-time sampling of external analog signals under the action of 1GHz clock generated by high-speed clock circuit. The built-in serial transceiver inside the FPGA chip receives the high-speed differential data stream in LVDS format, and uses the speed-down shunt module to perform a 1:4 speed-down shunting of the data stream for subsequent processing. The data from the DMUX directly enters the high-speed FIFO. The write clock of the FIFO is provided by the DMUX module. When the FIFO is full, a full flag will be generated. At this time, the master FPGA sends a read FIFO clock, and the data in the FIFO is read out at a slow speed, the collected data is read out slowly through the USB interface chip, and sent to the host computer for storage or display.

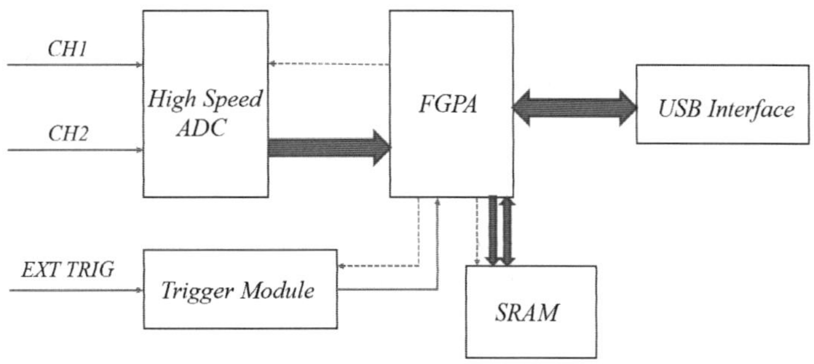

Fig. 6. Schematic diagram of data acquisition module

3.2 Test Verification of Acoustic Time Measurement Accuracy

In order to test and verify the sound-time acquisition accuracy of ultrasonic temperature measuring instrument, comparative experiments are carried out on three steel specimens with different thickness of 10mm and 20mm. The acoustic time T_1 is measured by oscilloscope DSOX3104T (the time measurement accuracy of oscilloscope DSOX3104T can reach 0.1ns), and compared with the acoustic time T_2 measured by the electromagnetic ultrasonic temperature measuring instrument, $\Delta T=T_2-T_1$ is the measurement error of the instrument. During the measurement, the ambient temperature remains constant, the electromagnetic ultrasonic probe and the specimen remain stationary, and the output of the power amplifier module is simultaneously connected to the input of the oscilloscope DSOX3014 and the electromagnetic ultrasonic probe. It can be seen from the data in Table 2~3 that the acoustic time measurement accuracy of ultrasonic temperature measuring instrument reaches 1ns.

Table 2. Acoustic time measurement of 10mm thick steel cylinder block

#	Results of EMAT (ns)	Results of DSOX3014	Absolute error(ns)
1	6182.9	6182.6	0.3
2	6185.3	6184.8	0.5
3	6183.4	6182.8	0.6
4	6182.0	6182.2	0.2
	Average value		0.4

Table 3. Acoustic time measurement of 20mm thick steel cylinder block

#	Results of EMAT (ns)	Results of DSOX3014 (ns)	Absolute error(ns)
1	12315.3	12316.0	0.7
2	12315.6	12316.0	0.4
3	12311.8	12312.0	0.2
4	12303.7	12303.0	0.7
	Average value		0.5

4 Reconstruction Experiments of Internal Temperature Distributions

In order to verify the reliability of the designed and developed ultrasonic nondestructive temperature measuring instrument, a verification experiment of the internal temperature field measurement of the steel structure is carried out and the experimental device is shown in Fig.7. A steel specimen with a thickness of 29mm is selected, the side of the specimen is approximately insulated and 5 thermocouples (TC1-TC5) are evenly drilled along its height direction. The lower surface of the specimen was the heating end, and the upper surface was the ultrasonic detection end. Under two different experimental conditions of heating plate temperature of 300℃ and 500℃, the transient temperature distributions inside the specimen are reconstructed by ultrasonic non-destructive temperature measuring instrument, and compared with the temperature measurement results of thermocouples.

In order to realize the ultrasonic detection of the transient temperature field inside the structure, while the lower surface of the steel is heated, the upper surface of the steel starts to excite ultrasonic shear wave detection. The ultrasonic non-destructive temperature measuring instrument collected and recorded the ultrasonic transit time and the temperature data of TC1-TC5 every 1s, and the heating time is set to 300s.

Fig.8 presents the comparison of the measurement results of the internal temperature of the steel structure in two different experimental conditions, respectively. It can be found that the ultrasonic measurement results are in agreement with those obtained by thermocouples. The average relative errors of temperature measurement are 4.06% and 5.00% respectively in the two different experimental conditions. Therefore, through experimental verification and analysis, the following conclusion can be drawn: The ultrasonic thermometer designed and developed in this work can effectively measure the internal temperature distributions of high-temperature solid structures.

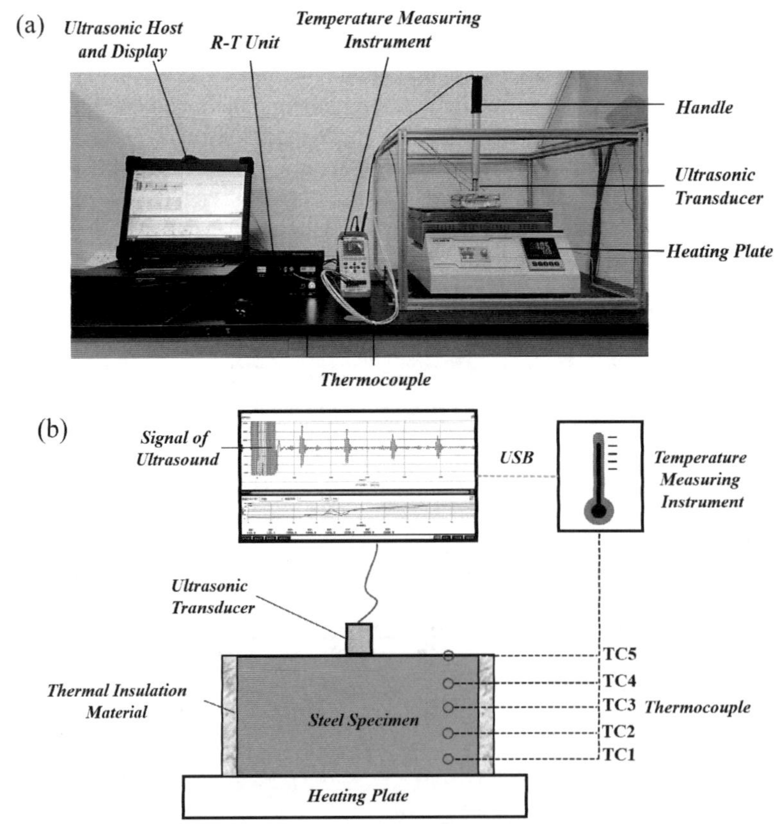

Fig. 7. Diagram of internal temperature field measurement experiment device

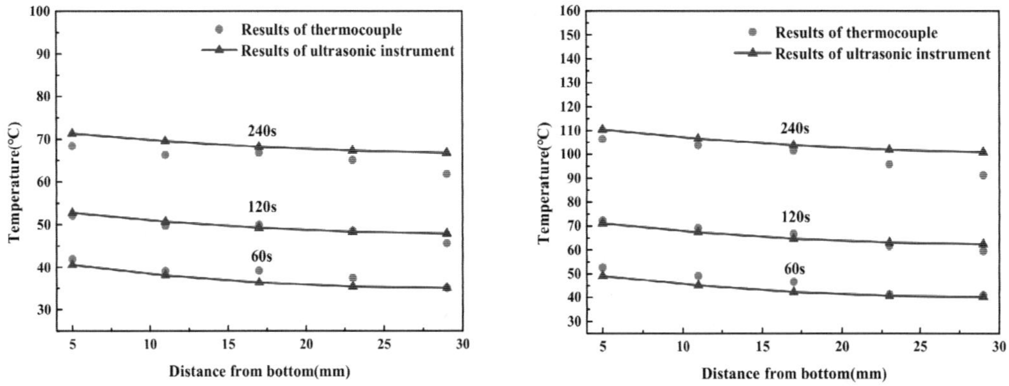

Fig. 8. Comparison of measurement results of internal temperature field

5 Conclusions

This study set out to design and develop a set of ultrasonic non-destructive temperature measurement instrument for the internal temperature measurement of high-temperature solid structures. The electromagnetic ultrasonic probe, electromagnetic ultrasonic excitation and receiving system, and high-speed acquisition module for acoustic time data are designed. In addition, the reliability of the ultrasonic non-destructive temperature measuring instrument is verified by carrying out internal temperature measurement experiments on steel structures under different experimental conditions. The summary is as follows:

(1) After the material selection and design of the permanent magnets of the electromagnetic ultrasonic probe, and the size optimization of the shell structure, the electromagnetic ultrasonic shear wave probe developed in this work can withstand a high temperature of 800 °C. The multi-waveforms of square wave, sharp pulse, sine wave and cosine wave can be selected, the multi-frequency and multi-channel excitation of 1~5MHz and the miniaturization of mechanical structure are achieved.

(2) The acquisition card with 1GSPS sampling rate combined with the phase-based acoustic time-to-zero-division method realizes the nanosecond acoustic time accuracy capture required by the ultrasonic temperature measurement technology. By using ultrasonic temperature measuring instrument to carry out ultrasonic detection of steel structures with different thicknesses, and compared with the acoustic time data collected by the oscilloscope, the analysis shows that the acquisition accuracy reaches 1ns.

(3) Under two different experimental conditions, the ultrasonic temperature measurement instrument designed and developed in this work was used to reconstruct the internal temperature distribution of the steel specimen, and the temperature measurement results of the thermocouple were compared. The experimental results show that the ultrasonic temperature measuring instrument can effectively measure the temperature distribution inside the structure. It has the advantages of non-contact measurement, good real-time performance and high precision. It provides a certain technical support for the non-destructive measurement of the internal temperature of solid structures and has important engineering application value.

References

[1] Yin W. Research On The Internal Temperature Measurement Method Of Power Battery Based On ERT. Published online 2013.

[2] Chernyshev, S.L., Lyapunov, S.V. & Wolkov, A.V. Modern problems of aircraft aerodynamics. Adv. Aerodyn. 1, 7 (2019). https://doi.org/10.1186/s42774-019-0007-6.

[3] Periyannan S, Balasubramaniam K. Multi-level temperature measurements using ultrasonic waveguides. Meas J Int Meas Confed. 2015;61:185-191. doi:10.1016/j.measurement.2014.10.050.

[4] J YY. Development of the Technology in Temperature Measurement. Metrol Meas Technol. 2009;4(29):62–66.
[5] Wang B, Niu Y, Qin X, Yin Y, Ding M. Review of high temperature measurement technology based on sapphire optical fiber. Meas J Int Meas Confed. 2021;184(July):109868. doi:10.1016/j.measurement.2021.109868.
[6] Manabu Takahashi II. Ultrasonic Sensing of Internal Temperature Distributions in Heated Materials. 2006;(2):1071-1074.
[7] Smith TB, Schetz JA, Bui TT. Development and ground testing of direct measuring skin friction gages for high enthalpy supersonic flight tests. In: 22nd AIAA Aerodynamic Measurement Technology and Ground Testing Conference. ; 2002:1-11. doi:10.2514/6.2002-3134.
[8] Gartenberg E, Oberts AS. Twenty-five years of aerodynamic research with infrared imaging. J Aircr. 1992;29(2):161-171. doi:10.2514/3.46140.
[9] Shi C, Hu B, Liang X. Reconstruction of Internal Temperature Distribution in Solid Structure by Ultrasound Thermometry. Jixie Gongcheng Xuebao/Journal Mech Eng. 2017;53(12):44-51. doi:10.3901/JME.2017.12.044.
[10] Wei D, Yang X, Shi Y, Xiao G, Du Y, Gui Y. A method for reconstructing two-dimensional surface and internal temperature distributions in structures by ultrasonic measurements. Renew Energy. 2020;150:1108-1117. doi:10.1016/j.renene.2019.10.081.
[11] Shi Y, Wei D, Zeng L, Qian W, Gui Y, Du Y. Research on reconstruction method for two-dimensional temperature field in solid using ultrasound. Zhongguo Kexue Jishu Kexue/Scientia Sin Technol. 2019;49(5):518-530. doi:10.1360/N092018-00119.
[12] Wei D, You-An S, Bi-Nan S, Ye-Wei G, Yan-Xia D, Guang-Ming X. Reconstruction of internal temperature distributions in heat materials by ultrasonic measurements. Appl Therm Eng. 2017;112:38-44. doi:10.1016/j.applthermaleng.2016.09.169.
[13] Shi Y, Wei D, Gui Y, Qian W, Zeng L, Qiu B. Research on parameter estimation in ultrasonic measurement of internal transient temperature distributions. Zhongguo Kexue Jishu Kexue/Scientia Sin Technol. 2016;46(5):509-517. doi:10.1360/N092015-00353.
[14] Shi W, Tong Y, Lu C, Chen Y, Zhu Y. An electromagnetic ultrasonic reconstruction method for internal temperature field of high temperature casting and forging. Yi Qi Yi Biao Xue Bao/Chinese J Sci Instrum. 2021;42(8):218-229. doi:10.19650/j.cnki.cjsi.J2107929.
[15] JinHong L. Ultrasonic thickness measurement method for high pressure and temperature tube by waveguide bar. Nondestruct Test. 2009;31(8):593-599.
[16] K C. Design of Ultrasonic Temperature Measurement System Based on TDC7200. Instrum Tech Sens. 2019;(08):79-82+105.
[17] WANG Y W, Design of portable ultrasonic guided wave industrial thermometer system[D]. North University of China, 2019.
[18] YUAN. Optimum Design and Experimental Study of Temperature-resistant and Pressure-bearing Ultrasonic Probe. Pipeline Tech Equip. 2019;(01):13-17+28.
[19] Tang Z, Sun X, Zhang P, Zheng Y, Lyu F. Research on composite electromagnetic ultrasonic transducer integrating thickness measurement and guided wave detection. Yi Qi Yi Biao Xue Bao/Chinese J Sci Instrum. 2020;41(9):98-109. doi:10.19650/j.cnki.cjsi.J2006440.
[20] Wei D, Shi Y, Hu B, Shou B, Gui Y. Experimental Study on Sensing the Internal Temperature Distributions of Structures by Electromagnetic Ultrasonic. Jixie Gongcheng Xuebao/Journal Mech Eng. 2019;55(6):93-99. doi:10.3901/JME.2019.06.093.

Detection Technology of Intentional Electromagnetic Co-channel Interference Source

Dongchu Su[1], Zengtian Chang[1], Jian Yang[1*], Bowei Chang[1], Zhijun Zhu[1], and Lu Qiang[1]

[1]63870 Troops, Huayin, Shaanxi, 714200, China

[*]Corresponding author's E-mail: ssyn_yangjian@163.com

Abstract: The characteristics and action mechanisms of intentional interference sources were analyzed. Then, a fast detection method of intentional co-channel interference was proposed based on such technologies as real-time spectrum analysis, digital phosphor display, and co-channel signal direction-finding. Next, two classical intentional interference scenarios were reproduced, and interference sources were detected fast through the proposed method.

1. Introduction

In recent years, spectrum-dependent equipment has been more frequently subjected to comparative testing in competitive bidding. Manufactured intentional electromagnetic interference behaviors tend to occur repeatedly at the comparative testing site, e.g., interference sources are added in the comparative testing of jamming equipment to enhance the jamming power. For another example, navigation interference signals are emitted to affect the positioning of unmanned aerial vehicles (UAVs) positioning during comparative testing. Traditional radio monitoring mainly refers to detecting and checking unintentional interference signals. Unintentional interference, which is generated by radio-frequency transmitting equipment operating in the adjacent frequency band, can be easily checked, and disposed of with unconscious and non-antagonistic features as well as single change patterns. In intentional interference activities, however, interference sources are concealed and started/stopped under the control of a remote-control unit, and interference signals present frequency overlapping with useful signals. Hence, intentional interference cannot be timely detected and alarmed by traditional radio monitoring means, which, moreover, are incapable of positioning and checking interference sources.

In this study, the potential interference means were analyzed based on the frequency use nature of test equipment. Furthermore, possible interference sources at a comparative testing site were timely detected and checked through real-time spectrum analysis, digital phosphor display, and same frequency signal direction-finding, expecting to put an end to malicious interference signals at the site and ensure the frequency use safety.

2. Intentional interference analysis

Intentional electromagnetic interference, namely interference sources are intentionally set, is characterized by the concealed deployment of jamming equipment, frequency overlapping between interference frequency and comparative testing equipment, and instantaneous trigger of interference signals. Consequently, the interference at comparative testing sites can be hardly detected or the detection may be neglected. Based on set-up objectives, interference sources can be divided into two types: counterparty suppression type and self-enhancement type.

2.1. Counterparty suppression type

As for the counterparty suppression type, one party imposes malicious interference with the normal frequency use of another party by transmitting high-power same frequency interference signals so that the interfered equipment fails to operate normally and is even destroyed. Suppression-type interference is typically featured by high signal power and frequency aiming. However, interference signals may be missed, or misjudged to be accidentally transmitted by legal frequency-dependent equipment since they are triggered instantaneously.

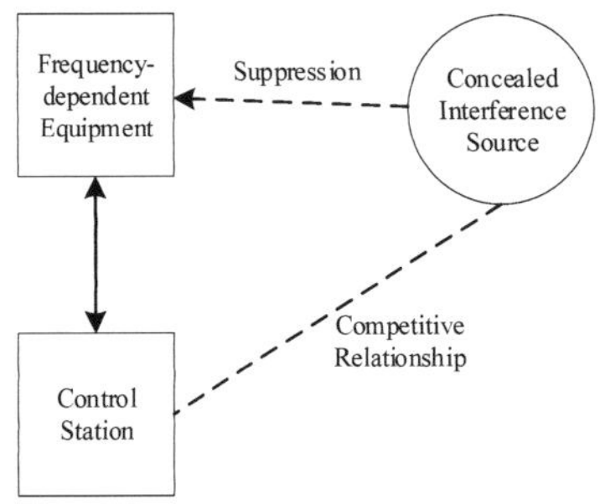

Figure 1. Schematic diagram of counterparty suppression-type interference

2.2. Self-enhancement type

When it comes to self-enhancement type interference, the tested interference unit is enhanced by adding high-power same-system interference sources in a concealed fashion to form higher-power interference signals, thus affecting the determination of the real interference performance for the tested interference unit. As a result, the measured interference performance index is virtually high. Self-enhancement type interference is typically characterized by high signal power and frequency and system sharing with the tested interference unit. The superposed frequency spectrum differs from the original one only in power, making it difficult to judge whether there are multiple interference sources.

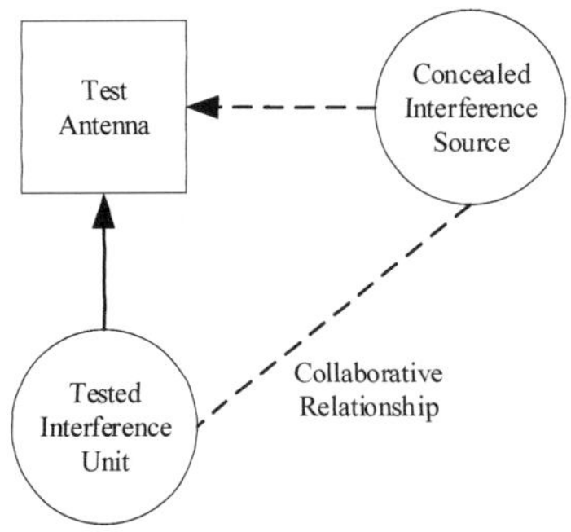

Figure 2. Schematic diagram of self-enhancement type interference

It can be known by analyzing the features of the above two interference types that the traditional radio monitoring means can discover interference at most but fails to position interference sources or judge the number of interference sources, accompanied by weaknesses in deterring malicious interference behaviors and preventing the repeated occurrence of interference signals. Therefore, it is necessary to add automatic direction-finding and positioning[1,7], automatic processing and warning, and means of identifying the number of interference sources for the traditional radio monitoring, thus forming the ability to respond to and react against interference signals fast.

3. Interference detection technologies

3.1. Same frequency signal direction-finding technology

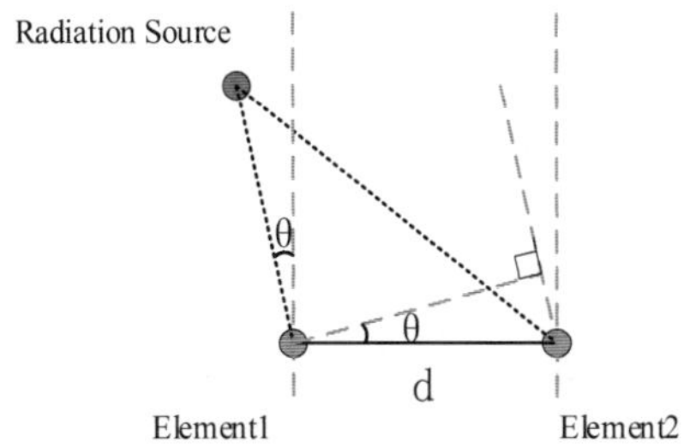

Figure 3. Direction-finding principle of phase method

The same frequency signal direction-finding technology refers to measuring the direction of electromagnetic waves through the phase method or amplitude method[4,5,8,9]. With the phase method in Figure 3 as an example, the distance from the receiving antenna Element 1 to Element 2 is assumed to be *d*, the included angle between echo direction and array normal is θ, and thus the signal received by Element 1 is as below:

$$r_1(t) = A\cos(2\pi f_0 t)$$

Then, the signal received by Element 2 is denoted as follows:

$$r_2(t) = A\cos[2\pi f_0(t-\tau)] = A\cos\left(2\pi f_0 t - \frac{2\pi d \sin\theta}{\lambda}\right)$$

Where λ stands for the signal wavelength.

On the above basis, the phase difference between $r_1(t)$ and $r_2(t)$ can be obtained as below:

$$\Delta\varphi = \frac{2\pi d \sin\theta}{\lambda}$$

Namely, the direction-finding angle θ is:

$$\theta = \arcsin\frac{\lambda\Delta\varphi}{2\pi d}$$

The blind source separation technology can realize the same frequency multi-signal direction-finding, mainly including four steps: (1) estimate the number of signal sources via a covariance matrix; (2) solve the separation matrix according to the estimated number of signal sources; (3) solve the hybrid matrix based on the separation matrix; (4) obtain the direction of signal sources. This technology separates signal sources one by one and then performs direction-finding operations for each signal source using a direction-finding method, thereby increasing the signal-to-interference ratio and improving the direction-finding resolution and accuracy of multiple same-frequency targeted signals at a low signal-to-noise ratio (SNR).

3.2. Digital phosphor display technology

As for the digital phosphor display technology, information to be displayed on a screen is regarded as a dot matrix[2,3,6]. The color shade or color temperature of each dot in this dot matrix on the screen is regulated according to its occurrence probability within a certain period. Then, the color shade is altered based on the occurrence probability of signals. Thanks to the digital phosphor display technology, a lot of radio frequency signals can be converted into a distinct view so that the instability and transient changes in radio frequency signals can be checked more easily. Moreover, signals concealed in the spectral background with different modulation systems can be discovered through a real-time spectrum analyzer. The comb-like spectral signals concealed beneath radar signals are displayed in Figure 4.

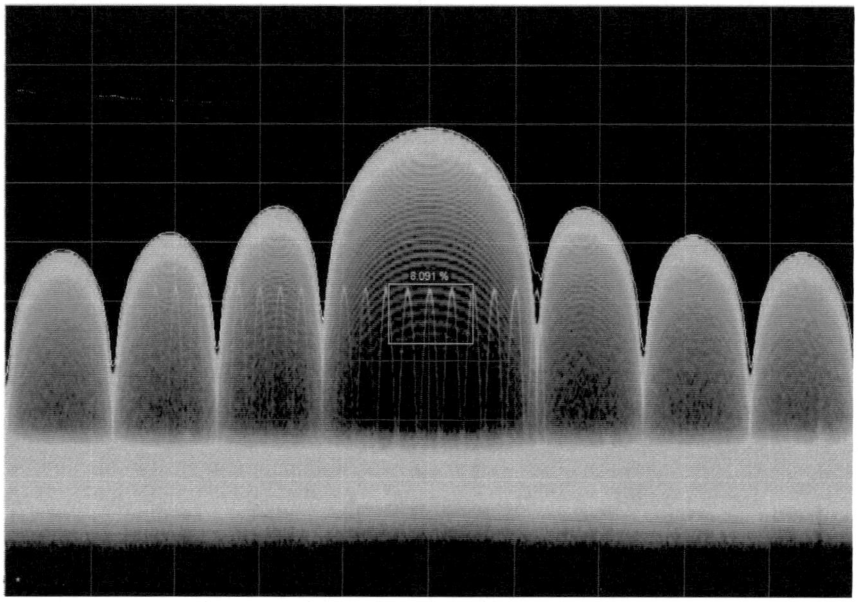

Figure 4. Digital phosphor display technology

3.3. *Time/azimuth amplitude method*

As for the time/azimuth amplitude method, the amplitude-time/azimuth change curves of a single signal source and multiple signal sources at different positions are determined through the rotation of receiving antennas. On this basis, the presence/absence of multiple signal sources is judged according to different features presented by such change curves.

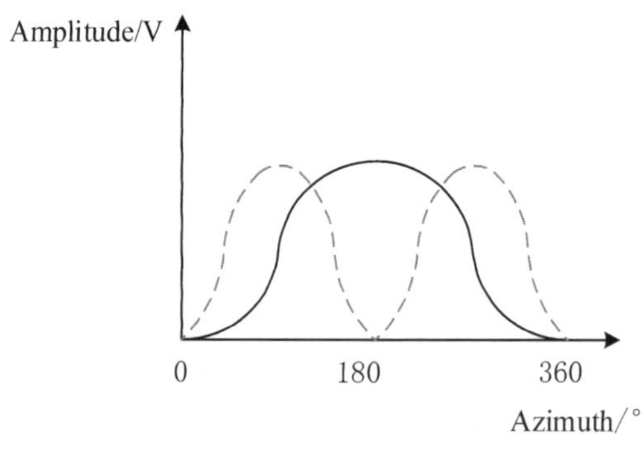

Figure 5. Azimuth amplitude diagram of a single signal source and multiple signal sources

In the case of a single signal source in an ideal space, the spatial field strength should decline evenly with the increase in the distance square. With the presence of multiple signal sources, multi-dot signal sources may be superposed at the same spatial point. Moreover, the spatial field strength will be distributed unevenly due to the varying operating distances, namely, at a long distance from a signal source, the measured spatial field strength may present a broken line-like decline and even increase.

Besides, the spatial field strength will also be affected by different azimuths of one signal source in the space. The above radiation characteristics are especially distinct on the azimuth amplitude diagram. As shown in Figure 5, the x-coordinate represents the azimuth of receiving antenna while the y-coordinate denotes the measured field strength. The solid line represents a single signal source, for which only one peak appears through the circular scanning of the receiving antenna. The number of peaks is more than one for the multiple signal sources represented by the dotted line. Based on this feature, whether the received signal is single or superposed by multiple signal sources can be distinguished, thus providing methodological support for discovering self-enhancement type interference.

The amplitude-time/azimuth change curves can be measured using test antenna rotation, mobile monitoring, and waterfall plot during the comparative testing to judge whether there are multiple interference sources. Multiple interference sources, if existing, can be disposed of by combining automatic direction-finding and digital phosphor display technology.

3.4. Intentional interference detection technology

Intentional interference detection is a process of detecting intentional jamming signals in comparative testing based on the same frequency signal direction-finding technology, digital phosphor display technology, and the time/azimuth amplitude method. This technology integrates the abilities of automatic direction-finding, positioning, and multi-interference-source discrimination, and it can effectively detect intentional electromagnetic interference while not affecting the environmental frequency spectrum at the comparative testing site.

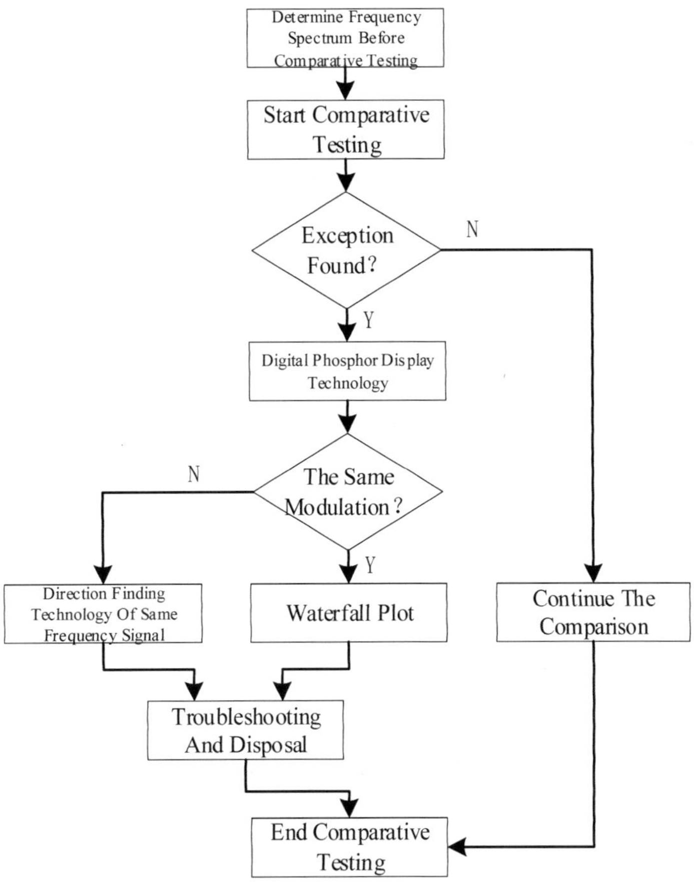

Figure 6. Intentional interference detection process at the comparative testing site

The intentional interference detection process at the comparative testing site is exhibited in Figure 6. Before comparative testing, the environmental frequency spectrum is firstly determined, and the spectral information under normal circumstances is recorded. During comparative testing, the digital phosphor display image can be observed immediately in case of abnormal spectral changes like suspicious interference to judge whether there are interference signals of different systems. If yes, the same frequency signal direction-finding technology is adopted to position, check and dispose of interference signal sources, and the comparative testing is ended, since the suppression effect can be exerted only when the interference power is greater than the power of useful signals. If same-system interference signals are judged, the waterfall plot is immediately viewed to acquire amplitude-time/azimuth change information and judge the presence/absence of multiple interference sources, followed by checking and disposition, and the comparative testing is finished.

The signal direction-finding, digital phosphor display, and waterfall plot functions are enabled during the whole comparative testing process for the sake of real-time spectrum monitoring. In addition, a warning message will be automatically sent in case of any abnormal spectral changes, reminding radio monitoring personnel to cope with transiently triggered interference signals.

During the same frequency signal direction-finding process, moreover, the trigger threshold is manually set to effectively reduce the false alarm rate. To be specific, the direction-finding trigger threshold is manually set according to the power of useful signals, and direction-finding will be triggered only when the signal power exceeds this threshold. As for positioning, interference sources can be positioned through intersecting position lines in the presence of two or more direction-finding stations. The greater the number of direction-finding stations, the higher the positioning accuracy, and the more efficiently the interference sources can be checked.

4. Verification of detection technologies

4.1. Verification of counterparty suppression-type interference detection

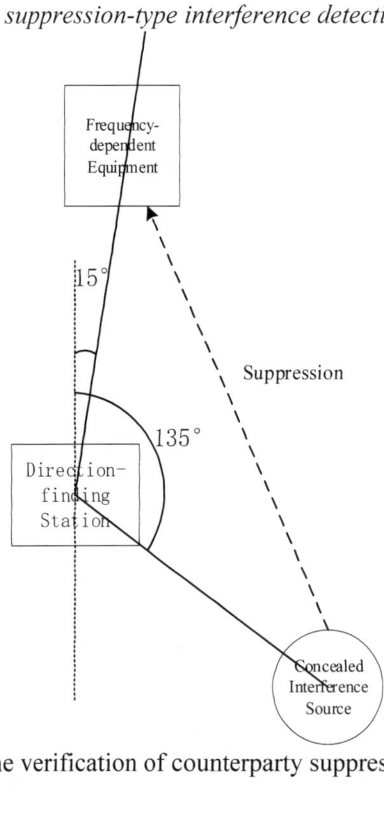

Figure 7. Scenario setting for the verification of counterparty suppression-type interference detection

As shown in Figure 7, the frequency-dependent equipment and concealed interference source were set to simulate the counterparty suppression-type interference in comparative testing, where the two are located at 15° and 135° directions from the reference line, respectively.

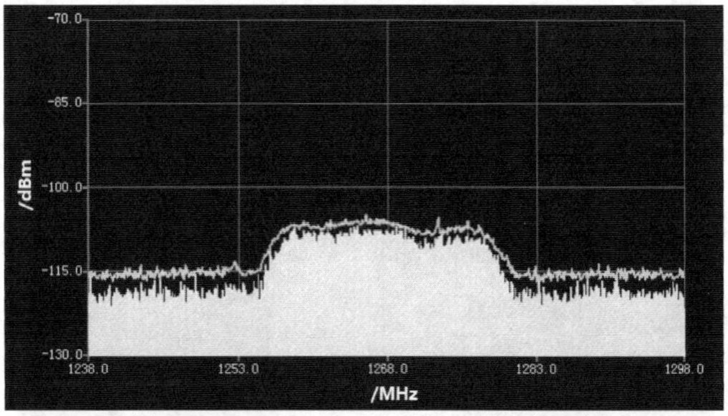

(a) Frequency spectrum of useful signals

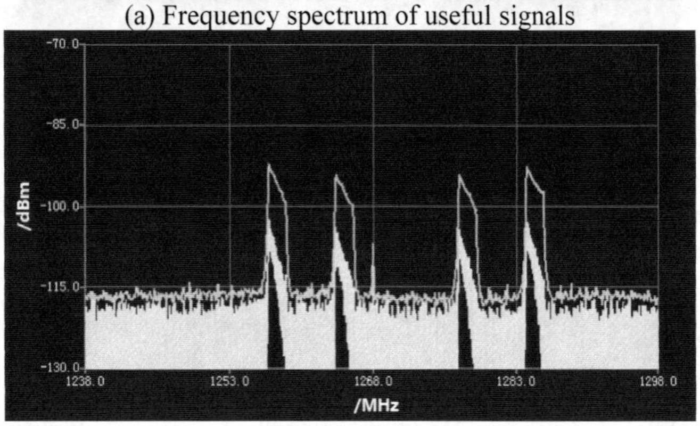

(b) Frequency spectrum of comb-like spectral interference signals

Figure 8. Settings of signal spectra

For the frequency-dependent equipment as shown in Figure 8(a), the center frequency and bandwidth of useful signals were set as 1,268 MHz and 20 MHz, respectively to simulate the navigation signals during the comparative testing of UAVs. The concealed interference source emitted a comb-like spectral signal, which consisted of four transiently triggered narrow-band signals, nearby 1,268 MHz to simulate navigation interference signals.

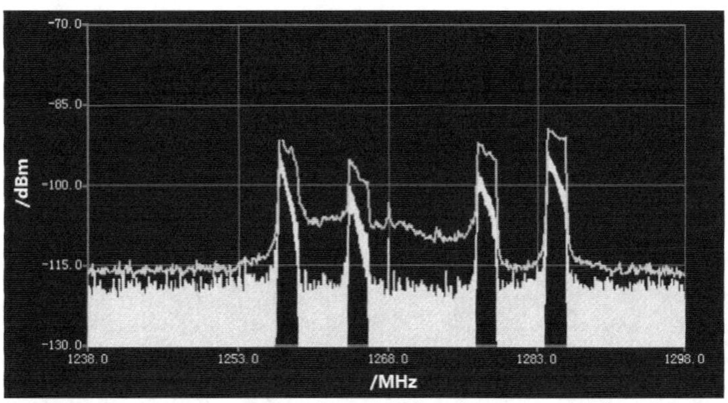

Figure 9. The frequency spectrum of useful signal superposed interference signal

In Figure 9, the power of the interference signal was greater than that of the useful signal, thus suppressing the useful signal.

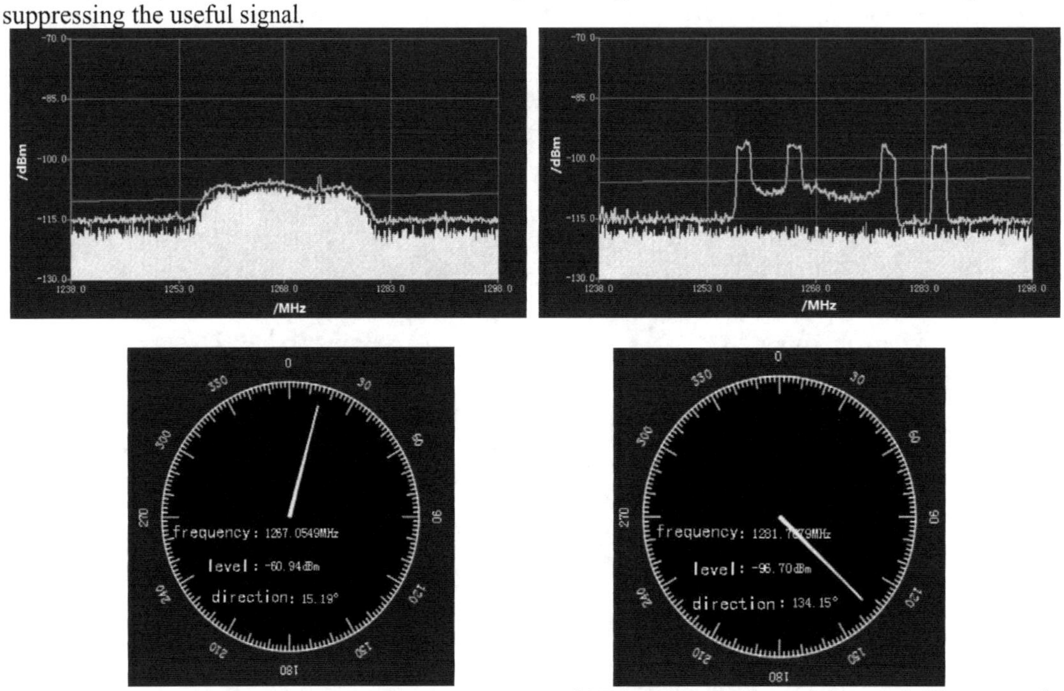

(a) Direct-finding of useful signal

(b) Direction-finding of the comb-like spectral interference signal

Figure 10. Signal direction-finding

In Figure 10(a), the direction-finding function was enabled to automatically direct at the azimuth (15.19°) of the useful signal. In Figure 10(b), a direction-finding threshold was set according to the power of the useful signal. When a transiently triggered interference signal appeared, the power was transiently greater than the direction-finding threshold, thus immediately triggering the azimuth information to turn to the azimuth (134.15°) of the interference signal. An alarm was emitted, reminding monitoring personnel to check. The ability of the automatic direction-finding technology to cope with transiently triggered interference was verified by this process.

4.2. Verification of self-enhancement type interference detection

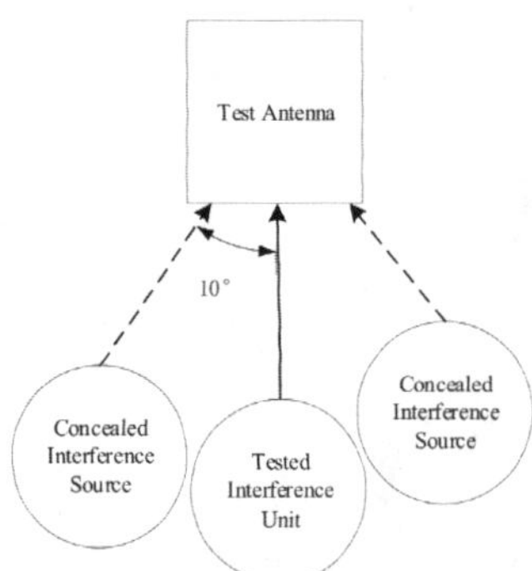

Figure 11. Scenario setting for the verification of self-enhancement type interference detection

A verification scenario (Figure 11) of interference detection was set to simulate the self-enhancement type interference appearing in the comparative testing of the interference unit, in which the test antenna was connected to a spectrometer, and the beam angle of the test antenna was set as 5°, and the circular scanning cycle as 4 s. The signal source emitted three interference signals to simulate the tested interference unit and two concealed interference sources. All interference signals were modulated from white noise with a bandwidth of 5 MHz. The included angle between each of concealed interference sources and the tested interference source was 10°.

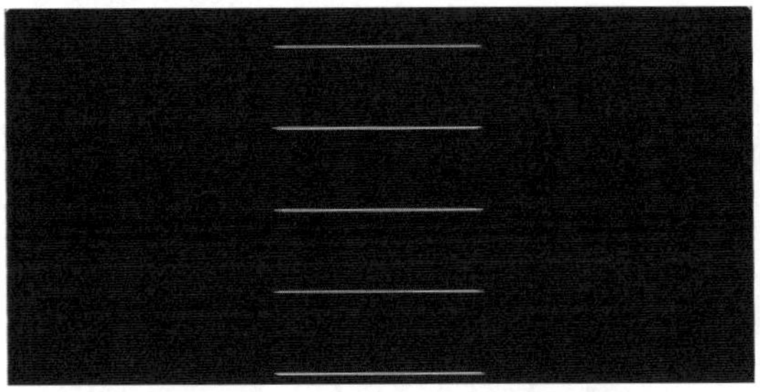

(a) Waterfall plot of a tested interference unit

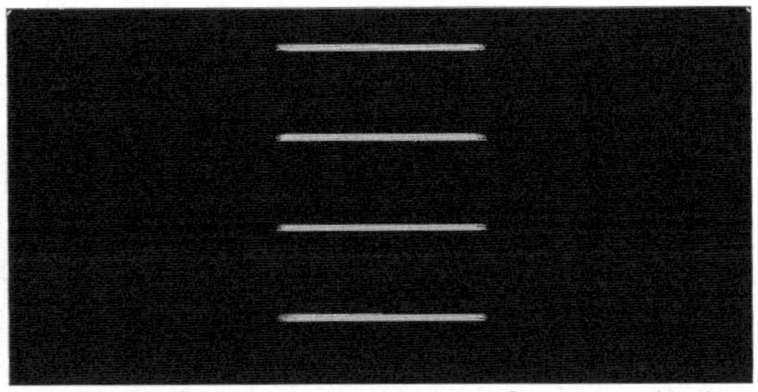

(b) Waterfall plot of tested interference unit plus two concealed interference sources

Figure 12. Circular scanning waterfall plot of receiving antenna (SWT = 20 ms)

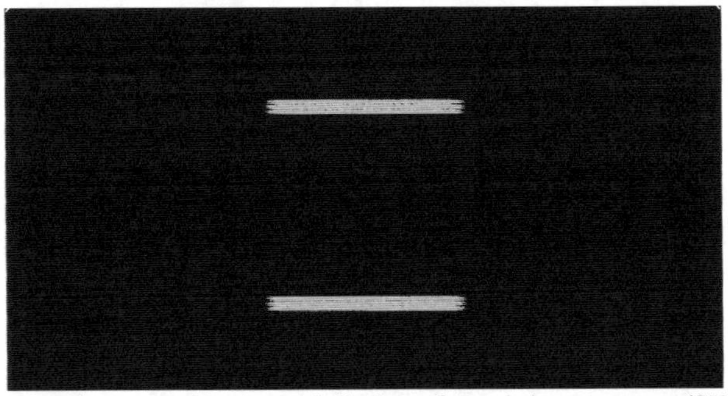

Figure 13. Circular scanning waterfall plot of receiving antenna (SWT = 3 ms)

The circular scanning waterfall plot of the receiving antenna is displayed in Figure 12(a). Therein, the signal peak appeared once every 4 s. Together with the output of two concealed interference units, the appearance time of signals was lengthened as exhibited in Figure 12(b). According to the waterfall plot in Figure 13, the scanning time was set as 3 ms, while the resolution was increased. It could be observed that three peaks emerged every 4 s, indicating the presence of three signal sources, thereby verifying the ability of this technology to handle self-enhancement type interference.

5. Conclusion

In this study, potential interference means were divided into counterparty suppression type and self-enhancement type according to the frequency use nature of test equipment. Then, the characteristics of the two interference types were analyzed as follows: The counterparty suppression type was featured by the transient trigger, narrow band, and same frequency. Sharing the modulation system with the tested interference unit, the self-enhancement type interference could not be distinguished from the tested interference signal in aspects of spectral shape and bandwidth. The traditional electromagnetic environmental monitoring lacks direction-finding means as well as the ability to cope with transiently triggered interference signals. Given this, an intentional interference detection technology consisting of automatic same frequency signal direction-finding technology, digital phosphor display technology, and azimuth amplitude method was proposed in this study. This technology could cope with transiently triggered signals, narrow-band same frequency signals, and multi-interference-source superposed signals. In the end, scenarios were set up to simulate the interference at the comparative testing site and verify the ability of the proposed detection technology to cope with the intentional electromagnetic interference at the comparative testing site. It is proved that the proposed detection technology is capable of real-time detection and checking of possible interference sources and putting an end to malicious interference behaviors at the comparative testing site.

References

[1] Yu Y M, Yao Y, and Cheng X H 2013 Introduction of TDOA Positioning Technology and Practical Application *J. China Radio* (11): 57-58+70.
[2] Mo F F 2019 On the Application of Digital Fluorescence Technology in Superheterodyne Spectrum Analysis *J.* Electronics World (21): 191-192. DOI:10.19353/j.cnki.dzsj.2019.21.103.
[3] Chou Z D, Wang B R, Zhang C, and Xu J H 2013 Design of digital phosphor in real-time spectrum analyzer *C* Assembly documents of 2013 National Conference Millimeter Waves, 1507-1510.
[4] Yu C L, Zhang Y F, and Wan F 2012 Actuality and development trend of passive location technical system and equipment *J. Journal of Air Force Radar Academy* 26 (02): 79-85.
[5] Li H W 2012 Studies on Target Localization and Tracking in Passive Coherent Location Radar *D.* Xidian University.
[6] Guo S J, Tang P F, Su S Y, and Chen Z P 2011 Design of real-time spectrum analysis applying DPX *J. Journal of Electronic Measurement and Instrument* 25 (09): 781-786.
[7] Zhu Q H 2007 Study on TDOA directional finding and location measurement *J. Telecommunication Engineering* (01): 53-56.
[8] Wang B C, Zhang G Y, and Hou H Q 2006 Research on passive location technology *J. Shipboard Electronic Countermeasure* (06): 20-26.DOI:10.16426/j.cnki.jcdzdk.2006.06.005.
[9] Shan Y H, Sun Z K, and Huang F K 2002 Passive location technology in continuous development *J. Aerospace Electronic Warfare* (01): 36-42.

Research on Online Monitoring and Analyze of Hydraulic Turbine Based on Low Cost, Fast and Internet of Things Technology

Liang Bai[1], Yin Zhang[1], Wuke Liang[1*], Jiawang Luo[1]

[1]Department of Energy and Power, School of Water Conservancy an Hydropower, Xi`an University of Technology, Xi`an, Shaanxi, 710048, China

[*]Liangwuke@vip.163.com

Abstract. In view of the problems such as difficulty in and high cost of vibration signal collection of rotating mechanical rotating shaft, this paper proposes a low-cost, fast, stable and reliable hydraulic turbine rotating wheel vibration signal online monitoring scheme specially designed to the imbalance and dynamic balance counterweight of hydraulic turbine rotating shaft. Three high-frequency wireless acceleration sensors are installed on the rotating shaft to monitor the vibration signals of the three directions in real time. Vibration signals are sent to the cloud platform to facilitate data storage. The acceleration signal is decomposed by WOA-VMD. The time domain signal shows the running track of the shaft to guide the dynamic balance counterweight of the rotor. The frequency domain signal is used to analyze the machine vibration in the shaft frequency domain and predict the health status of the shaft according to the frequency change. According to this system, completed the experiment of a hydraulic turbine experiment, proved the feasibility and rationality of the system, and has good application prospect and practical value.

1. Introduction

Vibration signals contain important information for mechanical condition detection and fault diagnosis. Accurate measurement of them helps in early warning of equipment failure and avoid equipment damage. For the vibration signal measurement, many scholars and researchers have proposed the mechanical method, electromagnetic method, optical method, computer vision method and image method. Among these measurement techniques are the contact type and the non-contact type, but for the vibration measurement of the moving parts, the wire contact type measurement mode appears helpless. In this case, non-contact sensors are required[1-8], Wang[9] et al. used an electrostatic induction sensor to measure the vibration of the rotating shaft. This principle is based on the induced charge output signal amplitude on the metal shaft surface, but this method has defects such as uneven induction charge distribution and unknown charge quantity. Read[10] et al. improved the electrostatic induction sensor to measure the vibration signal of the eccentric axis by installing a small electrostatic metal film on the shaft surface to measure the output current through the charge detection circuit. However, the ambient temperature and humidity can change the dielectric material between the electrodes, affecting the output result and causing the measurement error. Induction sensors are easily subject to ambient temperature, humidity, magnetic field and other factors, and scholars have designed different optical vibration measurement instruments. Ismail[11] et al. employed the laser Doppler oscillometer and the optical flow algorithm to measure the surface vibration of the silicone model, which can capture the model surface vibration at different frequencies, which has good robustness. The applicability of this method was demonstrated by testing by the rotor vibration measurement of wind turbines using laser Doppler vibration measurement by Gohler[12] et al. However, when the laser Doppler test, the measurement accuracy will decrease because of the measurement Angle amplification, and the accompanying noise will be amplified. Khalil[13] et al. introduced the angle analysis method to analyze the laser Doppler vibration, and conduct the accuracy and noise analysis of the vibration measurement in the three directions. The experimental results and the 3D accelerometer measurement results have verified the rationality of this method. Complex optical sensors can be designed to work in harsh environments and can perform high sensitivity and high noncontact displacement measurements. However, such devices are very expensive for conventional industrial applications, and the laser Doppler oscillometers are susceptible to rigid-body motion and asynchrony. Recently, vision-based measurements have developed rapidly with advances in computer vision and image sensor manufacturing technologies. Advantages of this technique include contactless measurements, massless introduction, and effectiveness. Wang[14] et al. Used digital image (DIC) technology to measure the frequency response function of the full-field vibration, taking the car hood as the research object, through the full-field vibration measurement on the surface of the hood, identified the mode parameters of the car hood, and predicted the characteristic vector between the frequency and the vibration shape. Peng[15] et al. Employed a high-speed camera to make a non-contact measurement of the high-speed maglev rotor vibration, based on the phase motion and video amplification technology, the high-frequency small amplitude microvibration measurement. Sarrafi[16] et al. used phase motion estimation to detect vibration damage of wind turbine blade, extracting resonance frequency and operation deflection shape to detect damage, etc., proving the feasibility of non-contact video measurement of structure damage. However, both DIC and video measurement will produce a large amount of highly heavy data, so image processing technology is needed to extract feature signals from the raw data, while bringing data storage and data processing problems and consuming a lot of storage equipment and computing resources. Therefore, it can not be widely used in the industry. Recently, wireless sensor networks have been widely used in the monitoring of rotating machinery, becoming the best alternative to traditional wired systems. Bengherbi[17] et al. developed a wireless sensor node based on FPGA to detect machine vibration system and fault diagnosis. The rationality of this method is verified, but this system consumes more power quickly and cannot run for a long time. Dos[18] et al. made use of the Internet of Things module to build a wireless microelectromechanical system (MEMS) acceleration sensor applied for large rotor vibration monitoring. The MEMS accelerometer combined with wireless communication can provide a feasible alternative which combines the traditional measurement with the Internet of Things, achieving the device data Internet access. The rapid development of the Internet of Things technology has provided a reliable technology for data transmission. It is necessary to monitor the state of the equipment to ensure the safe and stable operation of the equipment. Thus, this paper selects the Internet of things module ESP32 combined with the traditional vibration sensor, designs a set of wireless spindle vibration online monitoring system that conducts modal decomposition on the collected data and diagnosis on the equipment fault, improving the reliability of the equipment and avoiding causing heavy losses. This embedded system of the Internet of Things will solve the difficulties of data collection and transmission of rotating equipment in practical engineering, and help the staff to master the operation state of the equipment to ensure the safe and stable operation of the equipment.

2. Hardware selection and circuit design

2.1. Master controller

Data collection is carried out according to the requirements of vibration and analysis of the measured equipment, which can be collected by single channel or multi-channel. Generally, in order to improve the reliability of the data collected, multi-channel data collection is selected, and then the average value is taken as the signal sampling value at this time. In order to completely reproduce the original signal, the sampling frequency should be as high as possible, but it is limited by the hardware equipment, so the sampling frequency cannot be infinite. Considering that the wireless sensor should need to be installed on the rotating equipment, the overall size should not be too large, and be compact enough and easy to install while the power consumption should be low enough to meet the normal operation of the whole system, to ensure that the system can run for a long time. To meet the above requirements, we chose the low-cost, low-power consumption, and small-size Internet of Things module ESP32 as the main controller.

The ESP32 is a microcontroller that meets the above functions equipped with a Xtensa 32-bit LX7 dual-core processor. The maximum working frequency is 240MHz and the Flash is up to 16MB. In addition, external flash and external R AM can be connected through SPI / QSPI / OSPI and other interfaces and the chip includes a well-functioning Wifi subsystem, which is subject to the I EEE802.1b / g / n protocol. The chip comes with a 2-way 13-bit A DC acquisition circuit, a total of 20 channels exist, Practical multi-route ADC collection, four sets of SPI communication mode can expand the external RAM. At the same time, the ESP32 chip has a variety of unique hardware security mechanisms. Hardware encryption accelerators support the AES, SHA, and RSA algorithms. The RNG, HMAC, and digital signature module provide more security performance, and the perfect security mechanism enables the chip to meet the strict security requirements.

2.2. Acceleration transducer

The transducer is a piezoelectric acceleration sensor with compression structure, wide frequency sound, good environmental characteristics, light weight, and ease to install. The sensor model is: YK-YD500, sensitivity is 10mv / g, and the peak measurement range is 500g. The frequency response is 1Hz-12kHz, error is ± 3%, maximum lateral sensitivity is 5%, operating temperature range is -40~ + 120℃, transient temperature is 4mg / ℃ (0.3Hz), input impedance is less than 100 ohms, and the overall weight of sensor is 8g.

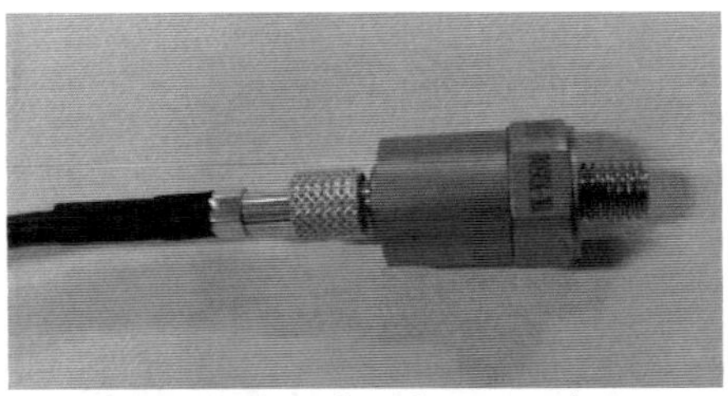

Figure 1. Acceleration transducer

2.3. ADC conversion

ESP32 integrates two 13-bit step-by-step approximation register ADC with a total of 20 simulated channel transfers. To achieve lower power consumption, the chip ULP coprocessor can measure the voltage in a sleep mode, and can awaken the C PU by setting a threshold or some other trigger mode. With the vibration sensor range level 5-15v, the ESP32 ADC channel cannot be measured directly. In order to ensure the safety of the ADC channel, an ADC front circuit is designed. The circuit is as shown in Figure 2.

The chip AD620 has the characteristics of small volume, low power consumption, low noise, low disorder, and high open-ring gain, which is suitable for measuring and amplifying weak signals. The output signal voltage of the vibration sensor is 5-15V, and the signal is compressed to 1-3V, which belongs to the reference voltage range. The signal differential input AD620 chip passes through the voltage tracer, and the pin 6 output signal is the ADC channel acquisition signal. The AD620 internal gain resistance is 24.7 **kΩ** and the gain equation is:

$$G=\frac{49.4}{RG}+1 \tag{1}$$

For the desired gain, the external control resistance value RG is:

$$RG = \frac{49.4}{G-1} \tag{2}$$

The voltage following circuit, the gain G is 1, so the control resistance should be as large as possible. In this circuit, R30 and R34 sum the external control resistance RG, $100\mathrm{k\Omega}$ differs nearly 2000 times from $49.4\mathrm{k\Omega}$, so RG can be considered infinite compared to the internal block. The output voltage is equal to the A D620 differential input voltage. The Multisim simulation signal is shown in Figure 3.

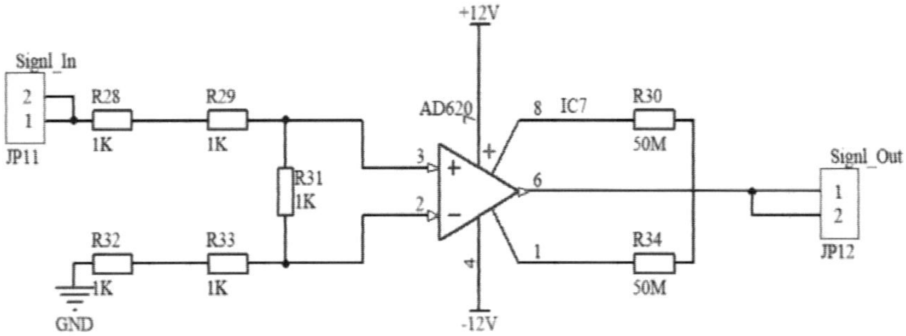

Figure 2. ADC front-end circuit

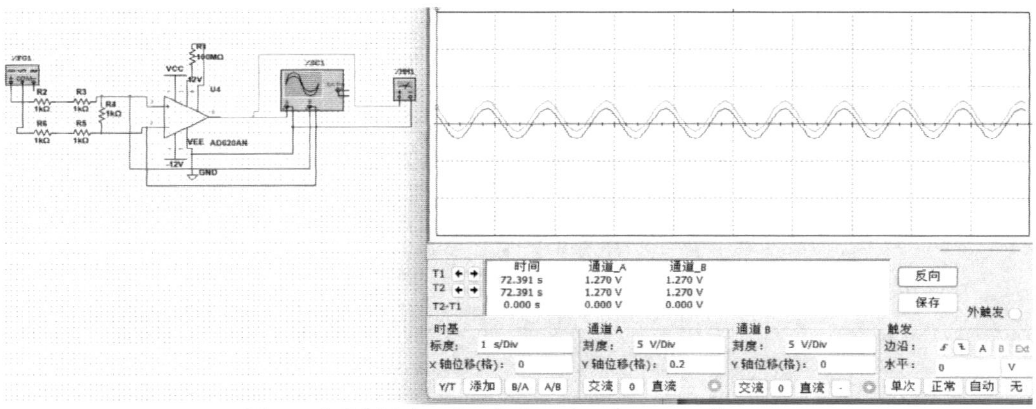

Figure 3. Multisim simulation signal output diagram

It can be clearly seen from the figure above that the output signal perfectly follows the input signal. At this time, the output signal can be directly sent to the ADC acquisition GPIO port. The data collected by E SP32 were connected to the cloud platform through the MQTT protocol and deposited in the database for subsequent studies.

The data acquisition and storage flow is shown in Figure 4

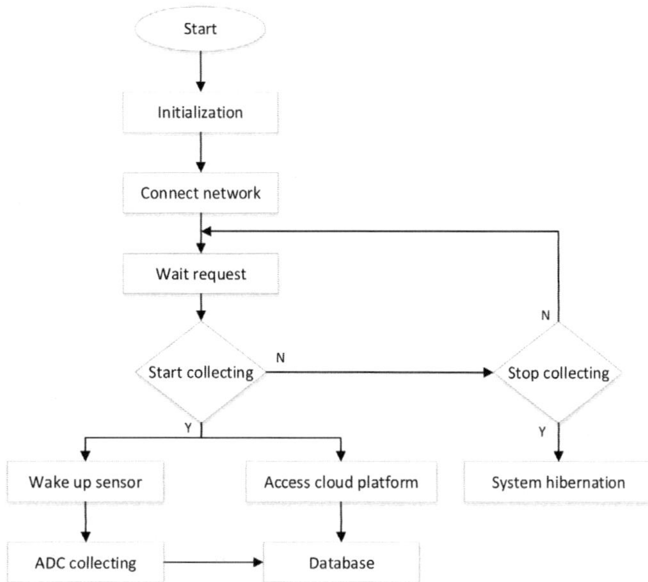

Figure 4. Flow chart of data acquisition and storage

3. The WEB software design

The WEB software system has data analysis and processing, data storage, data download, real-time waveform monitoring, remote control and other functions. Vibration signal is a complex signal with strong nonlinear overlap and multiple modes, and signal decomposition is needed to analyze each eigenmode. The VMD signal decomposition algorithm is selected to decompose the vibration signal. WEB software takes cloud platform data, decomposes real-time monitoring signals, and writes them into the local database to facilitate follow-up research by staff.

3.1. Master contro VMD decomposition principle

The VMD algorithm is to decompose the signal $f(t)$ into K modal functions u_k, each with a fixed central frequency ω_k and frequency band width. Establish constraint variational model expressions as follows:

$$\min_{\{u_k\},\{w_k\}} \left\{ \sum_k \left\| \partial_t \left[\left(\delta(t) + \frac{j}{\pi t} \right) * u_k(t) \right] e^{-j\omega_k t} \right\|_2^2 \right\} \tag{3}$$

$$s.t. \sum_{k=1}^{K} u_k = f(t) \tag{4}$$

In the formula: $\delta(t)$ is the impact function, K is the number of signal decomposition modes, u_k is the k mode component of the signal U decomposition, j the rotation operator, ω_k the central frequency of the mode function, Hz.

Introduce the penalty factor α and the Lagrange multiplier $\lambda(t)$ to make the above equality constrained variational problem equivalent to an unconstrained variational problem. The penalty factor α is to reduce the Gaussian noise interference and ensure the accuracy of signal reconstruction under strong noise, and the Lagrange multiplier $\lambda(t)$ guarantees the stringency of the constraints, so the augmented Lagrange expression is:

$$L(\{u_k\},\{\omega_k\},\lambda) = \alpha \sum_k \left\| \partial_t \left[\left(\delta(t) + \frac{j}{\pi t} \right) * u_k(t) \right] e^{-j\omega_k t} \right\|_2^2 + \left\| f(t) - \sum_{k=1}^{K} u_k(t) \right\|_2^2 + \left\langle \lambda(t), U(t) - \sum_{k=1}^{K} u_k(t) \right\rangle \tag{5}$$

Use the multiplier alternating direction algorithm (ADMM) to solve the unconstrained variational equation after the transition, through constantly iterating u_k, ω_k, λ, the augmented Lagrange saddle point is obtained as the optimal solution of the question. Due to the equivalence of the Fourier transform,

Parseval's theorem is used to solve it in the frequency domain, and the intrinsic mode u_k, frequency ω_k and Lagrange multiplier $\lambda(k)$ are even more advanced by equation (6) (7) (8).

$$u_{k,n+1}(\omega)=\frac{f(w)-\sum_{i\neq k}u_i(\omega)+\frac{\lambda(w)}{2}}{1+2\alpha(w-w_k)^2} \tag{6}$$

$$\omega_{k,n+1}(w)=\frac{\int_0^{\infty}\omega|u_k(\omega)|^2d\omega}{\int_0^{\infty}|u_k(\omega)|^2d\omega} \tag{7}$$

$$\lambda_{k,n+1}(\omega)=\lambda_{k,n}+\tau\left[f(w)-\sum_{k=1}^{K}u_{k,n+1}\right] \tag{8}$$

In the formula, $f(w)$ is the Fourier transform of $f(t)$, τ the update parameter, $\omega_{k,n+1}$ the central frequency of the current eigenmode function. The extended Lagrange saddle-point convergence constraints are as follows:

$$\sum_{k=1}^{K}\frac{|u_{k,n+1}(\omega)-u_{k,n}(\omega)|}{|u_{k,n}(\omega)|}<\varepsilon \tag{9}$$

In the formula, ε is a given positive value, which belongs to the calculation accuracy.

3.2. Client service software

The client software realizes the functions of user login, signal decomposition, waveform detection and remote control system for turning on signal acquisition. The GUI interface is developed with HTML, JavaScript, CSS3. The data adopts AJAX request to complete the data request without the page refresh, facilitating users to monitor the vibration of rotating parts in real time and display fault information. The client software architecture is as shown in Figure 6:

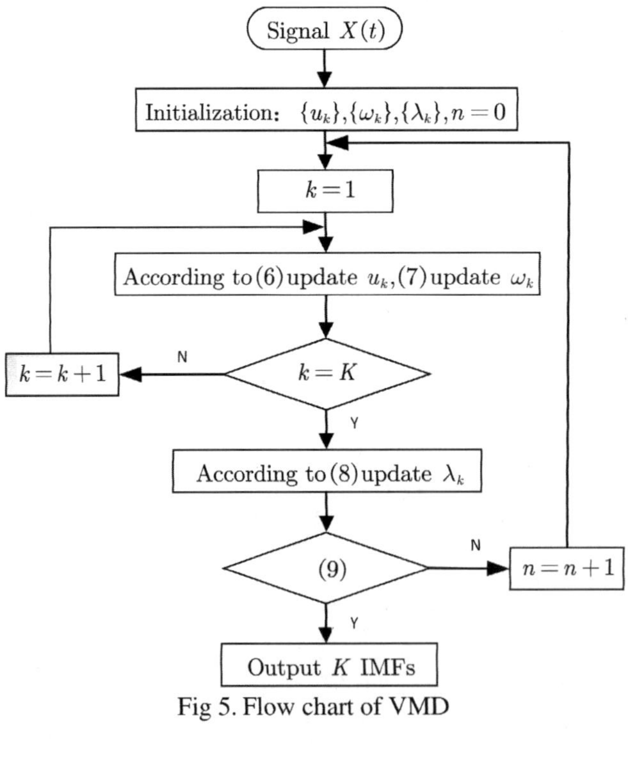

Fig 5. Flow chart of VMD

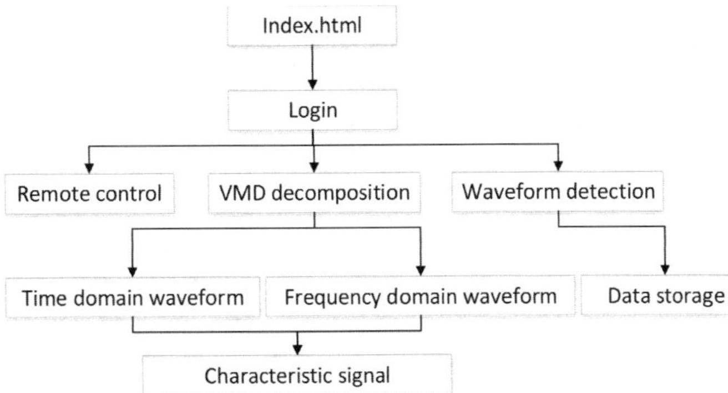

Fig 6. Client software architecture diagram

4. Experiments and data analysis

In order to effectively verify the rationality and validity of the proposed method, the dynamic balance experiment of the hydraulic turbine runner of a power station is conducted, collecting the turbine vibration signal at different speeds, studying the relationship between the turbine imbalance and spindle vibration, drawing the rotor trajectory, performing dynamic balance experiment of the rotor, and ensuring the rotor running coaxial. The sensor uses the magnetic seat installation and the turbine wheel. A total of three wireless vibration sensors are installed to collect the vibration signals from the three directions of the turbine. The sensor mounting diagram is shown in Figure 7 and the sensor frequency is sampled at 20kHz.

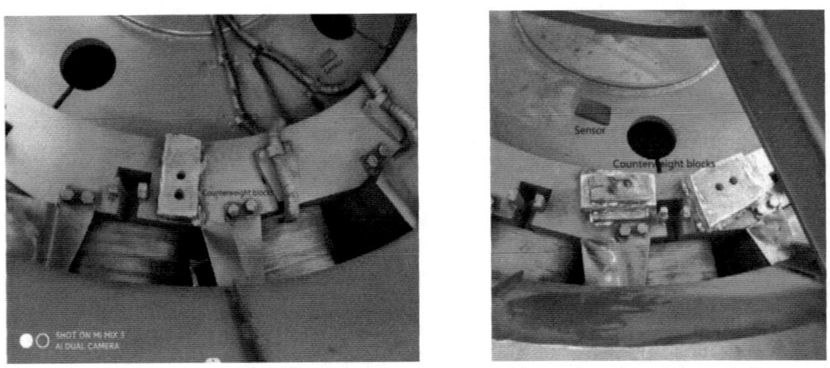

a) x-axis vibration sensor b)y-axis vibration sensor

Figure 7. Installation diagram of the wireless vibration sensor

Online monitoring is shown in Figure 8: the online monitoring interface has channel selection, enabling to select different channels to display the data of different sensors. This design has three channels to choose to measure the vibration irradiation values in x, y and z respectively. Sometimes the interface has the domain map and frequency domain map. When the frequency domain map is selected, the data shows the vibration frequency distribution through Fourier transform. The time domain and frequency domain diagram are shown in Figure 8: it can be seen from the frequency domain map that the frequency components are rich and the signal is complex, so the original signal mode is decomposed to obtain the eigenmode components.

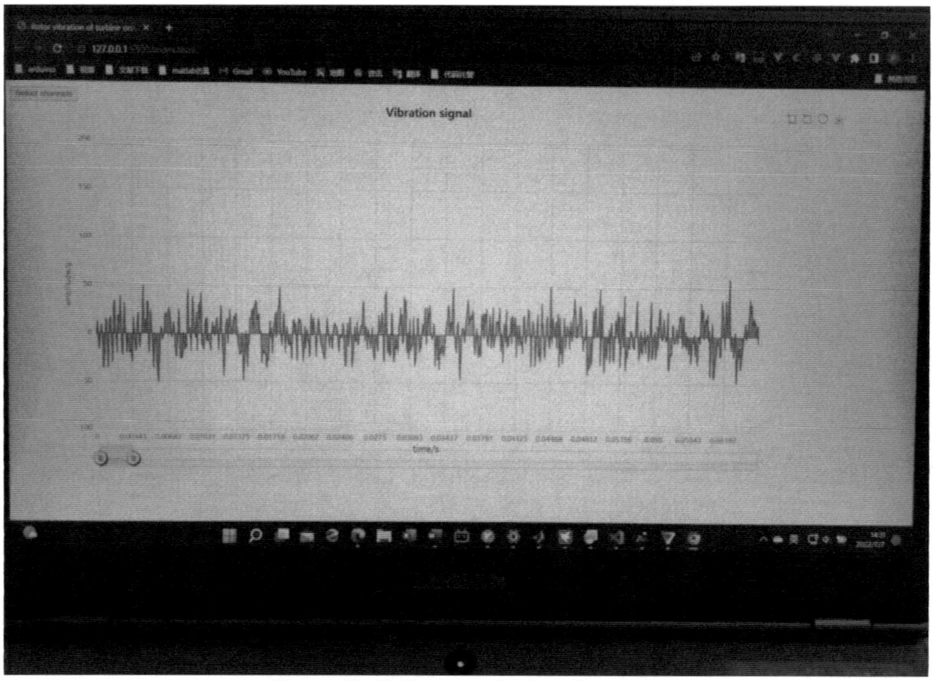

Figure 8. Online monitoring diagram

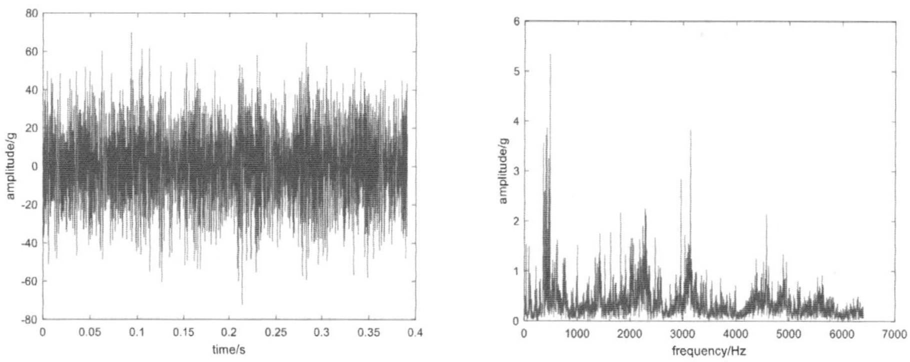

a) Time-domain map b) Frequency diagram

Figure 9. Time and frequency domain diagram of vibration signal:

In this paper, the signal envelope spectrum is taken as the fitness function to decompose layer k and penalty factor α as the optimization parameters, and the signal vibration mode of the water turbine is decomposed by using the whale algorithm. The decomposition convergence process is shown in Figure 10: the envelope spectrum is minimal when the iteration reaches the tenth time. At this time, the decomposition layer number k is 7, and the penalty factor α is 3500. According to this parameter, the original signal mode is decomposed, and the temporal frequency domain map of each mode component is shown in Figure 12. Rotor running track diagram 11 shows:

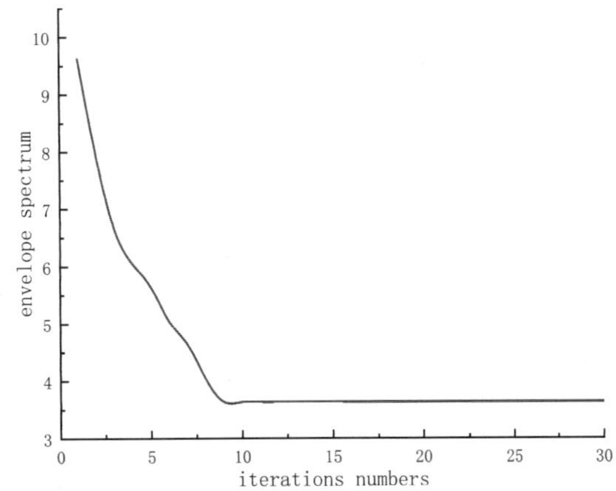

Figure 10. Decomposition and convergence

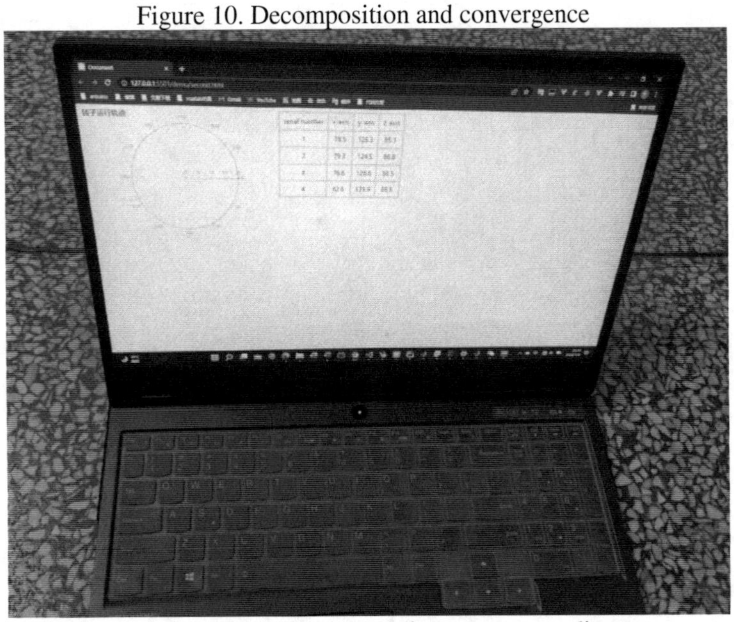
Figure 11. rotor running track diagram process diagram

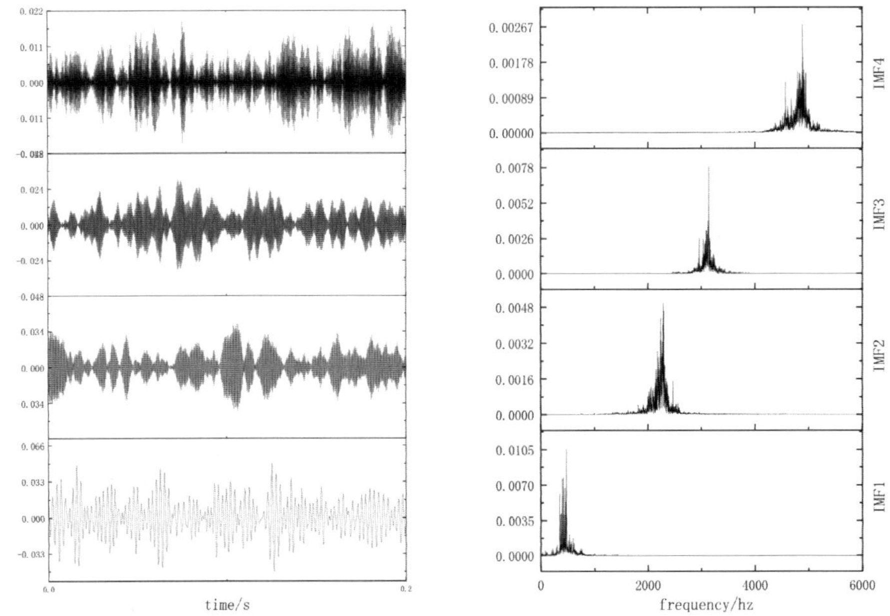

a) Time domain map b) frequency domain map

Figure 12. Time and frequency diagram of the rotor balanced vibration signal mode component

Figure 12 is a time-frequency diagram of the vibration acceleration signal at rotor balance, observed being strongly periodic from the time domain and having four frequency components. With increasing frequency, the irradiance is attenuated, which is consistent with the VMD decomposition principle. Figure 13 is a decomposition diagram of the rotor in the nonequilibrium vibration signal, which shows that the frequency of the high frequency part decreases and the vibration spoke value increases significantly. Due to the limited space, only the rotor dynamic equilibrium and non-equilibrium vibration signal decomposition is shown here, while the debugging process diagram is not unfolded.

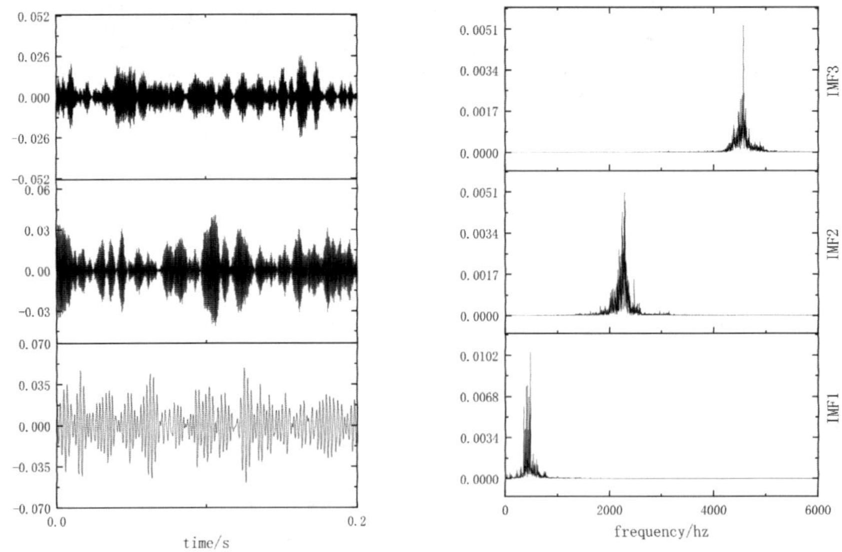

a) Time-domain map b) frequency-domain map

Figure 13. Time and frequency diagram of the nonequilibrium vibration signal mode component of the rotor

5. Conclusion

This paper proposes a low-cost, fast, stable and reliable online monitoring scheme for the vibration signal of the turbine runner. The system is specially designed to analyze the unbalance and dynamic balance weight experiments of the rotating shaft of the turbine. The vibration signal is decomposed by optimizing the VMD decomposition algorithm, the rotor dynamic balance is completed according to the running track, and the health state of the rotating shaft is predicted according to the frequency domain analysis. The main conclusions are as follows:

1) The acceleration sensor is combined with the Internet of Things module, and the measurement data is connected to the cloud platform, which realizes real-time online monitoring of runner vibration

2) When the rotor is unbalanced, the high-frequency part of the vibration signal begins to attenuate, the vibration amplitude increases, and the frequency changes, indicating that the rotating shaft is not concentric, and the rotor needs to be dynamically balanced.

5.1. Innovation

1) The online monitoring system designed in this paper is combined with wireless sensing to realize the signal measurement of the rotating part of the rotating machinery, which greatly reduces the cost compared with other non-contact measurement technologies.

2) The system is easy to install and the sensor is small in size, which is convenient for on-site experimental installation.

3) The signal is completely decomposed, and the modal components of the runner in different states can be obtained, which is convenient for fault analysis.

References

[1] Meng, D.J.; Miao, C.Y.; Li, X.G.; Li, J.; Shi, J.; Xu, W.; Yang, X.; Xu, D.G.; Liu, T.G.; Yao, J.Q. A vibration sensor based on Sagnac interferometer and fiber ring laser for fault diagnosis of bearing. Opt. Fiber Technol. 2021, 64, 102554.

[2] Goyal, D.; Dhami, S.S.; Pabla, B.S. Vibration Response-Based Intelligent Non-Contact Fault Diagnosis of Bearings. J. Nondestruct.Eval. Diagn. Progn. Eng. Syst. 2021, 4, 021006.

[3] Barusu, M.R.; Deivasigamani, M. Non-Invasive Vibration Measurement for Diagnosis of Bearing Faults in 3-Phase Squirrel Cage Induction Motor Using Microwave Sensor. IEEE Sens. J. 2021, 21, 1026–1039.

[4] Ghemari, Z.; Saad, S.; Khettab, K. Improvement of the Vibratory Diagnostic Method by Evolution of the Piezoelectric Sensor Performances. Int. J. Precis. Eng. Manuf. 2019, 20, 1361–1369.

[5] Feldman, J.; Hanrahan, B.M.; Misra, S.; Fan, X.Z.; Waits, C.M.; Mitcheson, P.D.; Ghodssi, R. Vibration-based diagnostics for rotary MEMS. J. Microelectromech. Syst. 2015, 24, 289–299.

[6] Amici, C.; Ragni, F.; Tiboni, M.; Pollet, J.; Buraschi, R. Quantitative Kinematic Assessment of the Sit-to-Stand Transition Using an IMU Sensor. In Proceedings of the 2021 24th International Conference on Mechatronics Technology (ICMT), Singapore, 18–22 December 2021, in press.

[7] Soto-Ocampo, C.R.; Mera, J.M.; Cano-Moreno, J.D.; Garcia-Bernardo, J.L. Low-cost, high-frequency, data acquisition system for condition monitoring of rotating machinery through vibration analysis-case study. Sensors 2020, 20, 3493.

[8] Lu, S.; Zhou, P.; Wang, X.; Liu, Y.; Liu, F.; Zhao, J. Condition monitoring and fault diagnosis of motor bearings using undersampled vibration signals from a wireless sensor network. J. Sound Vib. 2018, 414, 81–96.

[9] Wang L, Yan Y, Hu Y, et al. Intelligent condition monitoring of rotating machinery through electrostatic sensing and signal analysis[C]//2013 IEEE International Conference on Smart Instrumentation, Measurement and Applications (ICSIMA). IEEE, 2013: 1-4.

[10] Reda K, Yan Y. Vibration measurement of an unbalanced metallic shaft using electrostatic sensors[J]. IEEE Transactions on Instrumentation and Measurement, 2018, 68(5): 1467-1476.

[11] Ismail H M, Pretty C G, Signal M K, et al. Laser doppler vibrometer validation of an optical flow motion tracking algorithm[J]. Biomedical Signal Processing and Control, 2019, 49: 322-327.

[12] Göhler B, Lutzmann P, Scherer-Klöckling C, et al. Laser-Doppler-Vibrometrie an dynamischen Strukturen von Windenergieanlagen[J]. tm-Technisches Messen, 2020, 87(6): 392-403.
[13] Khalil H, Kim D, Nam J, et al. Accuracy and noise analyses of 3D vibration measurements using laser Doppler vibrometer[J]. Measurement, 2016, 94: 883-892.
[14] Wang W, Mottershead J E, Siebert T, et al. Frequency response functions of shape features from full-field vibration measurements using digital image correlation[J]. Mechanical systems and signal processing, 2012, 28: 333-347.
[15] Peng C, Zeng C, Wang Y. Phase-based noncontact vibration measurement of high-speed magnetically suspended rotor[J]. IEEE Transactions on Instrumentation and Measurement, 2019, 69(7): 4807-4817.
[16] Sarrafi A, Mao Z, Niezrecki C, et al. Vibration-based damage detection in wind turbine blades using Phase-based Motion Estimation and motion magnification[J]. Journal of Sound and vibration, 2018, 421: 300-318.
[17] Bengherbia B, Zmirli M O, Toubal A, et al. FPGA-based wireless sensor nodes for vibration monitoring system and fault diagnosis[J]. Measurement, 2017, 101: 81-92.
[18] Dos Santos Pedotti L A, Zago R M, Giesbrecht M, et al. Low-cost MEMS accelerometer network for rotating machine vibration diagnostics[J]. IEEE Instrumentation & Measurement Magazine, 2020, 23(7): 25-33.

Design of fretting signal acquisition system based on AD7177

Taiping Mo[1,a], Yuan Tang[1,b], Ying Su[1]

[1]School of electronic engineering and automation, Guilin University of Electronic Science and Technology, Guilin 541004, China

[a]mtp@guet.edu.cn, [b]1412012932@qq.com

Abstract. In Fretting is an important research direction in the detection of urban underground space structure. Traditional geophones have three issues with achieving fretting detection in the field, high cost, large size and low accuracy. A fretting signal acquisition system based on AD7177 is proposed to solve the above problems, which takes STM32 as the design core. A decision-making scheme that combines the signal acquisition system with the vibration sensor is introduced, which is used to settle the large size and high cost of the instrument. To improve the system accuracy, the AD7177 high-precision A/D conversion module and GPS are used for signal processing and positioning timing respectively. Finally, tests and field experiments were conducted in a state equipped with large storage space and WIFI transmission, which shows the system is convenient and successful. The experimental results show that the scheme is effective, the AD7177-based fretting signal acquisition system has high accuracy and small volume, which has a good ability to complete fretting detection.

1. Introduction

In the micro motion detection, which is no specific source the signal collected by the micro motion detection comes from people's daily activities and natural phenomena of nature, and the vibration signal comes from all directions that the internal information is rich [1]. At present, micro dynamic detection has been applied to many exploration fields such as engineering geological exploration and mine resource exploration, which has great engineering application value[2] .

The fretting detection method is based on the theory of stationary random process, the dispersion curve of the Rayleigh wave is extracted from the collected fretting signal and inverted to obtain the velocity structure of the underground medium [3]. The instruments and equipment used in fretting detection are all kinds of general seismic detection instruments, which no special equipment for fretting detection [4]. Most general seismic instruments can meet the performance indicators of fretting detection, but they are not designed for fretting detection. Therefore, that is still a large room for improvement in terms of portability and cost. [5] In this paper, to facilitate the use of equipment in fretting field detection, a portable fretting acquisition system with stable storage and data transmission is designed, which has real-time data detection, storage, and wireless communication [6].

2. General Design Scheme

The system adopts an integrated design, which is mainly composed of a geophone, ARM peripheral circuit, data acquisition module, power supply circuit, Bluetooth, WIFI, and other peripheral modules. the detector response signal is pre-processed by the pre-circuit, then through the data acquisition circuit, analog-to-digital conversion, the digital signal is transmitted to the STM32 processor and processed by the software module[7]. The central processor provides control instructions and chip pin control levels to the data acquisition module to set the sampling rate and sampling length required for data acquisition, and to give instructions for the start and end of acquisition. In addition to data acquisition modules, the system also needs to have a data storage unit and an interface with the host to communicate and download programs[8], the specific hardware design block diagram is shown in figure 1.

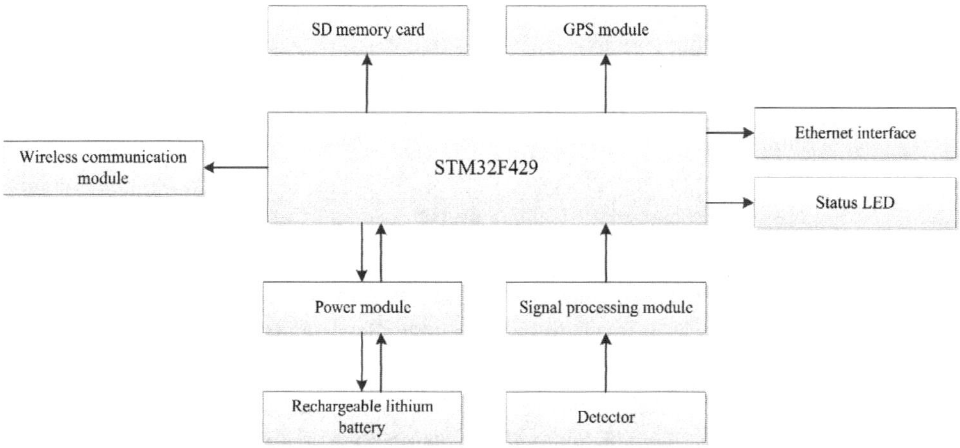

Figure 1. Hardware design block diagram

3. Hardware Structure

3.1 Signal processing module

In micromotion detection, the moving coil detector will transform the weak vibration signal into a weak electrical signal. The impedance matching, buffering, noise suppression, and amplification of the electrical signal are done by the signal processing module, finally, the analog-to-digital conversion of the analog signal to the digital signal. Due to the fretting signal is small, to accurately detect the weak vibration signal after the signal is filtered and conditioned. The programmable gain amplifier (PGA) is designed in the circuit, that the adjustable multiple is 1-128 times, the vibration signal can be amplified to the extent that AD can detect it to be ensured. After amplification of the signal by the AD7177 internal removal of harmonics and other interference signals[9].

3.1.1 Analog signal conditioning circuit design

Analog signal conditioning includes impedance matching, differential filtering, and harmonic suppression is shown in figure 2. The AD8211 has the highest CMRR relative to frequency among the current similar products. This feature makes the AD8211 extremely good at suppressing the interference of linear harmonics, thus helping us simplify the filtering requirements. Moreover, it has low noise and high precision and the input voltage noise is only 8nV√Hz, which is extremely suitable for fretting detection.

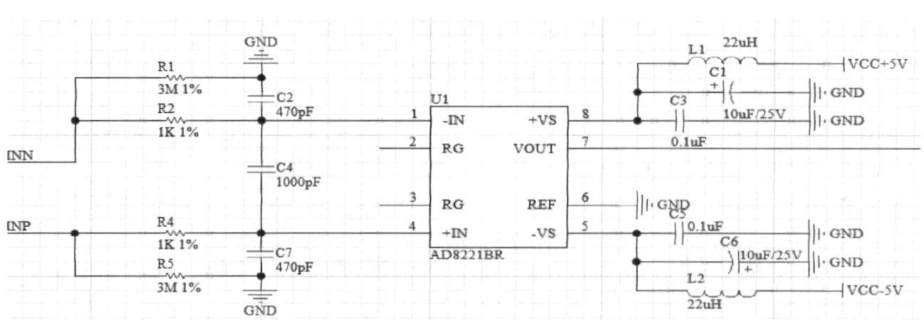

Figure 2. Signal conditioning circuitry

3.1.2 Programmable gain amplifier

Design uses the program-controlled amplification circuit with a magnification factor of 1-128 times, which role is to amplify the vibration signal measured to the full range of AD, the dynamic range of the overall magnification is increased. So that the performance of AD7177 will not be wasted and also can improve the signal-to-noise ratio. In the circuit, ADG1208 and AD8675 are selected as the core to design the gain of 2 adjustable amplifier circuits. AD8675 has ultra-low misalignment and drift and track-to-rail output similar to OP184 that even lower voltage noise. We only need to use STM32 microcontroller through the program of the ADG1208 A0, A1, A2 pin high and low level (0,1) input, which can obtain 1, 2, 4, 8, 16, 32, 64, 128 times gain. The schematic diagram of the programmable gain amplifier is shown in figure 3.

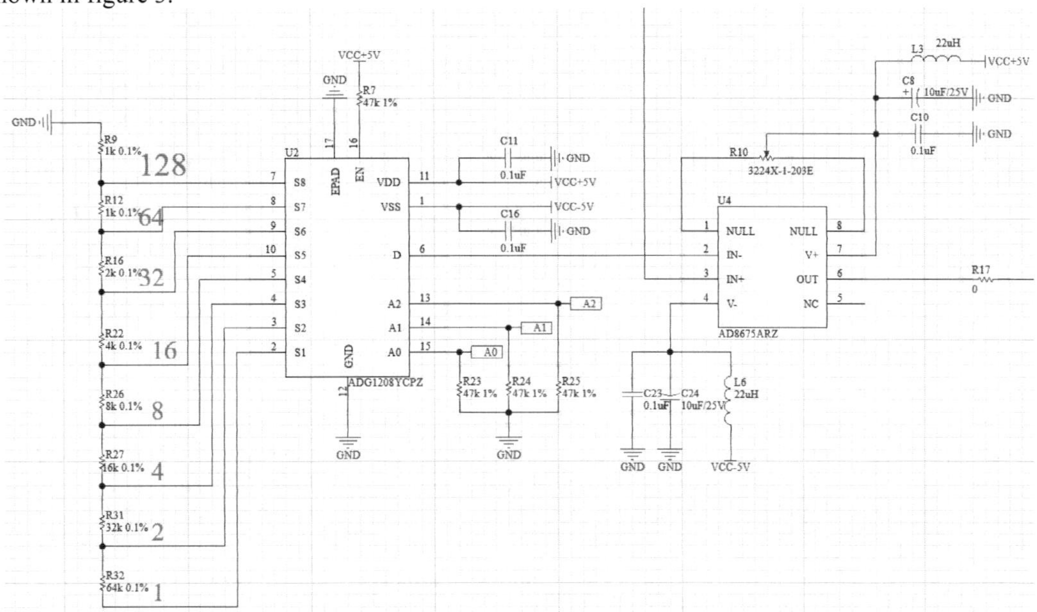

Figure 3. Programmable gain amplification circuit

3.1.3 Analog-to-digital conversion circuitry

The design of the analog-to-digital converter circuit uses the AD7177 as the core. The AD7177 is a 32-bit low noise, fast build, multiplexing, 2/4-channel (full difference/pseudo-difference) Σ-Δ analog-to-digital converter (ADC) suitable for low bandwidth input. Its output rate is fast and flexible, which ranging from 5SPS to 10kSPS. AD7177 also integrates analog and digital signal conditioning modules. The amplified signal is transferred to AIN0 and AIN1 pins of AD7177 after the single-ended differential circuit, which make the signal have a stronger anti-interference ability, then the AD7177 completes the conversion from an analog signal to a digital signal is shown in figure 4.

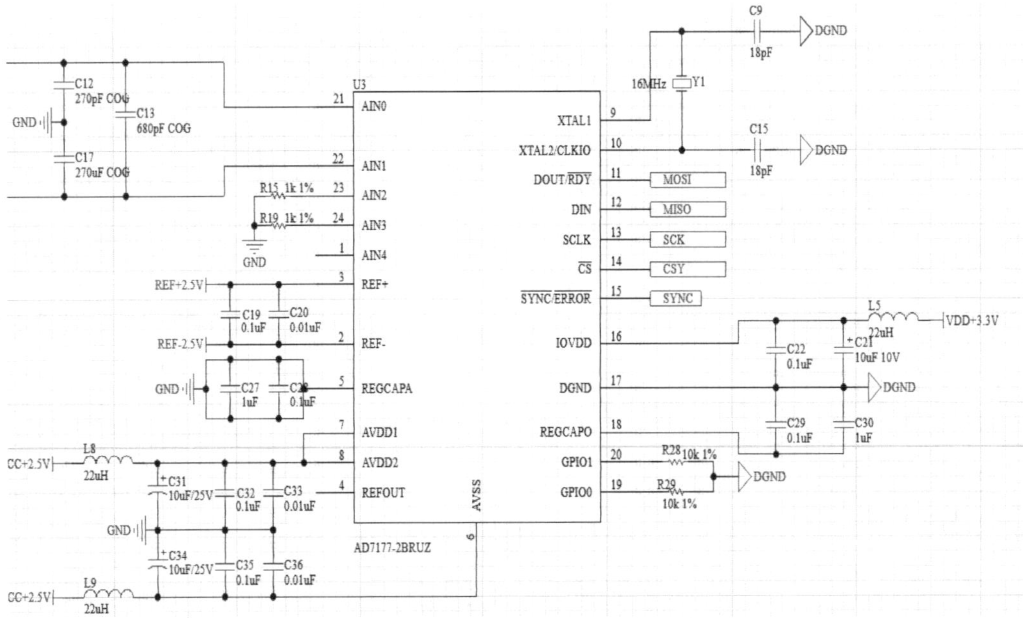

Figure 4. Analog-to-digital conversion circuitry

3.2 GPS synchronization module

Due to the characteristics of fretting detection, therefore often necessary to deploy geophone in a large area, which requires that each acquisition station of the fretting acquisition system should implement synchronous acquisition. Only when the data of the collection point is at the same time, the availability of data can be guaranteed. The time synchronization design adopted in this design is GPS synchronization. GPS synchronization technology has the advantages of high precision, simple layout, independent synchronization, and low cost, that also can provide the position information of geophone to facilitate the later processing of data. Figure 5 is the GPS module circuit diagram, ATGM332D is selected because that has the leading acquisition sensitivity in the industry and can receive six satellite navigation signals at the same time, realizing six-in-one multi-mode satellite navigation in a real sense[10]. Which is characterized by low cost and low power consumption and is extremely suitable for the design of a micromotion signal detection system.

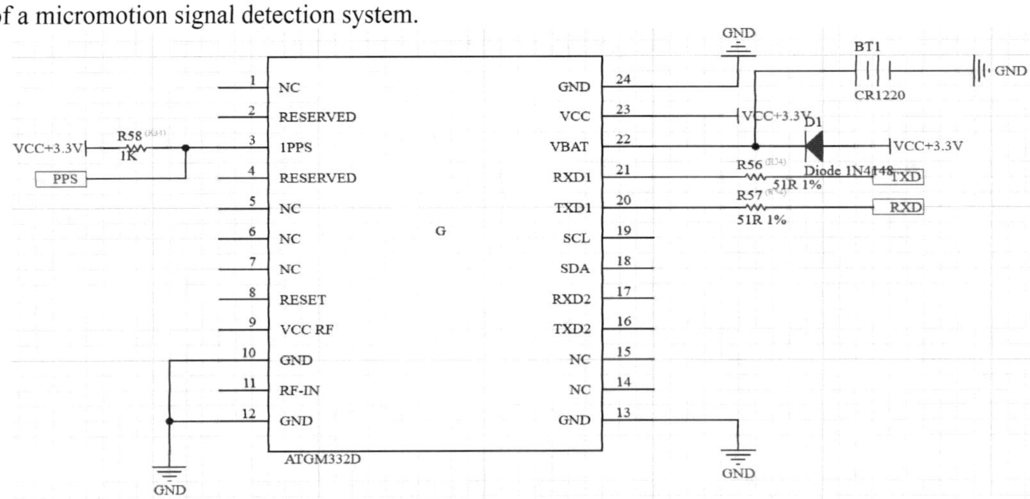

Figure 5. GPS module circuit

4. System software design

The development environment of the micromotion acquisition device is Keil and the software part of the whole system is designed in C language. Which mainly includes the writing of the main program, the setting of PGA and AD modules, data transmission, storage and GPS synchronization, etc. The main program is responsible for the configuration of the system clock, the configuration of DMA, the initialization of GPIO and some special ports, as well as the configuration of each software and hardware module. First, the sample rate, acquisition time, and signal amplification are set on the host computer software. Then, tasks are issued to the instrument, and the processor gives instructions to each module according to the received tasks. The collected data is converted to the actual voltage value of the geophone and stored in the SD card by the DMA controller. To facilitate long-term field acquisition, the storage is set to one hour to generate a file for subsequent detection and processing, and the first address of recorded data is used as the geophone identification code. The program flow is shown in figure 6.

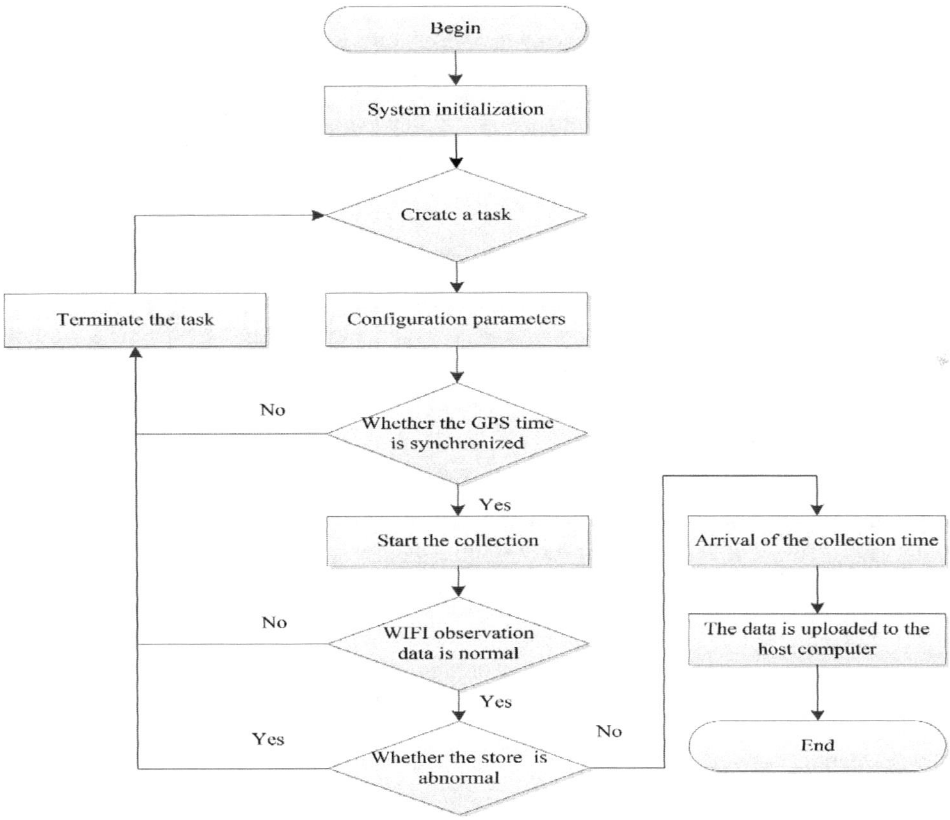

Figure 6. Software flowchart

5. System testing and practical application

5.1 Instrument self-noise

In the fretting acquisition system, the self-noise level of the instrument is a critical index. The lower the self-noise level of the acquisition instrument, the more real the signal it collects. In the test, the input end of the instrument is short-circuited, and the test result of short-circuit noise is shown in figure 7. The amplitude of the noise signal is far lower than that measured by the detector, which proves that the instrument has a good self-noise level.

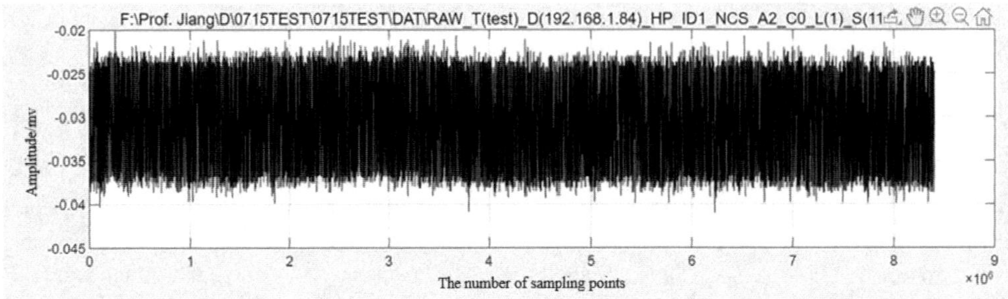

Figure 7. Instrument self-noise level

5.2 Experiment on the top of the urban bus tunnel (Taking Shenzhen Universiade Park as an example)

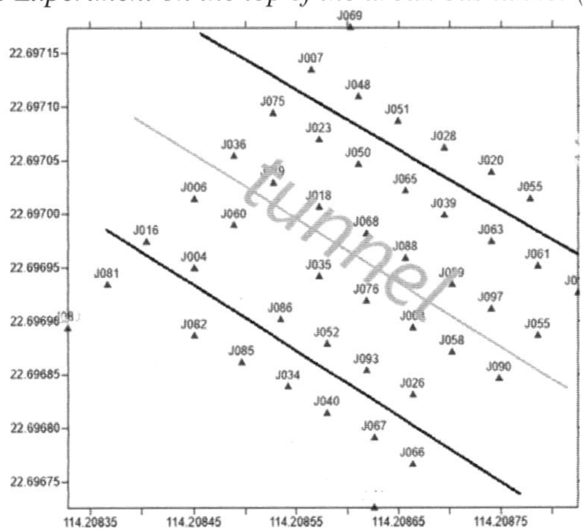

Figure 8. the observation points

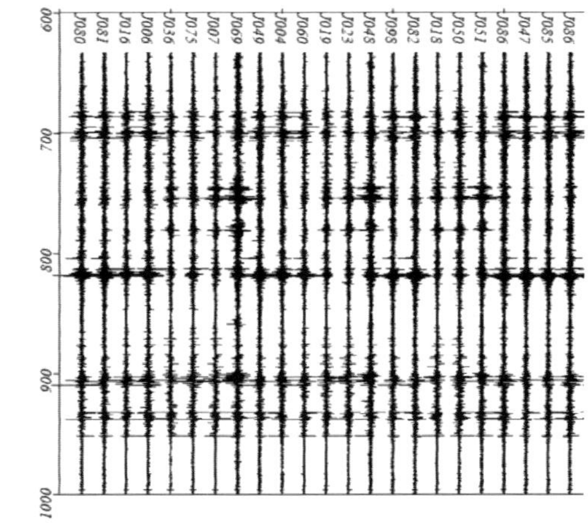

Figure 9. the noise data overview of some points

Figure 8 shows the observation points, and figure 9 shows the noise data overview of some points. The noise intensity distribution is determined based on the amplitude recorded by the noise observations. The experimental results show that the noise intensity at the top of the tunnel is large, the spread direction is consistent with the tunnel, and the range is slightly narrower than that of the tunnel. Therefore, the overall design of the instrument is correct and can be put into the field experiment of micromotion, which also lays a good foundation for the subsequent improvement and system upgrade.

6. Conclusion

In fretting detection, the performance of the fretting acquisition instrument determines the quality of data collected. The traditional cable exploration system is no longer suitable for fretting detection, and the cable-free system emerges as the times require, becoming a new direction to realize low-cost exploration. The components used in the micromotion acquisition device designed in this paper are cost-effective devices with low power consumption and low noise characteristics while ensuring the high performance and low cost of the whole instrument. To facilitate field construction, the whole instrument adopts a portable integrated design, which contains AD7177, a 32-bit high-precision AD chip, the accuracy of the data collected by the instrument is ensured, that is completely suitable for fretting detection. System can be used in the exploration of urban underground space with different types of geophone and has a good application prospect.

Acknowledgments

This work was supported by National Natural Science Foundation of China under Grant (61863007). Guangxi Natural Science Foundation under Grant (2020GXNSFDA238029). Guangxi Key Laboratory of Automatic Detecting Technology and Instruments (Guilin University of Electronic Technology): YQ21107

References

[1] Tian R, Lv S X, Lin J, Zhang L X :The key technology of single-channel wireless storage seismograph [J]. Chinese Journal of geophysics,2017,60(11):4273-4281.
[2] Li Z Y, Xian S, Mu J L: The modular design of solar cell energy acquisition and storage system [J]. Sensor and Microsystem,2022,41(01):85-88.
[3] Zhang Y Z,Xu C K,Zhou X F, Jin Z X :Design and Implementation of geophone based on STM32 Microcontroller [J]. Electronic Technique and Software Engineering,2021,(19):92-94.
[4] Yu H Q :Application of geophysical fretting technology in engineering geological prospecting [J]. Fujian Transportation Science and Technology,2021,(07):6-8+35.
[5] Wei J D, Cao G B, Liu B :Field measurement method of geophone performance parameters [J]. Oil Geophysical Prospecting, 2021, 56 (03) : 446-461+409-410.
[6] Tian R Y :Research on key technologies of wireless low-power nodal seismic acquisition system [D]. Jilin University,2021.
[7] Zou L :A quality Control method of SmartSolo node system [J]. Geophysical prospecting equipment,2020,30(05):335-337.
[8] Shi K, Yang F Q, Li Y F, Mo Y J, Qu X Y : Study of urban underground space structure using fretting detection[J]. Minerals & Geology, 2020,34(02):355-365.
[9] Feng W L, Li H Q, Tuo X G, Shen T : Design of Lightweight Small Piezoelectric Ceramic Seismic Source System[J]. Sensors and microsystems,2020,39(02):95-97+100.
[10] Liao C W, Deng T, Ding W, Wang H, Chen W M :Scheme and results of high precision clock synchronization technology for fretting data acquisition [J]. JGG,2009,29(06):153-156.

A Super-Resolution Altitude Measurement Technology for VHF Radar Based on Synthetic Steering Vector and Model Matching

Yi Le*[a], Ji Tang[a], Qiang Huang[b], Linhai Guan[a], Jianyang Zhou[a], Xiaoxuan Dong[c]

[a] Nanjing Glarun Defense Systems Co.,Ltd., Yuhuatai District, Nanjing, Jiangsu, P. R. China;

[b] Glarun Technology Co.,Ltd., Jiangning District, Nanjing, Jiangsu, P. R. China;

[c]Nanjing Research Institute of Electronic Technology, Yuhuatai District, Nanjing, Jiangsu, P. R. China;

[*] Corresponding author: leyi2133@163.com

Abstract: As radar is a kind of measuring device, the altitude measuring accuracy is a quite important core tactical indicator. In engineering practice, the altitude measuring accuracy deteriorates significantly in low elevation range, especially for low frequency radar such as VHF radar. This urgent problem to be solved is what we focus on in this article. After building a multipath reflection model, a method of altitude super-resolution measurement based on synthetic steering vector and model matching is proposed. Then some simulation analysis is finished, and some validation experiments are carried out on a practical VHF radar prototype. The results of theoretical simulation and prototype experiments show obviously improved altitude measuring performance compared with traditional measuring methods, verifying the correctness and engineering feasibility of the super-resolution altitude measurement technology proposed in this article. This study should have certain practical guiding significance for the theoretical and application research of related fields.

1. Introduction

The wavelength of VHF radar is generally 1 ~ 10m (working frequency of 30 ~ 300MHz), also known as meter-wave radar [1]. At the beginning of the development of radar technology, most of the early radars operated in this frequency band. However, with the growing demand for measuring performance, the VHF radar was once considered to be a kind of radar with poor performance due to its low measuring accuracy, and its development was limited to a certain extent.

However, in recent years, the development of stealth technology and anti-radiation missile (ARM) technology have attracted more attention from all countries in the world, posing a major threat to modern radar. In this case VHF radar has been refocused because of its potential performance in anti-stealth and anti-ARM [2][3]. For VHF radar, because the wavelength is comparable to the overall size of the aircraft, it works in the resonance region of the airframe. At the same time, the large wavelength of meter wave also requires the coating of aircraft absorbing materials to reach a thickness of more than ten centimeters, which is also difficult to achieve in reality. Therefore, the VHF radar has good anti-stealth performance [4]. In addition, the VHF radar also has the advantage in anti-ARM because if the ARM is to be able to intercept the VHF radar, its antenna size is required to be quite large, exceeding the size limit of the missile itself.

Because of such outstanding advantages, VHF band is becoming a hot spot in the development of a new generation of radar. But at the same time, the characteristics of meter-wave band determine that it is quite difficult to obtain satisfactory altitude measuring performance. That is because the beams of VHF radar are so wide that they hit the ground while working, causing the beam splitting phenomenon. In this case it will be strongly affected by the multipath interference effect caused by terrain factors such as ground or sea surface, which greatly reduces the altitude measuring accuracy at low elevation range, or even completely lose the ability of altitude measurement in serious cases. Therefore, improving the altitude measuring accuracy of VHF radar is a key problem to be solved urgently. This is also what we focus on in this article.

For low frequency radar such as VHF radar, due to its large beam width, in low-elevation range it receives the direct signal and the multipath reflection signal at the same time. Therefore, in fact the altitude measurement for VHF radar is a problem of DOA estimation. In this case the performance of the traditional altitude measuring methods such as amplitude and phase comparison methods decreases sharply, because they are based on the direct-propagation model. The problem is in essence caused by the mismatch of the model. Meanwhile, the direct wave and multipath reflected wave received by the radar in the low-elevation range are coherent, and the classical spatial spectrum estimation methods such as MUSIC algorithm are greatly limited in dealing with coherent sources.

In this article, the multipath reflection model is analyzed, and the synthetic steering vector based on the model is derived and introduced into the classical spatial spectrum estimation algorithm. Then the target's elevation angle is measured by searching the spectral peak within the elevation range using the spatial spectrum estimation method, so as to solve the problem of model mismatching and the limitation in coherent sources. The brief flow-process diagram is shown in Figure 1. Then some theoretical simulations and radar prototype experiments are carried out. It shows that the super-resolution altitude measurement technique based on synthetic steering vector and model matching has an obvious improvement on the altitude measurement for VHF radar, and its engineering feasibility is verified.

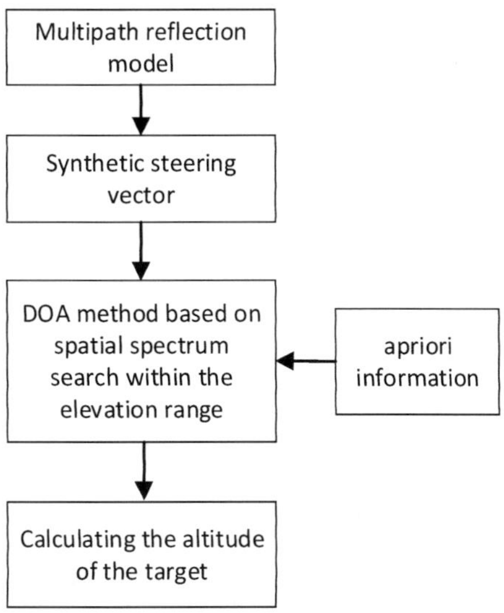

Figure 1. Flow-process diagram of the super-resolution altitude technique based on synthetic steering vector and model matching

2. Ii. Multipath reflection model of radar echo

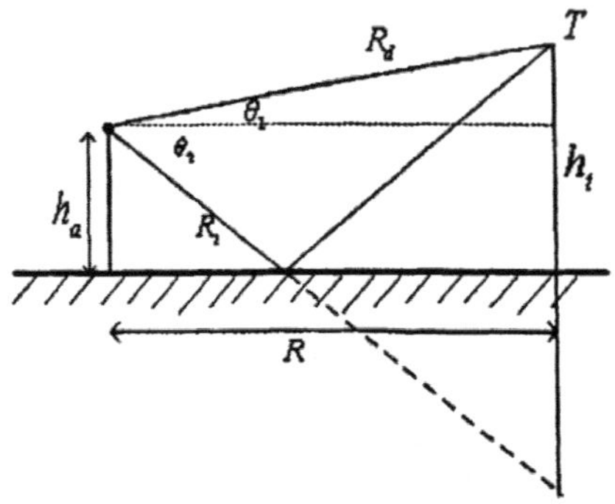

Figure 2. Multipath reflection model

The multipath reflection model of radar is shown in Figure 2 [5]. In the figure, h_a is the height of the center of the radar antenna, and H_t is the flight height of the target. R is the horizontal distance between the radar and the target, and Rd is the slope range between the radar and the target. $\theta_1$and θ_2 are the elevation angles of the target and its mirror image relative to the radar, and Ri is the multipath propagation distance generated by the ground. In addition to the signal directly reflected by the target, the echo signal received by the radar also includes the signal transmitting through multipath reflection.

If the direct received signal is taken as the reference, the reflection signal is:

$$s_i = \rho e^{j\alpha} \tag{1}$$

The total received signal is:

$$s = 1 + \rho e^{j\alpha} \tag{2}$$

Where ρ is the complex reflection coefficient, which is related to the frequency of the signal, the angle of incidence, the nature of the reflecting surface, etc.; α is the phase difference result from the propagation path of the reflected wave.

For a one-dimensional linear array composed of N units, according to the array signal synthesis principle, the data vector received by the array is:

$$s_l = [\mathbf{a}(\theta_1)\ \mathbf{a}(\theta_2)][1\ \rho \mathrm{e}^{j\theta}]^T s_d(t_l) + \mathbf{n}_l \tag{3}$$

In this formula $[a(\theta 1)\ a(\theta 2)]$ is the array flow vector, $s_d(t_l)$ is the complex envelope of direct wave signal, n_l is a complex Gaussian white noise uncorrelated with the signal.

$$\mathbf{a}(\theta_1) = [1,\ \exp(j2\pi(d/\lambda)\sin(\theta_1)),\ \ldots\ ,\exp(j2\pi((N-1)d/\lambda)\sin(\theta_1))]^T \tag{4}$$

$$\mathbf{a}(\theta_2) = [1,\ \exp(j2\pi(d/\lambda)\sin(\theta_2)),\ \ldots\ ,\exp(j2\pi((N-1)d/\lambda)\sin(\theta_2))]^T \tag{5}$$

Then equations (4) and (5) are the array steering vectors for the direct and reflected waves.

From the above multipath reflection model of radar, it can be seen that the radar received signal is a composite signal of direct wave and multipath reflection wave. They travel through different paths and arrive at the antenna at different times, resulting in different phases and amplitudes of the two signals, so the spatial envelope of the combined signal presents a modulated shape caused by multipath interference. The phase and amplitude differences of multipath signals are also related to the nature of the reflecting surface. The complex reflection coefficient ρ shows the different topographic factors.

It's obvious that in the presence of multipath reflection, if we use the traditional radar altitude measurement method, such as the method of multi-beam amplitude comparison, the angle measurement results will point somewhere between the real target and its mirror image, resulting in altitude measurement errors and affecting the measurement accuracy.

3. Super-resolution altitude measurement technique based on synthetic steering vector and model matching

From the above analysis of multipath reflection model, it can be seen that the key to altitude measurement of VHF radar lies in decomposing the signal source from the radar echo signal containing direct wave and multipath reflection wave, which is a problem of signal classification in essence. For many years, the eigen-decomposition based super-resolution spatial spectrum estimation algorithms have been extensively studied. The multiple signal classification (MUSIC) algorithm is an efficient processing method. It is worth noting that the classical MUSIC algorithm has a good directional resolution for incoherent signals, but the performance of MUSIC algorithm significantly deteriorates when dealing with coherent signal sources [6]. In fact, the direct wave signal and multipath reflected signal received by radar are typically coherent. Therefore, in this study, based on the MUSIC algorithm, we introduce the synthetic steering vector for model matching to achieve the estimation of coherent signal sources.

3.1 Spatial spectrum estimation method

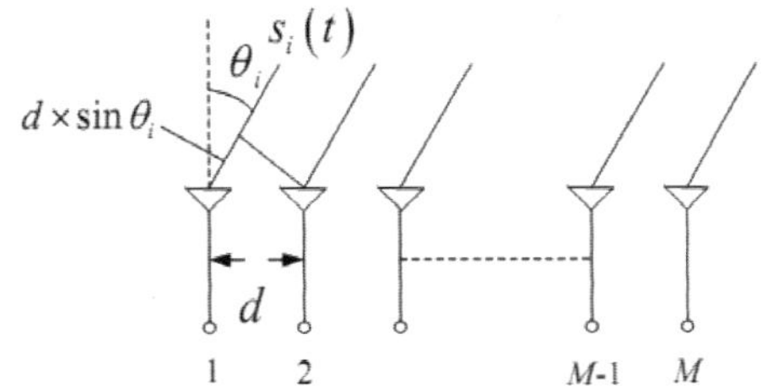

Figure 3. linear array model

Taking the linear array model shown in Figure 3 as an example. For a linear array with M units, d is the unit spacing distance, and θ_i is the incident signal angle of the ith signal source. The array received signal is

$$X(t)=[\mathbf{a}(\theta_1),\ \mathbf{a}(\theta_2),\ \dots,\ \mathbf{a}(\theta_3)]\mathbf{S}(t)+\mathbf{N}(t) \tag{6}$$

The steering vector of the i-th signal source is

$$\mathbf{a}(\theta_i)=[1,\ \exp(j2\pi(d/\lambda)\sin(\theta_i)),\ \dots,\exp(j2\pi((M-1)d/\lambda)\sin(\theta_i))]^T \tag{7}$$

Then the covariance matrix of the received signal is:

$$\mathbf{R}_x=E\{\mathbf{X}(t)\mathbf{X}^H(t)\}=\mathbf{A}\mathbf{R}_S\mathbf{A}^H+\sigma^2\mathbf{I} \tag{8}$$

According to the properties of the matrix, $\boldsymbol{AR}_S\boldsymbol{A}^H$ is a Hermitian positive semidefinite matrix, so R_x has M eigenvalues. Considering that noise power $\sigma^2>0$, therefore if there are N incoherent sources, with $N\le M$, then the M eigenvalues satisfy:

$$\lambda_1\ge\lambda_2\ge\dots\ge\lambda_N=\lambda_{N+1}=\dots=\lambda_M \tag{9}$$

Then the eigenvector matrix consisting of the eigenvectors corresponding to these M eigenvalues is:

$$\mathbf{U}_s=[\mathbf{u}_1,\mathbf{u}_2,\dots,\mathbf{u}_N] \tag{10}$$

$$\mathbf{U}_n=[\mathbf{u}_{N+1},\mathbf{u}_{N+2},\dots,\mathbf{u}_M] \tag{11}$$

Equations (10) and (11) are the signal subspace and the noise subspace of the array received signal, respectively. By searching the entire airspace for the following equation, a peak will appear at the angle where the source is located.

$$P = \frac{1}{\mathbf{a}^H(\theta)\mathbf{U}_N\mathbf{U}_N\mathbf{a}(\theta)} \tag{12}$$

For the altitude measurement problem of VHF radar, the steps of the altitude measurement algorithm based on the spatial spectrum estimation can be summarized as follows:

a) For the altitude measurement problem of VHF radar, the steps of the altitude measurement algorithm based on the spatial spectrum estimation can be summarized as follows:

b) Calculating the covariance matrix of the array receiving signal Rx. In practice, the covariance matrix can be estimated by using the data of pulses obtained from each pulse group of the radar.

c) Performing eigenvalue decomposition on the obtained covariance matrix.

d) The number of sources is estimated from the eigenvalue. For the altitude measurement problem of VHF radar, the multipath model usually appears as an obvious mirror target, so the number of sources can generally be taken as 2.

e) Find the noise subspace UN corresponding to the smallest eigenvalue.

f) Carry out spectral peak search using the formula (12) in the range of elevation angle.

The angle corresponding to the peak is the angle of the source. For the multipath model of VHF radar, the target angle is generally above the elevation angle.

3.2 Super-resolution altitude measurement method based on synthetic steering vector and multipath model matching

As mentioned above, the multipath reflected signal and the direct wave signal received by the meter-wave radar are coherent signals, and the performance of the MUSIC algorithm will deteriorate. In this case, a multipath reflection model represented by formulas (1) to (5) is introduced, and a synthetic steering vector is formed by taking the relationship between the reflected wave signal and the direct wave signal as prior information. The angle information of the target can be obtained by performing the one-dimensional MUSIC spectral peak search in the spatial domain based on the steering vector.

In combination with the form of the synthesized echo signal described by Equation (5), the steering vector of Equation (7) is rewritten as:

$$\mathbf{a}_{syn}(\theta_1,\theta_2) = \mathbf{a}(\theta_1) + \rho e^{j\alpha}\mathbf{a}(\theta_2) \tag{13}$$

It can be seen from the above formula that in the actual situation, the target distance and the height of the radar antenna are known. Therefore, according to the multipath reflection model shown in Figure 2, it can be seen that the angle of the multipath reflection wave and its phase difference α with the direct wave are the function of the target elevation angle [5], and the ground complex reflection coefficient ρ can also be determined by the priori model. Therefore, the steering vector of formula (13) is a function of only the target angle of θ_1.

The target angle can be determined by substituting the synthetic steering vector of Equation (13) into Equation (12) in the MUSIC spatial spectrum estimation method.

4. Simulation analysis and practical prototype experiment

In this study we use a VHF linear array radar as a prototype equipment. The super-resolution altitude measurement method based on synthetic steering vector and model matching is verified and analyzed.

The antenna array of the VHF radar is composed of 12 rows of vertically placed elements, and each row consists of 2 radiating elements. Each antenna element corresponds to a digital T/R channel. The radar operates in the meter-wave band.

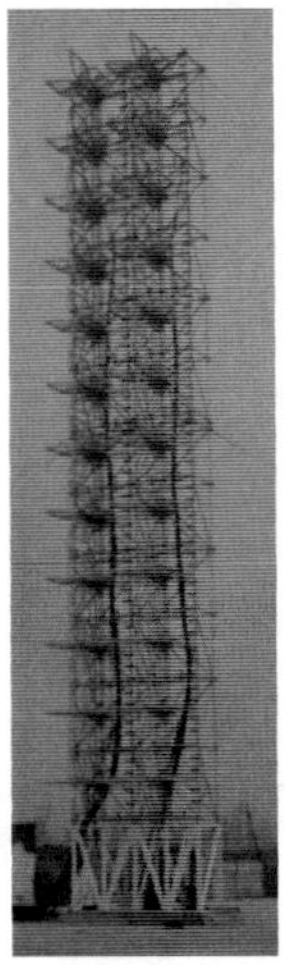

Figure 4. VHF linear array radar prototype

4.1 Simulation results

Using the flat terrain model, echo data received by the array is simulated and generated according to formula (3). The elevation angles of the target and its mirror image are 1.9 ° and -1.9 ° respectively, and the typical value of the reflection coefficient is -0.9. The super-resolution altitude measurement method based on synthetic steering vector discussed in this article is used to estimate the angle of the target, and the spatial spectrum search results of formula (12) are shown in Figure 5.

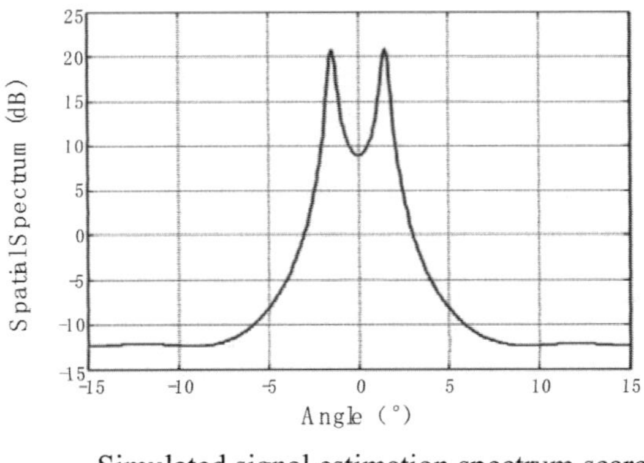

Figure 5. Simulated signal estimation spectrum search result

As can be seen, the spatial spectrum search result shows two peaks, corresponding to the angles of the target and its mirror image. The simulation results show that the proposed method can correctly and effectively find the elevation angles of the target and its mirror image.

4.2 Practical experiment results on the prototype

The VHF radar prototype is deployed in the real environment detecting the flying aircrafts in the air. An ADS-B equipment is used to obtain the true altitude information of the aircraft targets as the true value. The ADS-B targets are paired with the targets detected by the radar, so as to statistic the altitude estimation deviation and evaluate the actual performance of the proposed super-resolution altitude estimation method based on synthetic steering vector and model matching. By tracking batches of targets continuously we statistic the altitude measurement results, as shown in Figure 6 compared with the results of traditional amplitude comparison method.

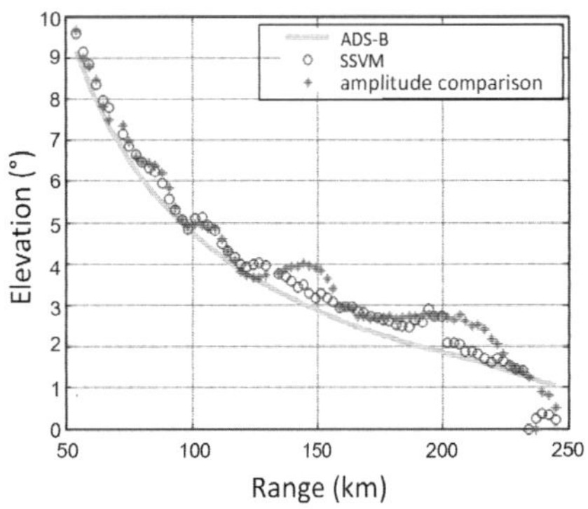

a) elevation measurement results of the prototype experiment

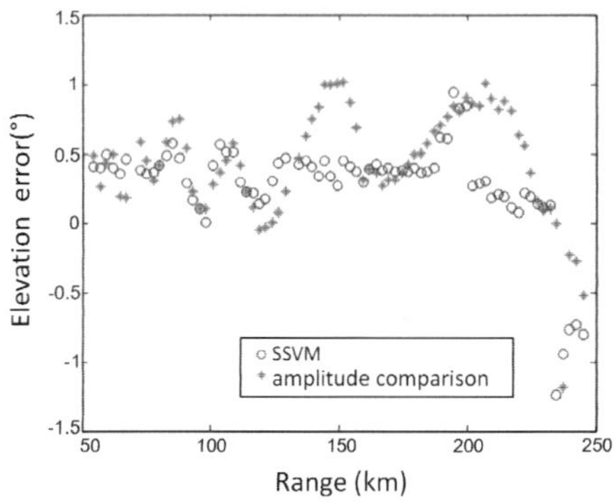

b) elevation error of the prototype experiment

Figure 6. Results of the prototype experiment compared with traditional method of amplitude comparison

As can be seen from the figures, compared with the traditional method of amplitude comparison, the method proposed in this paper has much better measuring performance in the low-elevation area (beyond 130km). By statistics, the altitude measurement accuracy of the proposed method in low elevation area reaches about 1/50 of the beam width, while that of the traditional method is about 1/10. The experimental results show that the altitude super-resolution technique based on synthetic steering vector and model matching has an obvious advantage.

5. Conclusions

Due to its unique advantages in anti-stealth and anti-ARM, more attention is being paid to VHF radar. The difficulty of altitude measurement in low-elevation area caused by its large beamwidth has become an urgent problem to be solved as a hot research topic. In this article, a super-resolution altitude measurement technique using synthetic steering vector and spatial spectrum estimation based on multipath model matching is proposed. The performance of the method is simulated and analyzed. Then an engineering verification experiment is carried out on a VHF radar prototype. The experimental results show a satisfactory altitude-measuring performance, realizing the super-resolution altitude measurement capability. Correctness of the method proposed is verified theoretically and experimentally, and its feasibility of engineering implementation is confirmed. This study should have positive and practical guiding significance for the theoretical and engineering research in related fields.

References

[1] A. Gu. "Development of meter-wave 3D radar", Modern Weaponry, 4, 50-51(2009).
[2] F. Y. Tao, "Introduction to Modern Meter-band Radar", Land-based Air Defence Weapons, 4, 34-38(2002).
[3] J. L. Li, Y. Z. Li, X. S. Wang, "Study on Anti-Stealth with meter-band Polarimetric Radar", Radar Science and Technology, 3, 321-327(2005).
[4] X. R. Xia, "The Development of Stealth Technology", Aerospace China, 1, 40-44(2002).
[5] M. I. Skolnik, Radar Handbook, Publishing House of Electronics Industry, Beijing, 2010.
[6] Y. L. Wang, Theory and algorithm of spatial spectrum estimation. Tsinghua University Press, Beijing, 2009.

Health state estimation of lithium ion batteries based on electrochemical impedance spectroscopy

Kelun Zhao*[a], Zhimao Ming[a], Jiajie Wu[b], Wei Sun[b]

[a]Guangzhou GRG metrology and test Co., Ltd, Guangzhou, China

[b]Huazhong University of Science and Technology, Wuhan, China

* Corresponding author: zhaokl@grgtest.com

ABSTRACT: The estimation method based on the electrochemical impedance spectrum can comprehensively describe the reaction process of Li-ion battery without causing damage to the battery, and is suitable for estimating the state of health of the battery. In this paper, the equivalent circuit model of Li-ion battery is established by analyzing the ac impedance spectrum of Li-ion battery and the internal reaction process of the battery, and corresponding each component of the model and the composite link with the curve segments in the impedance spectrum, and the model parameters are identified offline with the help of The Nyquist diagram of impedance spectrum. It is found that the charge transfer internal resistance and low-frequency impedance amplitude in the equivalent circuit increase monotonically with the aging of the battery. Then, the temperature compensation coefficient was introduced to solve the influence of temperature factors on the SOH estimation of lithium batteries, And the root mean squared error of SOH estimation before and after temperature compensation was reduced from 31.7 to 3.0, and the sum of squares for Error was reduced from 13054.7 to 19.8, which significantly improved the SOH estimation of batteries with different temperatures; and a model based on BP neural network was constructed, And the correlation coefficient of model prediction was 0.994. Solving the influence of production process differences on SOH estimation of lithium batteries, and the advantages and disadvantages of the two estimation methods are summarized.

1. Introduction

Lithium ion batteries have the advantages of long cycle life, environmental protection, low self-discharge rate, and are widely used in electric vehicles, energy storage systems and other occasions [1-3]. In response to the national call for energy conservation and environmental protection, lithium batteries need to be graded according to the State of Health (SOH), which requires a more accurate estimation of the battery's health status.

The traditional experience-based health state estimation of lithium-ion batteries is mainly realized by combining experimental data with prior knowledge. These methods are convenient to operate and simple in principle, but the estimation accuracy is low and the anti-interference ability is poor. With the development of machine learning algorithm, the SOH estimation method of lithium battery based on data drive appears [4-7], such methods have high data requirements, incomplete or uncertain data acquisition will have a great impact on the results.

Battery models are usually required to estimate battery health, Xiong Rui *et al* [8] Firstly, an electrochemical model was established based on the physical and chemical reaction process inside the battery. Then, parameters of the model were identified by an optimization algorithm, and five aging characteristics were quantified to estimate the health status of the battery. Omar, Monem [9] The study of et al. showed that high temperature and high rate current charge and discharge would reduce the cycle life of the battery, and thus established a cycle life model with battery internal resistance as the characteristic quantity to predict the SOH of the battery. Also need to build a battery equivalent circuit model, but this article USES the analysis method based on electrochemical impedance spectroscopy can be more comprehensive understanding of lithium battery internal reaction process information, and because of its incentive and response amplitude signals are small, can guarantee no damage inside the battery at the same time the more accurate estimates of the battery SOH [10].

There are many scholars who have carried out research in this aspect, Zhang Wenhua, Pei Feng [11] Based on the change of impedance spectrum, the change rule of charge transfer resistance with the health state of the battery was summarized, and the health state estimation of lithium battery was realized. Arijit Guha [12] Based on the experimental data of electrochemical impedance spectroscopy, a model based on capacity degradation and internal resistance growth was proposed to estimate SOH. Xu Xinmin [13] et al. found the monochromatic variation rule of impedance amplitude and battery SOH under the condition of relatively low frequency, and estimated SOH by fitting curve on this basis. Their research results are also reflected in this paper. In this paper, the estimation method based on impedance amplitude and the estimation method based on charge transfer resistance are compared and their advantages and disadvantages are found. In addition, their estimation methods do not consider the influence of temperature and other factors on SOH estimation, which is also one of the key points discussed in this paper.

2. Equivalent circuit model of lithium ion battery

2.1 Equivalent circuit model and impedance spectrum significance

Figure. 1 is the equivalent circuit model of lithium ion battery used in this study.

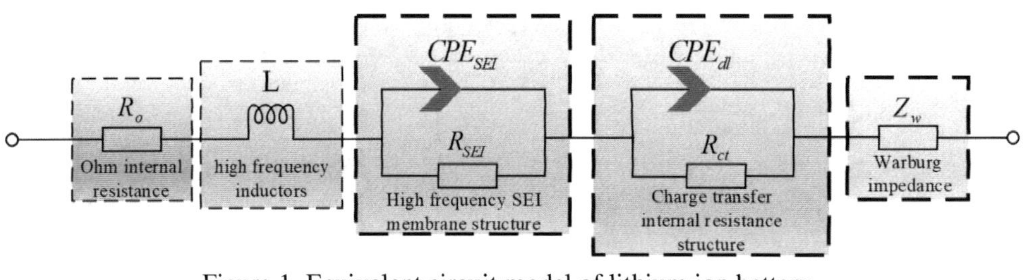

Figure 1. Equivalent circuit model of lithium ion battery.

Among them, the electrode reaction at high frequency presents sensibility externally, represented by the high frequency inductance L, and the SEI film structure contains the membrane impedance R of the SEI filmSEIAnd constant phase element CPESEI, describe the diffusion process of lithium ion through SEI film on the surface of electrode material. Charge transfer internal resistance structure contains charge transfer internal resistance RctAnd constant phase element CPEdlIs used to describe the charge conduction process in electrode reaction. Warburg impedance is generated by the diffusion process inside the battery, and is represented by Z_W [14].

The electrochemical impedance spectrum corresponding to the equivalent circuit model is shown in Fig. 2.The part below the real axis represents the high-frequency inductance, which can be ignored generally, and the value of the intersection between the curve and the real axis represents the ohmic internal resistance ROThe first high frequency arc corresponds to the high frequency SEI film structure, the second intermediate frequency arc corresponds to the charge transfer resistance structure, and the low frequency straight line corresponds to the Warburg impedance.

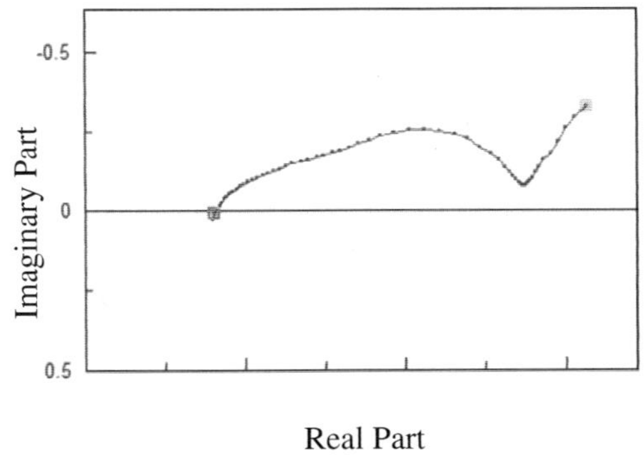

Figure 2. Nyquist diagram of electrochemical impedance spectroscopy of a cell.

2.2 Model parameter identification

Its impedance can be obtained from the equivalent circuit model as shown in (1).

$$Z = j\omega L + R_0 + \frac{R_{SEI}}{R_{SEI}T_{SEI}(j\omega)^{P_{SEI}} + 1} + \frac{\alpha}{\alpha T_{dl}(j\omega)^{P_{dl}} + (j\omega T_W)^{P_W}} \tag{1}$$

Among them, $\alpha = R_{ct}(j\omega T_W)^{P_W} + R_W \tanh[(j\omega T_W)^{P_W}]$, T_{SEI} and P_{SEI} is constant phase element CPE_{SEI}. The parameters T_{dl} and P_{dl} is constant phase element CPE_{dl}. The parameters R_W denotes Warburg impedance Z_W. The magnitude of theta T_W and P_W is parameters in Z_W.

Electrochemical impedance spectrum curve of the lithium battery exchange each segment has a composite elements in the equivalent circuit and the matching, regardless of the real axis under the high frequency part of the first of two pieces of circular arc with parallel structure of the R and CPE fitting, then straight with Warburg impedance matching part, this is piecewise local fitting, then to simulate global as initial value, Generally, the fitting effect of Warburg impedance will be poor after overall fitting, and the error may reach more than 20%, so the fitting result of Warburg impedance is generally not used.

3. Health status estimation of lithium ion batteries

3.1 Definition and influencing factors of SOH for lithium ion batteries

The health status of lithium ion battery can be represented by the SOH, SOH commonly used to define parameters of internal resistance, residual cycles, capacity, including internal resistance and the temperature of the battery and charged state, is not easy to measure, each battery remaining cycles have individual differences, the general definition of residual cycles will return to capacity, So it is appropriate to define SOH in terms of capacity [15], its definition is as formula (2).

$$\mathrm{SOH} = \frac{C_{\mathrm{now}}}{C_{\mathrm{new}}} \times 100\% \tag{2}$$

In which C_{now} represents the capacity released when the battery is discharged at a certain rate from the fully charged state to the cut-off voltage, C_{new} indicates the rated battery capacity.

In the actual cycle, the capacity of lithium-ion batteries will constantly lose, so the influencing factors of SOH need to be analyzed before the battery status assessment.

As shown in Fig. 3, the temperature is the main factor affecting the battery state of health, the battery when work in low temperature condition, its internal diffusion effect will be blocked, leading to the charge transfer resistance increases, the solidification happens inside the battery solution, electron transport slow, conductive performance degradation, low temperature lithium battery cathode analysis is serious, the diaphragm compatibility, which led to the decrease of the SOH; Battery after repeated charge and discharge cycle, its temperature rise, high temperature will destroy the balance of the battery inside, cause side effects, inside the battery of the SEI film thickness increases, at the same time high temperature accelerated the decomposition of electrolyte, resulting in a decline in the rate of the battery internal various ion penetration, making capacity attenuation, health status [16].Similarly, charge and discharge current ratio [17], charge and discharge cycle interval [18], accumulated discharge depth [19] Will affect the battery SOH.

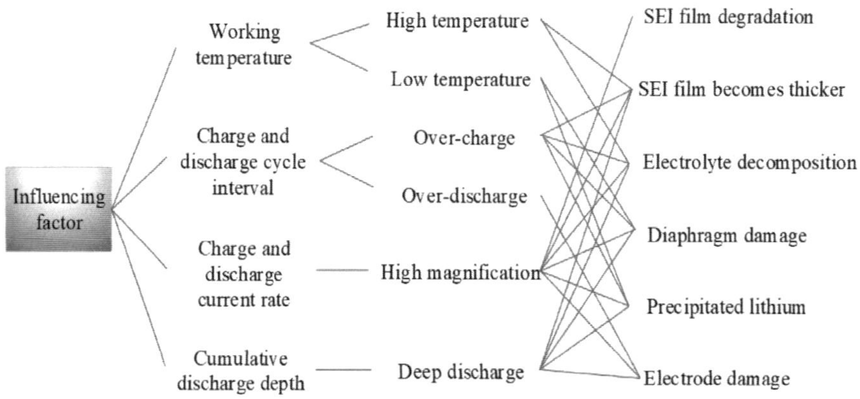

Figure 3. Influence mechanism of battery health status.

Is coupling between various influencing factors, and is not a simple one-to-one correspondence relationship, so if you want to quantitative analysis based on the various factors of lithium ion battery SOH change will be very tedious, estimate of the SOH accuracy will be low, but after knowing these lead to capacity attenuation factor, again by electrochemical impedance spectroscopy analysis, will be more clear.

3.2 Analysis of impedance spectrum aging of lithium ion battery

The data used in this paper are from experiments conducted by Ulrich Stimming of Faraday Research Institute and Alpha A. Lee of Cavendish Laboratory [20], 12 commercial lithium ion button batteries (45mAh Eunicell LR2032) were selected for the experiment. Among the 12 batteries, 8 were tested at room temperature of 25℃, 2 at 35℃, and 2 at 45℃. The variable of working state was temperature. Charge and discharge process: 1C(45mA) constant current to the cut-off voltage, then constant voltage charge to 4.2V, and then 2C(90mA) constant current discharge to 3V.For EIS test, electrochemical impedance spectroscopy (EIS) is measured with a small excitation current of 5mA. Charge and discharge cycles are conducted until the battery reaches the End of life (EOL).

Figure 4. Nyquist diagram of impedance spectrum of lithium battery with different SOH.

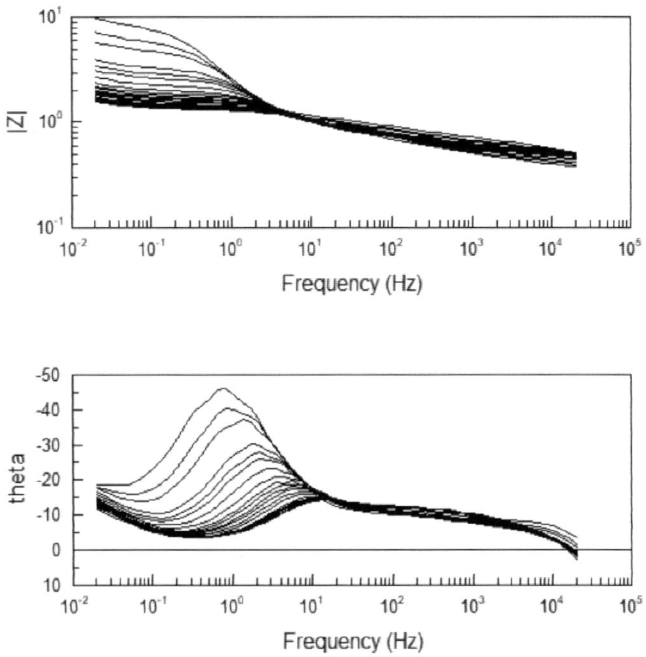

Figure 5. Bode diagram of impedance spectrum of lithium battery with different SOH.

The frequency range is 0.02hz to 20000Hz, and 60 frequency points are measured in each cycle. The real and imaginary parts, impedance amplitude and impedance phase Angle of the electrochemical impedance spectrum were measured at each frequency point. Fig. 4 and Fig. 5 can be drawn in ZView software according to some battery data.

Fig. 4 is Nyquist diagram of impedance spectrum of lithium battery with different SOH at 35℃, and Fig. 5 is Bode diagram of impedance spectrum of lithium battery with different SOH at 25℃. Five rules can be concluded by observing the two graphs:

(1) With the decrease of battery SOH, the high frequency part of Nyquist diagram of impedance spectrum hardly changes below the real axis, which indicates that the inductance L in the equivalent circuit model hardly changes, which is not suitable for SOH estimation.

(2) With the decrease of SOH, the ohmic resistance R of the battery$_{O}$Almost coincident, indicating that the ohmic internal resistance is not suitable for the estimation of SOH.

(3) With the decrease of battery SOH, the high frequency arc of impedance spectrum does not change significantly, indicating that R$_{SEI}$With CPE$_{SEI}$Not suitable for estimation of SOH.

(4) The mid-frequency arc of impedance spectrum Nyquist graph increases monotonically with the aging diameter of the battery, and the SOH of the battery can be estimated by the charge transfer resistance.

(5) As the battery life decays, the slope of the line in the low-frequency region of Nyquist diagram does not change significantly, but the amplitude and phase Angle of the low-frequency impedance in Bode diagram obviously change significantly, which can be used to estimate the SOH of the battery.

Based on the above analysis, we can select the charge transfer resistance and low frequency impedance amplitude to battery SOH estimation, using ac impedance spectrum in the ZView battery equivalent circuit model for parameter identification, to 25 ℃ under two of the same type battery charge transfer resistance for fitting, the results are shown in Table 1, fitting the result error is within 2%, conform to the requirements.

Table 1. Identification results of charge transfer resistance.

R_{ct1}/ Ω	error	SOH1	R_{ct2}/ Ω	error	SOH2
0.705	1.3%	91.9%	0.7	1.1%	90.44%
0.734	1.8%	86.4%	0.711	1.1%	86.56%
0.757	1.6%	84.4%	0.728	1.2%	83.47%
0.806	2.0%	77.3%	0.739	1.1%	78.52%
0.823	1.4%	70.5%	0.745	1.0%	72.92%
0.847	1.6%	65.8%	0.759	1.0%	69.41%
0.867	1.5%	60.8%	0.776	1.1%	63.49%
0.935	1.6%	50.8%	0.792	1.0%	58.29%
0.997	1.4%	44.8%	0.846	1.1%	50.84%
1.102	1.2%	37.7%	0.898	0.9%	45.20%
1.177	1.1%	32.7%	0.959	1.0%	40.05%

3.3 SOH estimation method of lithium battery based on charge transfer resistance

First, the charge transfer resistance is used to estimate the battery SOH. Since the battery SOH is less than 50%, the lithium battery is close to scrap, and can only be recycled or applied to places with low requirements on the battery, so when estimating the SOH of lithium battery, all data less than 50% are excluded, and two batteries of the same kind are selected at 25℃. Fig. 6 and Fig. 7 show the relationship between charge transfer resistance and SOH.

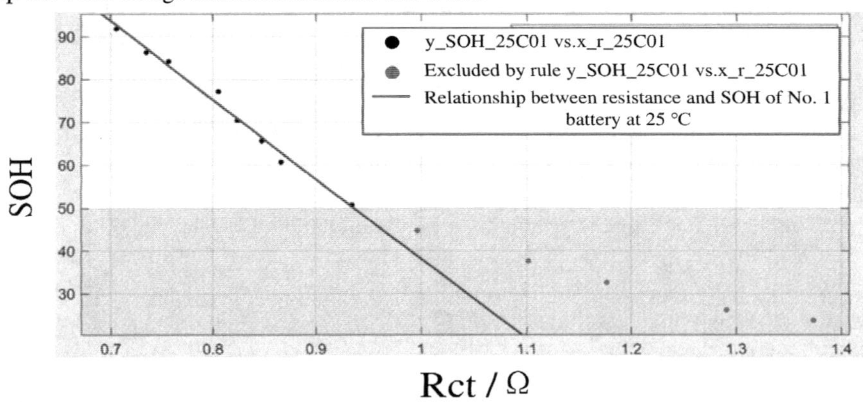

Figure 6. Relationship between charge transfer resistance and SOH of battery 1 at 25℃.

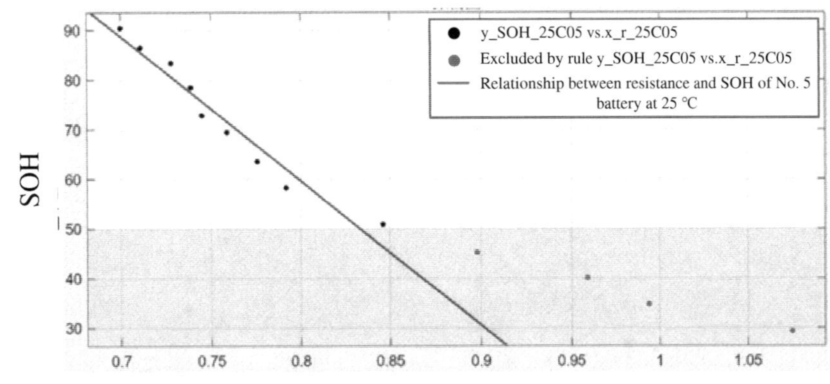

Figure 7. Relationship between charge transfer resistance and SOH of battery 2 at 25℃.

The fitting relation of battery 1 is shown in Equation (3), and that of battery 2 is shown in Equation (4).

$$SOH = -185.1x + 223.3 \quad (3)$$

$$SOH = -291x + 292.4 \quad (4)$$

Where, x represents the resistance value of charge transfer, and fitting effect can be achieved by fitting correlation coefficient R^2 And root mean square error RMSE, R^2 The closer it is to 1, the smaller the RMSE is, indicating the better the prediction model can describe the experimental data.The fitting curve R of battery 1^2=0.9871, RMSE=1.721, fitting curve R of battery 2^2=0.9541, RMSE=3.161, the estimation goodness of SOH meets the requirements, but it can be found that the SOH estimation formula of the same battery at the same temperature also has a great difference, which is related to the measurement accuracy of impedance spectrum data, the fitting accuracy of ZView software for equivalent circuit parameters, and the difference of battery production process. The SOH estimation curve of one battery cannot be used to estimate the health status of other batteries.

The BP neural network can be used to process the data of four groups of batteries. The internal resistance of charge transfer in each group of batteries is selected as the SOH of 95%, 90%, 87%, 84%, 80%, 75%, 70%, 65%, 60% and 55% respectively, and the SOH estimation model of these batteries can be trained to predict the SOH of these batteries.

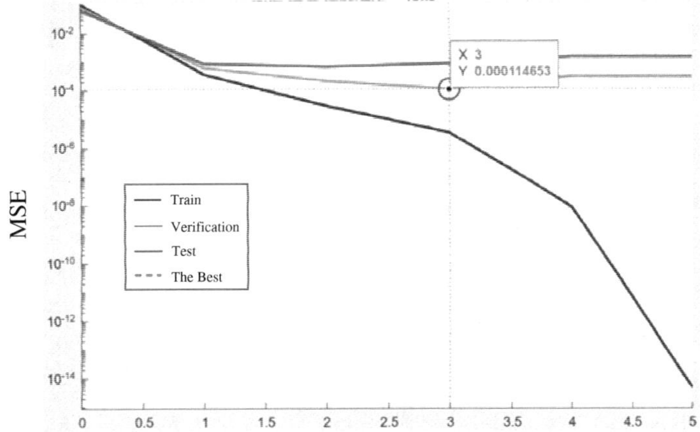

Figure 8. Estimated performance of battery SOH.

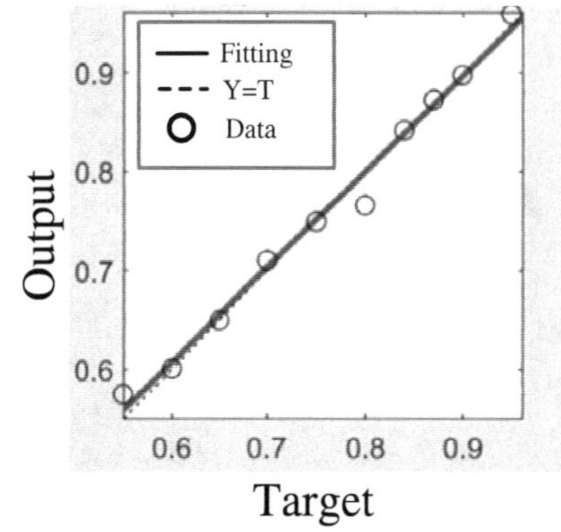

Figure 9. Fitting diagram of battery SOH estimation.

As can be seen from the performance diagram in Fig. 8, the verification result in the third round is the best. As can be seen from the fitting diagram in Fig. 9, the data fluctuates around the fitting curve, the overall trend is consistent, and the fitting coefficient is R2Close to 1, good fitting effect.

As shown in Fig. 10 is 45 ℃, 25 ℃, 35 ℃, the same kind of lithium ion battery in SOH = 90% of the health state of the electrochemical impedance spectra, from figure can be found in the aspects of the influence of temperature on the lithium ion battery is, curve and real axis intersection point represents the ohmic resistance fluctuation happens, semicircle diameter on behalf of the charge transfer resistance size has changed, The amplitude and phase Angle of the impedance also fluctuate, which greatly affects the estimation of the health state of lithium ion battery. Fig. 11 and Fig. 12 show the difference of health state estimation curves of the same kind of lithium batteries at 25℃ and 45℃, 35℃ and 45℃. The black data points in the figure represent the relationship between resistance and SOH in charge transfer of batteries at 25℃ and 35℃ respectively, and the green data points represent the relationship between resistance and SOH in charge transfer of batteries at 45℃ respectively. It can be found that the distribution of data points of different temperatures varies greatly.

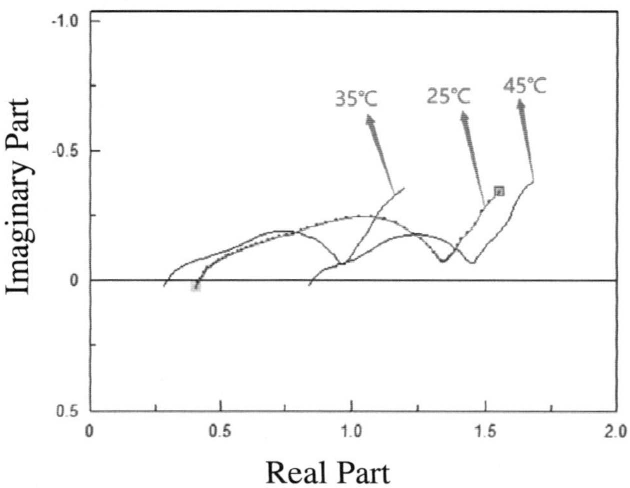

Figure 10. Nyquist diagram of electrochemical impedance spectra of lithium batteries at different temperatures.

Table 1. Identification results of charge transfer resistance.

R_{ct1}/ Ω	error	SOH1	R_{ct2}/ Ω	error	SOH2
0.705	1.3%	91.9%	0.7	1.1%	90.44%
0.734	1.8%	86.4%	0.711	1.1%	86.56%
0.757	1.6%	84.4%	0.728	1.2%	83.47%
0.806	2.0%	77.3%	0.739	1.1%	78.52%
0.823	1.4%	70.5%	0.745	1.0%	72.92%
0.847	1.6%	65.8%	0.759	1.0%	69.41%
0.867	1.5%	60.8%	0.776	1.1%	63.49%
0.935	1.6%	50.8%	0.792	1.0%	58.29%
0.997	1.4%	44.8%	0.846	1.1%	50.84%
1.102	1.2%	37.7%	0.898	0.9%	45.20%
1.177	1.1%	32.7%	0.959	1.0%	40.05%

3.3 SOH estimation method of lithium battery based on charge transfer resistance

First, the charge transfer resistance is used to estimate the battery SOH. Since the battery SOH is less than 50%, the lithium battery is close to scrap, and can only be recycled or applied to places with low requirements on the battery, so when estimating the SOH of lithium battery, all data less than 50% are excluded, and two batteries of the same kind are selected at 25℃. Fig. 6 and Fig. 7 show the relationship between charge transfer resistance and SOH.

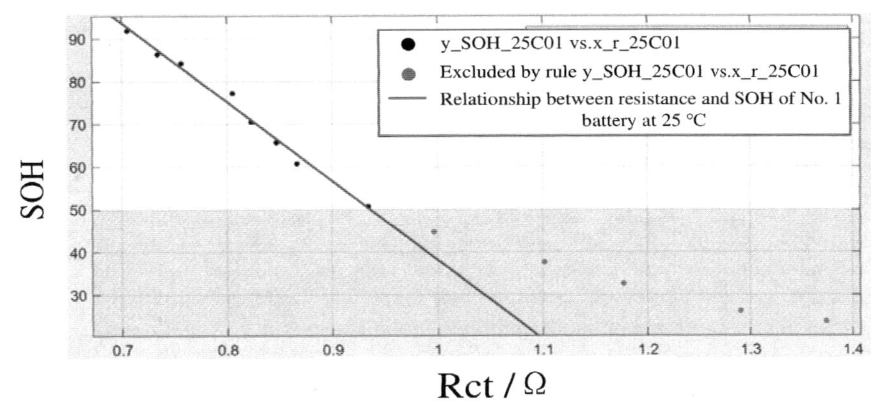

Figure 6. Relationship between charge transfer resistance and SOH of battery 1 at 25℃.

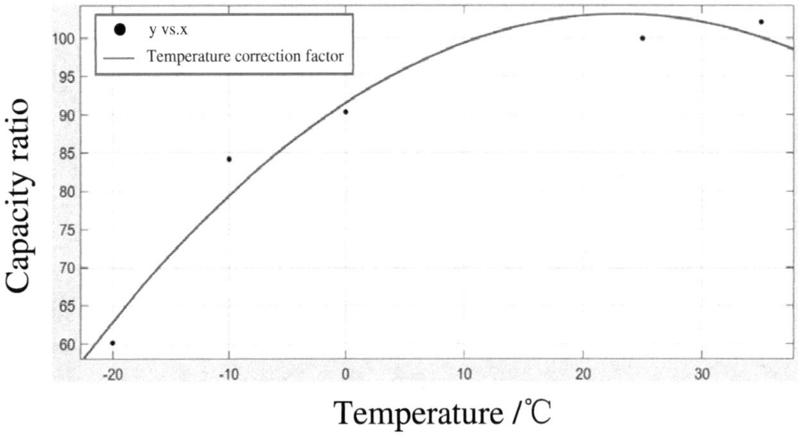

Figure 13. Temperature compensation function curve.

The temperature compensation function obtained is shown in Equation (5).

$$y = -0.0002x^2 + 0.01x + 0.9155 \tag{5}$$

SOH can thus be redefined as equivalent to the rated capacity C for any temperature is converted, as shown in Equation (6).

$$\mathrm{SOH} = \frac{C_{now}}{C_{\mathrm{new}} \times (-0.0002x^2 + 0.01x + 0.9155)} \times 100\% \tag{6}$$

Use the new SOH to redraw the relationship curve between charge transfer resistance and SOH, as shown in Fig. 14 and Fig. 15.

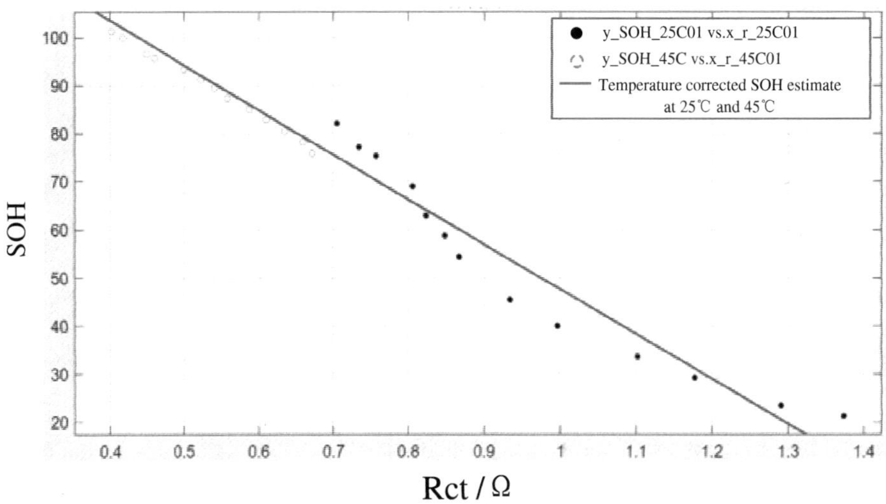

Figure 14. Battery health status curves after temperature compensation at 25℃ and 45℃.

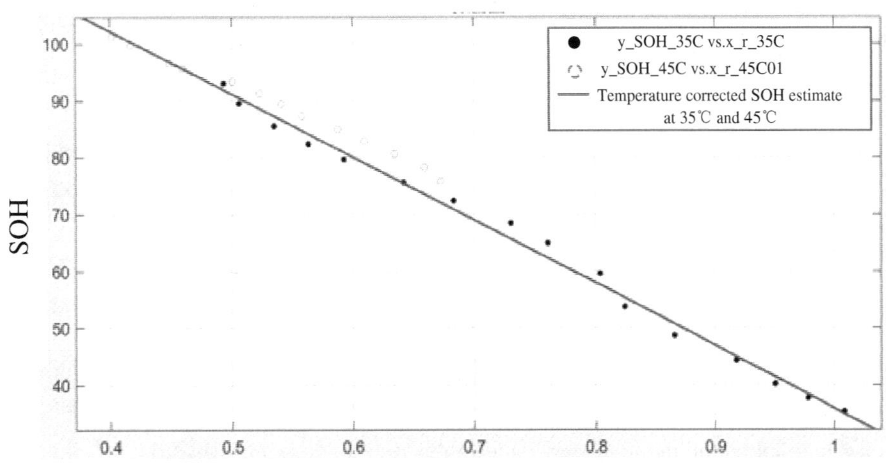

Figure 15. Battery health status curves of 35℃ and 45℃ after temperature compensation.

In the figure, black solid points are known fitting curves for SOH estimation of charge transfer internal resistance of lithium batteries at 25℃ and 35℃, while green virtual points are the relationship between charge transfer internal resistance of lithium batteries at 45℃ and SOH, which are the objects to be predicted. It can be intuitively found that the virtual points to be predicted are well distributed on the fitting curve.

The relationship between SOH estimation of charge transfer resistance at 25℃ and 35℃ after temperature compensation is shown in Equation (7) and Equation (8).

$$SOH = -0.9322x + 14.08 \tag{7}$$

$$SOH = -1.106x + 14.65 \tag{8}$$

The root mean square errors of curve fitting were 2.99076 and 1.52562, and the sum of squares of residual errors were 19.8 and 12.9, respectively, which were significantly improved compared with that before temperature compensation.

3.4 SOH estimation method of lithium battery based on low-frequency impedance amplitude

Through the previous analysis, the charge transfer resistance is used to estimate the SOH of lithium batteries, but in many cases, the model parameter R_{ct}It is very troublesome to obtain. Firstly, the impedance spectrum of lithium battery needs to be measured in the frequency range of 0.02Hz~20000Hz, and then the equivalent circuit parameter fitting of impedance spectrum can be obtained by simulation software such as ZView.This process is cumbersome, and requires special measuring equipment, expensive equipment, and charge transfer resistance R_{ct}Because of the error of software accuracy, this method is not the most suitable method for practical engineering application.

As can be seen from Fig. 5, in the Bode diagram of electrochemical impedance spectrum of lithium ion battery, impedance amplitude and impedance phase Angle also change with the aging degree of lithium battery, especially the impedance amplitude changes significantly at low frequency.

Plot the change relationship between impedance amplitude and SOH of two batteries of the same type at low frequency 0.02Hz at 25℃. It should be noted that SOH here is calculated from the capacity, and the capacity in the original data is obtained by the integration of current, which has no direct correlation with impedance amplitude. The graphs are shown in Fig. 16 and Fig. 17.

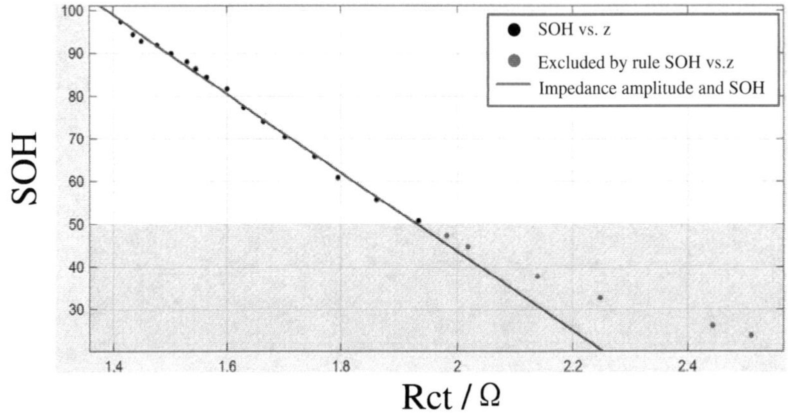

Figure 16. Relationship between impedance amplitude and SOH of battery 1 at low frequency (0.02Hz).

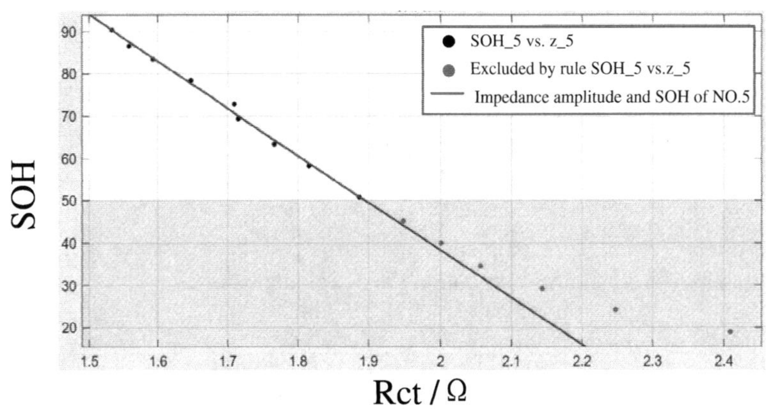

Figure 17. Relationship between impedance amplitude and SOH of battery 2 at low frequency (0.02Hz).

Can be seen from the diagram, along with the battery aging, electrochemical impedance of low-frequency impedance amplitude increases monotonously and increase obviously, the approximate linear relationship, the impedance of battery 1 amplitude normalization processing, which USES the current amplitude divided by the initial amplitude, the estimated curve can be used for this type of lithium battery SOH estimation, fitting a curve, shown in Fig. 18, The estimated function of the fitting is shown in Equation (9).

$$SOH=-130.5x+228.2 \tag{9}$$

Where x represents the normalized impedance change rate of the battery, and the fitting correlation coefficient R^2=0.9956, root mean square error RMSE=0.9973, goodness of fit meets the estimation requirements, and the results have high reliability, which can be used to represent the aging law of SOH.

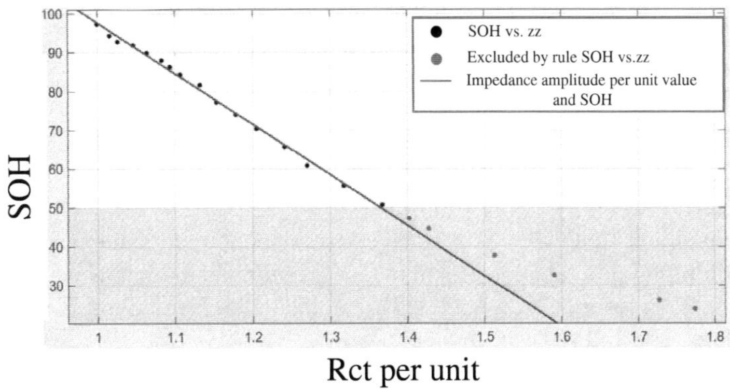

Figure 18. Relationship between amplitude of low-frequency impedance and SOH after normalization of battery 1.

Compared with the charge transfer resistance estimation method, using impedance amplitude at low frequency to estimate the health status of lithium ion battery has the following advantages:

(1) When charge transfer resistance is used to estimate SOH, the fitting coefficient and root mean square error of battery 1 health state curve are R respectively^2= 0.9871, RMSE = 1.721; When the low-frequency impedance amplitude is used to estimate SOH, the fitting coefficient and root mean square error of battery 1 health status curve are R respectively^2= 0.9956, RMSE = 0.9973.Therefore, it is more accurate to estimate the SOH of lithium ion battery by the low-frequency impedance amplitude.

(2) Greatly simplified the measurement requirements, only need to measure the frequency of 0.02Hz impedance amplitude rather than 0.02Hz~20000Hz impedance spectrum data, easier to practical operation.

(3) The results do not depend on the identification accuracy of equivalent circuit parameters by simulation software, and the impedance is easier to measure, the measurement accuracy is higher, the data is more reliable, and the results have greater application value.

However, SOH estimation based on impedance amplitude also has great defects. When the temperature is 45℃, the change of low-frequency impedance amplitude with SOH is no longer monotonous, and the method fails. Therefore, this method has limitations in practical application.

4. Conclusion

Aiming at the problem of health state estimation of lithium ion battery, an equivalent circuit model which can correctly describe the reaction process inside the battery was established in this paper, and the model was compared with the electrochemical impedance spectrum of lithium battery. Based on the corresponding relationship, parameter identification of equivalent circuit model was completed, and the final error met the requirements.

The model parameters and the battery capacity were analyzed, and the charge-transfer resistance and low-frequency impedance amplitude, which varied monotonically with the capacity, were selected and studied respectively.With the help of BP neural network, a model of estimating SOH of the same type of battery with some cells was established.The temperature compensation coefficient was introduced to estimate the SOH of lithium ion batteries at different operating temperatures. The final PREDICTED value and actual value RMSE of SOH were 2.99 and 1.53, which met the estimation requirements.

It is found that the amplitude of low frequency impedance increases linearly with the decrease of SOH.Considering the convenience and precision of measurement in engineering application, the curve of the relationship between 0.02Hz low-frequency impedance amplitude and SOH was fitted, and the fitting coefficient R of the final result was obtained2=0.9956, root mean square error RMSE=0.9973, which verifies the validity of the estimation method.

Acknowledgements
This paper is one of the stage results of the Guangdong Province Key Areas R&D Program Project Traction Battery System Testing and Evaluation Technology Research (2019B090908003).

References

[1] Kashiwagi T, Oda T. Energy policy & New National Energy Strategy in Japan[C]//2007 26th International Conference on Thermoelectrics. IEEE, 2007: 420-425.
[2] Yuan L X, Wang Z H, Zhang W X, et al. Development and challenges of LiFePO 4 cathode material for lithium-ion batteries[J]. Energy & Environmental Science, 2011, 4(2): 269-284.
[3] Hao Shiyu, Yin Huifei, Yang Ru, et al. International Journal of Electronic Measurement Technology, 2021, 40(09): 49-53.
[4] Liu Haotian, Wang Ping, Cheng Ze. Proceedings of the CSEE, 2021, 41(05): 1851-1860. (in Chinese)
[5] Cai L, Meng J, Stroe D I, et al. Multiobjective optimization of data-driven model for lithium-ion battery SOH estimation with short-term feature[J]. IEEE Transactions on Power Electronics, 2020, 35(11): 11855-11864.
[6] Zheng Xueying, Deng Xiaogang, Cao Yuping. Health Prediction of Lithium ion Battery Based on Energy-Weighted Gaussian Process Regression. Journal of Electronic Measurement and Instrument, 2020, 34(06): 63-69.
[7] Luo Fan, Huang Haihong, Wang Haixin. Rapid Prediction of Charge State and Health State of Retired Power Battery Based on Electrochemical Impedance Spectroscopy [J]. Chinese Journal of Scientific Instrument, 2021, 42(09): 172-180.
[8] Xiong R, Li L, Li Z, et al. An electrochemical model based degradation state identification method of Lithium-ion battery for all-climate electric vehicles application[J]. Applied energy, 2018, 219: 264-275.
[9] Omar N, Monem M A, Firouz Y, Lithium iron phosphate based battery -- Assessment of aging parameters and development of cycle life model[J]. Applied Energy, 2014, 113: 1575-1585.
[10] Cao Chunan, Zhang Jianqing. Introduction to Electrochemical impedance Spectroscopy [M]. Beijing: Science Press, 2002, 7.
[11] Zhang Wenhua, Pei Feng, Liu Ping, et al. Journal of Power Supply Technology, 2015, 39(01): 54-57.
[12] Guha A, Patra A. State of health estimation of lithium-ion batteries using capacity fade and internal resistance growth models[J]. IEEE Transactions on Transportation Electrification, 2017, 4(1): 135-146.
[13] Xu Xinmin, Wang Lian, Shi Huiling. Power Supply Technology, 2015, 39(12): 2579-2583.
[14] Barsoukov E, Macdonald J R. Impedance spectroscopy: theory, experiment, and applications[M]. John Wiley & Sons, 2018.
[15] Okoshi T, Yamada K, Hirasawa T, Battery Condition monitoring (BCM) for lead-acid batteries[J]. Journal of Power Sources, 2006, 158 (2) : 874-878.
[16] Ramesh S, Krishnamurthy B. A mathematical model to study capacity fading in lithium ion batteries: formation and dissolution reactions[J]. Journal of The Electrochemical Society, 2015, 162(4): A545-A552.
[17] Shi Wei. Study on Influencing Factors and Test Method of Life of Power Lithium-ion Battery Pack [D]. Beijing: Beijing Jiaotong University, 2014.
[18] Ouyang M, Ren D, Lu L, et al. Overcharge-induced capacity fading analysis for large format lithium-ion batteries with LiyNi1/3Co1/3Mn1/3O2+ LiyMn2O4 composite cathode[J]. Journal of power sources, 2015, 279: 626-635.
[19] Gao Fei, Yang Kai, Hui Dong, et al. Proceedings of the CSEE, 2013, 33(05): 41-45+8.
[20] Zhang Y, Tang Q, Zhang Y, et al. Identifying degradation patterns of lithium ion batteries from impedance spectroscopy using machine learning[J]. Nature communications, 2020, 11(1): 1-6.

Spacecraft hitch detection and health evaluation based on Multivariable Time Series

Xinye Tang[1*], Dezhong Jiang[1] and Botao Guo[1]

[1]China Academy of Launch Vehicle Technology, Bejing Aerospace Automatic Control Institude, Beijing, China

[*]E-mail: 17863109735@163.com

Abstract: In this paper, unsupervised method is studied to solve the problem that it is difficult for spacecraft to obtain tagged hitch data. Firstly, the incremental cross-correlation filtered attribute selection algorithm (ICF) is used to complete the selection of feature subsets of spacecraft multivariate time series (MTS);Then the unsupervised learning model of LSTM-SAE is trained on a large number of normal data; Finally, a complete hitch monitoring and health evaluation system is established and verified on the data of three fault degrees. The feature space of normal data is constructed, and the health state of spacecraft is measured by the reconstruction error of faulted data and the distance of feature space. It solves the problem of predicting and avoiding catastrophic failures of spacecraft under the conditions of lack of prior knowledge, unbalanced data distribution and incomplete failure modes.

1. Introduction

With the continuous increase of human space activities, the number of spacecraft in orbit is increasing, and the health status of spacecraft is also receiving more and more attention.

The existing fault diagnosis technology can be roughly divided into three categories: knowledge-based, model-based and data-based. Model based methods usually refer to quantitative model-based methods [1-2]. In recent decades, there have been some related studies. For example, reference [3] studied the model-based thruster fault detection and separation method, and carried out technical verification on aircraft systems such as X-38, mini AERCam and MIT spheres; Expert system is one of the earliest knowledge-based fault diagnosis methods for spacecraft. As early as the 1980s, NASA applied expert systems to spacecraft fault diagnosis [4]. Beijing Institute of control engineering applied the expert system to the satellite control system[5].The basic idea of data-based fault diagnosis is to learn the behavior mode of the system from historical data. From the perspective of deep learning, most current intelligent diagnosis methods can be regarded as shallow structure algorithms with nonlinear transformation, including HMM, SVM, logistic regression and traditional neural networks [6].

Spacecraft will produce a large amount of data in the process of testing and flying. This paper mainly studies the fault detection of multivariable spacecraft time series based on deep learning.

2. Attribute selection of multivariable time series of spacecraft

The parameters of spacecraft electrical system have strong correlation and similarity, and even some dimensions are almost the same in value and trend. Attribute selection is to select a group of the most representative and effective attribute subsets that contribute greatly to data mining. It not only reduces the time cost of the algorithm, but also improves the accuracy of subsequent knowledge discovery of the original data.

The goal of ICF algorithm is to reduce feature redundancy, which is measured by the cross-correlation parameters between paired variables. The cross-correlation function is a function of two discrete time series with respect to offset, which is calculated by sliding point product:

$$\emptyset x^1x^2(\tau) = \sum_{t=max\{0,\tau\}}^{min\{(T^1-1+\tau),(T^2-1)\}} x^1(t-\tau)\cdot x^2(t) \tag{1}$$

$\tau \in [-(T^1-1),\ldots,0,\ldots,(T^2-1)]$, T^1, T^2 is the length of time series. And introduce the normalized cross-correlation :

$$\overline{\emptyset}_{x^1x^2}(\tau) = \frac{\emptyset_{x^1x^2}(\tau)}{\emptyset_{x^1x^1}(0)\cdot\emptyset_{x^2x^2}(0)} \tag{2}$$

The cross-correlation value of the two time series is taken at the time when the sequence signals are best aligned, that is:

$$\overline{\emptyset}^*_{x^1x^2} = max_\tau|\overline{\emptyset}_{x^1x^2}(\tau)| \tag{3}$$

The ICF algorithm is divided into three steps:

The first step is to establish the characteristic redundancy matrix$R \in \{0,1\}^{D*D}$,If features i and j are pairwise redundant,$R_{ij} = 1$;Otherwise, $R_{ij} = 0$. For a given MTS data set, the calculation process of redundancy matrix R is as follows:

(1) For each sample, use formula (3) to calculate the maximum cross-correlation between each pair of univariate sequences. If $\overline{\emptyset}^*_{x_i^n x_j^n} \approx 1$, assume that features i and j are correlated on the nth sample, and increase the partial correlation count:$S_{ij} = S_{ij} + 1$(assume$S_{ij} = 0$ initially).

(2) Calculate the percentage of samples related to each pair of features in all samples,$S_{ij} = S_{ij}/N$.

(3) If features i, j are related on more than $\theta_p\%$ of the samples, set $R_{ij} = 1$.

(4) Set the diagonal of R to 0, that is $R_{ii} = 0$.

The specific value of $\theta_p\%$ needs to be verified by experiments.If $\overline{\emptyset}^*_{x_i^n x_j^n} > 0.9, \overline{\emptyset}^*_{x_i^n x_j^n} \approx 1$.

In the second stage, the initial feature set is preprocessed to delete the main noise.

In the third stage, feature selection rules are applied by using the information in R generated in the first two steps. Due to the length of article, the rules are not detailed here.The time series is reduced from thirteen dimensions to five dimensions by ICF algorithm.

3. Feature extraction of multivariable time series of spacecraft

3.1 Feature extraction network based on LSTM-SAE

3.1.1 LSTM-SAE network model

Autoencoder (AE) is one of the unsupervised deep learning methods. It is a typical single hidden layer neural network proposed by Rumelhart in 1986.It consists of an encoder and a decoder, in which the output of the encoder is used as the input of decoding. The raw data is sent to the encoder and converted into a compressed representation:

$$y = \varphi(Wx + b) \tag{4}$$

The decoder learns how to reconstruct the original input from the compressed representation:

$$g(x) = \theta(W'y + b') \tag{5}$$

The model should try to make the input and output close to or equal, for the sum of two sets of equal length sequences y_i y_i^p , the reconstructed loss function usually uses the mean square error:

$$MSE = \sum_{i=1}^{n} \left(y_i - y_i^p\right)^2 \tag{6}$$

Different from the simple cell structure of RNN, LSTM neural network is added with three independent gate structures in neurons, so that the state can add and lose information to achieve the purpose of long-term memory. It's basic cell structure is shown in the figure 1.LSTM-SAE combines Autoencoder with LSTM network, that is, input a sequence x to generate an output sequence.

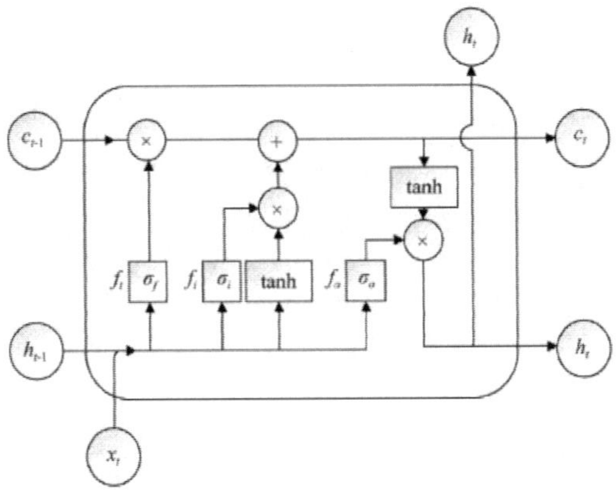

Figure 1. LSTM network structure diagram

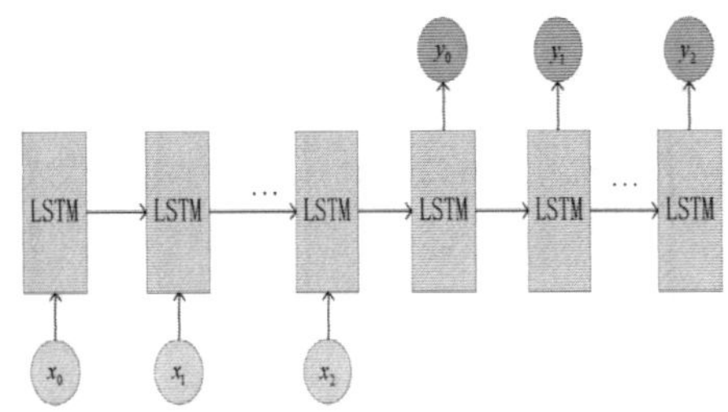

Figure 2. LSTM-SAE structure diagram

3.1.2 LSTM-SAE model training and evaluation

In the previous work, ICF algorithm was used to select five variables from the original 14 variables, and a total of fourteen features in time domain and frequency domain were extracted for each variable, with a total of 70 features in five variables. The length of each sequence is 30000, and every 30 data are divided into one sample. The final sample size is 30*70, and the sample set size is 1000. Divide the training set and verification set according to the proportion of [0.8,0.2].

The autoencoder network is composed of encoding layer and decoding layer, both of which are three-layer LSTM networks. In each epoch, the training process provides all health data samples for the model, and evaluates the effect of the model on the validation machine. Compare the training results of ICF algorithm and preprocessing data set with the training results of original data set.

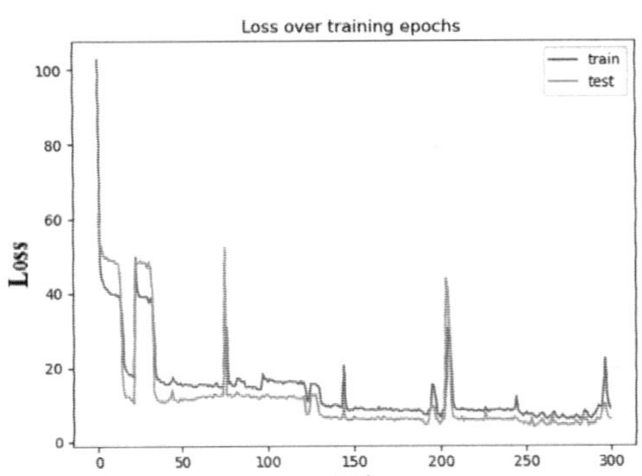

Figure 3. Training results of original data set

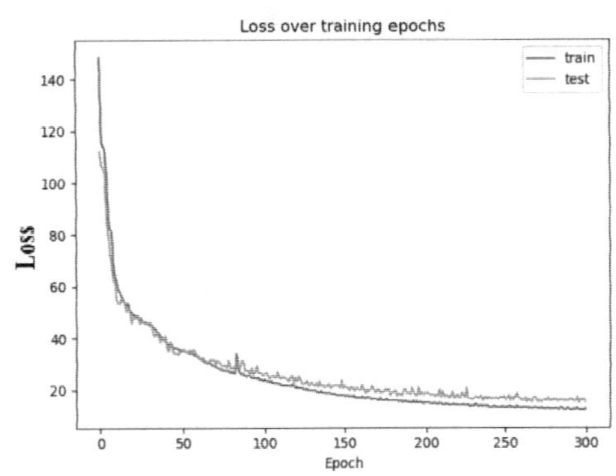

Figure 4. Training results after ICF algorithm

It can be seen from the figures that the convergence of the original data on the model is unstable, with obvious jitter; The data after ICF algorithm has better convergence effect on the model.

4. Fault monitoring and health evaluation

4.1 Evaluating indicator

Establish a feature space for the normal state data, quantify and calculate the degradation degree by measuring the difference between the measurement data and the normal state sample space. A is the feature space of the sample in the normal state, which is an expression of the normal working state of the system or equipment. σ_N is defined as the standard spacing mean of A.

$$\sigma_N = \frac{\sum_{i=1}^{n}\sum_{j=1}^{n}\left\|a_i - a_j\right\|}{n^2} \tag{7}$$

Let u_t be the data sample collected at time t and a_t is the feature vector extracted through the feature extraction network. σ_t is defined as the mean value of the external distance from the sample to the normal state space, which is used to characterize the distance between a_t and the normal state sample feature space A.

$$\sigma_t = \frac{\sum_{i=1}^{N}\left\|a_t - a_i\right\|}{N} \tag{8}$$

According to σ_N and σ_t, define the health degree of the percentage system:

$$H_t = \begin{cases} \frac{\sigma_N}{\sigma_t}\cdot 100, & \sigma_t \geq \sigma_N \\ 100, & \sigma_t < \sigma_N \end{cases} \tag{9}$$

4.2 Analysis of experimental results

There are test data of slight fault, medium fault and major fault of a certain model. Taking 5-volt voltage variable as an example, the original data of three degradation degrees are as follows:

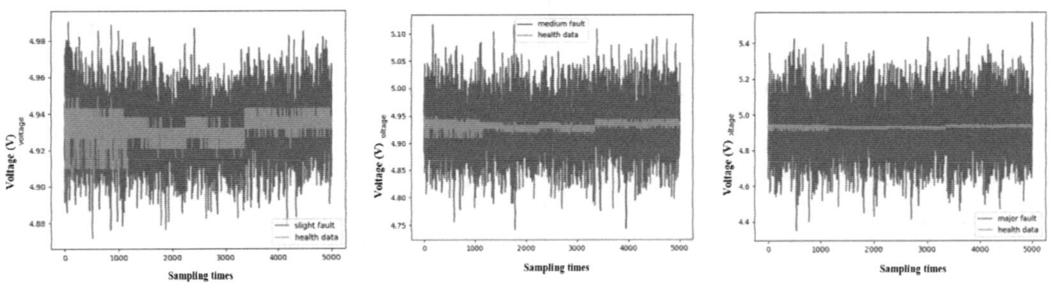

Figure 5. Original signals of three fault levels

The data of three degradation degrees after pretreatment are input into the trained model, and the reconstruction errors of different degradation degrees are compared. The reconstruction error diagram is shown in figure 6-8. The abscissa is the reconstruction error size, and the ordinate is the sample distribution.

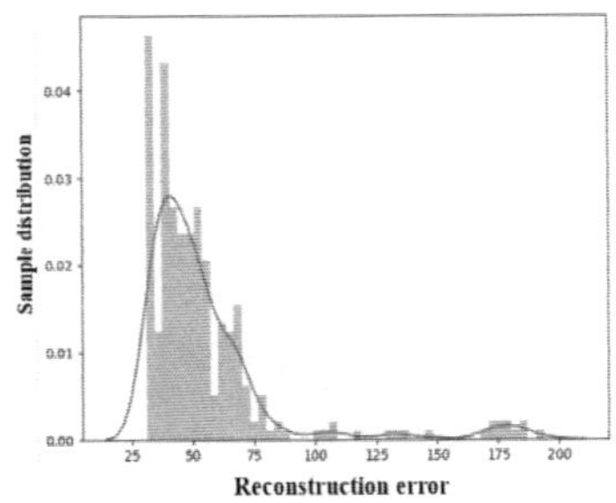

Figure 6. Reconstruction error of slight fault

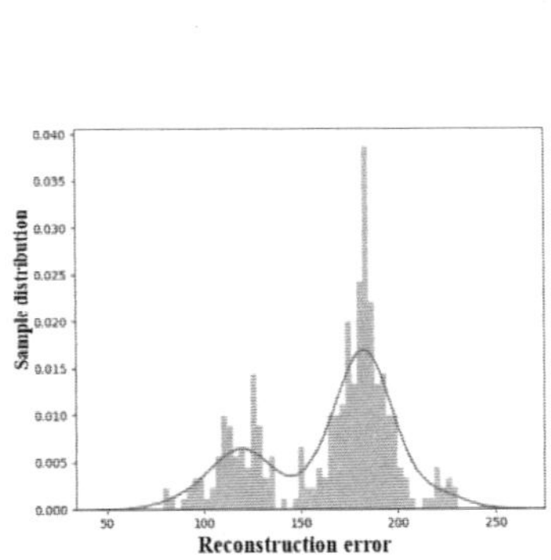

Figure 7. Reconstruction error of medium fault

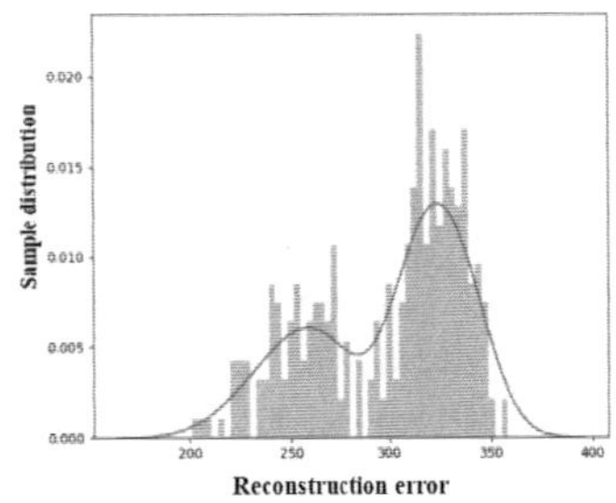

Figure 8. Reconstruction error of major fault

It can be seen that slight fault reconstruction errors are mainly distributed between 25 and 75, medium fault reconstruction errors are concentrated between 100 and 200, and major fault reconstruction errors are concentrated between 250 and 350. Different degrees of degraded data can be visually distinguished by reconstruction error.

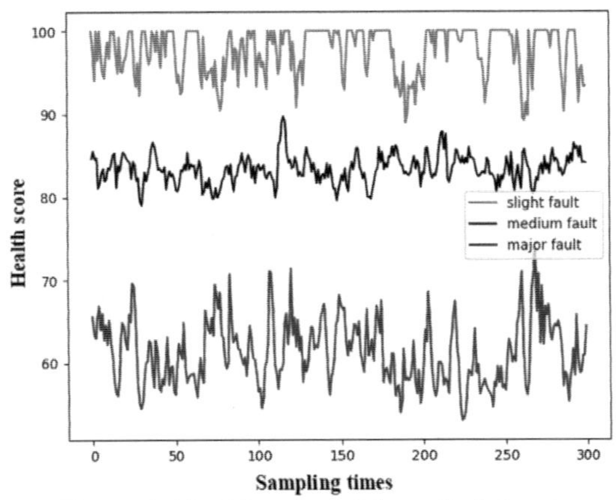

Figure 9. Health score of three fault levels

Finally, the health score is used to evaluate the data with different degrees of degradation more accurately. The result is shown in figure 9.

The results show that there is a clear distinction between the health of the three fault levels, and the more serious the fault, the lower the health score. Experiments show that the scheme of feature extraction and health evaluation in this paper has a good test effect.

5. Conclusion

Experiments show that the spacecraft fault monitoring and health assessment method based on multivariable time series proposed in this paper can effectively monitor the operating state of spacecraft. Compared with the traditional threshold setting method, this method can automatically learn and extract the characteristics of normal data without rich expert experience and domain knowledge; On the other hand, the threshold set by expert experience is often too broad to reflect the real working state of the system. It is of great significance to improve the reliability of products and systems and avoid catastrophic failures to carry out spacecraft anomaly monitoring technology.

Reference

[1] Xing Y , Wu H X , Wang X L , et al. Survey of fault diagnosis and fault-tolerance control technology for spacecraft[J]. Journal of Astronautics, 2003,24(3):221-226.

[2] Jiang L X , Hua-Wang L I , Yang G Q , et al. A survey of spacecraft autonomous fault diagnosis research[J]. Journal of Astronautics, 2009, 30(4):1320-1326.
[3] Wilson E,Sutter D. Motion-Based Thruster Fault Detection and Isolation[C]// Infotech@Aerospace,Arlington,USA,Sep 26-29,2005.
[4] Ali M , Scharnhorst D A , Ai C S , et al. A Flight Expert System For On-Board Fault Monitoring And Diagnosis[C]// International Society for Optics and Photonics,USA,Mar 26,1986.
[5] Ruoming A , Wang Jun. The Neural Network Expert System Applied to Power System Diagnosis[J]. Aerospace Control, 2012, 30(3):88-92.
[6] Widodo A , Yang B S . Support vector machine in machine condition monitoring and fault diagnosis[J]. Mechanical Systems and Signal Processing, 2007, 21(6):2560-2574.

Research on BSR evaluation method and positioning technology of battery module based on experimental bench

Yingling Huang[a], Bo Zhang[a], Kelun Zhao*[a]

[a]Guangzhou GRG metrology and test Co., Ltd, Guangzhou, China

* Corresponding author: zhaokl@grgtest.com

ABSTRACT: A battery module was taken as the research object, and the BSR test of the module was carried out based on the silent vibration table, and the BSR evaluation of the module was realized through the combination of subjective and objective mode. In order to achieve the accurate positioning of the abnormal noise of the module, the frequency point of the abnormal sound is locked based on the relative approach technology, and the playback and confirmation of the abnormal noise of the module is carried out based on the filter playback technology. Finally, acoustic imaging is used to locate the abnormal noise of the module. The research results show that the evaluation of BSR needs to be combined objectively and subjectively, and the rapid identification and location of abnormal noise of battery modules is realized by combining relative approach technology, filter playback technology and acoustic imaging technology.

1. Introduction

BSR is a kind of unpleasant high frequency noise caused by low frequency vibration [1-2]. BSR is an important indicator of vehicle NVH performance, which seriously affects ride comfort. Therefore, in the early stage of design on the relevant parts, the main engine plants raise the assessment requirements of BSR [3-4].

In the early stage of research and development, the BSR verification of automobile parts is mainly carried out by experimental bench. The test sample is installed on the silent vibration table through tooling, and then vibrates based on the given excitation spectrum, so as to test and evaluate the BSR [5-7]. At present, some main engine plants still simply adopt objective parameter quantitative assessment method, while some main engine plants simply adopt subjective evaluation method, both of which have their own advantages and disadvantages. This paper will take a battery module as an example to test and evaluate the BSR.

It is necessary to analyze the location of BSR and provide a basis for rectification after the BSR is found [8-9]. Shi Fuqiang et al. determined the source of BSR of door limiter and the rectification plan about BSR of door limiter through analysis of several broken cars and combining theoretical design and engineering experience [10]. Jiang Taotao et al. analyzed the possible factors of BSR from the perspectives of parts material, design size and part stiffness in view of BSR problem of sub-instrument panel, and finally determined the source of BSR based on actual inspection and simulation analysis [11]. Acoustic imaging technology has been widely used in sound source location in recent years. Shao Yuying et al. achieved accurate location of BSR of dry transformer based on beamforming algorithm and simulation calculation, and designed microphone array composed of 112 microphones for dry transformer [12]. Zhang Zhi et al. used planar acoustic array to collect acoustic signals on four surfaces of the engine, and then used cross-spectral matrix beamforming algorithm to conduct acoustic imaging on each surface of the engine at different speeds. The three-dimensional distribution of the sound source on the engine surface was realized by cross layer method, and the sound source was located to each part of the engine accurately. Acoustic imaging based on beamforming algorithm usually has a good localization effect for the maximum sound source, but the maximum radiated sound source is not necessarily generated by the BSR. This paper focuses on how to locate the BSR based on acoustic imaging technology.

2. Battery module bsr test and evaluation

This paper takes a battery module as the research object. The sample was installed on the silent vibration table with tooling, and a microphone was placed 30 cm away from the sample center surface for acoustic signal acquisition, as shown in Fig. 1. Then input the random excitation spectrum to the silent vibration table, as shown in Table 1. Finally, the silent vibration table was started to test the BSR of the module. At the same time, 200 mm away from the sample, the sample can be judged to have BSR.

Figure 1. Test bench layout.

Table 1. BSR tests excitation spectrum.

Frequency (Hz)	Acceleration (g^2/Hz)
7	0.00002
11	0.16
15	0.002
100	0.00005
118	0.000002
RMS	0.5g

At present, the objective parameters of BSR are mainly sound pressure level, frequency spectrum and loudness. The acoustic signals obtained from the test were analyzed, as shown in Fig. 2. The results of SPL and loudness were shown in Table 2.

Table 2. BSR acoustic test results.

Sample number	Sound pressure level (dBA)	N10 loudness (sone)	Subjective evaluation
23#	40.8	2.70	no BSR
24#	40.2	2.53	noticeable clicking sound
25#	41.2	2.66	no BSR

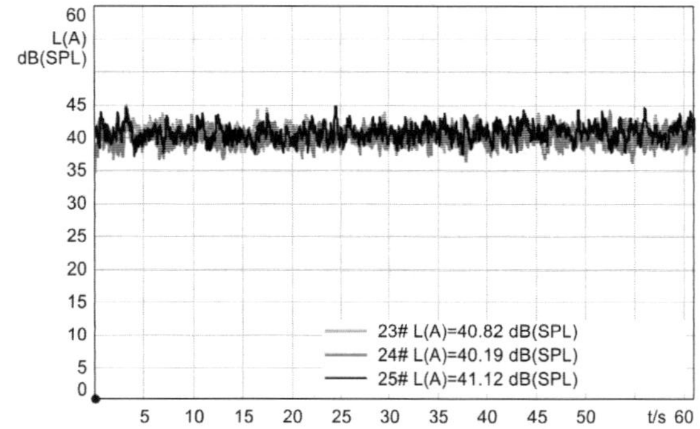

(a) Sound pressure curve

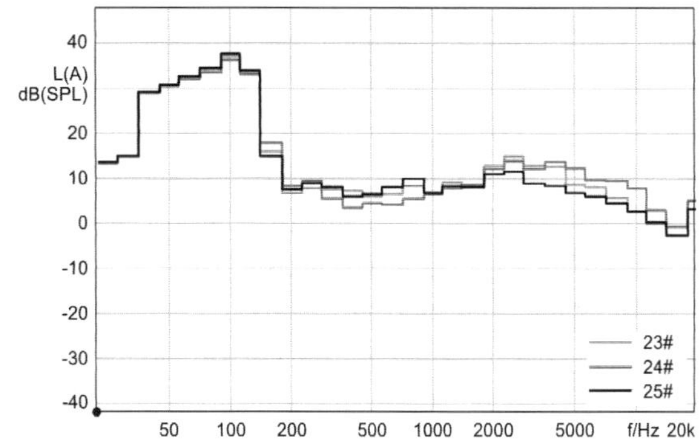

(b) Sound pressure spectrum

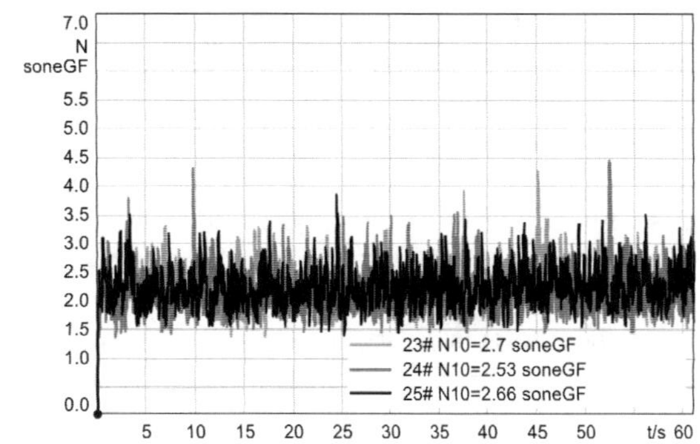

(c) Loudness curve

Figure 2. Acoustic curve.

As can be seen from Fig. 2, the acoustic curves of the three samples are basically consistent. It can be seen from Table 2 that there is no significant difference between SPL and loudness of the three samples. However, from the subjective evaluation point of view, the 24# sample has obvious abnormal sound. Simple objective parameter evaluation can not achieve a comprehensive assessment of abnormal sound of samples. Therefore, subjective evaluation is necessary in the test and evaluation of abnormal sound. The analysis and evaluation of abnormal sound can be carried out effectively by combining subjective and objective methods.

3. Bsr positioning technology of battery module

In the field of engineering technology, aiming at the BSR of parts, we mainly collect objective data for spectrum analysis and sound quality analysis, and then identify and locate BSR based on engineering experience. Traditional analysis methods and experience can only identify some common BSR, but also depend on the engineering experience of relevant personnel. So how to realize the rapid and accurate recognition and location of BSR, there is no effective technical means at present.

At present, the acoustic characteristics of BSR are obtained by conventional testing methods, and then the characteristics are compared based on the existing database of BSR, so as to identify and lock the source of BSR. The current technology relies on the existing BSR database. For the occasional BSR, if there is no relevant database, it is difficult to locate the BSR through feature comparison, and the existing technical means cannot achieve rapid and real-time BSR location.

Although acoustic imaging technology has been widely used in sound source location, the maximum sound source does not necessarily originate from BSR. It is necessary to further study how to apply acoustic imaging technology to BSR location. This paper combines relative volume analysis technology, filtering playback technology and acoustic imaging technology to identify and locate BSR and achieve rapid identification and visualization of BSR. The positioning technology is shown in Fig. 3.

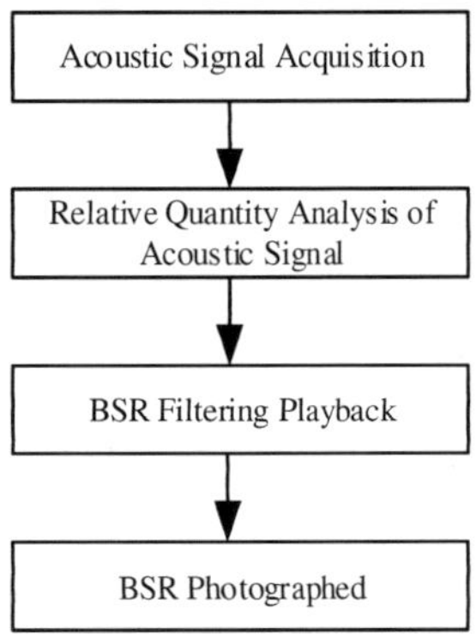

Figure 3. Positioning technology of BSR.

3.1 Relative quantity analysis and filtering playback

Relative quantity calculation is based on reference quantity. The relative quantity differs from the sound pressure level/loudness in that the reference used in this calculation is not a fixed value, but a signal in the time domain and/or frequency domain at a point in the sound signal. The relative quantity uses the sound itself signal to establish a series of reference quantities, through analyzing the actual signal and the change between these reference quantities, the abnormal sound in the sound is extracted. In the pretreatment stage, the collected time signal is analyzed by octave-range filter method, and the time signal is converted into a time-frequency signal. According to the characteristics of the analyzed signal, the appropriate octave-range bandwidth and time constant (time weighting) are selected. For the acoustic signal of the module, 1/12 octave-frequency range was used for analysis, FFT Size was set as 8192, and time constant T was set as 1024 ms.

Relative quantity analysis calculates an estimated value of the signal in the time domain or the frequency domain around a certain point through regression algorithm, and then compares the actual value with the estimated value. The larger the difference between these two values is, the more obvious the mutation is, that is, the stronger the BSR characteristics felt by the human ear.

Therefore, after completing the pretreatment stage, variable analysis is needed to calculate the difference between the estimated value and the actual value, which is used to calculate the magnitude of the mutation in the time domain/frequency domain. Firstly, linear regression algorithm is used to smooth the time domain/frequency domain data. The time-domain smoothing interval is 200ms, which is suitable for signal mutation in time. The smoothing interval in frequency domain is an octave, which is suitable for signal mutation in frequency. Based on ArtemiS SUITE, this paper adopts the time-domain + frequency-domain mixed mode for signal smoothing processing. Then the mutation part of the signal is extracted by calculating the difference between the actual signal and the smoothed result.

Then determine the time and frequency of abnormal sound. A relative quantitative analysis was conducted on the collected module acoustic signals, and the results are shown in Fig. 4.

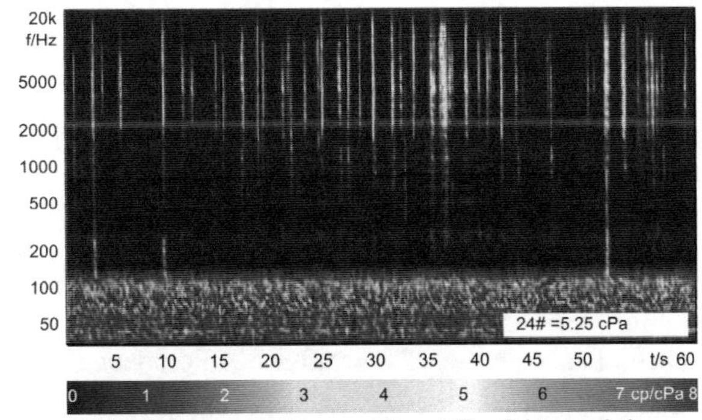

(a) Relative quantity results of the 24# module

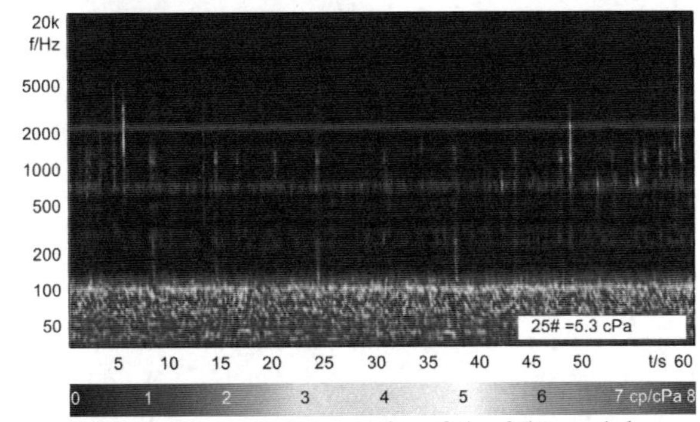

(b) Relative quantity results of the 25# module

Figure 4. Relative quantity analysis of battery module.

As can be seen from Fig. 4, the 24# sample has obvious BSR in 2000-8000Hz frequency band relative to the 25# sample, which may be the frequency band where the rattling sound occurs.

The possible frequency band or frequency point of BSR can be locked by relative quantity analysis. However, whether it is the actual BSR needs further confirmation. Generally, filter playback technology is used to confirm the abnormal frequency components. First, bandpass filtering is performed on the acoustic signals of the possible abnormal frequency band or frequency point, and only the acoustic characteristics of the frequency band are retained. Then the filtered acoustic signal is played back by using a two-ear playback earphone. Determine whether it is the perceived BSR through hearing, and then lock the frequency distribution of BSR. For the acoustic signals of the battery module, bandpass filtering was carried out in 2000-8000Hz frequency band, and 3 engineers used playback earphones to return the filtered acoustic signals. Compared with the actual perceived BSR, the frequency band of 2000-800Hz is finally locked as the frequency band where the rattle and click sound of the module is located.

3.2 Acoustic imaging and BSR localization

The basic principle of sound source location technology based on microphone array is to receive multiple channels of sound signals through the microphone. Through beamforming algorithm, the output of the grid point where the sound source is located is strengthened to form the "main lobe", and the output attenuation of other focused grid points is formed to form the "side lobe", so as to identify the sound source. Microphone array is composed of several microphone units arranged according to a certain spatial geometry law. Compared with the measurement method of single microphone, acoustic array has a series of obvious advantages in signal receiving and processing. Mainly in anti-reverberation, noise removal, speech separation ability is stronger, higher spatial resolution. The multi-channel signal processing method of microphone array enriches the processing domain of sound signal, and adds space domain beyond time domain and frequency domain, which makes the sound source location more accurate and adaptive.

Conventional beamforming algorithm is one of the most widely used sound source localization algorithms based on the principle of time-delay summation. Based on the spatial sound signals received by the sound array, phase alignment and summation of discrete grid focal points are carried out one by one. When the grid point is consistent with the sound source position, a maximum value will be output in this direction to achieve the purpose of the sound source location.

Based on the near-field spherical wave model, the expression of the signal received by the array can be obtained. Then the phase compensation is made by the delay of each element to grid point and reference element to grid point. Then the compensated signal is weighted and summed with the signal collected by the array. According to the standardization of microphone, the traditional beam output expression is:

$$Y(r,t) = 1/\mathrm{M}\sum_{l=1}^{M}\omega_l p_l(t-\Delta\tau_l) \quad (1)$$

Where, ω_l is the weighting coefficient of Microphone No. l, and the frequency domain expression of beamforming can be obtained after Fourier transform of the above equation:

$$Y(r,\omega) = 1/\mathrm{M}\sum_{l=1}^{M}\omega_l p_l(\omega)\exp(jw\Delta\tau_l) \quad (2)$$

Equation (2) gives the expression of the traditional time-delay summing beamforming algorithm. It can be found from spherical wave equation that the distance between array and sound source not only causes the phase difference between the received signal, but also causes the amplitude difference. However, this factor is not considered in the traditional algorithm. Therefore, a beamforming algorithm based on cross-power spectral imaging function is introduced. In this method, the cross-power spectrum generated by grid points on the sound source surface is solved, and the cross-power spectrum generated by different grid points is subtracted from the measured cross-power spectrum of the sound source data, and the grid point pointed to by the minimum difference is the location of the sound source. This method takes full account of the influence of amplitude attenuation, and can identify not only the position of sound source, but also the intensity of sound source.

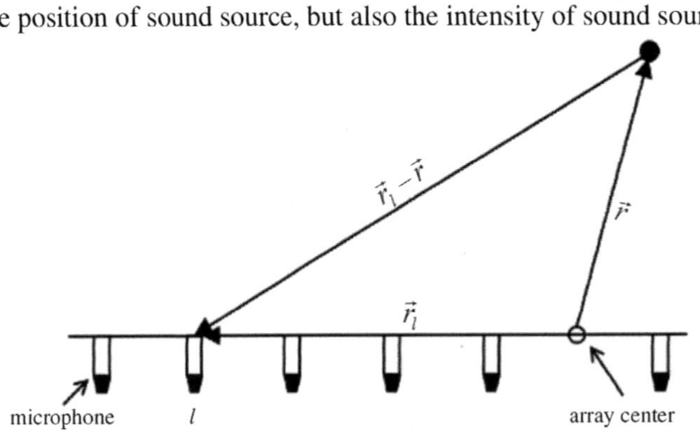

Figure 5. Beamforming computing model.

The near field beamforming algorithm based on cross-power spectral imaging function is still based on the near field spherical wave hypothesis. The location of the sound source is identified by scanning the focal point on the sound source surface, and its model is shown in Fig. 5. In the figure, $\vec{r}_l$ represents the position vector of the focal point relative to the center point of the array, and $\vec{r}_l$ represents the position vector of the No. l microphone relative to the center point of the array.

According to the hypothesis of near-field spherical wave, the sound pressure signal $p_l(\omega)$ received by Microphone No. l from a sound source somewhere in space can be expressed as:

$$p_l(\omega) = p_0\left[\exp\left(jk|\vec{r}_l - \vec{r}|\right)/|\vec{r}_l - \vec{r}|\right] \tag{3}$$

In the equation, $|\vec{r}_l - \vec{r}|$ represents the distance between the focus and microphone No. l, k represents the beam, and $v_l(\vec{r}) = \exp\left(-jk|\vec{r}_l - \vec{r}|\right)/|\vec{r}_l - \vec{r}|$ is the pointing vector:

$$p_l(\omega) = p_0 v_l(\vec{r}) \tag{4}$$

Under ideal conditions, the cross-spectrum generated between microphone No. l and microphone No. n at the sound field focusing model point can be expressed as:

$$G_{ln}^{mod} = p_l(\omega)p_n^*(\omega) = av_l^*(\vec{r})v_n(\vec{r}) \tag{5}$$

Where, a is the amplitude coefficient, essentially the square of acoustic wave amplitude. If the difference function is defined as the sum and variance of the cross-spectrum of the sound source and the cross-spectrum of the focal point is defined as the difference function, then:

$$D(a,\vec{r}) = \sum_{l,n=1}^{M}\left|G_{ln} - G_{ln}^{mod}\right|^2 = \sum_{l,n=1}^{M}\left|G_{ln} - av_l^*(\vec{r})v_n(\vec{r})\right|^2 \tag{6}$$

let $g = [G_{ln}]$, $h = |v_l^*(\vec{r})v_n(\vec{r})|$, Plug in to simplify things:

$$D(a,\vec{r}) = |g - ah|^2 = g^H g - a\left(h^H g + g^H h\right) + a^2 h^H h \tag{7}$$

In the equation, a reflects the real intensity of the sound source, and, $\vec{r}$ reflects the location information of the sound source, so that the focus with the smallest difference function is the real location of the sound source.

First, minimize the difference function with respect to a, let $g \cong ah$, then:

$$a = \left(h^H g\right)/\left(h^H h\right) \tag{8}$$

Substitute this equation into (7), then:

$$D(a,\vec{r}) = g^H g - \left(h^H g\right)^2/\left(h^H h\right) \tag{9}$$

Since g in the equation is the cross-spectrum of sound source, it comes from the signal received by array. Is independent of $\vec{r}$, so minimizing the difference function maximizes the cross-spectral image function related to $\vec{r}$. The cross spectrum function is defined as:

$$I^4(\omega,\vec{r}) = \frac{\left(h^H g\right)^2}{h^H h} = \frac{\left|\sum_{l,n=1}^{M} G_{ln}v_l^*(\vec{r})v_n(\vec{r})\right|^2}{\left|\sum_{l,n=1}^{M} v_l^*(\vec{r})v_n(\vec{r})\right|^2} \tag{10}$$

After the above equation is normalized through the channel, we get:

$$I^4(\omega,\vec{r}) = \frac{1}{M^2}\frac{\left(h^H g\right)^2}{h^H h} = \frac{1}{M^2}\frac{\left|\sum_{l,n=1}^{M} G_{ln}v_l^*(\vec{r})v_n(\vec{r})\right|^2}{\left|\sum_{l,n=1}^{M} v_l^*(\vec{r})v_n(\vec{r})\right|^2} \tag{11}$$

In the above equation, when $\vec{r}$ is consistent with the position of the sound source, the maximum output value is the position of the sound source, but the output dimension of $I^4(\omega,\vec{r})$ is the square of energy, so take the square root of the above equation and simplify it:

$$I^2(\omega,\vec{r}) = \frac{1}{M}\frac{\sum_{l,n=1}^{M} G_{ln} v_l^*(\vec{r}) v_n(\vec{r})}{\sum_{l,n=1}^{M} v_l^*(\vec{r}) v_n(\vec{r})} = \frac{1}{M}\frac{V^T G V^*}{M^T I W^*} \tag{12}$$

Where, $V = \left[v(\vec{r})\right]$ is defined as the turning vector of the focus, $W = \left|v(\vec{r})\right|^2$ as the constructed column vector, and I as the identity matrix. Equation (12) is the cross-spectral image function beamforming algorithm constructed in the form of energy. This algorithm can not only obtain the position of sound source, but also obtain more accurate intensity of sound source.

Acoustic imaging was calculated for acoustic signals in 2000-8000Hz frequency band of the battery module, and the results are shown in Fig. 6.

(a) Rough results of the 24# module (b) Precision results of the 24# module (c) BSR position of module 24#

Figure 6. Acoustic imaging results of the battery module.

It can be seen from Fig. 6 that abnormal sound in 2000-8000Hz frequency band of battery module mainly comes from the end face buckle, and there is design gap in the buckle. After pressing the buckle, the abnormal sound disappears. It shows that the identification of abnormal sound based on the combination of relative quantity analysis, filtering and playback and acoustic imaging is reliable and accurate.

4. Conclusion

Taking a battery module as the research object, the BSR test and location of the module vibration are studied based on the silent shaking table. The results show that:

(1) Simple objective parameter evaluation can not achieve a comprehensive assessment of abnormal sound of samples, so in the test and evaluation of abnormal sound, subjective evaluation is necessary, through the combination of subjective and objective can effectively analyze and evaluate abnormal sound.

(2) Combined with relative volume analysis technology, filtering playback technology and acoustic imaging technology for rapid identification and location of abnormal sound is reliable and accurate.

Acknowledgements

This paper is one of the stage results of the Guangdong Province Key Areas R&D Program Project Traction Battery System Testing and Evaluation Technology Research (2019B090908003).

References

[1] Z. Danhui, Y. Le, and C. Zhang, "Study on braking noise optimization in low frequency range based on a pure electric vehicle," Automotive Engineering, vol. 43(09), pp. 1411-1417, September 2021.

[2] H. Haibo, Z. Zhiwei, Z. Siwen, W. Yudong, Y. Mingliang, et al., "Optimization of automobile firewall acoustic package for multi-level goals," Journal of Southwest Jiaotong University, in press.

[3] C. Xiaopeng, "Study on control of squeak and rattle of automobile door system," Jilin University, June 2017.

[4] P. Yun, L. Dongqiang, and Z. Hua, "Research on abnormal noise control of automobile interior parts," Automobile Applied Technology, vol. 46(20), pp. 71-76, October 2021.

[5] K. Honggui, Y. Xuesen, and T. Chun, "Diagnosing of body noise and optimizing of body design," Journal of Hefei University of Technology (Natural Science), vol. 32(S1), pp. 48-50, November 2009.

[6] Z. Rong, Z. Ruili, M. Jianmin, and J. Xiangyu, "Engine abnormal noise positioning based on microphone array and SRC-PHAT," Vehicle Engine, vol. 06, pp. 60-65, December 2018.

[7] L. Shijie, M. Jinyan, L. Qi, and Z. Hongwei, "Analysis and Research on Squeak and Rattle of Automotive Console," Machinery Design & Manufacture, vol. 07, pp. 183-186+190, July 2020.

[8] D. Wei, F. Chao, and Y. Jun, "Reactor abnormal noise analysis based on sound source localization," Water Resources and Power, vol. 32(10), pp. 177-179, October 2014.

[9] H. Shoutong, Z. Wenzhong, C. Yi, and T. Weijun, "Research and application of abnormal noise monitoring and fault location technology of belt conveyor idler," Manufacturing Automation, vol. 44(02), pp. 217-220, February 2022.

[10] S. Fuqiang, and S. Chengwu, "Study on the noise problem of commercial vehicle door," Machinery, vol. 43(05), pp. 77-80, May 2016.

[11] J. Taotao, H. Zhe, and D. Mengdie, "Analysis of the Abnormal Sound of a Vehicle's Sub-instrument Panel," Auto Time, vol. 11, pp. 152-154, May 2022.

[12] S. Yuying, W. Xiao, P. Peng, Y. Guogang, and K. Nan, "Research on defect detection method of power equipment based on acoustic imaging technology," China Measurement & Test, vol. 47(07), pp. 42-48, July 2021.

Design of signal processing circuit for acoustic Doppler Current Profiler

Chuansheng Jiang[1,2,a], Jun Zou[1,2,b], Yanping Liu[1,2,c], Bing Li[1,2,d]

[1.] NARI Group Corporation/State Grid Electric Power Research Institute., Nanjing 211106,China

[2.] Nanjing NARI Water Conservancy and Hydropower Technology Co.Ltd., Nanjing 211106, China

[a]jiangchuansheng@sgepri.sgcc.com.cn; [b]zoujun@ sgepri.sgcc.com.cn, [c]liuyanping@ sgepri.sgcc.com.cn; [d]libing@ sgepri.sgcc.com.cn

Abstract: according to the measurement principle and performance requirements of acoustic Doppler velocity profile, a circuit for weak frequency signal processing is designed. The measurement principle, circuit structure, data comparison before and after signal processing and frequency measurement accuracy data analysis are introduced. The performance comparison and application of the instrument are carried out according to the relevant standards. The experimental results show that the signal processing circuit realizes the constant amplitude amplification of the wide dynamic range frequency signal, and the processed signal is converted by A / D and then measured by digital frequency to obtain a high frequency accuracy. In the application of channel flow measurement, the velocity profile can reflect the change trend of cross-section velocity and complete the accurate measurement of flow rate. It shows that the flow profile instrument based on the signal processing circuit can meet the requirements of on-line monitoring of flow velocity and flow of river and channel.

1 Introduction

River discharge test is one of the most important items of hydrology test. Rapid and accurate measurement of river flow has become a new requirement for national economic construction, water resources management, flood control and drought relief. At present, current meter method is mainly used in river flow measurement. Its basic principle is to lay several vertical lines on the flow measurement section and measure the water depth and the flow velocity at each point of the vertical line with the current meter to calculate the average velocity of the vertical line. The flow rate of the section can be calculated by calculating the area and the average velocity of the section[1,2].

The acoustic Doppler velocity profiler is a new flow measurement method developed and applied in recent years. Acoustic Doppler velocity Profiler is based on the Translation of English Acoustic Doppler Current Profiler, referred to as ADCP in English abbreviation. ADCP has the superiority of principle. It break through the traditional mechanical rotation sensor on the basis of the current meter. It uses ultrasonic transducer as sensor, and the transducer emits acoustic pulse wave. The acoustic pulse wave is backscattered by the backscattering body such as sediment particles and plankton which are not evenly distributed in the water body. The signal is received by ultrasonic transducer, and the flow rate is calculated by measuring doppler frequency shift.

ADCP can directly measure the flow velocity profile of the section with the characteristics of undisturbed flow field, short test duration and large speed range. ADCP flow measurement is the development trend of water conservancy industry at home and abroad. There is a large space in the domestic water conservancy industry market.

In this paper, a circuit for weak frequency signal processing is designed according to the measuring principle and performance requirements of acoustic Doppler current profiler. The circuit adopts the design idea of FM intermediate frequency system to realize the constant amplitude amplification of wide dynamic range frequency signal. In this way it greatly improves the detection range and measurement accuracy of the profiler.

2 System design

Doppler effect is the phenomenon that the frequency of the sound heard by the observer is different from that of the sound source due to the relative motion between the sound source and the observer. The physical principle is first discovered by Austrian physicist Christian Doppler in the 19th century. The typical Doppler effect occurs between the sound source and the observer[3]. In the static condition, when the observer is moving towards the sound source, the frequency of the sound wave he receives is higher than that he receives when he is stationary. When the observer is moving away from the source, he receives a lower frequency of sound than when he is stationary relative to the source[4].

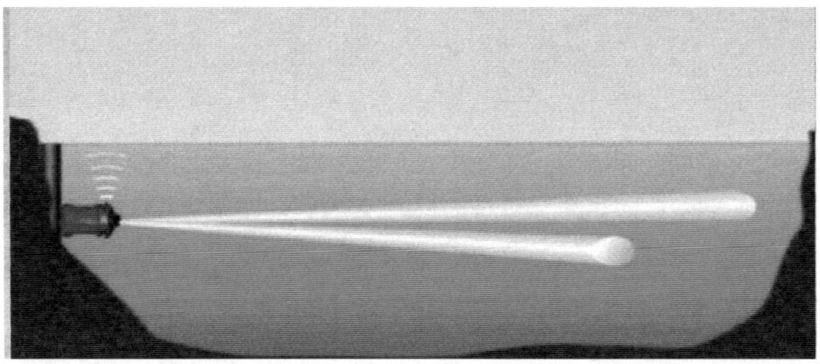

FIG. 1 Schematic diagram of measurement principle of 90° Acoustic Doppler velocity profiler

3 Signal processing circuit design

3.1 Circuit structure

The profillometer is mainly used in the field of flow monitoring in rivers and channels. The designed detection distance is 25 meters, the velocity range is ± 10 m/s, and the velocity accuracy is ±1% or ± 0.5 cm/s. The measuring circuit is the core of the velocity profiler. A measuring circuit is designed according to the measuring principle of the velocity profiler, as shown in Figure 2.

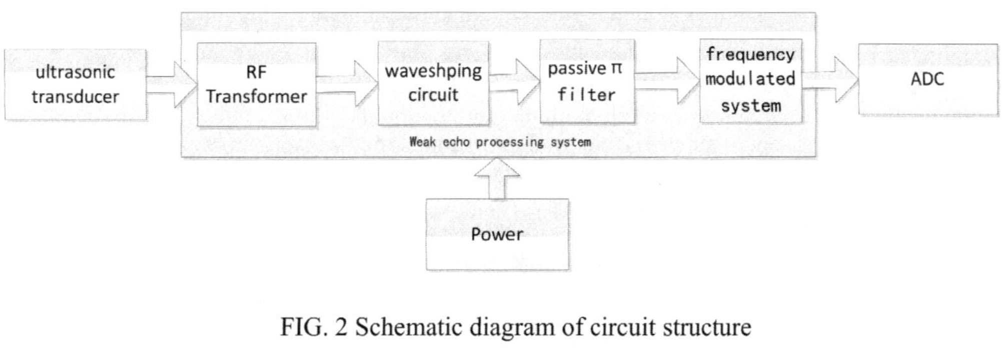

FIG. 2 Schematic diagram of circuit structure

The circuit consists of four parts, in which the weak signal processing system is the core of the whole circuit which realizes the signal shaping, filtering and amplification.

3.2 Difficulties in design

According to the instrument measurement principle, the ultrasonic signal spreads in water, and its attenuation is inversely proportional to the quadratic square of the distance. Therefore, the original signal is a continuous frequency signal with gradual attenuation, as shown in Figure 3.

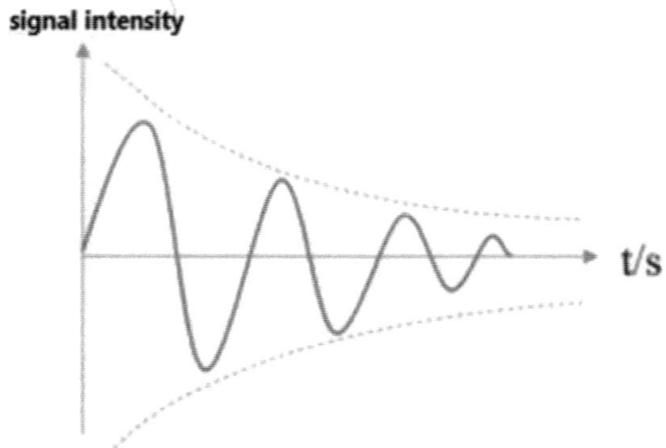

FIG. 3 Schematic diagram of ultrasonic signal propagation in water

The detection distance of the velocity profiler is up to 25 meters, and the signal intensity of the farthest section is less than 1/2500 of that of the nearest section. The dynamic range is beyond the processing capacity of the commonly used filter amplification and conditioning circuit, so the frequency modulation intermediate frequency system is adopted in this circuit for signal processing, as shown in FIG. 4.

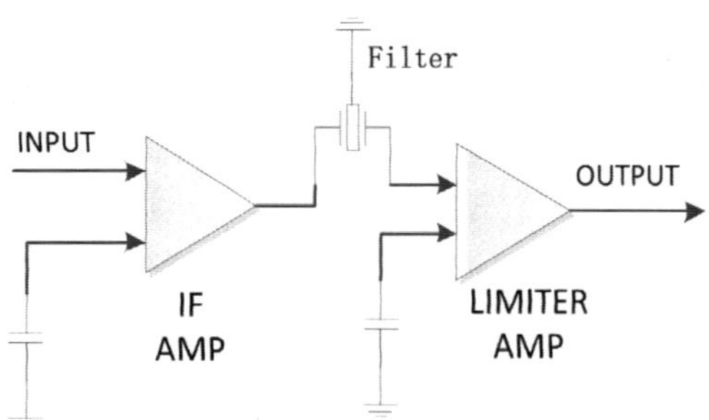

Figure 4. FM intermediate frequency system

In order to improve the signal-to-noise ratio, we choose the first level IF's gain of amplifier circuit according to the background noise signal strength and 25 meters underwater velocity signal intensity.In this way it makes the effective signal to the second limiting amplifier circuit of input threshold and below the input threshold noise signal at the same time. According to the instrument signal frequency, the ceramic filter is used to select the 1MHz intermediate frequency signal and attenuate the signal of other frequencies. Straight capacitors are used between the front and back stages to make the static working points of each amplifier independent. Impedance matching is required between the output and input circuits of the front and rear stages to suppress self-excited oscillations. After intermediate frequency amplification, the signal amplitude will be adjusted to about 0-3V through two stages of operational amplifier.

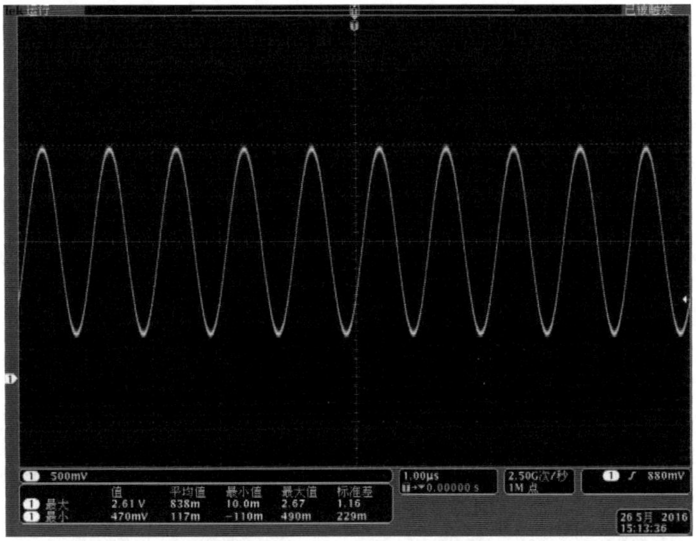

FIG. 5 Signal measured graph after processing

3.3 Signal digitization and frequency measurement

After the signal is processed by FM intermediate frequency circuit, it needs to be digitized through A/D conversion. According to Nyquist's theorem, the sampling rate must be at least twice the maximum analog signal bandwidth in order to recover the signal fully. Therefore, this circuit adopts a 12-bit sampling rate of 5MHz/s to collect signals of 1MHz frequency. The data collected is shown in Figure 6, and the actual effective bits are 11.5 bits.

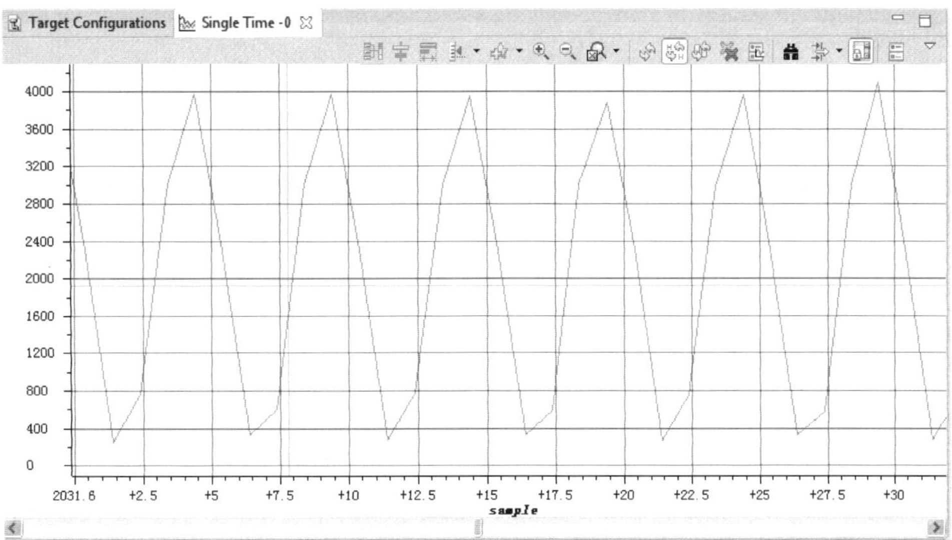

Figure 6 Signal collected by AD

The frequency measurement algorithm uses windowed FFT algorithm, and the accuracy of frequency measurement is ±0.67Hz after calculation. The basic formula of current measurement is:

$$v = \frac{f_r - f_t}{f_r + f_t} c = \frac{f_d}{2f_t + f_d} c \quad (1)$$

Where, v is water flow velocity, , $f_d = f_r - f_t$ (2)
is Doppler shift, and C is sound velocity in water. When there is an included Angle α between the direction of water flow and the direction of acoustic transmission, the radial flow measurement formula is:

$$v = \frac{f_d}{2f_t + f_d}\frac{c}{\cos\alpha} \quad (3)$$

According to Formula (3), if the direction of water flow is perpendicular to the direction of acoustic wave transmission (α= 90°), no Doppler effect will occur. Therefore, the ultrasonic wave emitted by the ultrasonic transducer has a certain Angle α with the direction of water flow, α is generally taken as $60^{\circ} \sim 70^{\circ}$.

4 Flow measurement comparison

In order to verify the actual use effect of the product, we choose a hydrology station for engineering application and flow test. According to The Code for Hydrology Survey (SL 195-2015), manual tests are selected at different water levels by combining the navigable ADCP with the artificial current meter method.

According to the analysis of the flow measurement data of the floating ADCP and the current velocity profiler, the systematic error is equal to mean relative error of 0.25%, the standard deviation Se is equal to 3.7%, and the random uncertainty is equal to twice of Se 7.4%.

The systematic error of single-line method for a class of stations is not more than ±1%, and the random uncertainty is not more than 8% according to the requirements of The Hydrological Data Reprogramming Code SL247-2012. The random uncertainty of flow measurement by ADCP can be increased by 2%. So the random uncertainty is not more than 10%. This flow test meets the requirement of alignment accuracy index.

5 Conclusion

According to the measuring principle and performance requirements of acoustic Doppler current profiler, a circuit for weak frequency signal processing is designed in this paper. The circuit adopts the design idea of FM intermediate frequency system to realize constant amplitude amplification of wide dynamic range frequency signal. Through the signal measurement, the signal amplitude is uniform and the waveform is smooth after the circuit processing. Performance test results show that the current velocity profiler with this circuit has the characteristics of long detection distance and measurement accuracy, which can better meet the requirements of online monitoring in application scenarios such as rivers and channels and it has broad application prospects.

References

[1]. ADCP Group,Division 3.Acoustic Doppler Current Profiler [J]. Marine Technology,1995,(3).

[2] Liu Yanping, Hua Tao, Zhou Bin. Research on H-ADCP Laboratory Detection Method [J]. Hydropower Automation and Dam Monitoring,2014,(4).

[3] Liu Yanping, ZOU Jun, Wei Shaohui. Measurement and Control Technology, 2018,(8).75-77,81.

[4] Wei Shaohui, LIU Yanping, Zou Jun. Development and Engineering Comparison measurement of Horizontal Stationary Acoustic Doppler Velocity Profiler [J]. Industrial Instrumentation & Automation Device, 2018,(6).112-116.[5] LIU Dezhu. Research on key Technology of Acoustic Doppler Velocity Measurement [D]. Harbin Engineering University,2010.1-114.

Research on fault diagnosis of gas steam boilers based on deep neural networks

Cong Deng*[ab], Xu-Feng Li[a], Peng Ji[a], Jing-Sheng Li[b], Qun Hou[b]

[a] Guangdong Institute of Special Equipment Inspection and Research, Foshan, China 528251;

[b] School of Mechanical and Automotive Engineering, South China University of Technology, Guangzhou, China 510640

[*] Corresponding author: davidcong78@vip.qq.com

ABSTRACT: Gas steam boilers are special equipment subjected to high temperature and pressure, it is particularly important to detect and deal with operation faults in time. Against the 4 types of common fault of gas steam boilers, the fault diagnosis model was established based on deep neural networks (DNN), and the model parameters were optimized and verified by experiments. The results show that the fault diagnosis model performs the best as Tanh of the activation function, 100 of Batch_Size, Adam of Optimizer, 10^{-5} of learning rate and 0.2 of Dropout since the output accuracy is above 92% and the model begins to converge with iterative times of 12. The accuracy of fault identification of reaches 100% in verification experiments and the model can still effectively recognize the dataset from the early stage which suggests that the fault diagnosis based on deep neural networks has high accuracy and provides a reliable solution for the safety supervision of special equipments.

1. Introduction

The concept of fault diagnosis technology was first proposed in the 1970s, and it has been widely used in the field of industrial production[1,2]. In the context of the continuous improvement of equipment automation and the Distributed Control System (DCS), fault diagnosis technology based on deep learning emerged and developed rapidly[3,4]. Compared with the traditional fault diagnosis technology, the fault diagnosis technology based on deep learning could effectively complete the fault pattern recognition in complex systems and gradually developed from single fault diagnosis to multiple fault classification diagnosis[5,6].

Logan et al.[7] developed an artificial neural network prediction system: Cerebellum Model Articulation Controller (CMAC), for continuous real-time fault diagnosis of large ships which achieved good results in practice. Tung et al.[8] applied neural networks to diagnose faults in dual-slot systems, which used the B-spline functions as the membership of fuzzy variables in the network to perform online training using the recursive least squares method and generated residual errors for fault diagnosis. Peng et al.[9] constructed a self-organizing competitive neural network model based on deep learning which realized the automated fault diagnosis of dynamometer cards. Jiang et al.[10] proposed an novel approach to derive the parity vectors that constructed optimized residual generators for linear and nonlinear systems which can significantly improve the sensitivity to small faults by evaluated thoroughly with a three-tank system, a numerical nonlinear example.

Gas Steam Boilers always work in a harsh environment of high temperature and high pressure, so major safety accidents are prone to occur if the failures in operation cannot be found and disposed of in time[11]. The application of fault diagnosis technology in boilers has attracted the attention of many engineers and researchers. However, there are few reports about the fault diagnosis based on deep learning involving boilers which is a typical nonlinear and high-complexity system. Due to the traditional fault diagnosis methods can hardly meet the actual requirements, the fault diagnosis model of gas steam boilers was established based on deep neural networks (DNN), and the model parameters were optimized and verified by experiments. The contributions of this paper lie in the following aspects.

- The data acquisition method of the faults of gas steam boilers is given and key features of the operation parameters is extractsed which is raleted to four typical faults of boiler system.
- An approach of parameters optimization is proposed to obtain the best performing fault diagnosis model through the comparison experiments.
- Through experimental verification, the fault diagnosis based on deep neural networks has good accuracy and applicability which provides a highly reliable solution to ensure the safe operation of boilers.

2. Objective and data acquisition

2.1 Fault information of gas steam boiler

2.1.1 The system of gas steam boiler

Figure 1 shows a schematic diagram of the gas steam boiler. The water-tube boiler is comprised of two systems, a combustion system, and a steam-water circulation system. The combustion system includes a preheater, burner, furnace, and other components: the gas and air are put into the burner in a predetermined ratio and burned in the furnace to generate heat. The steam-water circulation system mainly includes components such as a water drum, furnace tubes, convection tubes, steam drum, and downcomer: the preheated feed water is added to the water drum, and then the mixed hot water is heated to the mixture of steam and water which is sent to the upper drum through furnace tube or convection tubes where the stream flows away through the top outlet of the steam drum, and the water flows into the downcomer.

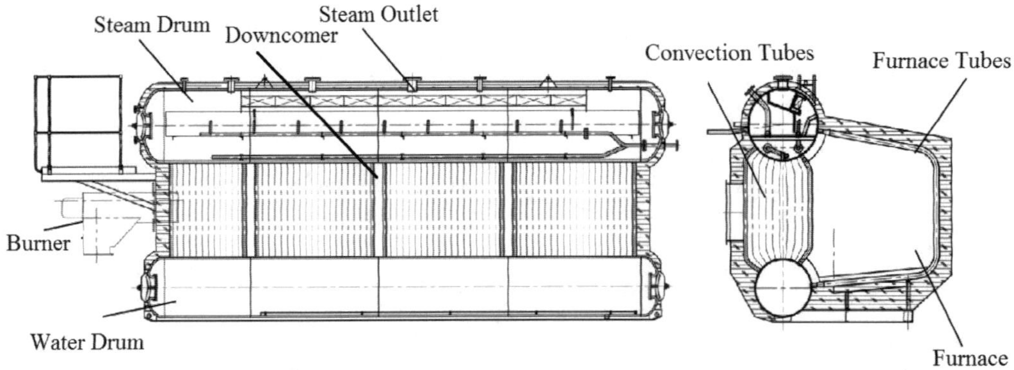

Figure 1. Schematic diagram of the experimental facility

2.1.2 Fault information

In the statistics of the fault of the gas steam boiler, 95% of the failures occurred in furnace tubes, upper drum, and preheater[12]. Common faults and characteristics in these parts are shown in Table 1.

2.2 Data acquisition

Due to the inappropriateness of obtaining fault data through the destructive test of boilers, Aspen Plus software was used to simulate the operation of the SZS4-1.25-Q gas steam boiler under four fault conditions. Nevertheless, the error between the data of the simulation model and the design parameters was less than 3%, which not only met the accuracy requirements but also could truly reflect the actual situation[13]. Table 2 shows some of the design parameters of the gas steam boiler.

Tab. 1. Common faults and characteristics[14]

Fault Type	Characteristics
Drum overfeed (DOF)	(i) the steam temperature of the outlet drops rapidly. (ii) the ratio of steam flow of outlet and feed water flow decreases. (iii) the export steam is charged with water. (iv) water strikes occur in the steam pipe.
Drum water shortage (DWS)	(i) the steam temperature of the outlet increases. (ii) the steam flow of the outlet is much larger than the feed water flow.
Furnace tubes leakage (FTL)	(i) the pressure of the furnace increases. (ii)gas temperature of outlet decreases. (iii) the flow of feed water increases; (iv) the water level of the steam drum decreases. (v) the steam temperature and pressure of the outlet decrease.
Preheater leakage (PHL)	(i) the pressure of the furnace increases. (ii) the water level of the steam drum decreases. (iii) the ratio of steam flow of outlet and feed water flow decreases. (iv)the gas temperature after the heater decreases. (v) the steam pressure of the outlet decreases.

Table 2. Design parameters of gas steam boiler

Parameters	Value	Parameters	Value
Rated evaporation/ t·h-1	4	exhaust gas temperature/℃	60
Rated working pressure/MPa	1.25	Fuel	Gas
Feed water temtepratrure/℃	20	Combustion efficiency	90.76%

Because of the performance of the above four faults, the following six characteristic parameters are used to diagnose the fault of boilers: the water level of a steam drum (WSD), the steam pressure of the outlet (SPO), the steam temperature of the outlet (STO), the ratio of steam flow of outlet and feed water flow (taken as absolute value) (RSF), the pressure of furnace (PF) and the gas temperature after the heater (GTH). By simulating a total of 11.2 minutes before and after the fault occurrence of each type, 672 sets of operation data were obtained, with the amount of 2,688 sets for four types. Figure 2 shows the variation of parameters with the fault of drum overfeed.

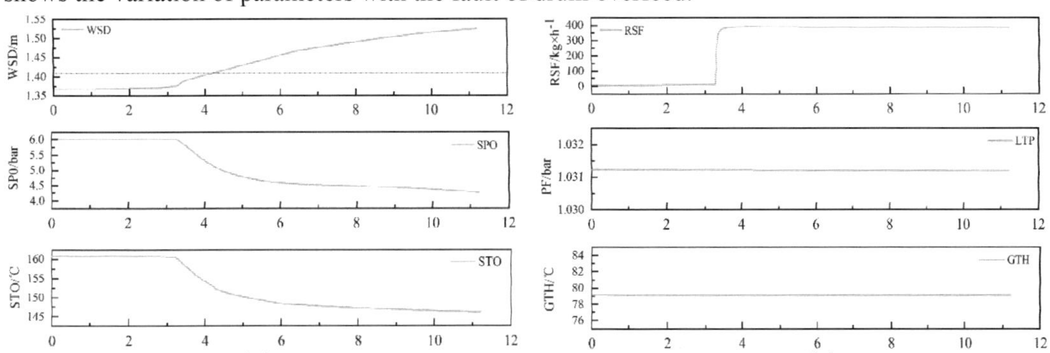

Figure 2. Parameters changes with the fault of drum overfeed

Nevertheless, the parameters with different units are inputted into the deep neural network model directly which could result in the inaccurate calculation. To eliminate the influence of dimensional units on the training model and improve the comparability of data sets, the preprocessing and normalization should be carried out before the import of continuous feature data. So the Z-score normalization method is used, and the conversion function is as follows:

$$x^* = \frac{x-\mu}{\sigma} \tag{1}$$

wherein μ is the mean value and σ is a variance of the sample data sets.

While the parameters were normalized, the binarization processing was used to the fault type, and then the data sets were randomly disrupted to reduce the risk of overfitting. Table 3 shows the first 5 groups of data in the preprocessed datasets.

Table 3. Partial preprocessed data

No.	WSD*	SPO*	STO*	RSF*	PF*	GTH*	PHL	DOF	DWS	FTL
106	-1.1496	-0.0680	-0.5523	0.0990	0.8177	-1.0326	0	0	0	1
1587	-0.4512	0.8101	1.7541	-1.6261	-0.8480	0.9931	0	0	1	0
666	0.0262	-1.6033	-0.6537	0.6630	1.0448	-0.8438	1	0	0	0
12	-0.9698	-0.0770	-0.1944	0.0686	0.8555	-0.9833	0	0	0	1
1041	1.6896	0.8514	-0.5880	0.7161	-1.0751	0.9964	0	1	0	0

3. Methodology

Neural networks are based on the extension of perceptron, while deep neural networks (DNN) can be understood as neural networks with many hidden layers. Hence DNN is sometimes called multilayer perceptron (MLP)[15]. Its topology structure consists of an input layer, hidden layers, and an output layer, as shown in Figure 3.

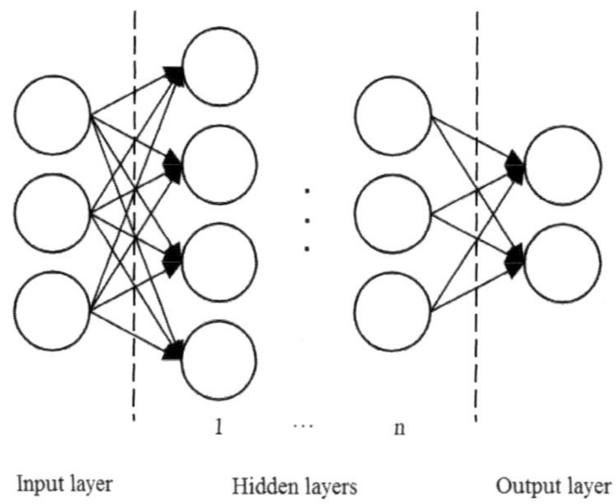

Figure 3. The structure of deep neural networks

The neurons within a layer are not connected and the neurons between contiguous layers are fully connected, i.e. any neuron in layer i must be connected to any neuron in layer i+1. The computational relationship between layer i-1 and layer i consists of a linear relationship and an activation function:

$$z_i = w_i x_i + b_i \tag{2}$$

$$y_i = \sigma(z_i) \tag{3}$$

$$x_{i+1} = \begin{cases} o_t & i=1 \\ y_i & 1 < i \le L \end{cases} \tag{4}$$

where z_i is the output vector of layer i that is not activated. w_i is the full connection matrix between layer i-1 and layer i. x_i is the input vector of layer i. b_i is the bias vector of layer i. y_i is the output vector after activated. o_t is the input feature vector.

In Equation (3), the following three functions are as activation functions:

LeakyreLu function:

$$\sigma(z)=\max(0,z)+ leak*\min(0,z) \tag{5}$$

Sigmoid function:

$$\sigma(z)=\frac{1}{1+e^{-z}} \tag{6}$$

Tanh function:

$$\sigma(z)=\frac{e^{z}-e^{-z}}{e^{z}+e^{-z}} \tag{7}$$

To make the analog output as accurate as possible, suitable parameters of wi and bi must be found. The optimization of a traditional machine learning algorithm is to use a suitable loss function to measure the output loss of the training samples and optimize the loss function to find its extreme value of minimization since the corresponding wi and bi are desired. In deep neural networks (DNN), the optimal extremum solution of the loss function can be done iteratively by algorithms such as GD and Adam, and the error will pass back for network parameter update by the backpropagation algorithm[16].

Assuming a multi-classification task with T hours of training data, the output layer of K categories, and the actual output vector of g_t, the output probability is:

$$P(g_t|x_t)=\prod_{j=1}^{K}\left(y_t^L(j)\right)^{g_t(j)} \tag{8}$$

The cross-entropy loss function is always used for network optimization in classification algorithms. According to the maximum likelihood estimation (MLE), the function is:

$$O=-\prod_{t=1}^{T}p(g_t|x_t) \tag{9}$$

Taking the logarithm to the Equation (9):

$$\ln O=-\ln\left(\prod_{t=1}^{T}p(g_t|x_t)\right)=-\sum_{t=1}^{T}\ln p(g_t|x_t) \tag{10}$$

The cross-entropy loss function can be defined as:

$$O=\sum_{t=1}^{T}\sum_{j=1}^{K}g_t(j)\ln\left(y_t^L(j)\right) \tag{11}$$

The error of the cross-entropy loss function is the gradient of the network parameters, and the error at layer i of the cross-entropy loss function is equivalent to the amount of inactivation at layer i. So the error of the output layer can obtain from the chain derivative rule:

$$\delta_t^i=\frac{\partial O}{\partial\alpha_t^L(j)}=\sum_{k=1}^{K}\frac{\partial O}{\partial y_t^L(k)}\frac{\partial y_t^L(k)}{\partial\alpha_t^L(j)} \tag{12}$$

where:

$$\frac{\partial O}{\partial y_t^L(k)}=-\frac{g_t(k)}{y_t^L(k)} \tag{13}$$

$$\frac{\partial y_t^L(k)}{\partial \alpha_t^L(j)} = c_{kj} y_t^L(k) - y_t^L(k) y_t^L((j), j), kj = \begin{Bmatrix} 1 & k = j \\ 0 & k \neq j \end{Bmatrix} \tag{14}$$

The error of the hidden layers is defined as:

$$\delta_t^L(j) = y\left(\sum_{j=1}^{s_{l+1}} W_{ji}^l \delta_t^{l+1}(j) \sigma'(\alpha_t^l)\right) \tag{15}$$

$$\frac{\partial O}{\partial W^i} = \frac{\partial O}{\partial \alpha_t^i} \frac{\partial \alpha_t^i}{\partial W^i} = \delta_t^i (x_t^i)^T \tag{16}$$

$$\frac{\partial O}{\partial b^i} = \frac{\partial O}{\partial \alpha_t^i} \frac{\partial \alpha_t^i}{\partial b^i} = \delta_t^i \tag{17}$$

Then the parameters of the hidden layers can be updated by using the loss function optimization algorithm:

$$W^i = W^i - \alpha \frac{\partial O}{\partial W^i} \tag{18}$$

$$b^i = b^i - \alpha \frac{\partial O}{\partial b^i} \tag{19}$$

The iterative formulas of GD and Adam algorithm are as follows:
GD algorithm:

$$w = w - \alpha * dw \tag{20}$$

Adam algorithm:

$$\begin{cases} v = \beta_1 v + (1 - \beta_1) dw \\ v = \beta_2 s + (1 - \beta_2) dw^2 \\ w = w - \alpha \dfrac{v}{\sqrt{s + \varepsilon}} \end{cases} \tag{21}$$

where: β_1 is a constant value of 0.9. β_2 is a constant value of 0.999. ε is a small constant value of 10^{-8}.

4. Fault diagnosis experiment of deep neural networks

4.1 Parameters optimization

More appropriate parameters need to be solved to improve the applicability of the fault diagnosis based on deep neural networks and the process of solution is essentially an optimization problem, which can be optimized by comparison experiments. So the comparison experiments of fault diagnosis parameters were carried out, and the input layer of the deep neural networks was set to 6-dimension, and the hidden layer was set to 20 layers where the number of neurons in each layer was 24, and the output layer was 4-dimension according to the 6 types of characteristic parameters and 4 types of faults. Meanwhile, the loss function was set to cross-entropy loss function and the activation function of the output layer was set to Softmax which could map the output to a real number between 0 and 1, and normalize them to ensure that the sum of the probability of various fault is 1[17].

Then, there were five parameters need to be optimized experimentally that were the activation function of the hidden layer, Batch_Size, Optimizer, learning rate, and Dropout. The networks performance of all combination of those five parameters has been evaluated by the cross-entropy loss and the accuracy, and 11 representative combination of parameters are picked and presented here to reveal the process of optimization. The 11 representative settings of optimized comparison experiments are shown in Table 4.

Table 4. The settings of optimized comparison experiments

NO.	Activation function	Batch_Size	Optimizer	Learning rate	Dropout
1	Leakyrelu	300	GD	0.00001	0
2	Sigmoid	300	GD	0.00001	0
3	Tanh	300	GD	0.00001	0
4	Tanh	400	GD	0.00001	0
5	Tanh	200	GD	0.00001	0
6	Tanh	100	GD	0.00001	0
7	Tanh	100	Adam	0.00001	0
8	Tanh	100	Adam	0.0001	0
9	Tanh	100	Adam	0.001	0
10	Tanh	100	Adam	0.00001	0.2
11	Tanh	100	Adam	0.00001	0.5

4.1.1 Optimized activation function.
Figure 4 shows the effect of the comparison experiments with the three activation functions, Tanh, Sigmoid, and Leakyrelu. It can be found that the cross-entropy loss gradually decreases and stabilizes at 0.18, and the accuracy gradually increases and stabilizes at 90.91% when Tanh is used as the activation function of the hidden layer.

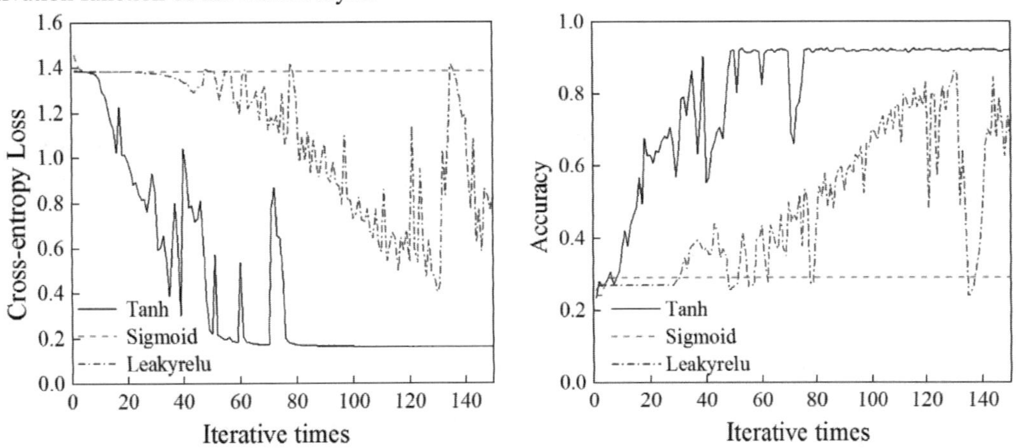

(a) The cross-entropy loss (b) Accuracy

Figure 4. The comparison experiments with three activation functions

4.1.2 Optimized Batch_Size.
Batch_Size is the number of selected samples to train at one time which may have an impact on the degree of optimization and the training speed of the model[18]. Four sizes of 100, 200, 300, and 400 are selected for iterative network optimization based on Tanh of the activation function in Figure 5. It is found that the cross-entropy loss and accuracy are better when the Batch_Size is 100 and 200 which

stabilizes at 0.16 and 0.17, 91.56%, and 92.03% respectively. However, the fluctuation range of accuracy is smaller at the low iterative times when Batch_Size is 100.

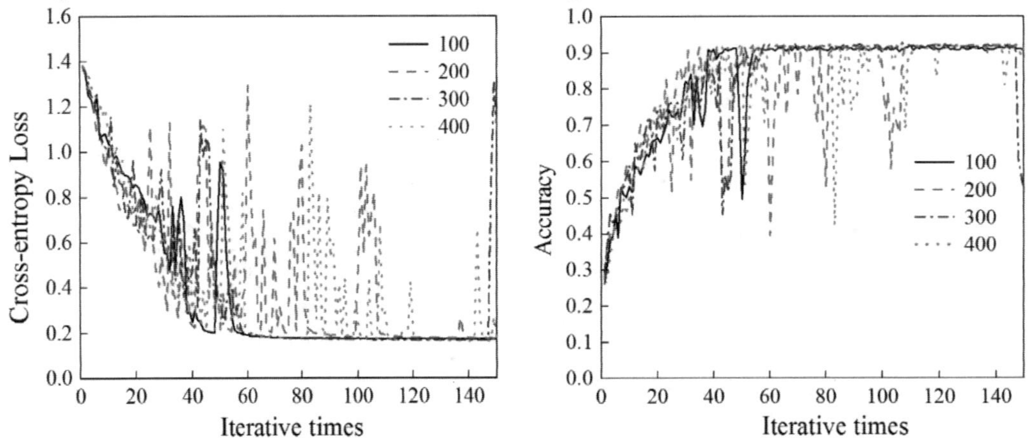

(a) The cross-entropy loss (b) Accuracy

Figure 5. The comparison experiments with different Bitch-Size

4.1.3 Optimized Optimizer.

The Optimizers of Adam and GD are used in the comparison experiments on the strength of Tanh as activation function and 100 as Batch_Size, and the results are shown in Figure 6. The cross-entropy loss and accuracy tend to be stable and equal as the iterative times' increase. However, it can be observed that the optimizer of Adam requires fewer iterative times to reach stability which suggests that Adam is more suitable for deep neural network modeling for fault diagnosis.

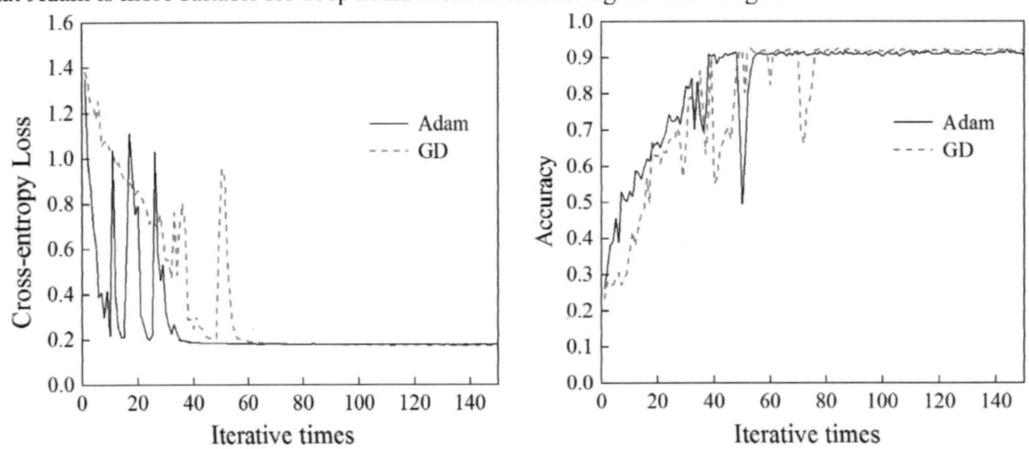

(a) The cross-entropy loss (b) Accuracy

Figure 6. The comparison experiments with different Optimizers

4.1.4 Optimized learning rate.

Based on the above-optimized parameters, three learning rates of 0.00001, 0.0001, and 0.001 are chosen for iterative network optimization, and the results are shown in Figure 7. It can be seen that the cross-entropy loss gradually decreases to 0.18 and accuracy increases and reaches 91.90% when the learning rate is 0.00001, which suggests that 0.00001 of the learning rate is optimized in this comparison experiment.

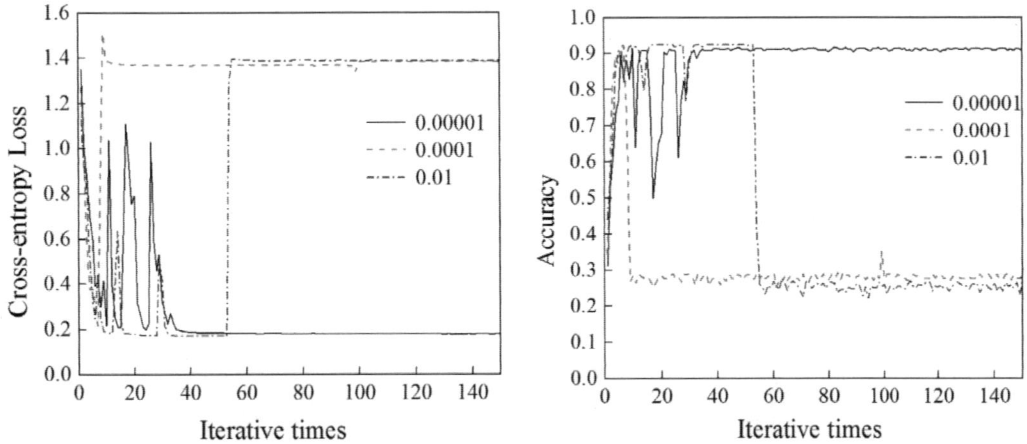

(a) The cross-entropy loss (b) Accuracy

Figure 7. The comparison experiments with different learning rate

4.1.5 Optimized Dropout.

In the context of frequent overfitting in deep neural networks, Dropout is proposed to temporarily discard some neurons at random in a certain proportion during the training of the network, thus weakening the dependency between neurons and improving the generalization ability of the network[19].

Figure 8 shows the effect of the comparison experiments with different Dropouts of 0 and 0.2 and 0.5 based on Tanh as activation function and 100 as Batch_Size and Adam as optimizer and 0.00001 of the learning rate. Similarly, the cross-entropy loss decreases, the accuracy gradually improves and both tend to be stable with different Dropout through 150 iterative times. It also can be found that accuracy is better with a Dropout of 0.2 even though the degree of network optimization is roughly the same with a Dropout of 0.2 and 0.5.

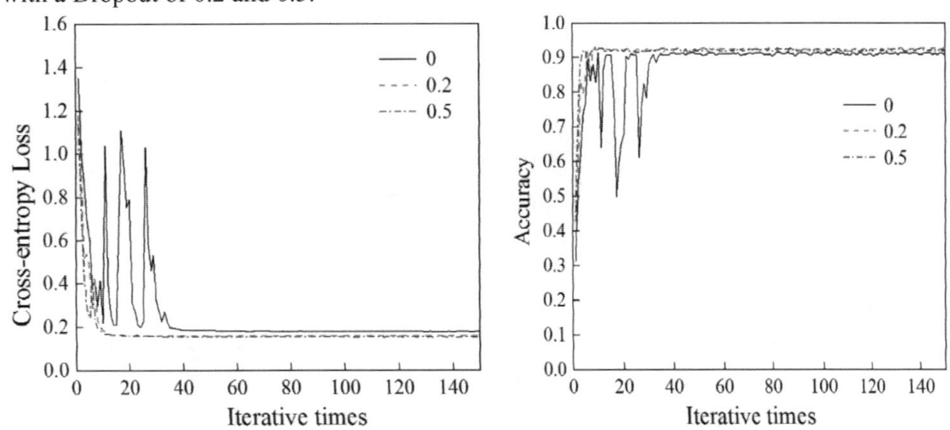

(a) The cross-entropy loss (b) Accuracy

Figure 8. The comparison experiments with different Dropout

4.1.6 Results of parameter optimization.

Up to now, the optimal parameters of the deep neural networks for fault diagnosis are found, and the output results of the cross-entropy loss and accuracy of the final optimized experiment are shown in Figure 9. The output accuracy is above 92% and the system begins to converge with iterative times of 12, which demonstrates that the fault diagnosis algorithm of the gas steam boiler system under the optimized parameters is very stable and usable.

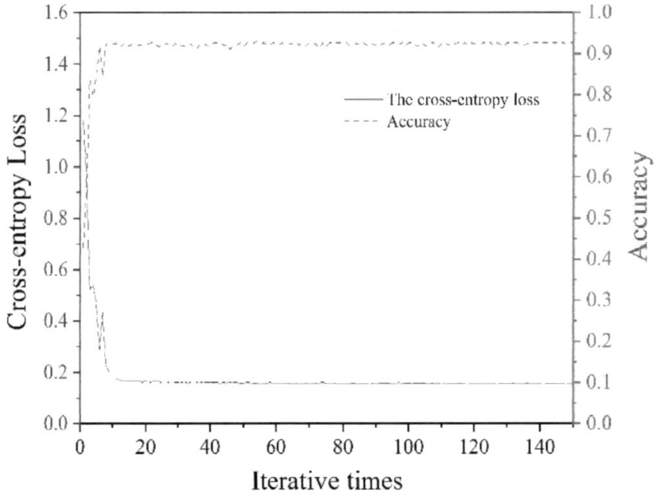

Figure 9. Results of parameter optimization

4.2 Experimental verification of fault diagnosis system

2,688 sets of operation data were used to build the networks. Besides, 80 sets of new data (20 sets of each fault), acquired by the experimental platform and simulation model which was shown in figure 10, were used to verify this fault diagnosis system based on deep neural networks. The curves in Figure 11 represent the output probability of the test data in each fault type under the deep neural networks model and the largest output probability of the fault represents the final diagnosis result.

The accuracy of fault identification reaches 100% for the 80 sets of experimental verification of the fault diagnosis model and the output probability of the fault generally has high directivity. The reason why the probability of PHL is 0.55 and DOF is 0.31 in the 20th set of data Figure 11 (d) is that the dataset comes from the early stage of PHL fault which leads to confusion. Overall, the fault diagnosis based on deep neural networks has high accuracy and applicability for the diagnosis of four types of operational faults which has high practical application value.

(a) The experimental platform

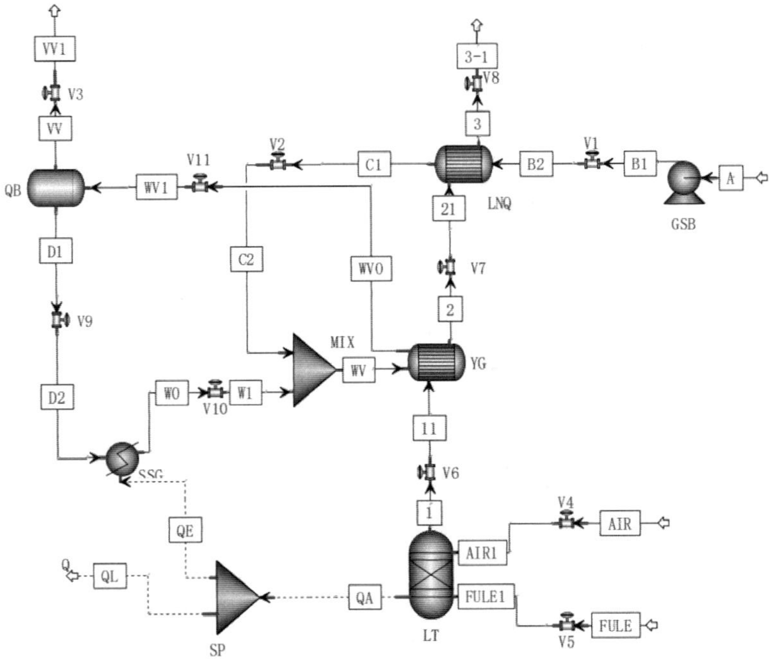

(b) The simulation model

Figure 10 Gas steam boiler of SZS4-1.25-Q

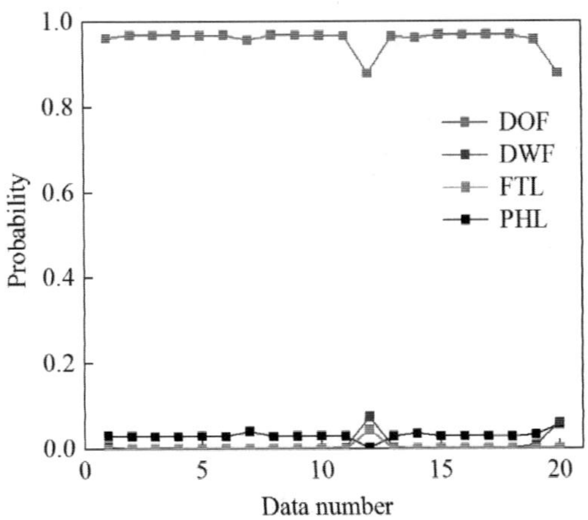

(a) Drum overfeed (DOF)

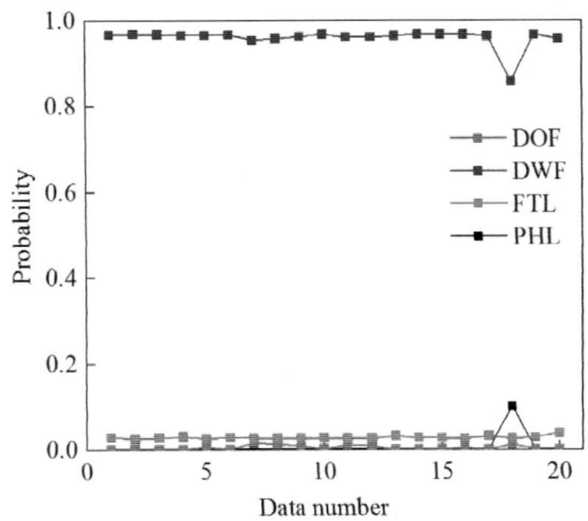

(b) Drum water shortage (DWS)

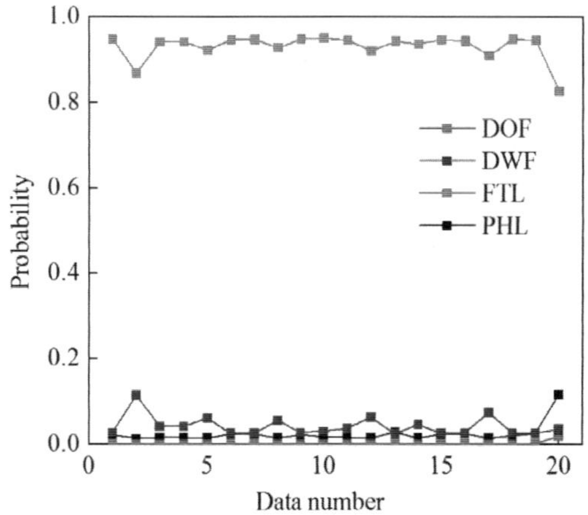

(c) Furnace tubes leakage (FTL)

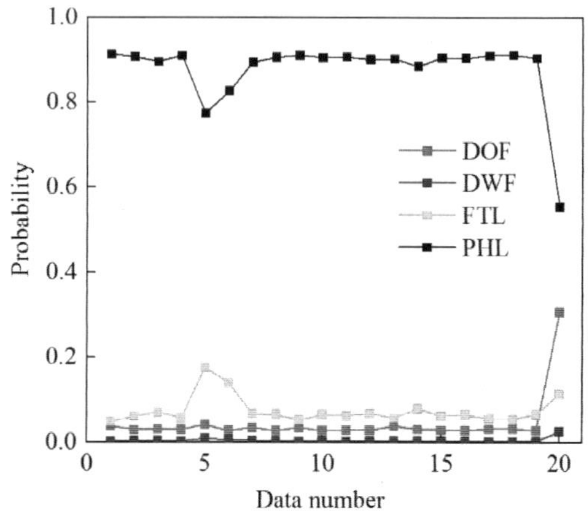

(d) Preheater leakage (PHL)

Figure 11 Output Probability with various faults

5. Conclusion and future perspective

This paper is focud on the fault diagnosis of gas steam boilers based on deep neural networks correlating four types of boiler faults with eight operating parameters. By designing an applicable deep neural network model and using comparison experiments for parameters optimization, an optimal fault diagnosis model is constructed. As Tanh of the activation function and 100 of Batch_Size and Adam of Optimizer and 10^{-5} of learning rate and 0.2 of Dropout, the fault diagnosis model performs the best which begins to converge with iterative times of 12 and the output accuracy is above 92%. The accuracy of fault identification reaches 100% for the 80 sets of experimental verification and the model can still effectively recognize the dataset from the early stage of fault which suggests that the fault diagnosis based on deep neural networks has high accuracy and applicability.

In the future study, the fault diagnosis model will be extended to more general types of faults and their corresponding mechanisms. There are reasons to believe that fault diagnosis based on deep neural networks will eventually be used in in the actual operation of boilers, providing a reliable solution for the safety supervision of special equipments.

Acknowledgments

We gratefully acknowledge financial support from the Science and Technology Project of Guangdong Institute of Special Equipment Inspection and Research (2020CY-03), the Science and Technology Project of State Administration for Market Regulation (2021MK087).

References

[1] Bunks, C., Dan, M. and Ani, T., " Condition-based maintenance of machines using hidden Markov models," Mechanical Systems & Signal Processing 14(4), 597-612 (2000).
[2] Baruah, P. and Chinnam, R. B., "HMMs for diagnostics and prognostics in machining processes," International Journal of Production Research 43(6): 1275-1293 (2005).
[3] Liu, H., Wang, Y. Y. and Chen, W.G., "Three-step imputation of missing values in condition monitoring datasets," Transmission & Distribution 14(16):3288-3300 (2013).
[4] Liu, J., Yan, S., Zeng, D., Hu, Y. and Lv, Y., "A dynamic model used for controller design of a coal fired once-through boiler-turbine unit," Energy 93:2069–2078 (2015).
[5] Oko, E. and Wang, M., "Dynamic modelling validation and analysis of coal-fired subcritical power plant," Fuel 135:292-300 (2014).
[6] Li, G., Sui, H., Xu, C. and et al, "Steady state modeling and simulation of an industrial combustor-style fluidized catalytic cracking regenerator," Asia-Pacific Journal of Chemical Engineering 8(4): 507-518 (2013).

[7] Logan, K. P., "Real-Time Simulation of Marine Diesel Engines for Fault Diagnosis," Marine Technology 39(1):21-28 (2002).
[8] Tung, M. H. and Chan, C. W., "Online fault diagnosis of nonlinear systems based on neurofuzzy networks," Pergamon Press 12: 67-72 (2005).
[9] Peng, X., "Application of BP neural network and self-organizing competitive neural network to fault diagnosis of suck rod pumping system," Acta Petrolei Sinica 27(2): 107-110 (2006).
[10] Jiang, Y. and Yin, S., "Optimized design of parity relation based residual generator for fault detection: data-driven approaches", IEEE Transactions on Industrial Informatics, 17(2): 1449-1458 (2020).
[11] Bunks, C., Dan, M., Al-Ani, T., "Condition-Based Maintenance of Machines Using Hidden Markov Models," Mechanical Systems & Signal Processing 14(4): 597-612 (2000).
[12] Yao, Y., Jiang, L., Deng, B. and et al, "Heat transfer analysis of stationary bed materials in a CFB boiler after a sudden power failure," Fuel Processing Technology 211: 830-839 (2020).
[13] Meng, Y., Wu, X., Oladejo, J. and et al, "Application of Machine Learning in Industrial Boilers: Fault Detection, Diagnosis, and Prognosis," Chem. Bio. Eng. Reviews 8(5): 535-544 (2021).
[14] Wang, S. L., Hu, H. L., Zhang, X. L., "Simulation of gas steam combined cycle based on Aspen Plus," Turbine Technology 47(6): 4I7-420(2005).
[15] Yan, H., Jiang, Y., Zheng, J. and et al, "A multilayer perceptron-based medical decision support system for heart disease diagnosis," Expert Systems with Applications 30(2):272-281 (2006).
[16] Guo, C. W., Jia, L. W.; Cheng L. and et al, "Parameter estimation of fractional uncertain differential equations via Adams method," Nonlinear Analysis 27:1-5 (2022).
[17] Li, H., Lyu, Y., Duan, G. and et al, "Improving finger vein discriminant representation using dynamic margin softmax loss," Neural Computing and Applications 6: 1-13(2021).
[18] Copelli, S., Torretta, V., Maestri, F. and et al, "Batchsize and Topological Criteria: a Combined Approach to Safely Optimize Hazardous Polymerization Processes," Chemical Engineering Transactions 33: 273-277 (2013).
[19] Srivastava, N., Hinton, G., Krizhevsky A. and et al, "Dropout: A Simple Way to Prevent Neural Networks from Overfitting," Journal of Machine Learning Research 15(1):1929-1958 (2014).

Research on stress wave wood nondestructive testing technology

Xiwen Wei[1,a], Chunxiao Du[2,b], Shuzheng Xu[3,c], Cong Tian[1,d], Xinnian Yang[1,e], Lina Hu[1,f] and Peng Pang[1,g]

[1] School of Electrical and Information Engineering, Heilongjiang University Of Technology, Jixi, China

[2] Harbin Veterinary Research Institute, CAAS, Harbin, China

[3] China United Network Communications Co., LTD. Nanjing Software Research Institute, Nanjing, China

[a]wxw198806@163.com, [b]duchunxiaoeasy22@163.com, [c]xeus5212@163.com, [d]zwqkt@126.com, [e]yangxinnian123@163.com, [f]jxdxhln@163.com, [g]pp171026@163.com

Abstract. As a kind of nondestructive testing technology, stress wave technology has the advantages of long propagation distance, strong anti-interference ability and convenient use, and has become one of the important means to detect wood properties and defects at home and abroad.This paper introduces several stress wave wood testing equipment commonly used in the field of wood non-destructive testing at this stage, and analyzes its characteristics. The research status of stress wave wood nondestructive testing technology is analyzed from three aspects: mechanical property testing, internal defect testing and factors affecting stress wave propagation velocity. Finally, the existing problems and development trends of stress wave technology in wood nondestructive testing are discussed to provide reference for further research in this field.

1. Introduction

With the world's attention to forest resources, wood non-destructive testing technology was proposed in the early 1960s, and then gradually emerged in developed countries such as the United States and Germany, and achieved certain results. Non-destructive testing can be performed on the growth characteristics of standing wood, physical and mechanical properties of logs and wood samples, and can also detect defects existing in logs and wood. In the 21st century, science and technology have made greater progress, which has accelerated the development of wood non-destructive testing technology in its research field [1]. According to different working principles and detection methods, there are dozens of non-destructive testing methods for wood. Among them, the most commonly used are X-ray detection, ultrasonic detection, infrared detection, nuclear magnetic resonance detection and stress wave detection [2]. Compared with other nondestructive testing technologies, stress wave nondestructive testing technology has long propagation distance, strong anti-interference ability, and is not limited by the shape and size of the tested wood. Besides, it does not damage the tested wood, with low development cost of stress wave nondestructive testing equipment. It has strong portability, no harm to the human body, simple sensor arrangement and fast detection process. As one of the technologies with the highest applicability and widest application in the non-destructive testing technology of wood at home and abroad, it is often used for the detection of physical and mechanical properties of wood and internal defects of wood [3-4].

The stress wave detection technology uses the impact stress wave generated after the object is hit, to detect the characteristics of the propagation in the object. It is a measurement techniques based on the relationship between the propagation speed of the stress wave in the wood and the density and elastic modulus of the tested wood [5-6]. At the same time, it can also detect the internal and external defects of the wood according to the difference of the propagation speed of the stress wave in the different strength parts of the wood. When the stress wave propagates inside the wood, the speed of the stress wave when it encounters defect areas such as decay, cracks, and holes is different from that when it passes through the intact parts [7], as shown in figure 1.

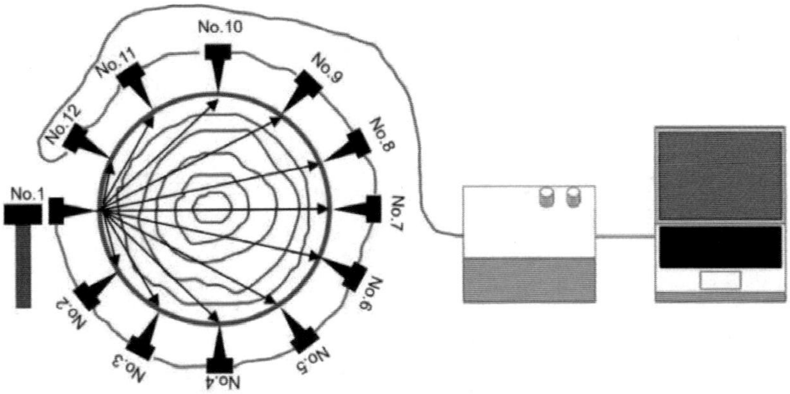

Figure 1. Stress wave nondestructive testing of wood cross section

The propagation time T of stress wave between any two sensors is recorded, and the time matrix can be expressed by equation (1), and the distance d between any two sensors is given, the stress wave propagation velocity matrix V is obtained, as shown in equation (2).

$$T = \begin{vmatrix} t_{11} & t_{12} & \cdots & t_{1n} \\ t_{21} & t_{22} & \cdots & t_{2n} \\ \cdots & \cdots & \cdots & \cdots \\ t_{n1} & t_{n2} & \cdots & t_{nn} \end{vmatrix} \quad (1)$$

$$V = \begin{vmatrix} 0 & d_{12}/t_{12} & \cdots & d_{1n}/t_{1n} \\ d_{21}/t_{21} & 0 & \cdots & d_{2n}/t_{2n} \\ \cdots & \cdots & \cdots & \cdots \\ d_{n1}/t_{n1} & d_{n2}/t_{n2} & \cdots & 0 \end{vmatrix} \quad (2)$$

2. Stress wave wood nondestructive testing equipment

At present, most of the stress wave nondestructive testing equipment used in my country is from abroad, and its main brands are the Fakopp series produced by the Hungarian FAKOPP company and the Arbotom series produced by the German RINNTECH company. Fakopp 2D tree imaging evaluation system is a stress wave wood detection technology based on stress wave transfer path and imaging algorithm. It is mainly used to detect live standing trees and up to 32 sensors can be set to obtain tomographic images of wood cross-sections. The measurement time accuracy reaches 2μs, as shown in figure 2. Arborsonic 3D acoustic tomography system is a more advanced version than

Fakopp 2D, adding inversion algorithm and 3D visualization, and can fit multiple 2D tomograms into 3D images. In 2015, Liao et al. [8] used Fakopp stress wave tester and micro-drill resistance instrument to screen the internal defects and sizes of cylindrical wooden components of ancient buildings on site.

The main function of Arbotom is to assess the internal health status of trees or wood, and to accurately locate internal defects, and the detection results are intuitive and clear. In theory, there is no upper limit on the number of Arbotom sensor probes, so Arbotom can be applied to trees of any diameter at breast height, and can be used to detect internal defects of wood and judge the quality, stability and availability of tree stems, as shown in figure 3. One of the features of Arbotom is that it can be adaptively adjusted according to the shape of the section of the tree under test. After the system software adjusts the simulation diagram according to the cross-sectional shape of the tested tree, the measurement result is more realistic and the error is smaller. Northeast Forestry University Wang's research group used the Arbotom stress wave wood detection equipment to measure and analyze a large number of living standing wood samples of different tree species such as larch, white birch, and ash. A large number of research results have been achieved in many aspects such as stress wave propagation in non-frozen wood and quantitative detection of sample defects [9].

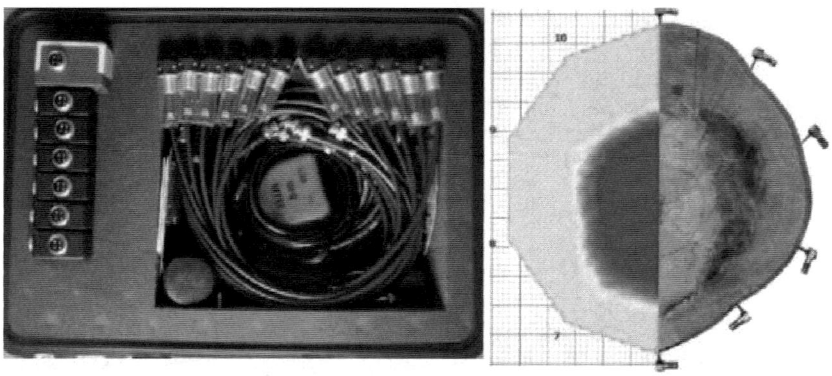

Figure 2. Fakopp device and detection effect

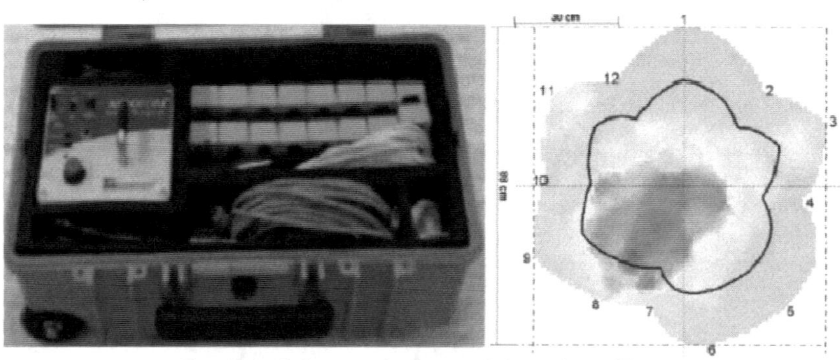

Figure 3. Arbotom device and detection effect

Some non-destructive testing equipment for wood based on stress waves has also been developed in China. In 2013, An et al. developed a stress wave nondestructive testing equipment based on a digital signal processor (DSP), which has the advantages of fast detection speed and low cost. In 2018, Xu et al. [10] designed a non-destructive testing equipment for wood based on a charge-type accelerometer, which can detect cracks, voids and other damage inside the wood by calculating the transmission time of the stress wave inside the wood. Most of the above studies are still in the laboratory stage, and there is still much room for improvement in terms of detection accuracy, imaging clarity, and comprehensive functionality. In recent years, the TOP-900 and F3D stress wave tree non-destructive testing systems produced by Zhejiang TUOPU YUNNONG Company can quickly detect the internal defects of trees with high imaging accuracy. It has been widely used in ancient and famous trees, street trees, precious wood trading and forestry scientific research.

3. Research status of stress wave wood nondestructive testing technology at home and abroad

3.1. Testing of physical and mechanical properties of wood by stress waves

The physical and mechanical properties of wood can be detected by stress wave technology, and the accuracy of stress wave technology detection is proved by comparing with traditional physical and mechanical properties testing methods [11]. In 2016, Madhoushi et al. [12] used stress wave nondestructive testing technology to measure the transverse stress wave propagation velocity and dynamic modulus of elasticity (MOEd) in the DBH direction of 14 Northeast cottonwood trees, as well as the speed and MOEd of the longitudinal stress wave propagation in the height direction of 3 logs. The results show that the longitudinal MOEd of trees and logs is larger than the lateral MOEd. In 2020, Llana et al. [13] established stress wave propagation velocity and tree species and bending elastic modulus models through non-destructive testing of logs such as common alder, European white birch and American sycamore, regardless of tree species density. The model provides a good estimate of the mechanical properties of hardwood thinned small-diameter logs and is mostly used to pre-classify materials in forests.

The application of stress wave technology in the evaluation of physical and mechanical properties of trees started relatively late in China, but with the rapid development of stress wave detection technology, Chinese forestry scientists began to gradually conduct in-depth research, and tried to establish the mathematical model on the relationship between stress waves and various properties of wood. In 2015, Liu et al. [14] analyzed the propagation law of stress waves in healthy trees, and established a mathematical model of the ratio of the direction angle in the tree cross section to the velocity v_T in the propagation direction and the radial velocity v_R. The results show that the model is in good agreement with the theoretical model, the linear regression has a high degree of fit (R^2=0.95), and the model is not affected by changes in tree species. In 2019, Yang et al. [15] proposed a method based on shock stress wave test to evaluate the quality of hardwood logs, using moment analysis and wavelet transform to extract two acoustic evaluation parameters—time center and damping ratio. The results show that compared with the traditional sound velocity classification, the method increases the occupancy rate of high-grade sheets in the high-quality group by about 26%, while the occupancy rate of high-grade sheets in the low-quality group decreases by more than 15%.

3.2. Detection of internal defects of wood by stress wave

Establishing a model of the propagation velocity of stress waves in wood can not only obtain the physical and mechanical properties of wood, but also judge the internal defects of wood. It plays an important role in evaluating wooden structure buildings and predicting the properties of panels, which can promote the development of stress wave wood nondestructive testing technology to a certain extent. In 2017, Zhang et al. [16] used a total of 40 trees of 4 tree species in the Botanical Garden of Zhejiang A&F University as experimental samples to measure the propagation velocity of stress waves between points on longitudinal sections with different angles. The results show that the propagation velocity between any two points on the same section increases with the increase of the angle in the direction, and the propagation velocity between any two points on different longitudinal sections is related to the angle between the longitudinal section and the radial section. In 2020, Wei et al. [17] used *birch*, *ash*, *elm*, and *larch* logs as test samples to analyze the relationship between the stress wave propagation velocity and the direction angle in the longitudinal section, and the relationship between the propagation velocity and longitudinal section angle in the cross section, as shown in figure 4 and figure 5. And establish a multiple regression model of the propagation velocity and direction angle and longitudinal section angle, as shown in table 1 and table 2. The results show that the model has a good guiding role for the detection of internal defects in trees. The position and size of internal defects in logs can be accurately detected by using the model combined with two-dimensional imaging technology.

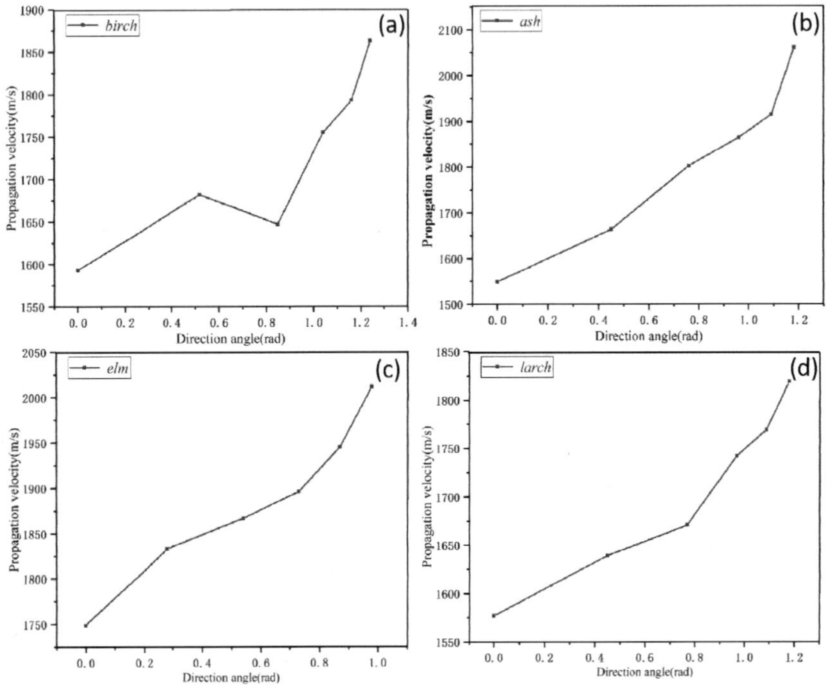

Figure 4. Stress wave propagation velocity diagram in different direction angles

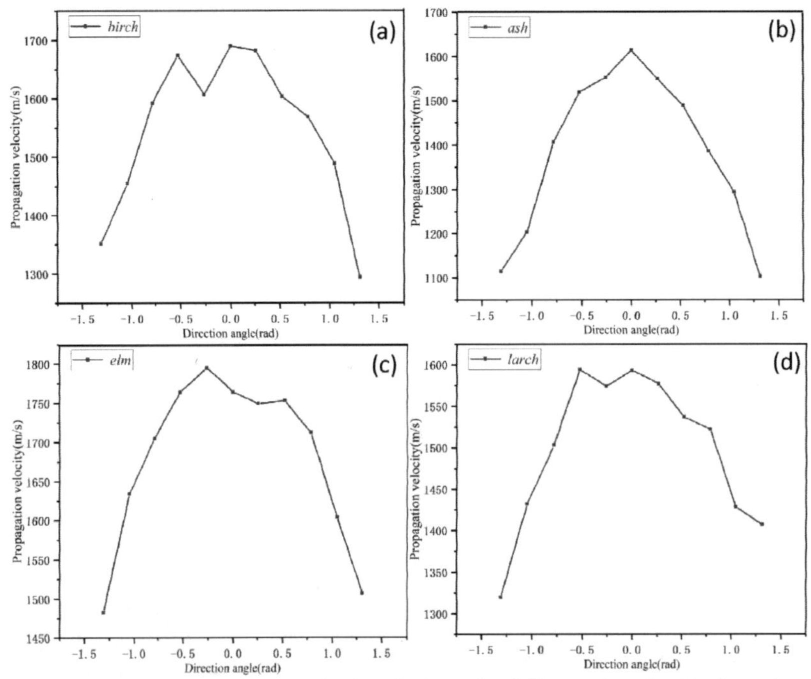

Figure 5. Stress wave propagation velocity diagram in different longitudinal section angles

Table 1. Fitting formula of stress wave propagation velocity of different tree species under different directions.

Tree species	Fitting formula	R^2	p
Birch	$y = 153.09x^2 + 1595.67$	0.83	<0.01
Ash	$y = 324.91x^2 + 1577.44$	0.95	<0.01
Elm	$y = 230.54x^2 + 1782.99$	0.92	<0.01
Larch	$y = 161.20x^2 + 1586.85$	0.97	<0.01

Table 2. Fitting formula of stress wave propagation velocity of different tree species under different longitudinal section angles.

Tree species	Fitting formula	R^2	p
Birch	$y = -203.58x^2 - 10.11x + 1685.46$	0.93	<0.01
Ash	$y = -283.01x^2 + 6.1x + 1578.68$	0.97	<0.01
Elm	$y = -166.29x^2 - 1.58x + 1793$	0.95	<0.01
Larch	$y = -136.65x^2 + 12.72x + 1592.42$	0.92	<0.01

With the vigorous development of the computer field in recent years, researchers have begun to try to use some algorithms to realize the internal stress wave imaging of wood [18]. In 2015, Chen et al. [19] proposed a 3D stress wave imaging method for wood internal defects based on the TIDW algorithm, and initially realized the 3D stress wave imaging of wood internal defects. Du et al. [20] proposed a three-dimensional stress wave imaging method based on TKriging in 2018, and compared it with the TIDW algorithm. The results show that the average imaging accuracy of the TKriging algorithm is 89.1%, which is about 10% higher than that of the TIDW algorithm. In 2021, Wei et al.[21] based on the propagation velocity model, proposed a method named twelve-directional stack imaging (TDSI) for the internal defect detection of larch wood. Figure 6 is an imaging contrast diagram of wood section with 6cm defect diameter based on TDSI and CDIM. The results show that the TDSI has a good effect on quantitative detection of defects, and the imaging precision of defects is more than 90.99%.

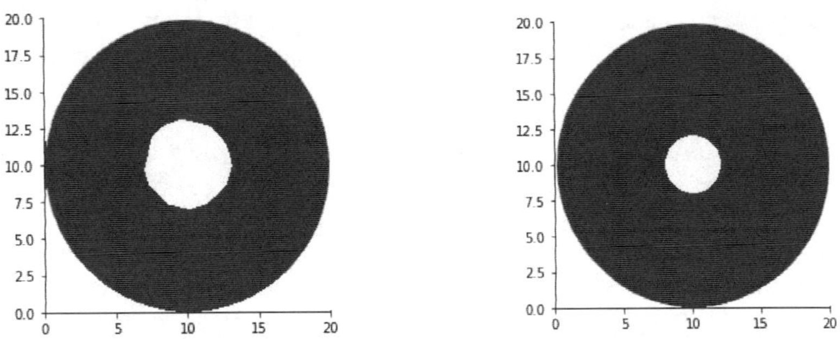

(a) TDSI (b)CDIM

Figure 6. Image comparison of TDSI and CDIM

3.3. Influencing factors of stress wave propagation in wood

With the continuous development of stress wave wood nondestructive testing technology, forestry technicians found that the propagation speed of stress wave in wood is affected by many factors, such as tree species, temperature, moisture content and compressive strength. Therefore, exploring the influence of various factors on the propagation velocity of stress waves in wood has become the main topic. Xu et al. [22] studied the influence of moisture content, temperature and propagation path on the propagation velocity of stress waves in wood from 2011 to 2014. The results show that at different temperatures, when the moisture content is below 32%, the propagation velocity decreases greatly with the increase of moisture content, and vice versa; below 50%, the propagation velocity decreases linearly with the increase of temperature. When the moisture content is above 50%, the propagation velocity also shows a downward trend with the increase of temperature, but there is an obvious jump near 0°C, as shown in figure 7; when the stress wave propagates in the cross section of the log, the path is arc-shaped and travels the fastest along the radial direction.

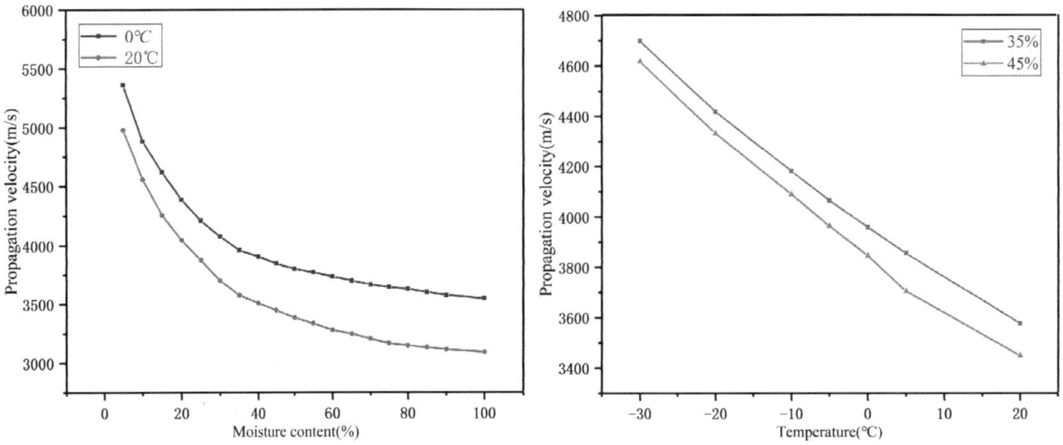

Figure 7. Effect of moisture content and temperature on stress wave propagation velocity

In 2019, Chen et al. used stress wave technology to analyze the influence of factors such as main tree species, DBH, different climate distribution areas, and different orientations of living trees on the propagation velocity of living trees in coastal defense forests in Hainan Island. The results show that the stress wave propagation velocity of Casuarina japonica has a very significant positive correlation with the diameter at breast height, but the stress propagation wave velocity is basically stable when the diameter at breast height is greater than 16 cm, that is, the density of trees increases with the increase of tree age, and it is basically stable after reaching mature forest. There are some differences in the propagation velocity of stress waves among different tree species. Casuarina japonica has the highest velocity, followed by Acacia japonica, and Acacia bigleaf is the lowest. Different climate distribution areas and different orientations of Casuarina japonica have no significant effect on the propagation velocity of stress waves.

4. Problems and prospects of stress wave wood nondestructive testing

At this stage, the stress wave nondestructive testing technology needs to be further studied in the following aspects.

(1) The existing stress wave wood nondestructive testing system has basically realized the convenience of testing, but the system supporting analysis and imaging software still cannot achieve high integration, and the equipment needs an external computer during the testing process. Therefore, the embedded stress wave wood nondestructive testing system still needs continuous innovation and research and development.

(2) The establishment of a stress wave propagation velocity model in wood can promote the development of stress wave wood nondestructive testing technology to a certain extent, and is of great significance to the development of stress wave imaging technology. At present, the stress wave propagation velocity model applied to the detection of wood defects has great limitations. Most of them are a single function relationship between the velocity and the angle in a certain cross section. There are few studies on the law of stress wave propagation in the log space, and no scholar has constructed a regression model of stress wave propagation velocity in wood space.

(3) Factors such as moisture content, temperature, and light have a great influence on the propagation speed of stress waves in wood. At present, scholars have established mathematical models between the propagation speed of stress waves and moisture content, temperature, etc., to determine the moisture content of wood. However, no one has quantified the influence of moisture content and introduced it into the spatial propagation velocity model, and has not yet established a regression optimization model for the propagation velocity of stress waves inside the wood.

(4) At present, studies on tomography and 3D imaging of internal defects in wood using stress wave technology have existed, and the research on imaging algorithms is mainly based on tomography, which can roughly reflect the position and size of internal defects in wood. Some algorithms are not mature enough, and the amount of data collected by stress waves is not sufficient. Especially when performing 3D imaging of stress waves, a complete 3D imaging system cannot be established, resulting in low accuracy and poor stability of reconstructed images. Therefore, new imaging algorithms need to be studied urgently to provide strong technical support for the realization of 3D high-precision reconstruction of internal defects in logs.

5. Conclusion

As one of the non-destructive testing methods of wood, stress wave technology has only been developed for a few decades, but the research results are relatively rich. 3D imaging of internal defects can be realized by using existing stress wave wood testing equipment, and the development of traditional stress wave wood testing technology has reached a bottleneck. However, in recent years, the rapid development of the computer field and the rise of artificial intelligence have also brought new possibilities for the non-destructive testing technology of stress wave wood. Various algorithms have also been applied to the imaging and physical modeling of wood. Wood non-destructive testing technology is gradually forming a system, which provides a new technical support for the development of wood testing technology.

Acknowledgments

Thanks for the support of the Natural Science Foundation of Heilongjiang Province of China (LH2020C091), the Heilongjiang Province College Students Innovation and Entrepreneurship Training Program(202211445027), and the 2022 Special Foundation Project of Fundamental Scientific Research Professional Expenses for Undergraduate Universities in Heilongjiang Province .

References

[1] Cheng L, Dai J, Yang Z et al. (2020) Variation of Larch Wood Property Indexes Based on Nondestructive Testing Data. Bioresources, 15, 2906-2923.

[2] Zhang Q, Dai Y, Li J et al. (2020) A review of the nondestructive testing of wood based on acoustics. Science & Technology Review , 38, 95-103.

[3] Nowak T, Karolak A, Sobotka M et al. (2019) Assessment of the Condition of Wharf Timber Sheet Wall Material by Means of Selected Non-Destructive Methods. Materials, 12.

[4] Voessing KJ, Gaal M, Niederleithinger E (2018) Air-coupled ferroelectret ultrasonic transducers for nondestructive testing of wood-based materials. Wood Science and Technology, 52, 1527-1538.

[5] McKinley P, Kamke FA, Sinha A et al. (2018) Analysis of Adhesive Penetration into Wood Using Nano-X-Ray Computed Tomography. Wood and Fiber Science, 50, 66-76.

[6] Dackermann U, Crews K, Kasal B, et al. (2014) In Situ Assessment of Atructural Timber Using Stress-wave Measurements. Materials and Structures, 47, 787-803.

[7] Erdene-Ochir T, Ishiguri F, Nezu I et al. (2020) Utilization potential of naturally regenerated Mongolian Betula platyphylla wood based on growth characteristics and wood properties. Silva Fennica, 54.

[8] Liao C, Zhang H, Li D et al. (2013) Screening method of internal defects in cylindrical wood members of ancient architectures. Journal of Beijing Forestry University, 35, 123-126.

[9] Gao S, Wang L, Wang Y et al. (2010) Comparisons of Stress Wave Propagating Velocities in Frozen State and in Normal Temperature State of Standing Trees. Scientia Silvae Sinicae, 46, 124-129.

[10]Xu L, Qi H, Wang Q et al. (2018) Design of Nondestructive Testing System for Wood Based on Charge Acceleration Sensor.Shanxi Electronic Technology, 9-11+15.

[11]Shang D, Duan X, Yang Z et al. (2008) Stress Wave Nondestructive Testing Technology and Its Application in Ancient Wood Architectures' Protection. World Forestry Research, 21, 44-48.

[12]Madhoushi M, Daneshvar S. (2016) Predicting the Static Modulus of Elasticity in Eastern Cottonwood (Populus Deltoides) Using Stress Wave Non-destructive Testing in Standing Trees. European Journal of Wood and Wood Products, 74, 885-892.

[13]Llana DF, Short I, Harte AM (2020) Use of non-destructive test methods on Irish hardwood standing trees and small-diameter round timber for prediction of mechanical properties. Annals of Forest Science, 77.

[14]Liu G, Li G, Sun Y et al. (2015) A stress wave propagation velocity model of standing trees. Journal of Zhejiang a&f University, 32, 18-24.

[15]Yang Y, Qu Y, Xu F et al. (2019) Quality evaluation of hardwood log based on impact stress wave testing. Journal of Vibration and Shock, 38, 126-134,150.

[16]Zhang C, Feng H, Li G et al. (2017) Stress wave propagation velocity model for different angles in a longitudinal section of standing trees. Journal of Zhejiang a&f University, 34, 926-933.

[17]Wei X, Sun L, Xu S et al. (2020) Quantitative detection of log defects based on stress wave propagation velocity model. Journal of Beijing Forestry University, 42, 143-154.

[18]Li X, Yang M, Li WL (2022) Three-Dimensional Reconstruction and Predictive Simulation Algorithm of Forest and Fruit Wood Borer Galleries Based on Two-Dimensional Images and Different Influencing Factors. Agronomy-Basel, 12.

[19]Chen F, Feng H, Du X et al. (2015) Three Dimensional Stress Wave Imaging Method of Wood Internal Defects Based on TIDW. Chinese Journal of Sensors and Actuators, 28, 1625-1633.

[20]Du X, Feng H, Hu M et al. (2018) Three-dimensional stress wave imaging of wood internal defects using TKriging method. Computers and Electronics in Agriculture, 148, 63-71.

[21]Wei X, Xu S, Sun L et al. (2021) Propagation Velocity Model and Two-dimensional Defect Imaging of Stress Wave in Larch (*Larix gmelinii*) Wood. Bioresources, 16, 6798-6812.

[22]Xu H, Xu G, Wang L et al. (2014) Construction of Stress Wave Time Isolines on Log Cross Section and Analysis of Its Effect Factors. Scientia Silvae Sinicae, 50, 95-100.

A Remote Monitoring and Maintenance System for Industrial Robots

Jian Zhou[1,3], Xiao-jian Ding[1,4], Li-Jun Sun[1*], Yun-Fan Yang[1], Yuan-hang Wang[1], Wen-Wei Liu[1,2]

[1] China Electronic Product Reliability and Environmental Testing Research Institute, Guangzhou 511370, China.

[2] The Xi'an Saibao Industrial Technology Research Institute Co.Ltd, Xi'an 710299, China.

[3] Guangdong Provincial Key Laboratory of Electronic Information Products Reliability Technology, Guangzhou 511370, China

[4]The Ministry of Industry and Information Technology Key Laboratory of general quality technology and application of intelligent manufacturing equipment, Guangzhou 511370, China

* Corresponding Author: sunlj@ceprei.com.

Abstract: Technologies such as external sensors, data acquisition, trend analysis, and cloud services make remote operation and maintenance of equipment status monitoring more convenient and lower maintenance costs. In this paper, an industrial robot operation and maintenance system is designed. First, the online monitoring of the robot is realized by collecting the control parameters, state parameters, fault information and external sensor monitoring information of the industrial robot. Then, the health baseline of the industrial robot is established based on the historical operation data. The online data is compared with the baseline data to realize the online assessment of the robot's health status. Finally, a visualization platform for remote operation and maintenance of the robot is built to realize the online display of parameters, health trend tracking and fault status early warning, which is convenient for maintenance personnel to take timely countermeasures.

1. Introduction

Industrial robots are known as "the jewel in the manufacturing crown". The research and development, manufacture and application of industrial robots are an important indicator of a country's technological innovation and high-end manufacturing level. At present, China is the world's largest consumer of industrial robots. With the explosive growth of robot market ownership, industry maintenance and safety issues are becoming more and more prominent. According to incomplete statistics, the maintenance costs and downtime losses caused by equipment failures have accounted for 30% to 40% of the production costs of enterprises.

The "Four Families" of industrial robots have developed their own operation and maintenance diagnostic systems for maintenance and support problems. Taking ABB Robotics as an example, the concept of "remote service" is proposed by the company. The failure of the industrial robot itself will be warned, and its service engineers can view all the failure information of the field equipment through the remote service platform, and analyse the data within a few minutes [1]. The remote monitoring service solution different from ABB is designed by Japan's FANUC Robot, which realizes the remote monitoring, remote diagnosis, and maintenance of the robot without adding equipment such as "service box". The user's device can always upload its own operation data to the server located in the data centre through the Internet, and the user can log in to the relevant web page on any networked device to check the robot's operation status [2]. In terms of academics, relevant scholars have carried out a lot of research. In 2003, Chen Yong, a scholar from Southeast University in China, designed an Internet-based remote monitoring platform for industrial robots, which uses information fusion technology to process data from multiple sensors on the automotive production line [3]. In 2006, Japanese scholar Suzuki K et al designed a residual generator for fault diagnosis of the servo system, and embedded the model into the industrial robot remote monitoring system for evaluation [4]. In 2008, Changfeng Li, a scholar from Harbin Institute of Technology in China, realized remote real-time monitoring and control of industrial robots based on TCP IP network [5]. In 2009, Finnish scholar Sallinen M et al. proposed a web-based remote monitoring and maintenance system for industrial robots, which enables robots to operate more efficiently and healthily [6]. In 2015, Lee J et al. realized the health management of industrial robots based on the temperature and torque signals of the robot, and developed a supporting remote monitoring system to predict early failure characteristics [7]. In 2017, Jianming Xu, a scholar from Zhejiang University of Technology in China, developed a remote monitoring system based on industrial robots, which realized local monitoring and remote web-side monitoring [8]. In 2020, Zefeng Wu, a scholar from Nanjing University of Aeronautics and Astronautics in China, used various sensors in the robot system to collect the data of the system, and then uploaded it to the cloud server through Ethernet, to realize the remote monitoring of industrial robots [9]. Guiwu Fu, a scholar from Foshan Institute of Science and Technology in China, developed a C/S architecture industrial robot remote monitoring system based on the SDK communication secondary development interface [10].

Aiming at the characteristics of complex working conditions, large amount of data, and wide distribution of industrial robots, this paper proposes a remote operation and maintenance system for robots based on the health assessment and early warning method. The rest of this paper is organized as follows: Section 2 presents the architecture of the industrial robot operation and maintenance system. Section 3 presents the implementation of the remote operation and maintenance system. Section 4 gives the design content of the system power module. Section 5 gives the case effect of the industrial robot operation and maintenance system. Section 6 summarizes the full text.

2. The architecture of the system

In order to carry out systematic monitoring, evaluation, diagnosis and prediction, a remote operation and maintenance system for industrial robots is built. The system architecture is shown in Figure 1.

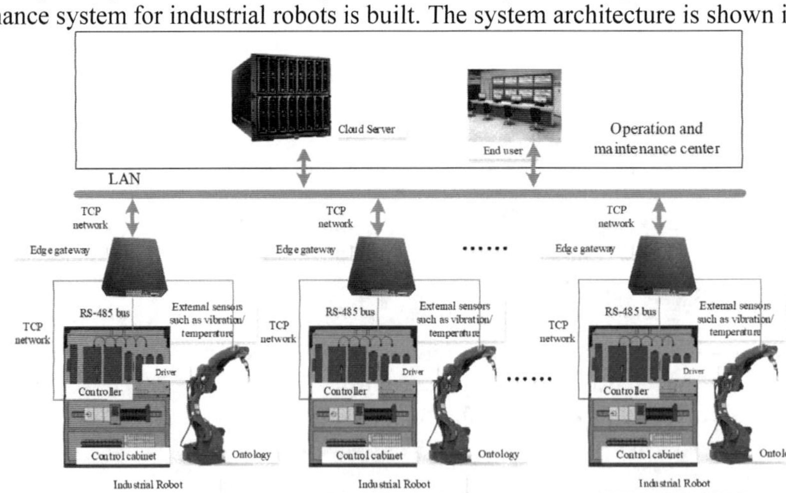

Figure 1. Architecture of the system.

The system is mainly composed of edge gateways and maintenance centre. Among them, the edge gateway is in the core unit, and one edge gateway corresponds to one industrial robot. On the one hand, it realizes information exchange with the robot controller based on TCP IP network communication, and obtains the internal status data of the driver based on the RS-485 bus interface; Real-time waveform acquisition and signal feature extraction. The centre of operation and maintenance is configured with a cloud server, and the remote operation and maintenance visualization platform software is deployed to realize functions such as data analysis, data storage, data display, health assessment, trend tracking, alarm and early warning.

3. System implementation scheme

The industrial robot includes two working modes: online operation mode and remote commissioning mode. In the online operation mode, the robot performs normal production line operations, and the edge gateway obtains the robot's working condition and collects data under the same working condition of the robot. In the remote debugging mode, engineers can obtain the debugging control right of the robot under test on site through the remote real-time monitoring network of the system. The operation of the robot is controlled by the commissioning control commands set in advance by the robot control system, and the status data of the robot in the commissioning state is obtained, and the test results are given. The system data flow is shown in Figure 2.

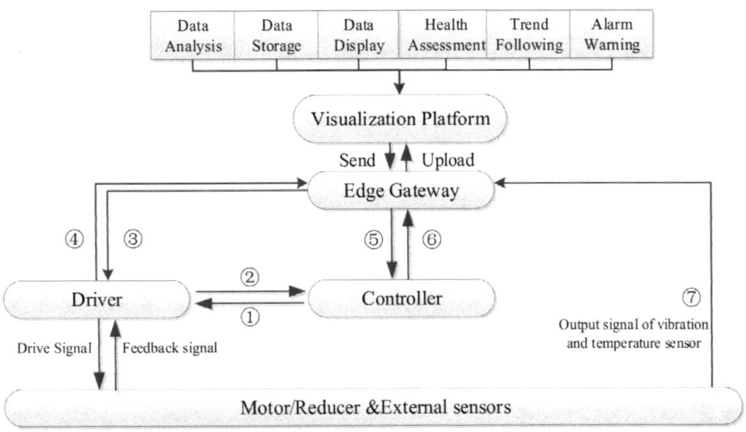

Figure 2. Schematic diagram of system data flow.

The specific workflow of each part is as follows:

①Pulse signal plus+/plus-; direction signal Dir+/Dir-; enable signal;

②Driver A/B/Z signal; alarm signal;

③Send: data collection instruction (sampling enable and collection duration, etc.);

④Upload: data package (real-time power, command position, feedback position, feedback speed, feedback torque and alarm code, etc.);

⑤Send: mode switching command (teaching mode/remote commissioning mode); servo enable; alarm clear; back-to-origin control; teaching mode operation command (single-axis forward and reverse jog, single-axis bidirectional uniform speed operation, Single-axis bidirectional uniform speed change operation, speed setting); control command (start, stop); program selection command (program number set in advance); speed selection command (30%, 50%, 75%, 100%);

⑥Upload: running status table (robot serial number, running program number, driver status, controller status, motor running circles, driver running time, robot controller running time, etc.); single-axis limit status; fault information table, etc.;

⑦Read: External sensor signal (vibration/temperature, etc.).

4. Design of functional modules

The functions of the system are mainly divided into data storage, baseline management, weight setting, health calculation, equipment monitoring station, alarm management, etc.

4.1. Module for data storage

Data storage is the most basic functional module of the system. All data of the system will be stored in this module, mainly including real-time working status information of equipment and equipment failure warning data information. After comparative analysis of various databases, MySQL database is to be selected. This database is a relational database management system. It has the characteristics of fast speed, reliability, open source and small size, and has been widely used. According to the needs of the monitoring system, data tables such as static information, status data, and fault details are mainly created. The static information table is used to complete the storage of user information, basic equipment information, common faults and solutions, and the state data table mainly completes the data storage of real-time status. The fault and warning information table mainly completes the storage of equipment fault details, including fault code, fault type, and the time when the fault occurs.

4.2. Module for Baseline Management

The health baseline for each parameter is determined based on the following methods:

——For event-type parameters, such as alarm status, defective products, and impact event parameters, once they occur, they will fail, and the health baseline is a fixed value;

——For real-time parameters, such as temperature, vibration, current, etc., all belong to real-time data, the mean of the test data is the healthy baseline, the distance from the healthy baseline is the standard deviation, and the standard deviation is the degradation baseline; the distance from the healthy baseline is 3 times the standard deviation is Fault baseline.

As shown in Figure 3, the system regularly saves the running data of the robot, and the historical data under the normal state of the robot can be selected as the baseline control value, that is, a healthy baseline is generated.

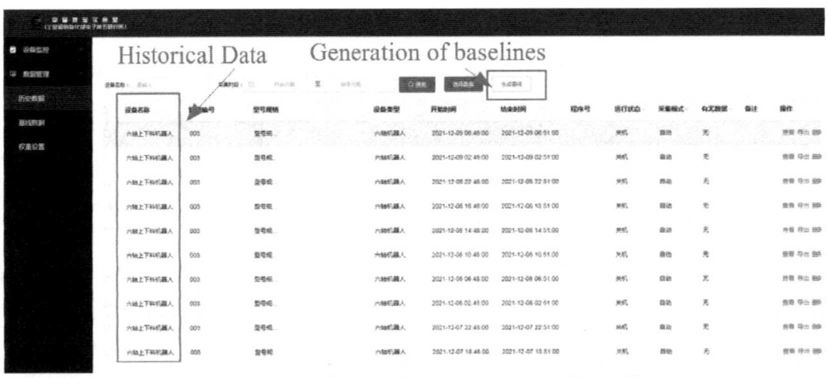

Figure 3. Generation and management of baselines.

As shown in Figure 4 , the generated health baseline is stored according to the number of the robot and each axis, and the baseline data information can be visually observed through visual charts. After the current baseline is activated, the robot's health status is evaluated based on that baseline.

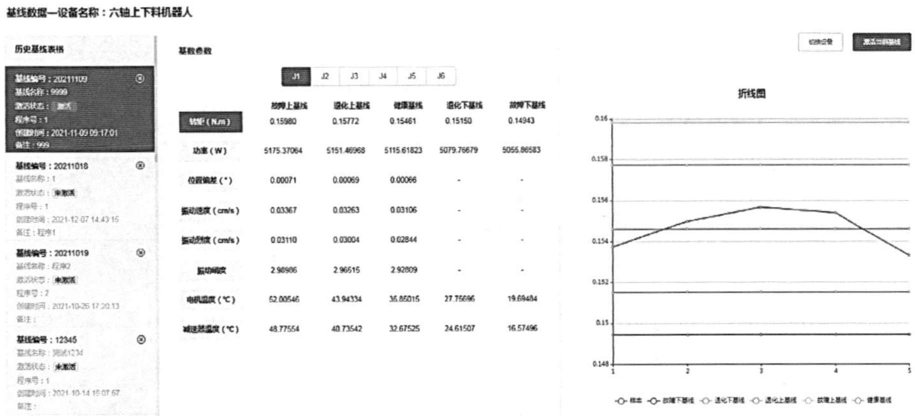

Figure 4. Activation of healthy baseline.

4.3. Module for Health Assessment

The health status of each axis of the robot is comprehensively evaluated through parameters such as torque, power, position deviation, vibration, and temperature. The evaluation method is as follows:

$$H_i = \sum_{k=1}^{N_i} \left(w_{i,k} * H_{i,k} \right), \quad s.t. \sum_{k=1}^{N_i} w_{i,k} = 1 \tag{1}$$

In formula (1), $H_{i,k}$ is the health score of the k_{th} parameter of the i_{th} axis, N_i is the number of all real-time health parameters of the i_{th} node, $w_{i,k}$ is the weight of the k_{th} real-time health parameter of the i_{th} node among all parameters. The sum of the weight parameters is 1, and the setting of the weight can be determined by experience or other methods. The more important the parameter, the greater the corresponding weight.

The calculation method of the health degree of the whole robot is based on the comprehensive evaluation of the health status of each axis. One of the simplest evaluation methods is as follows:

$$Health = \sum \upsilon_i H_i \tag{2}$$

In formula (2), υ_i is the weight ratio of the i_{th} axis, H_i is the health score of the i_{th} axis.

4.4. Module for Data Display

The function of this module is to visually compare the baseline data of each parameter of the robot. As shown in Figure 5, through the health of the whole robot and its joints, its health status is intuitively and clearly reflected, and real-time health status tracking and evaluation is performed.

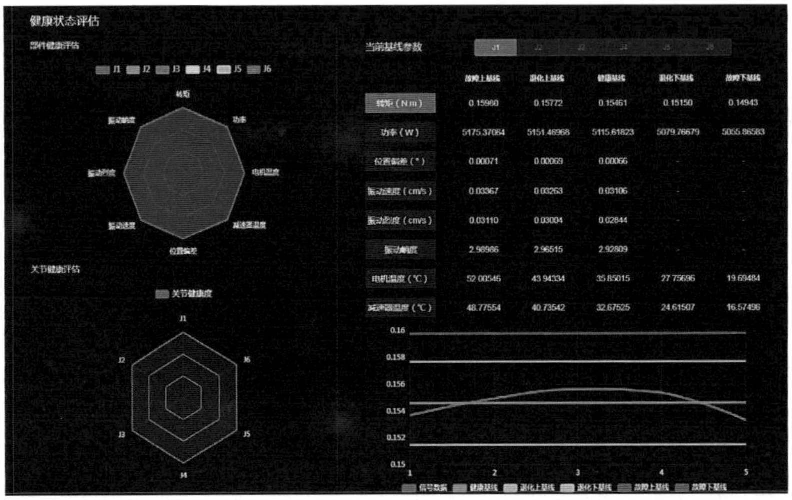

Figure 5. The effect of status display.

4.5. Module for Alarm Management

The module is shown in Figure 6. Alarm information such as alarm equipment, alarm type, and alarm code can be recorded in detail by this module.

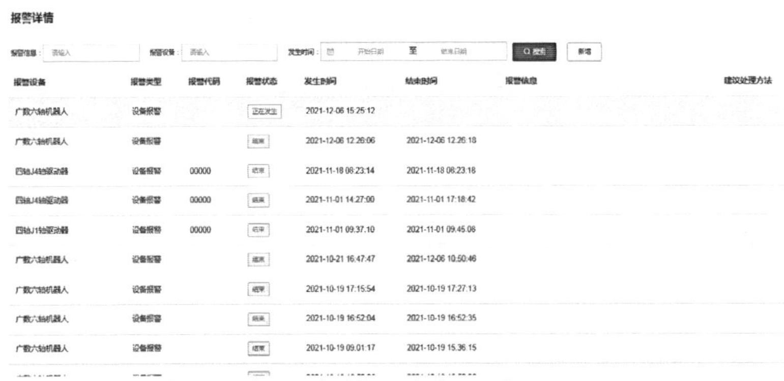

Figure 6. Alarm information.

4.6. Application

Taking a six-axis industrial robot as the research object, experiments are carried out to verify the feasibility and effectiveness of the robot remote operation and maintenance system. As shown in Figure 7 (a), the robot is set to operate under fixed working conditions on the production line. The operation and maintenance system collects 5 minutes of data every 2 hours, and collects 1 week of data in total. A total of 80 sets of data are shown in Figure 7 (b).

(a) Industrial robot. (b) Details of the data.

Figure 7 Robots for testing and data recording.

The first set of collected data under this working condition is used to generate and activate the baseline. The effect of baseline generation is shown in Figure 8 (a). According to the actual working conditions and working characteristics of the robot, the weights of the parameters of the robot are allocated, as shown in Figure 8 (b).

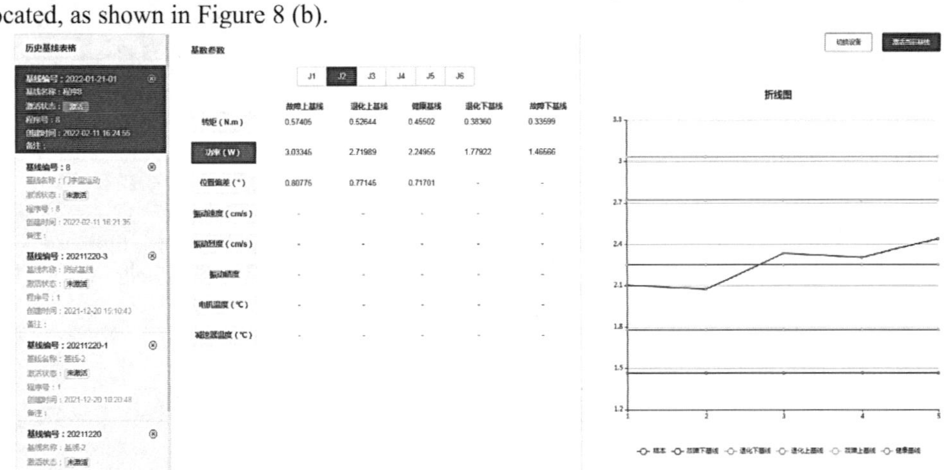

(a) Process of Baseline Generation.

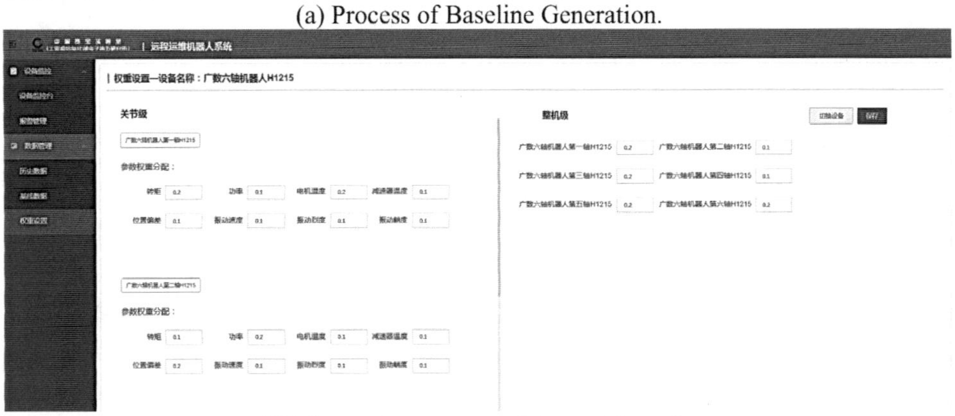

(b) The setting process of the weight ratio.

Figure 8 The process of health score generation.

Through real-time data collection and analysis, the background algorithm is invoked to evaluate the status of each joint parameter of the robot. And the details of the robot's health status are obtained as shown in Figure 9, which can intuitively reflect the health status of the robot's joints and the whole machine, and track and study industrial robots. Health trends change, health diagnosis and predictive maintenance for robots.

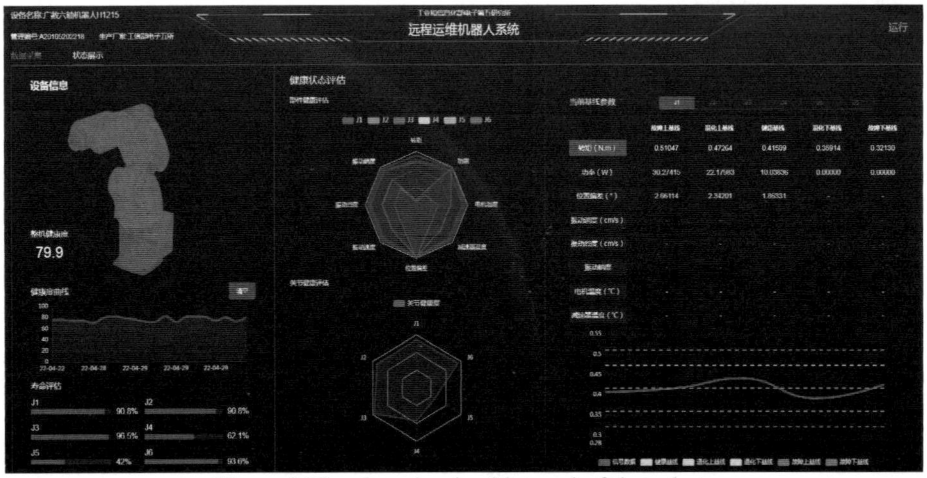

Figure 9 The changing health trend of the robot.

5. Conclusions

The system is suitable for the operation and maintenance of industrial robots. The edge gateway node realizes signal collection, including important parameters inside the robot controller (such as tracking error, feedback position, feedback torque and feedback speed, etc.) and the external equipment installed for important robot objects. Sensor signals (vibration/temperature, etc.), and the raw data is processed locally according to the model, and then the results and related data are uploaded to the cloud. The cloud stores and displays the data, tracks the long-term historical data of the robot, realizes the health status assessment of the robot based on the status baseline analysis, and provides a reference for the health score, which is helpful for safe operation.

Acknowledgments

This project is supported by the Key Research and Development Program of Shaanxi Province (Grant No. 2021GY-335), the Science and Technology Program of Guangzhou under No. 202201010298 and No. 202103010002 and the Ministry of Industry and Information Technology Project (Grant No. TC210804R-1).

References

[1]. Wu, Z., Wang, H., Zhang, Y., & Ai, M. (2021). Abb robot data collection based on dynamic link library. JPhCS.
[2]. FANUC Inc. (2012). Fanuc Robots' Ether Net/IP Protocol.
[3]. Yong,C, Xianzhong, D. (2003). Construction of a reconfigurable multi-agent system platform based on mobile robots. Journal of Southeast University: Natural Science Edition (z1), 6.
[4]. Suzuki, Keita & Fujii, Takao. (2006). Fault diagnosis of parameter change type faults and a fault tolerant system design for servo systems. Proceedings of the American Control Conference. 2006. 10.1109/ACC.2006.1657596.
[5]. Changfeng, L, Dongsheng, Q. (2008). Research on remote control of parallel robots based on Tcp/Ip protocol. Machine Tools and Hydraulics, 36(2), 3.
[6]. Sallinen, M., Heikkila, T., & Koskinen, J. (2009). Sensor based flexibility for robotics in

manufacturing applications. IEEE.
[7]. Lee, J., Ardakani, H. D., Yang, S., & Bagheri, B. (2015). Industrial big data analytics and cyber-physical systems for future maintenance & service innovation. Procedia CIRP, 38, 3-7.
[8]. Jianming, X & Xiangfei, P. (2017). Research on Industrial Robot Monitoring System Based on Socket Communication. Computer Measurement and Control, 25(7), 4.
[9]. Zefeng, W, Chenggang, L & Yong, S. (2020). A remote monitoring system for industrial robots based on cloud servers. Machinery Manufacturing and Automation, 49(6), 4.
[10]. Guiwu, F & Ying, T. (2020). Design of industrial robot remote monitoring system based on SDK communication interface. Machine Tool and Hydraulics, 48(17), 5.
[11]. Rizky, S. (2004). Panduan belajar sql server. Sql Database.
[12]. Ji, H. J., & Lim, H. (2016). Real-time bearing fault probability estimation using onsite learning for industrial robots. Isr: International Symposium on Robotics. VDE.
[13]. Eski, I., Erkaya, S., Savas, S., & Yildirim, S. (2011). Fault detection on robot manipulators using artificial neural networks. Robotics and Computer-Integrated Manufacturing, 27(1), 115-123.
[14]. Meng, F., Gong, J., Yang, S., Huang, L., & Tang, X. (2021). Study on tribo-dynamic behaviors of rolling bearing-rotor system based on neural network. Tribology International, 156, 106829.
[15]. Maclennan, Tang, J., Crivat, Z., & Bogdan. (2008). Data Mining with Microsoft SQL Server 2008. Wiley Publishing.

Reliability evaluation methods for unmanned aircraft systems with multi-rotors

Ming-Xi Cai[a,b], He-Long Wu*[a,b], Jian-Feng Yang[a,b]

[a] China Electronic Product Reliability and Environment Testing Research Institute Guangzhou, China;

[b] Guangdong Provincial Key Laboratory of Electronic Information Products Reliability Technology, Guangzhou, China;

[*] Corresponding author: wuhelong1314@163.com

Abstract: According to the characteristics of unmanned aircraft systems (UASs) with multi-rotors, this paper realizes the reliability evaluation. Referring to the relevant standards and specifications for reliability evaluation of electronic products and communication products, the methods of sampling, test plan, test conditions and index evaluation related to reliability evaluation of Referring to are proposed. Based on the exponential distribution, the reliability index and its calculation method are put forward, and the evaluation method for non-failure data is also studied. The method is applied and the reliability level of UASs with multi-rotors is obtained. This method meets the actual market demand and provides a practical method for UASs reliability evaluation.

1. INTRODUCTION

With the progress of aviation technology and wireless transmission, the application of multi rotor UASs has increased significantly, and the development and dealers of UASs with international cutting-edge have emerged in China. Drones are used more and more, and they are changing from professional level to consumer level. It can be seen that the UASs industry has entered a period of vigorous development.

However, with the development of UASs, some safety accidents have gradually occurred, so the reliability of multi rotor UASs has become a hot spot. Lv et al. Studied the system reliability of UASs and proposed an estimation method for the reliability test of samples [1]. Zhang Rui et al. Studied the reliability allocation method of UASs system [2]. Zhou Jinniu judged the reliability of UASs Based on fuzzy comprehensive evaluation method [3]. Based on the data used by domestic UASs in the outfield, Zhang Wenjin and others used the method of establishing mission reliability block diagram to successively establish the mission reliability models of subsystems and systems, and completed the prediction of mission success probability (MCP) [4]. But this method requires the accumulation of a large amount of data. Prescott D.R. et al. Proposed the method of using BDD for task planning of multi-stage mission system taking UASs as an example [5]. Therefore, most of the current reliability evaluation methods rely on the failure data of the underlying parts, and the evaluation is based on the reliability analysis method. Some reliability evaluation methods for sample tests also rely on a large number of samples, and a large number of failures are required. Most of them focus on the data analysis method, lacking a systematic test design method.

Therefore, this paper presents a reliability evaluation method for multi rotor UASs, determines the sampling, test scheme, test conditions and index evaluation methods, and puts forward the non-failure data processing method.

2. TEST METHODS

The goal of the reliability evaluation methods is to describe the reliability level of the system operation of UASs with multi-rotors as a whole, namely the mean operating time between failures (MTBF) and the mean operating time to first failures (MTTF). According to the life test method, UASs with multi-rotors reliability evaluation test adopts the sampling test method. Sampling from qualified batches of drone products produced within the last two years. The number of samples should be no less than 3 units. The censoring time of the test should be no less than 10 times the battery life of the product.

The test conditions are as follows:

a) Generally, the reliability field tracking test should be carried out after the UAS works for 3 endurance times. If it is necessary to track early failures and failures during normal use, the reliability field tracking test can be carried out after the UAS product is put into operation;

b) The mission profile and environmental profile should be determined in advance according to various information of the UAS life cycle;

c) One or a series of typical programs can be compiled according to the information of the mission profile, so as to simulate the mission profile in the test field test;

d) The test environment should be arranged according to the environmental profile of the UAS;

e) During the test, the simulation program of the task profile and the simulation environment of the environment profile shall be applied at the same time to fully expose the system failure;

f) If the accelerated method is used for the test, the following requirements should be met: 1) The accelerated method should have obvious acceleration; 2) The accelerated method should not change the failure mode and failure mechanism; 3) The acceleration factor should be determined.

3. RELIABILITY INDEX EVALUATION

The reliability indexes of UASs with multi-rotors include MTTF and MTBF.

3.1 Point estimation of MTTF

The calculation formula for the point estimation of MTTF is as follows:

$$m = \frac{T_0}{N} \tag{1}$$

Where T_0 is the sum of working time before the first failure of all samples, in hours (H); N is the total number of samples.

3.2 Point estimation of MTBF

The calculation formula for the point estimation of MTBF is as follows:

$$\mathrm{m} = \frac{\sum_{i=1}^{N} t_i}{r} = \frac{\sum_{i=1}^{N} t_i}{\sum_{i=1}^{N} r_i} \tag{2}$$

Where r is the total number of failures; t_i is the accumulated working time of the ith UAS in the evaluation period, in hours (h); r_i is the cumulative failure number of the ith UAS in the evaluation period.

3.3 Non-failure data analysis of MTBF

If the UASs products do not failure in the fixed time test, the lower confidence limit of MTBF is:

$$m_L = \frac{T}{-ln(1-\alpha)} \tag{3}$$

Where T is the total test time of all samples for the fixed time test, in hours (h); α is the confidence level.

3.4 Interval estimation of MTBF

The interval estimation method of MTBF is as follows:

$$m_L = mC_L \tag{4}$$

$$m_U = \mathrm{m}C_U \tag{5}$$

Where m_L is the lower confidence limit of MTBF; m_U is the upper confidence limit of MTBF; C_L is the lower confidence limit coefficient of MTBF for the fixed time test (see Table 1); C_U is the upper confidence limit coefficient of MTBF for the fixed time test (see Table 1).

Table 1. Confidence coefficient of two sides (or one side) when MTBF is calculated in the fixed time test

r	40% (two sides)		60% (two sides)		80% (two sides)		90% (two sides)	
	70% (one-sided confidence lower limit)	70% (one-sided confidence upper limit)	80% (one-sided confidence lower limit)	80% (one-sided confidence upper limit)	90% (one-sided confidence lower limit)	90% (one-sided confidence upper limit)	95% (one-sided confidence lower limit)	95% (one-sided confidence upper limit)
1	0.410	2.804	0.334	4.481	0.257	9.491	0.211	19.417
2	0.553	1.803	0.467	2.426	0.376	3.761	0.317	5.658
3	0.630	1.568	0.544	1.955	0.499	2.722	0.387	3.659
4	0.679	1.447	0.595	1,742	0.500	2.293	0.437	2.930

4. APPLICATION

In this application, three sets of UAS are used, which have been processed and passed the appearance acceptance and functional performance test. The test adopts the timed censoring test scheme, and the cumulative relevant test time of each sample should be 10 times the product endurance time.

4.1 Test profile

Make the UAS fly circularly according to the following eight stages until the UAS reaches the endurance time and returns when the power is insufficient , as shown in Figure 1.

Stage 1: take off the UAS from the ground and reach a vertical height of 10m above the ground in about 20 seconds;

Stage 2: hover the UAS at a height of 10m above the ground for 20s;

Stage 3: test the maximum ascent speed of UAS, and fly the UAS at the maximum ascent speed;

Stage 4: test the maximum horizontal speed of UAS, and fly the UAS at the maximum horizontal speed;

Stage 5: raise the UAS to a vertical height of 100m above the ground;

Stage 6: hover the UAS at a height of 100m above the ground for 10s;

Stage 7: test the maximum descent speed of UAS, and land the UAS at the maximum descent speed;

Stage 8: UAS landing flight.

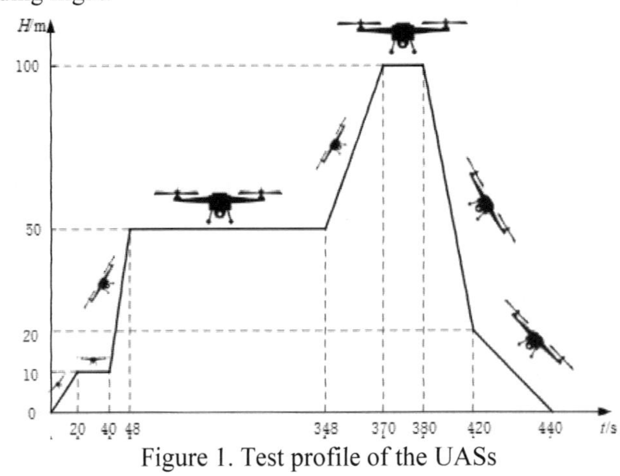

Figure 1. Test profile of the UASs

4.2 Test data

From October 12, 2021 to June 2, 2022, field flight test will be carried out in Mohe. There are 3 test samples, and each UAS has been tested for 12 times the endurance time. The operation process is recorded as follows:

Table 2. Reliability test operation record of UAS 01

No.	Date	Runtime		Downtime	Recovery time
		Start time	End time		
1	2021.10.22	9: 00	9: 25		
2	2021.10.25	9: 00	9: 22		
3	2022.01.26	9: 00	9: 10		
4	2022.05.05	9: 00	9: 03	9: 03	9: 30
5	2022.05.07	9: 00	9: 30		
6	2022.05.10	9: 00	9: 25		
7	2022.05.13	9: 00	9: 25		
8	2022.05.17	9: 00	9: 30		
9	2022.05.21	9: 00	9: 25		
10	2022.05.26	9: 00	9: 25		
11	2022.05.30	9: 00	9: 30		
12	2022.06.02	9: 00	9: 30		

Table 3. Reliability test operation record of UAS 02

No.	Date	Runtime		Downtime	Recovery time
		Start time	End time		
13	2021.10.22	9: 00	9: 30		
14	2021.10.25	9: 00	9: 30		
15	2022.01.26	9: 00	9: 30		
16	2022.05.05	9: 00	9: 30		
17	2022.05.07	9: 00	9: 28		
18	2022.05.10	9: 00	9: 30		
19	2022.05.13	9: 00	9: 25		
20	2022.05.17	9: 00	9: 25		
21	2022.05.21	9: 00	9: 30		
22	2022.05.26	9: 00	9: 25		
23	2022.05.30	9: 00	9: 23		
24	2022.06.02	9: 00	9: 30	9:20	9:22

Table 4. Reliability test operation record of UAS 03

No.	Date	Runtime		Downtime	Recovery time
		Start time	End time		
25	2021.10.22	9: 00	9: 25	9:22	9:24
26	2021.10.25	9: 00	9: 26		
27	2022.01.26	9: 00	9: 25		
28	2022.05.05	9: 00	9: 25		
29	2022.05.07	9: 00	9: 26		
30	2022.05.10	9: 00	9: 25		
31	2022.05.13	9: 00	9: 23		
32	2022.05.17	9: 00	9: 22		
33	2022.05.21	9: 00	9: 20		
34	2022.05.26	9: 00	9: 22		
35	2022.05.30	9: 00	9: 20		
36	2022.06.02	9: 00	9: 23		

4.3 Failure conditions and treatments

During this test, the tested product has three failures , as shown in Table 5.

Table 5. Summary of failure conditions and treatments

Sample No.	Date of failure	Failure phenomenon	failure cause	treatment
01	2022.05.05	UAS out of control, causing blade damage.	Failed to receive GPS signal	Replace the damaged blades, and the UAV flight test is normal
02	2022.06.02	When the UAV rises to 100m, the prompt of the remote controller drops and the signal is weak.	Signal link disturbed	After returning home, the UAV conducted flight test again, and there was no abnormality
03	2021.10.22	When the UAV flies to 50m, the flight data is not displayed, indicating that the aircraft loses control and starts to return	Signal link interruption	After returning home, the UAV conducted flight test again, and there was no abnormality

All faults are caused by external factors, and the UAS itself has no relevant faults, so all the three failures are not counted.

4.4 Reliability evaluation index calculation

Since the samples are new products and do not require preventive maintenance, only the MTBF is evaluated. Since there is no fault in the UAV product at the time of the timed censoring test, the point of the mean time between failures is estimated as follows when the confidence is 80%:

$$m_L = \frac{T}{-ln(1-\alpha)} = \frac{15.02}{-ln(1-0.8)} = 9.33h$$

5. CONCLUSION

This paper presents a reliability evaluation method for multi rotor UAV. Based on the relevant standards and specifications of reliability evaluation of electronic products and communication products, this method puts forward the methods of sampling, conditions, schemes, data analysis and so on. It gives the process and method of reliability evaluation applicable to UASs with multi-rotors in an engineering manner, and realizes the reliability evaluation under limited sample size. At the same time, the MTBF estimation method under fault free data is proposed, which effectively realizes the reliability evaluation of UASs with multi-rotors in practical engineering application.

Acknowledgment

The authors are supported by the Key-Area Research and Development Program of Guangdong Province under No.2020B0404020002, the Special Fund of Guangdong Province for promoting high-quality economic development under No. JQR217000010, the Development fund for the Fifth Electronics Research Institute of Ministry of Industry and Information Technology under No. 19Z14, Ministry of industry and information technology project （TC210804R-1） .

References

[1] Xiaolin Lv,Wenlai Zhu, "Research on reliability-estimating method for the UAV," Journal of Astronautic Metrology and Measurement, 2012, 32(6):3.

[2] Rui Zhang, Bo Lee, "Reliability allocation method of UAV system," Journal of Nanjing University of Aeronautics & Astronautics,041.0z1(2009):106-108.

[3] Bo Lee, "Mission-Oriented Reliability Modeling and Prediction for Unmanned Aerial Vehicle Swarm," National University of Defense Technology, 2018.

[4] Wenjin Zhang, Rui Kang, Shengkui Zeng, "Calculation of reliability index based on mission completion probability for UAV system." JOURNAL OF BEIJING UNIVERSITY OF AERONAUTICS AND ASTRONAUTICS, 29.12(2003):4.
[5] Prescott, D. R. , et al. "A reliability analysis method using binary decision diagrams in phased mission planning." Journal of Risk and Reliability 223(2009).

The study and development of UAV digital twin system

Jianfeng Yang[1,2,3], Chuangmian Huang[*1], Jingru Wang[4]

[1] China Electronic Product Reliability and Environment Testing Research Institute Guangzhou, China;
[2] Guangdong Provincial Key Laboratory of Electronic Information Products Reliability Technology Guangzhou, China;
[3] Guangdong Provincial Engineering Technology Research Center of UAV Reliability and Safety, Guangzhou, China;
[4] College of Aerospace Engineering, Nanjing University of Aeronautics and Astronautics, Nanjing, China;

[*] Corresponding author: huangchuangmian@163.com

Abstract. To develop the digital twin system of unmanned aircraft vehicle (UAV), a multi-information fusion and time-varying simulation technology based on the testing data is proposed. And a UAV digital twin system is developed. First, the system composition and the working principle of the UAV digital twin system are introduced, respectively. Second, the Ko displacement theory is applied to realize the integration of the acquired data and the finite element simulation model. Then, the model reduced method is introduced, which can reduce the scale of the finite element model and improve the calculation efficiency. Finally, the experimental verification is carried out. The test results and simulation calculation data show that the relative errors between the proposed calculation method and the test results are within 30%, which meets the requirements of engineering application. The test results show that the proposed UAV digital twin technology is feasible, providing a reference case for the development of UAV digital twin technology, which has practical engineering significance and military value.

1. Introduction

The Digital Twin (DT) is a new paradigm which is proposed by National Aeronautics and Space Administration (NASA) and U.S. Air Force [1]. The background of the emergency of the Digital Twin Paradigm is due to the need for NASA and the U.S. Air Force to develop a new generation of aircraft with lighter weight, higher loads, longer flight time, and more extreme service conditions over longer time periods than the present generation. Current approaches for certification, fleet management and sustainment are largely based on statistical distributions of material properties, heuristic design philosophies, physical testing and assumed similitude between testing and operational conditions and will likely be unable to address these extreme requirements. In order to solve this problem, a shift from the technological paradigm is required, and the digital twin technology is thus born.

After years of development, many scholars have published papers on digital twins. Tao et al. [2] proposed a five-dimension DT for complex equipment and a new approach for health management driven by DT for complex equipment. Dong et al. [3] reviewed five key technologies for modeling and simulation of airframe DT and discussed the state-of-the art of proposed technologies as well as future research directions. Li et al. [4] described the concept of digital satellite. At present, the company that can represent the advanced technology level of the United States is SpaceX, but there is almost no public information about the DT satellites of SpaceX. In recent years, Lockheed Martin, SpaceX, ANSYS, Parametric Technology Corporation, France Dassault, Germany Siemens, etc. have carried out a lot of research and exploration in DT technology, and achieved great results in design and manufacturing. However, the literature on equipment-level digital twin technology is rarely published [5, 6].

In this study, a multi-information fusion and time-varying simulation technology based on the testing data is proposed, and a UAV digital twin system is developed. Based on the existing simulation technology, the proposed technology reduces the degree of freedom and improves the calculation efficiency by applied the model reduced method. The method achieves the purpose of time-varying simulation by taking the collected data as the boundary condition of the simulation model.

2. Composition and working principle of the UAV digital twin system

The UAV digital twin system consists of 6 major modules, namely: real-time simulation, flight plan, flight parameters, parameter setting, data management, and UAVs management. The real-time simulation module is used to display the flight status of UAV (position, attitude, flight speed, etc.), remaining power, GPS signal strength, communication packet loss rate, communication delay, health level and structure stress, strain, temperature, vibration in real-time value etc. The display interface of this module includes 3 modes, namely: stress cloud mode, perspective mode, and rendering mode. The flight plan module is used to set the UAV's flight route and take-off and landing waypoints. By setting the UAV's flight route in the flight plan module, one-key take-off and landing of the UAV can be realized. The flight parameter module is used for changing the flight route of UAV by commands. It can also perform pre-takeoff inspection, collision alarm, detailed display of flight status parameters, etc. The parameter setting module can set the warning parameters, simulation type, simulation parameters and connection parameters. The data management module can record and reply the flight process. The UAVs management module is used to manage the related information of the UAV in the digital twin system.

(a) real-time simulation module. (b) flight parameters module.

Figure 1. The UAV digital twin system.

The working principle of the UAV digital twin system: The UAV involved in this study is a vertical take-off and landing UAV. Its flight control system is based on the open source code. So the code on the ground station of the flight control system can be transplanted to the digital twin system, which can be used to control the flight of the UAV. The UAV digital twin system developed in this study can obtain the flight parameters of the flight control system and testing data acquired by the airborne temperature-vibration-strain signal acquisition system (ATVSSAS). The obtained test data is disseminated by radio and is analyzed by using the ground workstation to be displayed on the graphical interface. Due to the limited number of channels of the ATVSSAS, it is impossible to monitor the entire structure of the UAV. Therefore, the digital twin system adopts the multi-information fusion and time-varying simulation technology based on the test data. In order to use the acquired strain data as the boundary condition of the finite element simulation model, and to solve the strain of other non-measurement points, Ko displacement theory [7] is adopted. For the purpose of real-time simulation, the model reduced method is used to shorten the calculation time.

3. Real-time simulation technology of the UAV digital twin system

3.1 Ko displacement theory

The Ko displacement theory has corresponding displacement theoretical formulas for different structure types. The following takes the cantilever beam as an example to illustrate the Ko displacement theory principle.

Differential equations based on the classical beam theory are given by :

$$\frac{d^2y}{dx^2}=\frac{M(x)}{EI} \tag{1}$$

in which y is the vertical displacement, x is the span-wise coordinate, $M(x)$is the bending moment, E is the Young modulus, and I is the moment of inertia. The bending moment, $M(x)$, in equation (1) may be related to the associated bending strain, $\varepsilon(x)$, at the bottom or top fiber through the outermost fiber bending stress equation, $\sigma(x)=[M(x)\cdot c]/I$:

$$M(x)=I\frac{\sigma(x)}{c}=EI\frac{\varepsilon(x)}{c} \tag{2}$$

in which c is the beam half depth. In light of equation (2), beam bending equation (1) can be written in terms of $\varepsilon(x)$ as:

$$\frac{d^2y}{dx^2}=\frac{\varepsilon(x)}{c} \tag{3}$$

For variable cross-section beams, replace c with the relation function $c(x)$ between c and x:

$$\frac{d^2y}{dx^2}=\frac{\varepsilon(x)}{c(x)} \tag{4}$$

The relationship between displacement and strain can be obtained by integrating the above equation twice.

By dividing the variable-section beam into n equal-length parts for discretization and placing n+1 strain sensors on the surface of the beam, and then integrating twice based on above equations, the deformation displacement y_i of each strain sensor is obtained.

$$y_i=\frac{(\Delta l)^2}{6}\sum_{j=1}^{i}\frac{1}{c_{j-1}}\{[3(2j-1)-(3j-2)\frac{c_{i-j+1}}{c_{i-j}}]\varepsilon_{i-j}+(3j-2)\varepsilon_{i-j+1}\}+y_0+(i)\Delta l\,tan(\theta_0) \tag{5}$$

Based on the above theory, Ko et al. extended it [8]. By distributing multiple 'strain lines' to reconstruct the twist angle of the structure. Taking the wing as an example, the following is the calculation method of the twist angle.

Two monitoring lines are laid on the surface of the structure, and n+1 strain sensors are placed equidistantly on each 'strain line'. The measured strains are $\varepsilon_i$and ε_i'. The displacements y_i and y_i' of each point on two lines are obtained by using the Ko displacement theory. The twist angle ϕ_i for each cross section can be derived from the displacement values:

$$\phi_i=\sin^{-1}(\frac{y_i-y_i'}{d_i}) \tag{6}$$

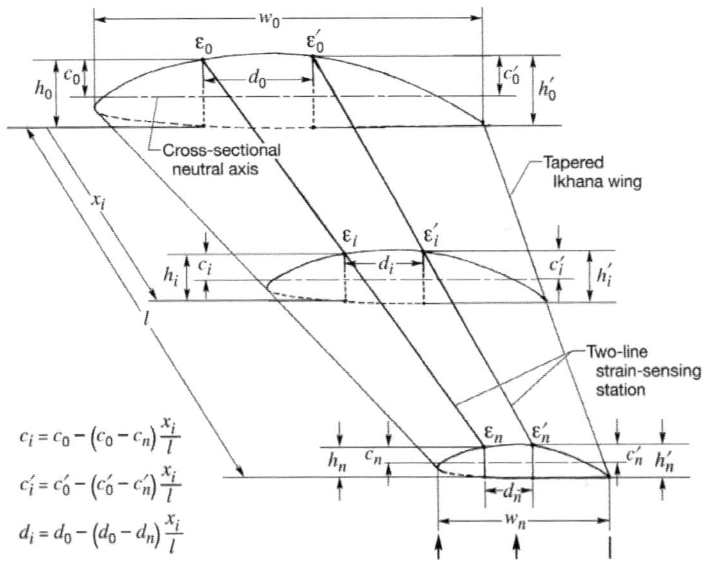

Figure 2. Deformation Reconstruction via Ko Displacement Theory.

3.2 Model reduced method

The idea of model reduced method is to condense the subsystems of a complex system into a super-element, and then assemble multiple super-elements [9]. For each super-element, its behavior characteristics are described only by the matrix generated by the boundary or external degrees of freedom.

After generating the corresponding matrices and applying multi-point constraints and single-point constraints, the following equation can be obtained as:

$$K_{ff}U_f = P_f \tag{7}$$

According to internal nodes (to be agglomerated by reduction) and external (boundary) nodes (retained nodes for analysis after condensation), the following expression can be obtained as:

$$\begin{bmatrix} K_{oo} & K_{oa} \\ K_{oa}^T & \overline{K_{aa}} \end{bmatrix} \begin{Bmatrix} U_o \\ U_a \end{Bmatrix} = \begin{Bmatrix} P_o \\ \overline{P_a} \end{Bmatrix} \tag{8}$$

Extracting the above equation and left-multiplying it by $[K_{oo}]^{-1}$ gives:

$$\{U_o\} = -[K_{oo}]^{-1}[K_{oa}]\{U_a\} + [K_{oo}]^{-1}\{P_a\} \tag{9}$$

Let $G_{oa} = -K_{oo}^T K_{aa}$ (boundary transformation) and $U_o^o = -K_{oo}^T P_o$ (fixing boundary displacement), the following expression can be obtained:

$$U_o = U_o^o + G_{oa}U_a \tag{10}$$

Substituting the expression for U_o into the following equation:

$$K_{oa}^T[G_{oa}U_a + U_o^o] + \bar{K}_{aa}U_a = \bar{P}_a \tag{11}$$

The following expressions can be obtained as:

$$G_{oa} = -K_{oo}^T G_{aa} + \overline{K_{aa}} \tag{12}$$

$$P_a = G_{oo}^T P_o + \bar{P}_a \tag{13}$$

Solving for the remaining structure yields the boundary displacement:

$$U_a = K_{aa}^{-1} P_a \tag{14}$$

There are two types of super-element configuration for a model: one is a single-layer super-element configuration, whose external nodes of all super-elements are internal nodes of the remaining structure [10]. In one word, all super-elements are directly connected to the remaining structure (Figure 3), the tree structure is a single-layer tree; the second is a multi-layer super-unit structure, that is, a super-unit will be connected to another super-unit, the tree-like diagram is A multi-level tree (Figure 4)

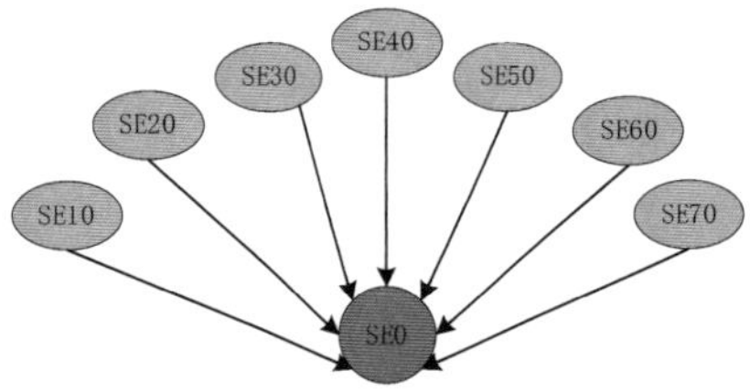

Figure 3. The configuration of a single-layer super-element.

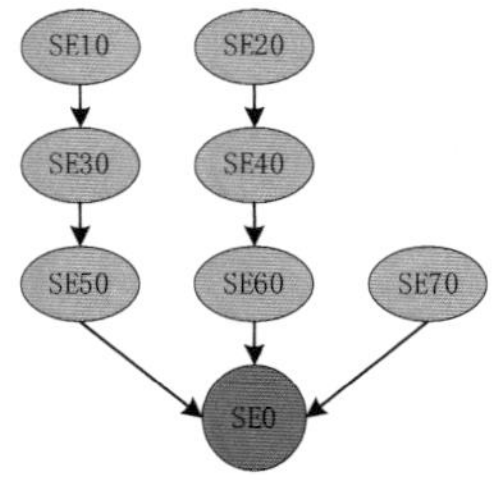

Figure 4. The configuration of a Multilayer super-element.

4. verification

In order to more accurately calculate the displacement of other non-measurement points on the structure, it is necessary to arrange multiple 'strain lines' on the structure. As shown in Figure , the green line is the highest point of the airfoil, and the red line is the installation position of the strain gauge. 10 strain sensors were evenly distributed on each line, the distance between the two strain sensor lines above and below the wing tip was 2.6 cm, and the distance between the two strain sensor lines above and below the wing root was 3.57 cm. The two strain sensor lines (red lines) are equidistant from the airfoil highest point line (green lines).

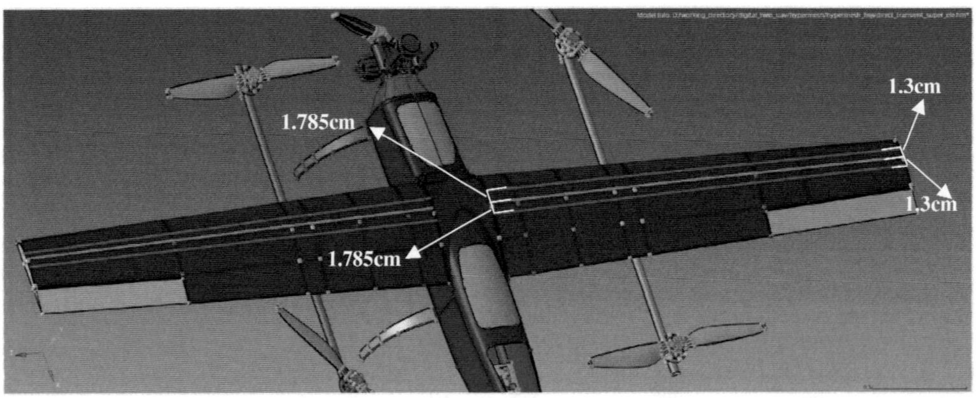

Figure 5. Schematic diagram of the installation position of the strain sensor.

To verify the effectiveness of the presented method, the fluid-structure interaction simulation method was used to calculate the strain value of the UAV in flight state. The strain values of the UAV strain sensors obtained by using the fluid-structure interaction simulation is used as the strain measurement values for verification. Figure shows the strain values.

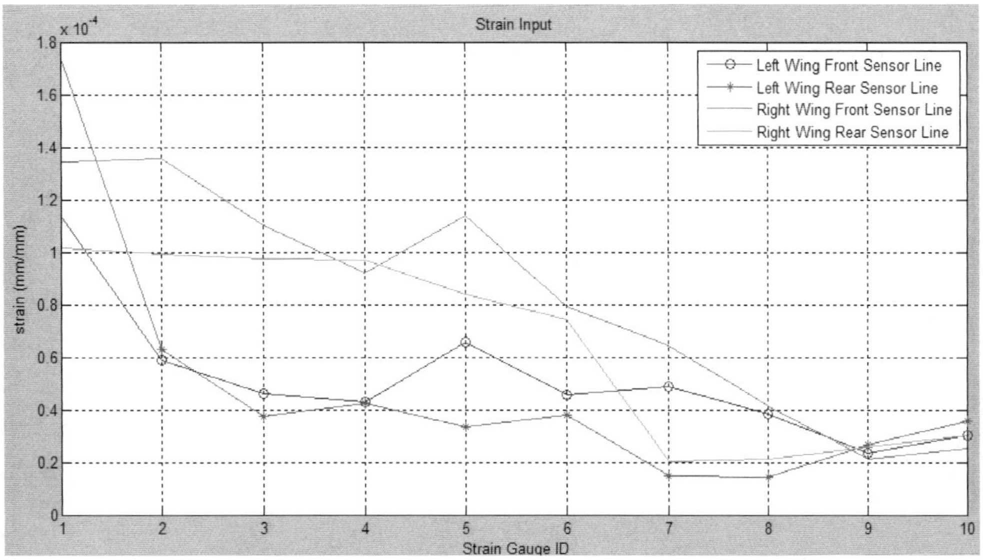

Figure 6. The strain values of the UAV strain sensors.

Using the above data, the displacement reconstruction is carried out through the Ko displacement theory, and the displacement value of each strain sensor position is obtained, as shown in Figure .

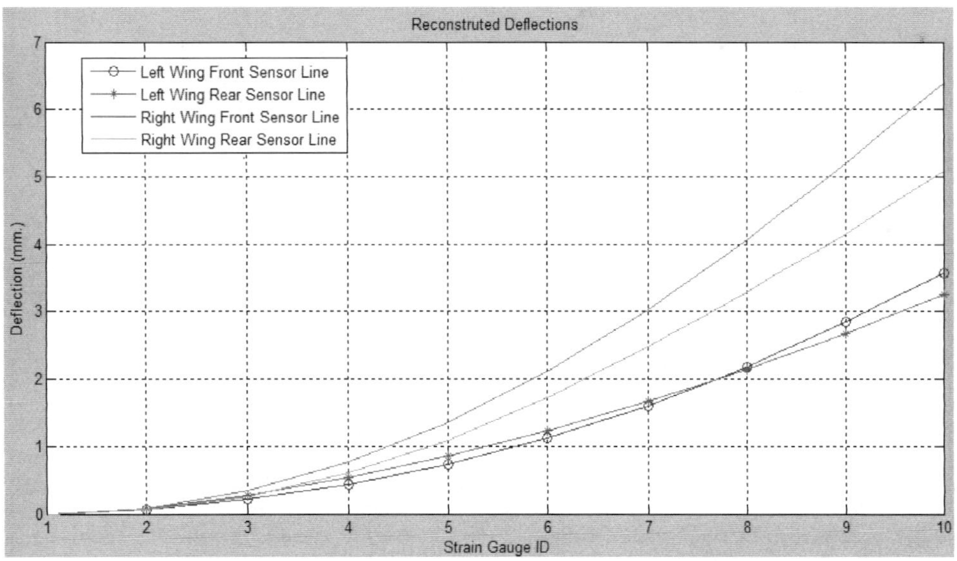

Figure 7. Deformation reconstruction results by using Ko displacement theory.

Use the displacement values obtained by the Ko displacement theory to calculate the stress and strain of the UAV structure. In order to analyze the accuracy of the presented method, the strain values of the finite element model corresponding to the installation positions of the strain sensors in the simulated structure is extracted and compared with the measured data of the strain sensors. Table 1 are the comparison results:

Table 1. Comparison between Nastran simulation results and strain gauge measurements.

		Sensor 1	Sensor 2	Sensor 3	Sensor 4	Sensor 5	Sensor 6	Sensor 7	Sensor 8	Sensor 9	Sensor 10
Front strain line of left wing	Test ($\times10^{-5}$)	11.29	5.87	4.61	4.28	6.56	4.57	4.88	3.84	2.35	3.02
	FEM ($\times10^{-5}$)	9.16	7.00	5.96	5.22	7.24	3.38	6.26	4.98	2.81	3.79
	Relative Error (%)	18.84	19.30	29.39	21.74	10.24	-26.06	28.35	29.62	19.68	25.55
Rear strain line of left wing	Test ($\times10^{-5}$)	17.24	6.30	3.73	4.23	3.35	3.81	1.46	1.44	2.65	3.57
	FEM ($\times10^{-5}$)	13.50	6.59	4.72	5.14	3.96	4.82	1.59	1.76	3.08	4.61
	Relative Error (%)	21.93	4.68	26.70	21.72	18.30	26.42	8.32	21.92	16.37	29.18
Front strain line of right wing	Test ($\times10^{-5}$)	13.44	13.56	11.04	9.19	11.39	7.95	6.43	4.17	2.11	2.51
	FEM ($\times10^{-5}$)	11.90	12.20	7.87	7.09	8.56	9.11	5.08	5.18	2.58	3.15
	Relative Error (%)	11.48	10.26	28.75	22.84	24.84	14.70	21.02	24.19	22.43	25.41
Rear strain line of right wing	Test ($\times10^{-5}$)	10.20	9.91	9.76	9.71	8.39	7.41	2.02	2.10	2.59	3.04
	FEM ($\times10^{-5}$)	7.25	7.46	7.02	9.00	5.99	8.29	2.40	2.57	3.31	3.32
	Relative Error (%)	28.58	24.73	28.06	7.23	28.54	11.83	19.29	22.45	28.10	9.16

It can be seen from the comparison results that the relative errors between the measured values and the simulated values are less than 30%. The possibility of causing this error is that the material properties in the finite element model are the lay-up materials in an ideal state based on the lay-up theory, and the materials used by the UAV are affected by factors such as production conditions, resulting in the lay-up of the layers. The unevenness, or discontinuity of the layup in some positions, cannot match the layup material in the ideal state, resulting in large errors in some parts.

5. Conclusion

This paper proposes a digital twin technology for UAV, develops a UAV digital twin system, and then describes the system composition and technical principles of the UAV digital twin system. The core algorithms in the digital twin system which are Ko displacement theory and model reduced method are explained in detail. The flight test verification was carried out, the test results show that the relative errors between the method proposed in this paper and the test results are less than 30%, which meet the needs of engineering applications. The methods and technical ideas proposed in this paper have certain significance for the development of UAV digital twin technology, and the developed digital twin system has strong engineering significance and military value.

Acknowledgments

The works was supported by the Key-Area Research and Development Program of Guangdong Province under No.2020B0404020002, the Special Fund of Guangdong Province for promoting high-quality economic development under No. JQR217000010, the Development fund for the Fifth Electronics Research Institute of Ministry of Industry and Information Technology under No. 19Z14, Ministry of industry and information technology project under No. TC210804R-1.

References

[1] Glaessgen, E.H. and Stargel, D.S., 2012.The Digital Twin Paradigm for Future NASA & U.S. Air Force Vehicles. In: The 53rd Structures, Structural Dynamic, and Materials Conference. Hawaii.
[2] Tao F., Zhang M., Liu Y., and Nee. A.Y.C., Digital twin driven prognostics and health management for complex equipment. CIRP Annals – Manufacturing Technology, 2018, 67(1): 169-172.
[3] Zhou X., Zhao F.B., et al., Key technology for modeling and simulation of airframe digital twin. Acta Aeronautica et Astronuatica Sinica, 2021, 42(3):023981.
[4] Dong Y.F., Li Z., and Lei M., Research on Concept of Digital Satellite. Aerospace Shanghai (Chinese & English), 2021, 28: 1-12.
[5] Tao, F. and M. Zhang, Digital Twin Shop-Floor: A New Shop-Floor Paradigm Towards Smart Manufacturing. IEEE Access, 2017. 5: 20418-20427.
[6] Sun H. J., Zhang W. J., and Zhang L. Y., Analysis on the intelligent development of aerospace manufacturing in advanced military enterprises in the United States and Europe. Satellite Applications, 2019, 6: 26-31.
[7] Ko, W.L., W.L. Richards, and V.T. Tran, Displacement Theories for In-Flight Deformed Shape Predictions of Aerospace Structures. NASA/TP-2007-214612, 2007.
[8] Ko, W.L., W.L. Richards, and V.T. Fleischer, Applications of Ko displacement theory to the deformed shape predictions of the doubly-tapered Ikhana Wing. NASA/TP-2009-214652, 2009.
[9] XU J., ZHOU L., CHEN P. L., Module granularity optimization analysis model based on digital twin. Computer Integrated Manufacturing Systems (in Chinese), 2019, 25(6):1419-1431.
[10] Tao F., Liu W.R., Liu J.H., Digital twin and its potential application exploration. Computer Integrated Manufacturing System, 2018,24 (1), pp. 1-18

Intelligent defect detection for Power Patrol Inspection

Chuangmian Huang[a], Riguang Lin[b], Lihao Wu*[b]

[a]The Fifth Electronics Research Institute of MIIT, Guangdong Provincial Key Laboratory of Electronic Information Products Reliability Technology, Guangzhou, China,511370;

[b]Guangzhou City University of Technology, Huadu District, Guangzhou, China, 510800

[*] Corresponding author: wulh@gcu.edu.com

ABSTRACT: In this paper, the insulator string images captured by Using unmanned aerial vehicles (UAVs) in the power inspection are taken as the research object. The image processing and deep learning methods are used to label the images, and the self-explosion fault of the identified and segmented insulator strings is identified and located.
In the process of semantic segmentation of insulator string images, the given image data set is enhanced, and images are exchanged with image processing algorithms. Based on SegNet, each pixel in the picture is classified, and then the self-encoding is used for reference, and the image is first encoded and then decoded. The coding layer before the model uses the VGG-16 network to extract smaller pictures, so as to meet the requirements of semantic segmentation, and the decoding layer after the model obtains the classification probability value of each point and classifies each point. The experimental results show that the SegNet algorithm can realize the recognition and segmentation of insulator string images under complex conditions with high accuracy.
In order to realize the identification and location of the self-exploding fault position of the insulator string, the training model can be used to detect the position of the self-exploding insulator in the image. First, the data of target detection is preprocessed, and then the data set transformation method of rotation transformation and image mirroring is used; then the SSD model is selected to identify and locate the self-explosion position of the insulator. The SSD model combines the advantages of the Faster R-CNN model and the YOLO model. It runs faster and has higher detection accuracy, which meets practical applications. The experimental results show that the SSD algorithm is more targeted to the insulator data set, and can improve the speed and accuracy of insulator identification and positioning.

1. Introduction

The insulator string is a key device in high-voltage transmission lines. Its main function is to support and fix the current-carrying conductor, so that a good insulation is formed between the current-carrying conductor and the ground. However, because the insulator string is exposed to the outdoors for a long time, it is prone to failure, so it is the monitoring object in Power Patrol inspection. In order to ensure the safe and reliable operation of the transmission line, the power grid operation department needs to regularly inspect [1], repair and maintain the transmission line and substation system to ensure the elimination of faults or hidden dangers. With the rapid development of China's economy, higher requirements have also been placed on the safe operation of infrastructure such as power transmission network equipment. As one of the important means to ensure the normal operation of the transmission and distribution network, overhead line inspection has always been faced with problems such as wide distribution of network cables, complex facility layout, high intensity of line inspection operations, long periods, and complex and harsh natural environment in some areas [2, 3]. The traditional power grid inspection method is manually inspected, but there are some problems with this method. For example, the labor intensity is high, the work efficiency is low, the manual input data volume is large, the inspection quality cannot be guaranteed, the data input is prone to errors, and in dangerous areas, it will endanger the lives of workers.

With the development of science and technology in China, especially the rapid development of unmanned aerial vehicles and intelligent robots, the required information is collected in real time through various information sensing devices. A large number of on-site pictures of power equipment and lines are taken to replace worker inspections. The basic work process is: demarcate the working area of the drone, set the inspection time, shoot the drone in the designated area, and analyze the drone shooting. The picture of the problem, and the problem is marked and fed back to the dispatch center, and then the dispatch center arranges the corresponding workers to repair according to the problem.

However, due to the large number of pictures taken by UAV (more than 300 images taken by a single elevated tower) and large size (4096 * 2160), it takes 5-10 minutes to manually mark a picture, which is a huge workload. At the same time, the relevant personnel performing the marking work are very prone to eye fatigue, resulting in missing and wrong marks. In view of the above situation, consider using the methods of image processing and machine learning (deep learning) to label the pictures. With the development of artificial intelligence, the deep learning algorithm can automatically learn the characteristics of big data, realize the identification and segmentation of insulator strings in the picture, and identify the self-explosion identification and positioning of insulator strings, so as to automatically process the line patrol pictures taken by these UAVs [4-8].

Based on the application of deep learning algorithm and computer vision, this paper realizes the recognition and segmentation of insulator string and the detection of insulator self-explosion fault under complex background. The recognition accuracy is high and the speed is fast. The power inspection can only detect defects, which is conducive to improving the intelligence of power system operation and maintenance, and is of great significance to ensure the security and stability of power system.

2. Related work

2.1 insulator string images segmentation and self-explosion identification

The main contents of this paper are: insulator string segmentation and insulator self -explosion identification and marking. Using data mining method to solve the following two problems.

For problem one, because the picture taken by the UAV is large, generally (4096 * 2160), the insulator string occupies a small part of the picture, so it is necessary to design an image segmentation algorithm to segment the area where the insulator string is located. Then mark the mask image of the marked insulator string bead area in each image of the given sample. The mask area is only for the entire insulator string. The comparison between the original image and the mask image is shown in Figure 1.

Figure 1. The left is the original image, and the right is the mask image

For problem two, according to the segmented image, the location of the insulator is preliminarily identified, the insulator string is segmented, and then the self-exploding insulator recognition model is constructed according to the Ground Truth of the marked samples. Finally, the training model is used to detect the position of self-exploding insulator in the image, and the Bounding Box is used to mark it. The self-explosion mark shall include two intact insulators around the self-explosion position. The self-explosion position mark of insulator is shown in Figure 2.

Figure 2. Example of location markers

2.2 Deep Learning for insulator string recognition and defect detection

The development of deep learning has greatly improved the accuracy of image recognition. In recent years, some methods based on CNN networks have been proposed for insulator string recognition and defect detection. In [9], the multipatch feature of insulator is extracted by CNN and then used to train SVM classifier. The SSD model [11] for insulator string recognition and defect detection is used by

Miao et al. [10]. In [12], An insulator defect detection method based on multilevel perception is proposed by Jiang et al, which uses SSD as the detection model. Faster R-CNN [14] is proposed for insulators string detection, and the U-net [15] is proposed for defect detection in [13]. A compact end-to-end neural network is designed Chang et al. [16], which is trained in the framework of conditional generative adversarial networks. A CNN architecture for insulator string recognition and defect detection is proposed by Tao et al. [17]. Meanwhile, a data augmentation method is used to address the scarcity of defect images. In [18], an insulator string segmentation network Up-Net is proposed by Sampedro et al. However, these methods mainly rely on popular deep learning recognition, and do not fully consider the characteristics of insulator string images, which will lead to low recognition accuracy.

3. Symbol description

Table 1. Symbol description

Symbols	**Meaning**
width	The Width of image
height	The Length of image
(x, y)	Original coordinates
(x′,y′)	Transformed coordinates
(x0, y0)	Center of rotation
top	Ordinate of the highest point after rotation
down	Ordinate of the lowest point after rotation
left	Abscissa of the leftmost point after rotation
right	ordinate of the rightmost point after rotation
newwidth	The Width after rotation
newheight	The Height after rotation
θ	Rotation angle
p	Correct value
q	Estimate
N	The default number of boxes that match the real object box
c	Confidence level of prediction box
g	Location information of real box
α	Weight parameter
Lioc(x, l, g)	Position loss function
Lconf (x, c)	Confidence loss
A	GroundTruth area
B	The region obtained by algorithm segmentation

4. Data preprocessing

4.1 Data analysis

The resulting number of original pictures is 40, as shown in Figure 3. Generally, the amount of data required for deep learning training is very large, and 40 pictures obviously cannot meet the training requirements. Therefore, the dataset needs to be augmented. The purpose of data augmentation is to increase the number of images.

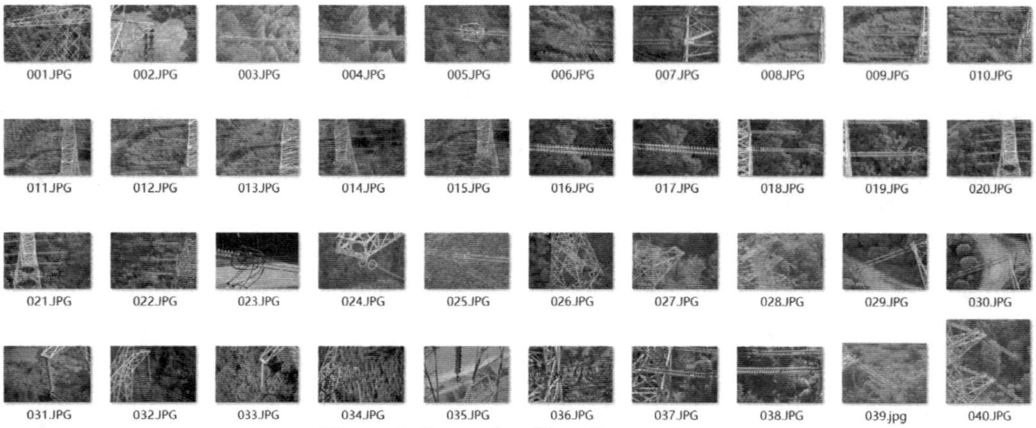

Figure 3. Example of location markers

4.2 Semantic segmentation data preprocessing

Image data set enhancement is to transform images with image processing algorithms and add noise to increase the number of training images. The commonly used methods are affine transformation, cutting, rotation, adding noise, cropping image, and generating antagonistic neural network to enhance the data set. In this experiment, the original image and the mask image are rotated up and down, left and right, and rotated 30 degrees, 45 degrees, 90 degrees, 135 degrees, 180 degrees, 225 degrees, 270 degrees and 315 degrees respectively by rotating and mirroring the image. Increase the number of pictures to 400 through these four methods.

The transformation principle is as follows.

The formula for swapping left and right is as equation (1) and equation (2).

$$x = width - x_0 - 1, \tag{1}$$

$$y = y_0 . \tag{2}$$

The formula for the up and down swap is as equation (3) and equation (4).

$$y = height - y_0 - 1, \tag{3}$$

$$x = x_0 . \tag{4}$$

The principle of rotation is as follows. And the formula for rotation is as equation (5) and equation (6).

$$newwidth = right - left, \tag{5}$$

$$newheight = top - down . \tag{6}$$

It is easy to get the left, top coordinates of the image after rotation (the coordinate system with the rotation center as the origin). Therefore, after the rotation, the coordinate system should be converted to take the upper left corner of the image as the coordinate origin. The following formula can be obtained from the transformation relationship.

$$x' = x + left, \tag{7}$$

$$y' = -y + top . \tag{8}$$

The matrix expression is as equation (9).

$$[x' \quad y' \quad 1]=[x \quad y \quad 1]\begin{bmatrix} 1 & 0 & 0 \\ 0 & -1 & 0 \\ left & top & 1 \end{bmatrix}. \tag{9}$$

The resulting rotation formula is as follows.

$$[x \quad y \quad 1]=[x' \quad y' \quad 1]\begin{bmatrix} 1 & 0 & 0 \\ 0 & -1 & 0 \\ -x_0 & y_0 & 1 \end{bmatrix}\begin{bmatrix} \cos\theta & -\sin\theta & 0 \\ \sin\theta & \cos\theta & 0 \\ 0 & 0 & 1 \end{bmatrix}\begin{bmatrix} 1 & 0 & 0 \\ 0 & -1 & 0 \\ left & top & 1 \end{bmatrix}. \tag{10}$$

Some processed images of the original images are as follows.

Figure 4. Original image

Figure 5. Rotation transformation

Figure 6. Swap up and down

Some processed images of the mask images are as follows.

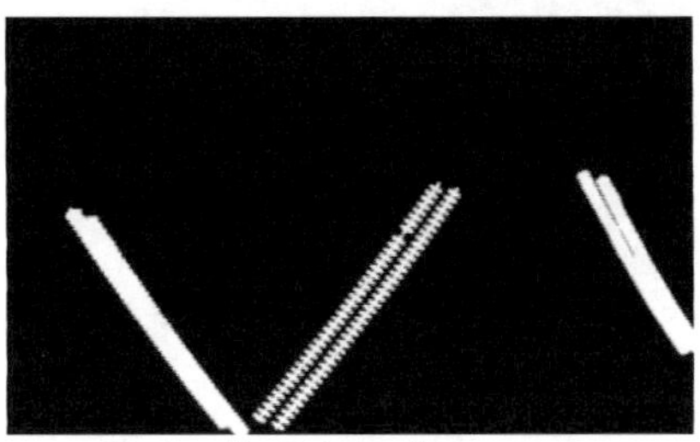
Figure 7. Original image

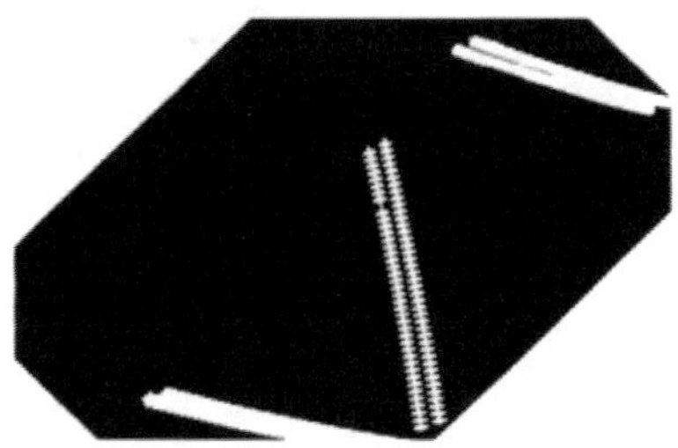
Figure 8. Rotation transformation

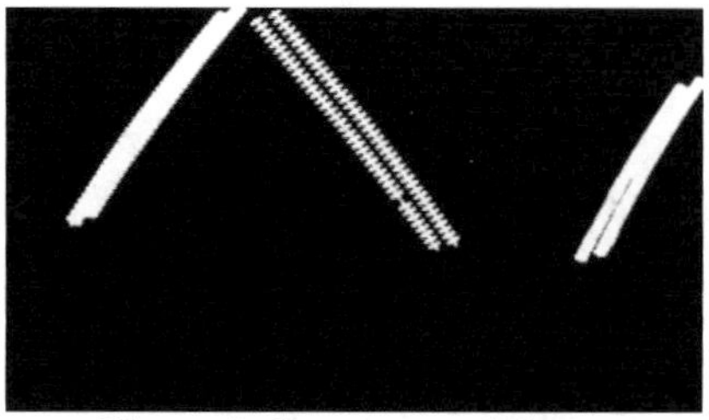

Figure 9. Swap up and down

4.3 Target detection data preprocessing

Object detection uses data set transformation and semantic segmentation. Here, the image is also rotated up and down, left and right, and rotated by 20 degrees, 40 degrees, 60 degrees, 80 degrees, 90 degrees, 100 degrees, 120 degrees, 140 degrees, and 160 degrees in the way of rotation transformation and image mirroring. Different from the rotation angle of semantic segmentation dataset, it improves the generalization ability of the model.

It is worth noting that when the image processing algorithm is used for the original image, the coordinates in the XML tag file corresponding to the original image should be transformed correspondingly. The algorithm principle is the same as the rotation principle and the modulation principle of semantic segmentation data preprocessing.

Some processing results of the original image are as follows.

Figure 10. Original image

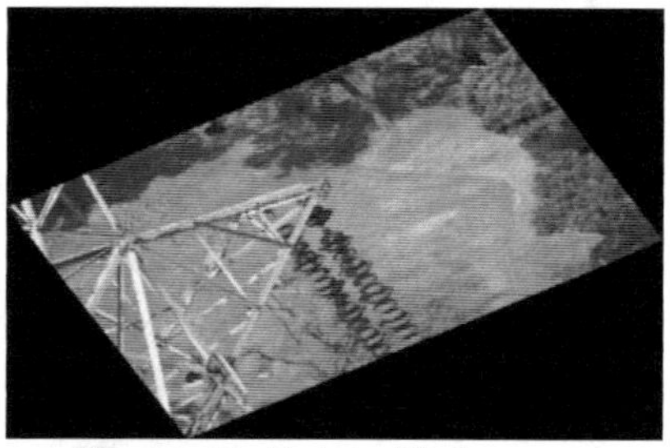

Figure 11. Rotate 30 degrees

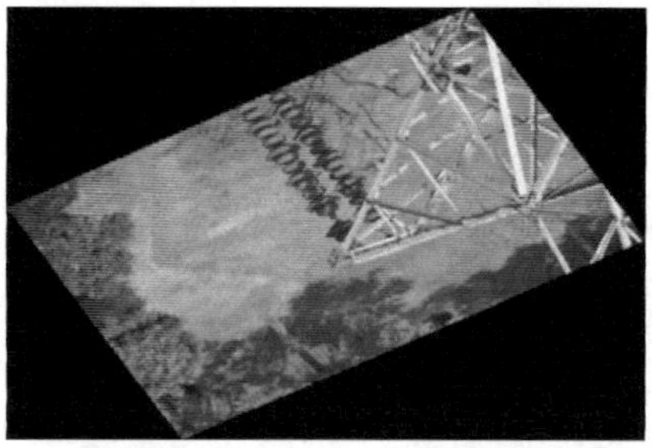

Figure 12. Rotate 210 degrees

The processing results of XML tags are as follows.

```
- <object>
  - <item>
      <name>1</name>
    - <bndbox>
        <xmin>1780</xmin>
        <ymin>1232</ymin>
        <xmax>1966</xmax>
        <ymax>1463</ymax>
      </bndbox>
    </item>
  </object>
</outputs>
<time_labeled>1577179089915</time_labeled>
<labeled>true</labeled>
- <size>
    <width>3936</width>
    <height>2624</height>
    <depth>3</depth>
  </size>
</doc>
```


Figure 13. XML file of the original image


```
- <object>
  - <item>
      <name>1</name>
    - <bndbox>
        <xmin>2215</xmin>
        <ymin>2098</ymin>
        <xmax>2377</xmax>
        <ymax>2299</ymax>
      </bndbox>
    </item>
  </object>
</outputs>
<time_labeled>1577179089915</time_labeled>
<labeled>true</labeled>
- <size>
    <width>3936</width>
    <height>2624</height>
    <depth>3</depth>
  </size>
</doc>
```


Figure 14. XML file for rotating 30 degrees image


```
- <object>
  - <item>
      <name>1</name>
    - <bndbox>
        <xmin>2344</xmin>
        <ymin>1941</ymin>
        <xmax>2506</xmax>
        <ymax>2143</ymax>
      </bndbox>
    </item>
  </object>
</outputs>
<time_labeled>1577179089915</time_labeled>
<labeled>true</labeled>
- <size>
    <width>3936</width>
    <height>2624</height>
    <depth>3</depth>
  </size>
</doc>
```


Figure 15. XML file for rotating 210 degrees image

5. Semantic segmentation algorithm

5.1 SegNet network

SegNet is used to classify each pixel in the image and identify the category of each pixel to realize image segmentation. SegNet is a semantic segmentation network obtained by modifying vgg-16 network based on FCN. There are two versions of SegNet: SegNet and Bayesian SegNet.

The SegNet network model borrows the idea of self-encoding, encoding the image first, and then decoding it. Its core is mainly composed of the encoding (Encoder) stage and the decoding (Decoder) stage. Encoding stage: The process of converting images into features and generating low-resolution images as feature maps. Decoding stage: The process of converting feature maps into image labels and mapping low-resolution images to pixel-level labels.

5.2 SegNet's network framework process

In the encoding phase, vgg-16 network is used to obtain the high-dimensional features of the input image through Convolution layer and Pooling layer. In the corresponding decoding stage, the Convolution layer and Upper sampling layer of the decoding stage correspond to the encoding stage one by one. The Upsampling layer can recover the characteristics of some images, and the Convolution kernel of the convolution layer is obtained through network training. After the decoding phase, the feature map with the same size as the input original map is obtained. Finally, the Softmax layer is used to check the maximum value of the feature depth corresponding to each pixel position as the category, and the final segmentation map is obtained.

 IOP Publishing
2366 (2022) 012039 doi:10.1088/1742-6596/2366/1/012039

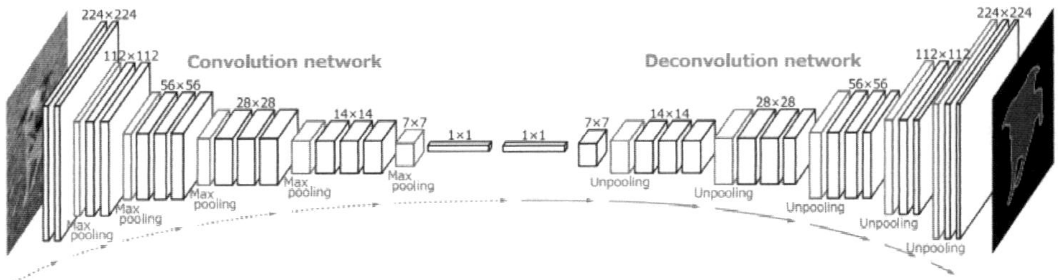

Figure 16. The SegNet network uses Convolutional layers, Batch Normal zation layers, ReLU activation Pooling, and Upsampling Softmax layers. The left half of the figure is the encoding (Encoder) stage, and the right half is the decoding (Decoder) stage

The left half shown in Figure 16 is the encoding phase of SegNet. In the encoding phase, vgg16 network is used to obtain the high-dimensional features of the input image through the convolution layer and the pooling layer; The right half of Figure 16 is the decoding stage. The convolution layer and the upper sampling layer of this stage correspond to the encoding stage one by one. The upper sampling layer recovers some image features. The convolution kernel of the convolution layer is obtained through network training. After the size of the feature image is consistent with the size of the input original image in the decoding stage, the softmax layer is used to check the maximum value of the corresponding feature depth of each pixel position as the category [19].

5.3 SegNet upsampling layer

Take max-pooling as an example, because in the downsampling operation stage of pooling, not only the feature of the largest pixel value in the window is extracted, but also the position of the largest pixel feature in the original image is recorded. According to the position of pooling-indices relative to the pooling filter in upsampling, put the data back to the corresponding position and do convolution training. The maximum pixel feature position recorded by Pooling is used to map the feature to the corresponding position in the original image, and the rest of the images are filled with zeros. The Softmax classifier does pixel-by-pixel classification.

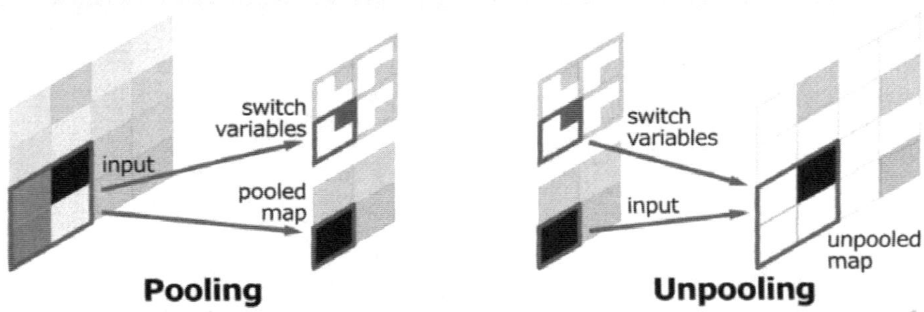

Figure 17. The left of figure is the Pooling operation, the goal is to downsample the input image and record the location of the downsampling; the right of figure is the Unpooling operation, the goal is to upsample the image to restore the image size

6. Target detection algorithm

6.1 Faster R-CNN

Fast RCNN is a two-stage method, which is the basis for many target detection algorithms to improve. Moreover, it is improved based on the existing RCNN and fast RCNN methods. The most striking point of the algorithm is the introduction of the region proposal network (RPN) network that can carry out "end-to-end" training. The main reason why the concept of RPN is proposed is that it is time-consuming to extract candidate regions with traditional methods. The main purpose of RPN is to speed up the extraction of candidate regions. The algorithm flow of fast RCNN is shown in Figure 18. First, CNN network is used to extract the feature map of the input image, and then RPN network is used to generate candidate regions based on the feature map. Finally, combined with the candidate regions generated by RPN and the feature map extracted by CNN network, the positioning and recognition of the target region are completed through the full connection network layer [20].

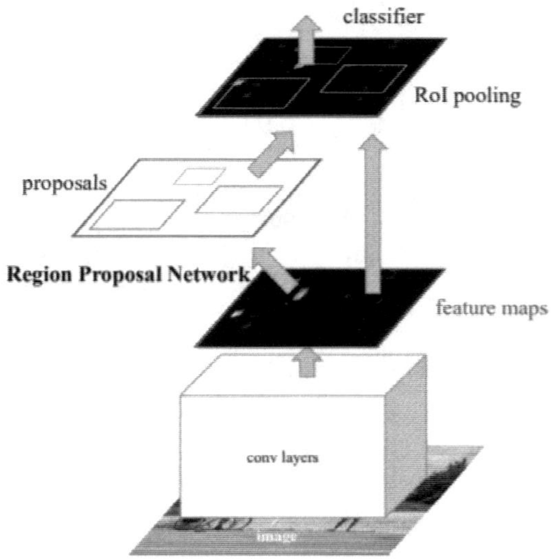

Figure 18. Fast RCNN flow chart

6.2 SSD

The most obvious difference between SSD (single shot multi-box detector) network and fast RCNN network is that SSD does not use RPN network to generate candidate boxes. The whole network only needs one step to output the results, which greatly improves the speed of text detection. In the process of neural network operation, different convolution layers can be used to output characteristic maps of different scales, and then they can be detected separately. Finally, the results can be integrated by non-maximum suppression algorithm.

The network structure of SSD is shown in Figure 19. Its workflow can be divided into the following four steps [11].

(1) The input image is first processed by the deep convolution network, which is similar to the fast RCNN. Taking max-pooling layer as the dividing line, we divided this section of structure into five groups composed of two or three CNN units, and named their output characteristic maps conv_1 to conv_11 from front to back.

(2) After layers of convolution, the feature map becomes smaller and smaller, which means that the information of small objects contained in the current feature map will be less and less, and gradually only the features of large objects will be left. In order to get the feature information of objects with different sizes, we can consider detecting objects with different sizes. For example, take the characteristic diagrams output by the six CNN cell groups conv_4 to conv_11. For each point in the feature map, there are several anchor boxes as a priori, but SSD network sets different sizes, numbers and proportions of anchor boxes for each feature map with different scales.

(3) Then, the candidate boxes whose confidence on all six scales exceeds the threshold are counted, and then all of them are mapped to the original graph. Finally, through the non- maximum suppression algorithm, the final prediction boundary box can be obtained.

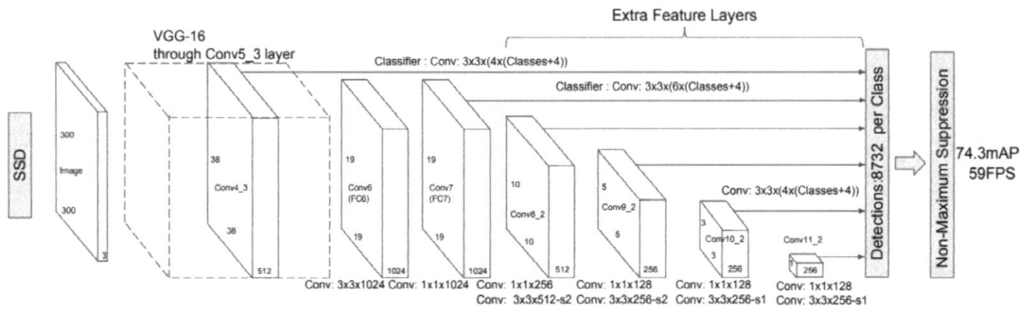

Figure 19. SSD structure

6.3 YOLO

Faster R-CNN has high detection accuracy, but its speed is far from meeting the real-time requirements. The object detection algorithm based on deep learning uses the idea of regression. Yolo does not choose to train the network based on the candidate region algorithm, but directly adopts the full graph training mode. Its advantage is that it can distinguish the target object and the background area more conveniently and quickly. Its disadvantage is that it can improve the detection speed while sacrificing the accuracy [21].

YOLO's network framework process is as follows.

(1) Scale the input image to a fixed size, pass the original image through the full convolution neural network, classify the feature matrix by Softmax function, and filter by using the maximum suppression NMS.

(2) The input image is divided into S*S grids, and the corresponding grids are detected. Each grid needs to predict B Bounding-boxes, each Bounding-box corresponds to 5 prediction parameters, including 4 location information [x, y, w, h], and a confidence value for prediction. The confidence value represents the confidence P of the target contained in the predicted bounding box and the prediction accuracy IOU of the bounding box.

7. Experimental design and analysis

7.1 Experiment platform

The software and hardware configuration of the experimental environment is shown in the table 2.

Table 2. Experimental configuration

CPU	**graphics card**	**Memory**
Ryzen3600	2080 Ti	16G
operating system	**Python**	**CUDA**
Ubuntu18.04.3	3.6.9	10.0.130
Tensorflow	**Keras**	**Scipy**
1.13.1	2.1.5	1.2.1

The python library used is as follows.

NumPy: NumPy（Numerical Python）is an open-source numerical computing extension to Python. This tool can be used to store and process large matrices much more efficiently, supporting arrays and matrix operations of a large number of dimensions.

SciPy: SciPy is a convenient, easy-to-use Python toolkit designed for science and engineering. It includes statistics, optimization, integration, linear algebra modules, Fourier transforms, signal and image processing, ordinary differential equation solvers, and more.

Matplotlib: Matplotlib is a 2D graphics library for Python.

OpenCV: OpenCV (Open-source Computer Vision Library) is a cross-platform computer vision library based on open-source distribution, which implements many common algorithms in image processing and computer vision.

TensorFlow: TensorFlow is an open-source software library for numerical computation using dataflow graphs.

Keras: Keras is an open-source artificial neural network library written in Python that can be used as a high-level API for TensorFlow, Microsoft-CNTK, and Theano

PyTorch: PyTorch is the python version of torch, a neural network framework Open sourced by Facebook, specifically for GPU-accelerated deep neural network (DNN) programming. Torch is a classic tensor library that operates on multidimensional matrix data, and is widely used in machine learning and other math-intensive applications.

7.2 Data Sources

(1) Select 40 original pictures of self-explosion defects, standard mask pictures based on the original pictures, and BoundingBox labels of self-explosive insulators given by the question maker.

(2) Mirror transformation and rotation transformation were performed on the 40 original images given by the question maker. The dataset based on semantic segmentation and the dataset based on object detection are processed separately, and 400 pictures for semantic segmentation and 420 pictures for object detection are obtained.

7.3 Experimental evaluation index

(1) Dice coefficient

The insulator string segmentation is evaluated by the Dice coefficient, and the similarity of the two samples is usually calculated. The formula is as follows.

$$dice(A,B) = \frac{2|A \cap B|}{|A| + |B|}. \tag{11}$$

The value range of Dice coefficient is [0, 1]. The closer the value is to 1, the more consistent the predicted results are with those marked by professionals.

(2) IOU

IOU is used for the evaluation of insulator self-explosion area. IOU represents the overlap between the candidate bound and the original ground truth bound, that is, the ratio of their intersection and union. The figure and formula are as follows.

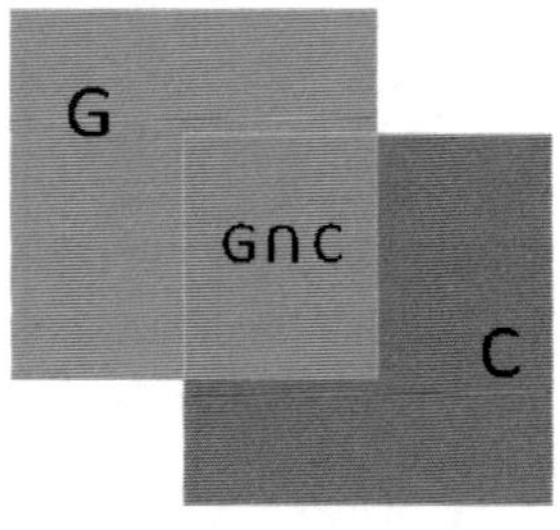

Figure 20. IOU

$$IOU = \frac{G \cap C}{G \cup C} \times 100\%. \tag{12}$$

7.4 Experimental Design

7.4.1 Semantic segmentation model experiment

(1) Training process

First load the vgg16 trained model, and train on the basis of the standard trained model, you will get better results. Because the standard trained model will have better weight and strong generalization ability. Open the txt file containing the training set and labels, and put all the file names of the labels corresponding to the training set in a TXT file in advance. This can facilitate the subsequent data reading functions, batch reading of pictures and labels, and also facilitate the subsequent data processing. The lines of the read TXT file are disordered, making the read pictures disordered. The obtained data set is divided into 10% data as test set and 90% data as training set. Set the way to save the model. Save the model every three iterations. Set the way to reduce the learning rate. If the loss rate of the test set does not decrease for three times, the learning rate will decrease and continue training. Set whether to stop early. When the loss rate of the test set does not decrease all the time, it means that the basic training of the model is completed and can be stopped. Related parameter settings: the number of pictures read in each training (batch-size) is set to 4, the number of iterations (epoch) is 15, the number of iterations at the beginning is 0, and the learning rate continues to decline. After setting the parameters, you can start training.

(2) Load dataset

First, according to the image file name in the loaded TXT file, read the image and label in the folder where the image is stored and the folder where the label is stored. Read (batch- size) pictures at a time. For the loaded pictures, first transform them into 4116 * 416 pictures by size, then normalize the pixels in the pictures, and finally add them to the training set to control their training through the training function. For the read label, that is, the mask image. It also converts its picture size to 416 * 416, and then converts it to an array. By judging the value of its pixels, it is classified and transformed into a matrix with only 0-1 elements. This matrix is a two-dimensional matrix. The row represents the position of the pixels and the list shows the category. When a pixel belongs to a certain class, the position of the corresponding class is 1, and the positions of other classes are 0. Finally, it is added to the label of the training set to control its training through the training function.

7.4.2 Target detection model experiment

(1) Training process

First, load the pre trained weights, a priori boxes and txt files containing indexes, and train on the basis of the standard trained models and a priori boxes, and you will get better results. Because the standard trained model will have better weight and strong generalization ability. Then open the txt file consisting of the file names of the training set and the labels, and put all the file names of the labels corresponding to the training set in a TXT file in advance. This can facilitate the batch reading of pictures and labels and the subsequent data processing. The third step is to scramble the lines of the read TXT file so that the read pictures are out of order. The fourth step divides the data set obtained in the third step: 10% of the data is used as the test set, and 90% of the data is used as the training set. Step 5: set the way to save the model and save it once for three iterations; Set the method of decreasing the learning rate. If the loss rate of the test set does not decrease for 3 times, the learning rate will decrease and continue training; Set whether to stop early. When the loss rate of the test set does not decrease, it means that the basic training of the model is completed and can be stopped. Step 6: relevant parameter setting: the number of pictures read per training (batch- size) is set to 4, the number of iterations (epoch) is 20, the number of iterations at the beginning is 0, and the learning rate continues to decline. Three training with different learning rates are set for rough training and fine training respectively. First, the coarse training with high learning rate can be selected for training. When the melody reaches a certain level, the fine training with low learning rate can be selected. After setting the parameters, the training can begin.

Journal of Physics: Conference Series **2366** (2022) 012039 doi:10.1088/1742-6596/2366/1/012039

(2) Loading of datasets

The data read from TXT is segmented first, and the loaded image information and classification information are saved respectively, and then stacked into the training set. Then, according to the image read by the index, it is trimmed, scaled and selected again. This operation can enhance the robustness of the model, and can also encode the picture for SSD detection.

(3) Data storage

The storage form of the training set is the same as that of the VOC2007 dataset. There are 3 folders: Annotations folder, JPEGImages folder, ImageSets folder. The Annotations folder saves the xml files of all labels; the JPEGImages folder saves all the original images, and the order of these images corresponds to the label files; the ImageSets folder saves some txt files about training, testing, and indexing.

7.5 Analysis of experimental results

For the semantic segmentation model, the number of iterations is set to 20. Some training processes are as follows.

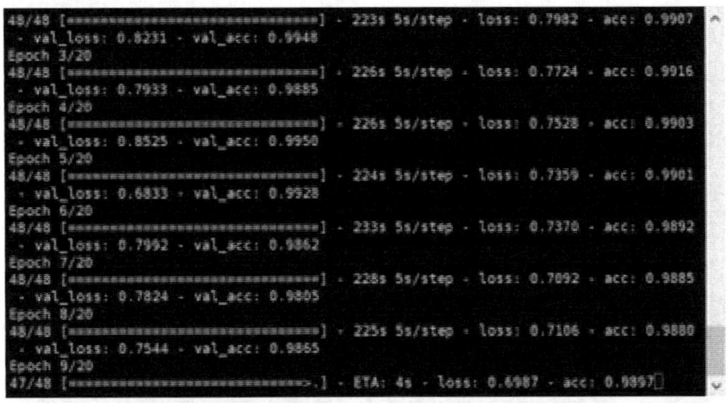

Figure 21. Semantic segmentation training 1

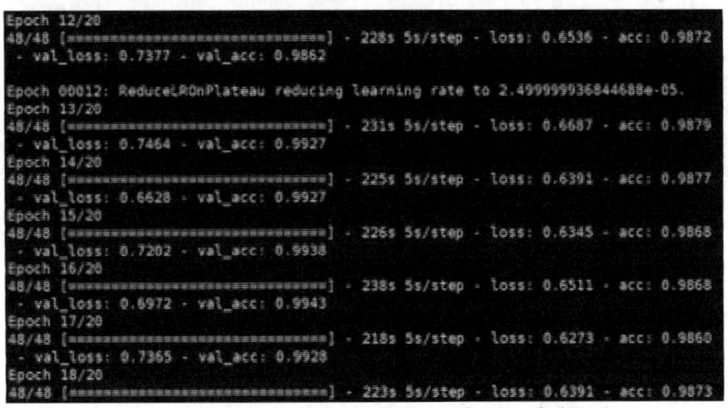

Figure 22. Semantic segmentation training 2

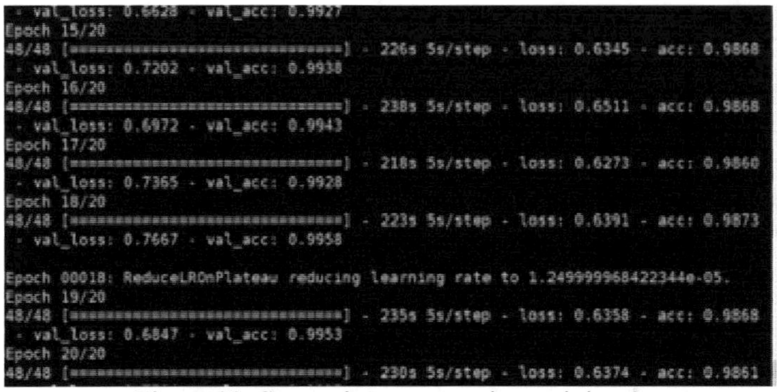

Figure 23. Semantic segmentation training 3

From the overall loss rate and accuracy of the model, as well as the loss rate and accuracy of the model on the test set, the model seems to be well trained. The overall accuracy and test set accuracy are as high as 99%. The overall loss rate decreased from 0.7982 to 0.6374. The loss rate of the test set fluctuates around 0.7.

Take a test picture to test the model, and find that the effect is not ideal. The main reason is that there are too few pictures for training, and the pretreated pictures have great correlation, which makes the model severely over fit. It should be processed by image segmentation and generating antagonistic neural network to obtain more data sets.

For the target detection model, set the number of iterations to 20 and the learning rate to 10 (negative 4). When the loss of training decreases to 1, the learning rate will be automatically adjusted to 10 (minus five); When the training change drops to 0.5, the learning rate will be automatically adjusted to 10 (minus six). Part of the training process is as follows.

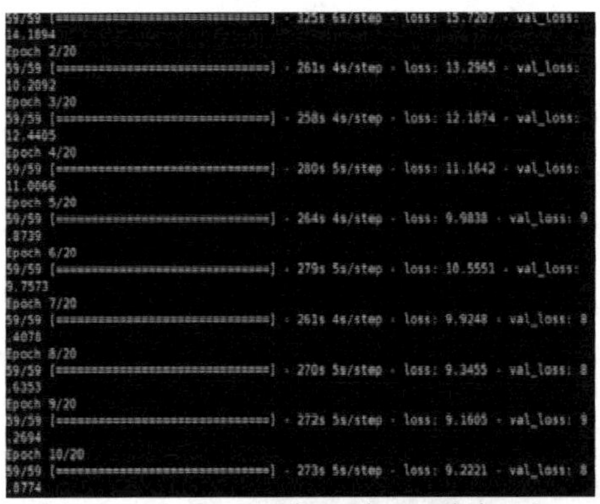

Figure 24. Target detection training 1

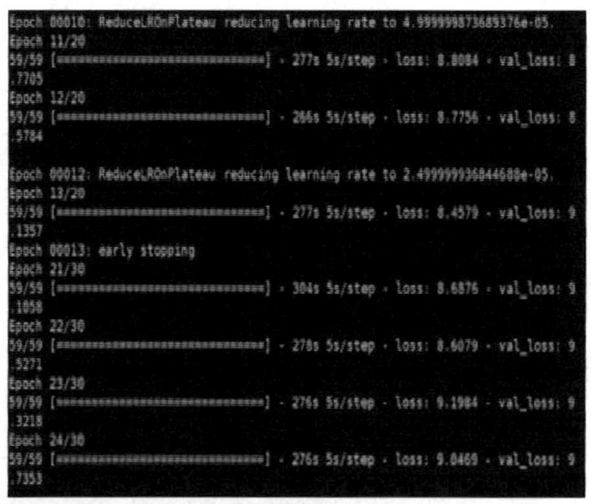

Figure 25. Target detection training 2

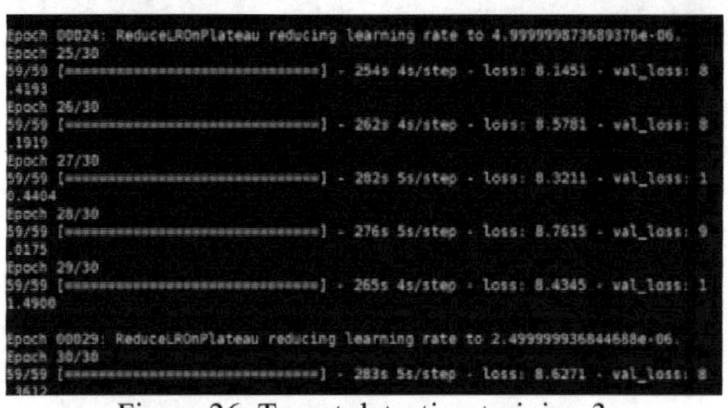

Figure 26. Target detection training 3

During the training, it can be found that the total loss of the model at the beginning is as high as 15.7207 and 13.2965. When the number of iterations reaches 3, the loss change rate of the model begins to become gentle; When the number of iterations reaches 12, the loss is 8.7756. At this time, the loss change rate of the model is less than 1, and the learning rate of the model drops to 2.5 times 10 (negative 5); When the number of iterations reaches 13, the fine training mechanism is triggered. This mechanism is triggered when the model training reaches a certain degree, making the model training more complete; When the number of iterations reaches 29, the loss of the model is 8.4345. At this time, the loss change rate of the model is less than 0.5, and the model learning rate is automatically adjusted to 2.5 times 10 (negative 6). At this time, the model is close to convergence, but the loss of the model is large.

Take a test picture and let the model test. It is found that there is no prediction box in the result. The possible reason is that the loss of the model is large. The model fails to identify the insulator, resulting in the convergence of the model, but the loss is large. The main reason may be that there are too few pictures for training, and the pretreated pictures are of poor quality for the training of the model. So that the model can't converge to a good result, we should use a variety of image processing methods to process the image, such as increasing noise, image transformation, stretch compression, color change, channel separation, etc. Sufficient data sets with good quality should have more satisfactory results.

8. Conclusion

The main contents of this paper are: the identification and segmentation of insulator string in the image taken by UAV, the detection of insulators in the image and the identification and marking of self-exploding insulators. Through the application of deep learning algorithm and computer vision, the insulator string recognition and segmentation under complex background and the detection of insulator self-explosion fault are realized. At the same time, several deep learning algorithms are proposed.

Firstly, this paper preprocesses the data. Because the given images can't meet the training requirements, the data set needs to be enhanced. The preprocessing of semantic segmentation data and target detection data respectively, and how to deal with the data set.

Secondly, SegNet model is used to identify and segment insulator strings. This paper introduces some concepts of this model, and focuses on the construction of SegNet model. It describes the model structure, model label, model loss function, the classification of each pixel in the image, and identifies the category of each pixel to realize image segmentation. The experimental results show that the SegNet model proposed in this paper can segment the image semantically. Its network model structure is easy to understand and the network training is more comprehensive. It can also realize the recognition and segmentation of insulator strings under complex background, and has high accuracy.

Thirdly, in order to realize the identification and location of insulator string self-explosion fault, the self-explosion insulator position in the image is detected by using the training model. The concepts and structures of Faster R-CNN and Yolo in the deep learning algorithm are introduced. SSD model combines the advantages of Faster R-CNN and Yolo. Therefore, SSD model is proposed to identify and locate the position of insulator self-explosion. Firstly, the SSD backbone network feature is used to extract part of the feature map; Then the prior frame is extracted from the feature map and adjusted; Then, the prior frame and SSD prediction results are transformed into the types of objects in the prediction frame; Finally, in order to be more accurate, the loss function is used to extract the results, which reduces the network leakage rate and recognition rate. The experimental results show that SSD algorithm can improve the speed and accuracy of insulator identification and location.

Acknowledgment

This work is supported by the Open Foundation of the Guangdong Provincial Key Laboratory of Electronic Information Products Reliability Technology, and Basic and Applied Basic Research Foundation of Guangdong Province(202002030228).

Riguang Lin contributed equally to this study

References

[1] PENG Chuang, ZHANG Hongmin, WANG Yongping. Fast Detection Method of Insulator String Image Basedon YOLOv3 [J]. Insulators and Surge Arresters, 2022(01):151-156.
[2] HUANG Xinbo, ZHANG Huiying, ZHANG Ye, LIU Xinhui, XING Xiaoqiang, YANG Luya. Composite insu- lator images segmentation technology based on improved color difference [J]. High Voltage Engineering, 2018, 44(8): 2493－2500. Van Derlofske, J. F., "Computer modeling of LED light pipe systems for uniform display illumination," Proc. SPIE 4445, 119-129 (2001).
[3] B. Yan, Q. Chen, R. Ye, and X. Zhou, "Insulator detection and recognition of explosion based on convolutional neural networks," International Journal of Wavelets, Multiresolution and Information Processing, vol. 17, no. 02, Article ID 1940008, 2019.
[4] P. I. Obi and J. P. Iloh, "Various types OF insulators used IN power system for safe operations OF the transmission lines," International Journal of Innovative Engineering, Technology and Science, vol. 1, no. 1, pp. 20–27, 2017.
[5] Z. Zhao, N. Liu, and L. Wang, "Localization of multiple in sulators by orientation angle detection and binary shape prior knowledge," IEEE Transactions on Dielectrics and Electrical Insulation, vol. 22, no. 6, pp. 3421–3428, 2015.
[6] S. Zhu, Q. Gao, Y. Lu, and D. Sun, "Identification and location of insulator string based on frequency-tuned," Transactions of China Electrotechnical Society, vol. 33, no. 23, pp. 5573–5580, 2018.
[7] Y. Tiantian, Y. Guodong, and Y. Junzhi, "Feature fusion-based insulator detection for aerial inspection," in Proceedings of the 2017 36th Chinese Control Conference (CCC), pp. 10972–10977, IEEE, Dalian, China, July 2017.
[8] Y. Zhai, H. Cheng, R. Chen, Q. Yang, and X. Li, "Multi-saliency aggregation-based approach for insulator flashover fault detection using aerial images," Energies, vol. 11, no. 2, p. 340, 2018.
[9] Z. Zhao, G. Xu, Y. Qi, N. Liu, and T. Zhang, "Multi-patch deep features for power line insulator status classification from aerial images," in Proc. Int. Joint Conf. Neural Netw. (IJCNN), Vancouver, BC, Canada, 2016, pp. 3187–3194.
[10] X. Miao, X. Liu, J. Chen, S. Zhuang, J. Fan, and H. Jiang, "Insulator detection in aerial images for transmission line inspection using single shot multibox detector," IEEE Access, vol. 7, pp. 9945–9956, 2019.
[11] Liu W, Anguelov D, Erhan D, et al. SSD: Single shot multibox detector[C]. European conference on computer vision. Springer, Cham, 2016:21-37.
[12] H. Jiang, X. Qiu, J. Chen, X. Liu, X. Miao, and S. Zhuang, "Insulator fault detection in aerial images based on ensemble learning with multi-level perception," IEEE Access, vol. 7, pp. 61797–61810, 2019.
[13] Z. Ling et al., "An accurate and real-time method of self-blast glass insulator location based on faster R-CNN and U-Net with aerial images," CSEE J. Power Energy Syst., vol. 5, no. 4, pp. 474–482, Dec. 2019.
[14] S. Ren, K. He, R. Girshick, and J. Sun, "Faster R-CNN: Towards real-time object detection with region proposal networks," IEEE Trans. Pattern Anal. Mach. Intell., vol. 39, no. 6, pp. 1137–1149, Jun. 2017.
[15] O. Ronneberger, P. Fischer, and T. Brox, "U-Net: Convolutional networks for biomedical image segmentation," in Proc. Int. Conf. Med. Image Comput. Comput. Assist. Interv., 2015, pp. 234–241.
[16] W. Chang, G. Yang, J. Yu, and Z. Liang, "Real-time segmentation of various insulators using generative adversarial networks," IET Comput. Vis., vol. 12, no. 5, pp. 596–602, Aug. 2018.
[17] X. Tao, D. Zhang, Z. Wang, X. Liu, H. Zhang, and D. Xu, "Detection of power line insulator defects using aerial images analyzed with convolutional neural networks," IEEE Trans. Syst., Man, Cybern., Syst., vol. 50, no. 4, pp. 1486–1498, Apr. 2020.

[18] C. Sampedro, J. Rodriguez-Vazquez, A. Rodriguez-Ramos, A. Carrio, and P. Campoy, "Deep learning-based system for automatic recognition and diagnosis of electrical insulator strings," IEEE Access, vol. 7, pp. 101283–101308, 2019.
[19] V. Badrinarayanan, A. Kendall and R. Cipolla, "SegNet: A Deep Convolutional Encoder-Decoder Architecture for Image Segmentation," in IEEE Transactions on Pattern Analysis and Machine Intelligence, vol. 39, no. 12, pp. 2481-2495, 1 Dec. 2017, doi: 10.1109/TPAMI.2016.2644615.
[20] Ren S, He K, Girshick R, et al. Faster r-cnn: Towards real-time object detection with region proposal networks[C]. Advances in neural information processing systems. 2015: 91-99.
[21] Redmon J, Farhadi A. Yolov3: An incremental improvement[J]. arXiv preprint arXiv:1804.02767, 2018.

Test Technology of Digital Prototype for Electric Resistance Equipment and Its Application in Simulation Platform

Zhijun Zhu [1], Lu Qiang [1], Jian Yang [1*], Siwei Luo [1], Dongchu Su [1]

[1]63870 Troops, Huayin, Shaanxi, 714200, China

[*]Corresponding author's E-mail: ssyn_yangjian@163.com

Abstract. The concept and technical principle of the digital prototype are introduced, and the test content and method of the digital prototype for reactance equipment are proposed. For typical communication countermeasure equipment, the content and essence of the digital prototype are discussed in detail. How to carry out the acceptance test and how to apply it in the simulation platform in the future are also discussed. A simple digital prototype, prototype test platform, and simulation platform are constructed to verify the concepts and methods proposed in this paper.

1. Introduction

A digital prototype is a kind of computer simulation model that can replace physical tests with model simulations in the industrial development stage to greatly improve research efficiency and reduce cost. The qualified digital prototype is called digital equipment in the equipment acceptance stage. All kinds of simulation platforms can carry out drilling and effect deduction with the help of digital equipment.

Digital equipment will be submitted to the simulation platform for drills and deductions. The fidelity of digital and physical equipment in real world performance directly affects the credibility of the results of drills and deductions. A digital prototype is a highly complex software model. The model is prone to be idealized and performs better than the actual installation. Therefore, conducting a comprehensive test on the digital prototype is necessary. The digital prototype test is to compare the response of the digital prototype to the simulation instruction with the response of the physical prototype to the operation instruction, so as to ensure the high consistency of the performance index between the model and the actual device.

This paper introduces the technical principle and test method of the digital prototype of electric resistance equipment. A simple digital prototype, prototype test platform, and simulation platform are constructed as an example of typical communication countermeasure equipment. The basic methods and processes of digital prototype tests and digital equipment applications are introduced.

2. Technical principle of the digital prototype

2.1 Basic concepts

A digital prototype is a complete set of computer simulation systems to verify the comprehensive performance of equipment in advance according to the design scheme and parameters in the equipment development stage. The purpose of the digital prototype is to call all kinds of computer simulation platforms and replace the actual equipment to complete certain performance verification. Its response to the simulation instructions is as consistent as possible with the physical response to the operation instructions. The digital prototype has the same function list and performance index as the physical prototype and has great advantages in terms of operating convenience and maintenance cost. However, its design and implementation are easily idealized, resulting in insufficient fidelity with the physical prototype, which affects the credibility of the simulation results to a certain extent. The comparison between the digital prototype and the physical prototype is shown in Table 1.

Table 1. Comparison of digital prototype and physical prototype

	physical prototype	digital mock-up
mode of occurrence	physical devices	computer file
user	person	testing software, simulation software, etc
environment facing	true natural and electromagnetic environment	the description of the digital environment by simulation software
credibility and cost	participate in the drill results are 100 percent credible but costly	participate in the drill results in high credibility, low cost

2.2 Technical principle

In essence, the digital prototype is a software package that encapsulates the processing logic of various instructions and provides call interfaces to external software (such as simulation software) according to the protocol. The external software loads the digital prototype and sends instructions to it. The digital prototype processes the instructions with the established logic according to its attributes, current status, and other information and outputs the results to the external software. The digital prototype should realize each function item and determine each performance index compared to the physical prototype.

The core of digital prototype construction is to achieve the following two elements:

(1) Business logic

Business logic is the processing of external instructions. In the execution of business logic, the digital prototype should consider its attributes, the current state, the environment, and other factors and give the response consistent with the physical prototype in the real environment. The complexity of business logic determines the fidelity of the prototype.

(2) Interactive interface

The digital prototype will eventually be delivered to various simulation platforms, so the simulation platform should standardize the software interaction interface. Each type of digital prototype realizes the interface to ensure the compatibility of the prototype and the platform. It is understood that there is no clear digital prototype interface standard at present.

This paper argues that the prototype interaction can learn from virtual instrument technology and adopt the SCPI (standard instruction set of the programmable instrument) form; that is, through the transfer of text format commands, the prototype is responsible for command parsing and execution, which simplifies the interface function call operation between the platform and the prototype, and solves the compatibility problem of the test and simulation platform for each unit digital prototype.

2.3 Characteristics of Digital Prototype of Electrical Resistance Equipment

Electronic countermeasure equipment is a complex electronic system based on radio frequency and microwave technology, mainly composed of reconnaissance and interference subsystems. Reconnaissance and interference capabilities are greatly affected by environmental impacts, such as reconnaissance sensitivity of the complex electromagnetic environment and interference distance of low-temperature environments. Therefore, the actual indicators of electronic countermeasure equipment usually deviate from their nominal indicators. In addition, due to large radiation power characteristics and spurious effects, the influence of a complex electromagnetic environment caused by reactance equipment itself on the electronic equipment and even false interference also exist.

It is difficult to predict the nonlinear characteristics of various RF devices in reactance equipment[1]. The fidelity of a digital prototype based only on theoretical or empirical models is far from enough. It is necessary to iteratively optimize the model parameters combined with a large number of measured data. In the design and implementation of the digital prototype of reactance equipment, the electromagnetic environment effect of actual equipment should be fully considered, including the influence of complex electromagnetic environment on performance indicators and their influence on friend equipment[2,3].

3 Digital prototype testing technology

The biggest risk of replacing the physical prototype with the digital prototype lies in its insufficient consideration of the real environment, and the possible performance defects of the equipment are ignored in the simulation process, which affects the evaluation of the deductive results. Therefore, it is necessary to fully test and modify the digital prototype to ensure that its performance on the virtual platform is highly consistent with that of the physical prototype in the real environment. The comparison relationship between the digital prototype and the real prototype is shown in the figure 1.

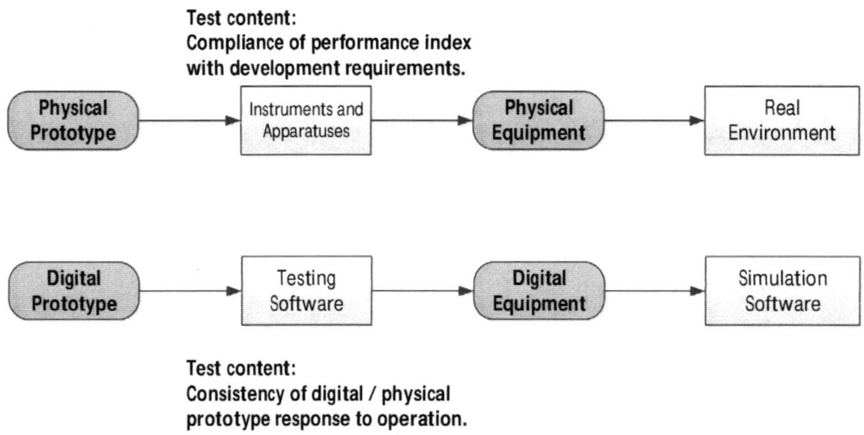

Figure 1. Digital/physical prototype test control relationship

3.1 Performance index test based on test platform

A digital prototype has the same function performance as the physical prototype, and its function performance is reflected in its response to computer instructions. Digital prototype testing is to use professional software to implement prototype loading, instruction sending, response query, and other operations, and judge the fidelity of the prototype by comparing the response of the digital prototype and the physical prototype.

For the reactance equipment, different from the physical prototype test by sending and receiving real electromagnetic signals, the digital prototype interacts with the test platform software by signal description rather than real signals. A signal description is a numerical description of signal parameters, including frequency, bandwidth, amplitude, style, direction, and other information[4].

The test platform software defines the signal description words and calls the corresponding interface of the prototype to execute instructions. The prototype is processed by business logic, and the response information is displayed by the test software. Taking the reconnaissance index test as an example, the interaction process between the test platform and the prototype is shown in Figure 2.

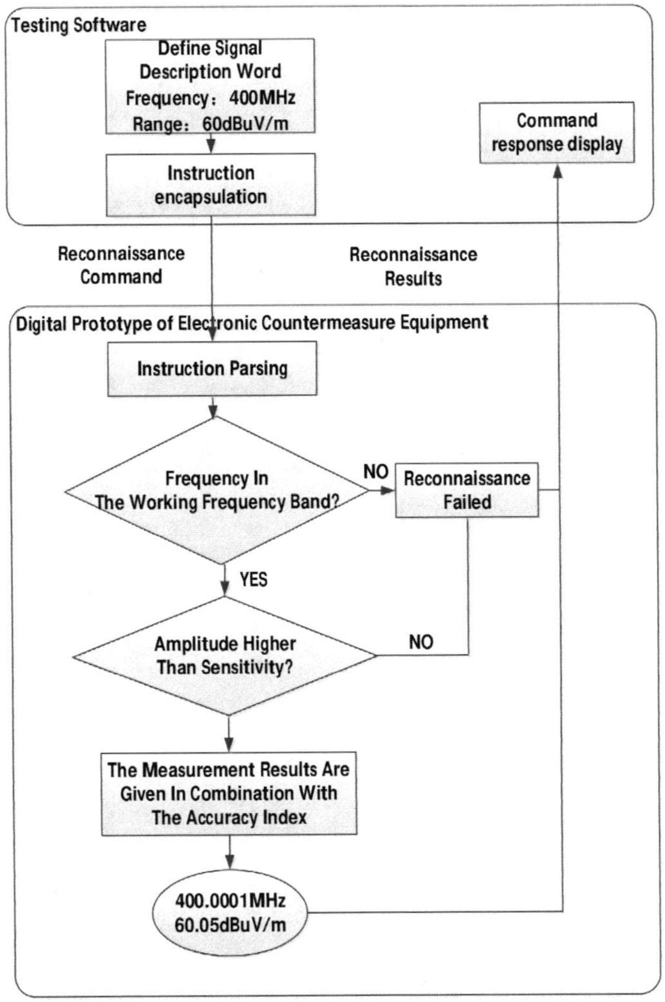

Figure 2. Interaction flow of reconnaissance index test

Digital prototypes with qualified performance indicators (highly consistent with physical prototypes) will be submitted as digital equipment.

3.2 Application Testing Based on Simulation Platform

The simulation platform is the end user of digital equipment, which calls various digital equipment to complete tasks such as drills and activity deduction. In order to ensure the compatibility of digital equipment and simulation platforms, the equipment interface is stipulated by the platform, and the specified software interface must be realized in the construction stage of digital equipment.

The simulation platform software usually performs environmental simulation, digital equipment deployment, instruction issuance, and effect evaluation. The simulation platform first transmits the digital environment (including natural environment and electromagnetic environment) to digital equipment, and the latter completes the adaptability judgment combined with its ability index. Digital equipment processes and responds to various instructions, and the simulation platform evaluates the effect and renders the scene according to the response. The interaction between platform and digital equipment is shown in Figure 3.

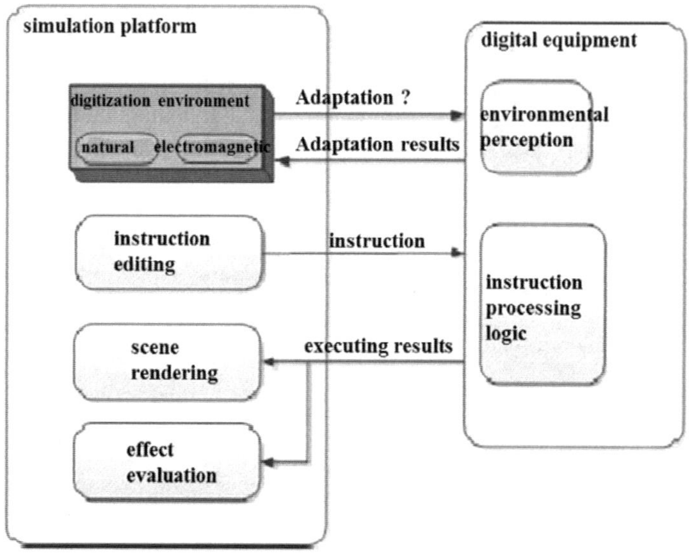

Figure 3. An interactive process between simulation platform and digital equipment

4 Digital Prototype Construction and Test Case

4.1 Digital prototype of typical communication countermeasure equipment

Typical communication countermeasure equipment has the functions of monitoring and direction finding, signal recognition, and interference deployment, and the corresponding technical indicators include reconnaissance sensitivity, direction finding accuracy, and maximum transmitting power[5][6]. A simple prototype with basic reconnaissance and interference functions is constructed. The prototype consists of two files:

(1) Library file. It contains the compiled business logic code and a three-dimensional model file. The three-dimensional model file is used to render the prototype in the simulation software.

(2) ConFigureuration file. It is used to set the working state parameters of the prototype to avoid the high coupling between the prototype and the test software. After the prototype is loaded, the state of the prototype can be dynamically updated by manually modifying the configuration file. This paper uses an XML (extensible markup language) file to store the prototype parameters.

The structure of the prototype is shown in Figure 4.

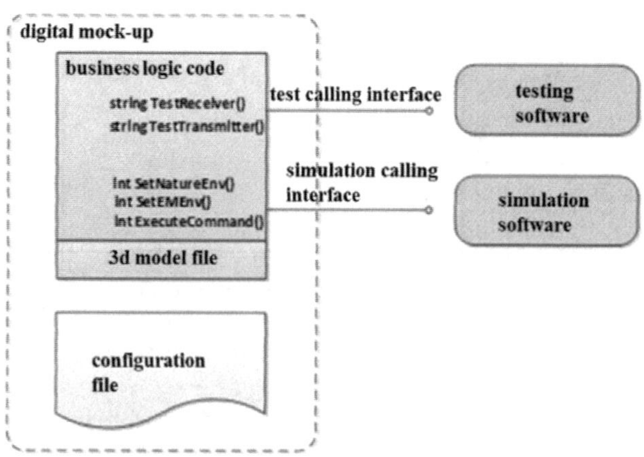

Figure 4. Structure of digital prototype file

4.2 Effect demonstration of test software

A prototype test software was designed and developed to realize the test function of reconnaissance and interference indicators. After loading the prototype file, the response of the prototype was searched by sending the signal description word to test the consistency of the digital prototype and the physical prototype. The software runs as shown in Figure 5.

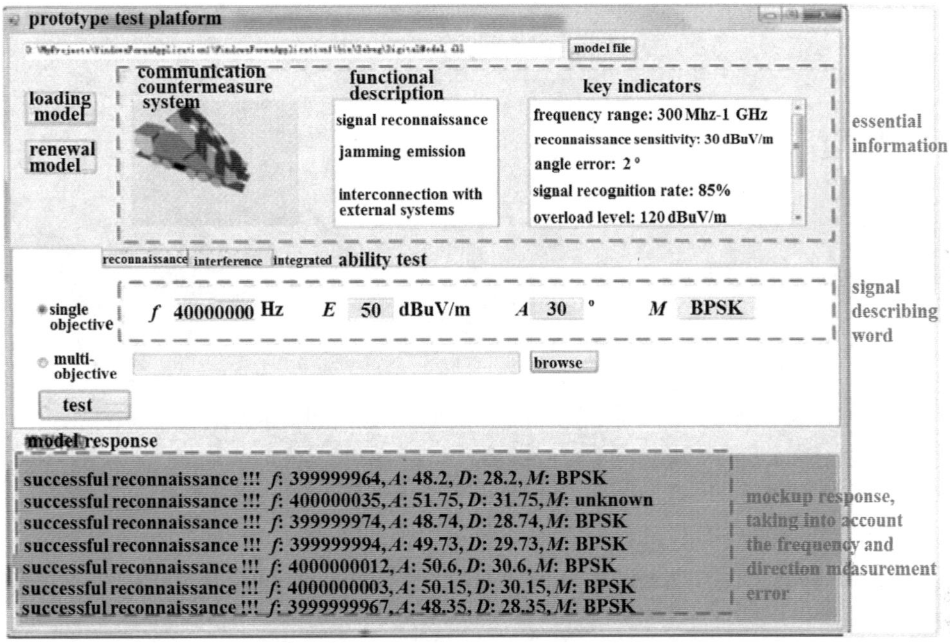

Figure 5. Example of prototype test platform

4.3 Effect demonstration of simulation software

Simulation software is designed and developed to realize the functions of environment simulation, digital equipment deployment, instruction delivery, scene rendering, etc. Setting scenario: positions A and B are deployed with four and two communication stations, and communication countermeasure vehicles are deployed to the designated location, performing reconnaissance and jamming tasks against position B.

The simulation software performs the following operations on digital equipment:

(1) Environmental perception and compatibility judgment

The simulation software firstly realizes the digitization of the environment (this paper focuses on the electromagnetic environment), which calculates the electromagnetic information of each point in space according to the deployment of electromagnetic radiation sources. The equipment perceives electromagnetic information and judges the compatibility of the electromagnetic environment according to its characteristics, such as receiver overload caused by strong signals and stray interference with our equipment. The digital electromagnetic environment is rendered by visualization (field strength thermal diagram), as shown in Figure 6.

Figure 6. Digital effect of the electromagnetic environment

(2) Electromagnetic signal reconnaissance and interference

Equipment perceives electromagnetic information, gives reconnaissance results combined with its indicators, and gives A and B identification combined with pre-stored protection frequency information; interference is carried out on the specified target of position B, and the interference effect (antenna main lobe range, theoretical coating cover distance, etc.) is given. The application effect of digital equipment in the simulation platform (reconnaissance and interference) is shown in Figure 7.

Figure 7. Application effect demonstration of digital equipment in the simulation platform

5 Conclusion

In this paper, the digital prototype and its test method are studied, and the author's understanding of the technical principle of digital prototype, prototype test method, and digital equipment application is expounded. Taking communication countermeasure equipment as an example, a simple prototype system of digital prototype, prototype test platform, and simulation platform is constructed, and the whole process of prototype construction, test, and application is introduced.

Accurate detection of targets under a complex electromagnetic environment is the premise for the effectiveness of reactance equipment, and the influence of electromagnetic environment effect caused by reactance equipment on friend equipment cannot be ignored. When designing the digital prototype of reactance equipment, the electromagnetic environment effect characteristics should be fully considered. In the process of prototype testing, the consistency between the electromagnetic environment effect characteristics and the actual equipment should be emphatically evaluated.

Reference

[1] Liu Delong, Chen Xu, Zhang Chen. (2021) Overview of digital prototype construction of electronic warfare system [J]. Ship Electronic Warfare,44 (5): 12-16.

[2] Hu Changming, Guard Zhong, Wang Changwu, et al. (2017) Exploration and Practice of Digital Prototype for Complex Electronic Equipment Structure [J]. Electronic Mechanical Engineering,33 (6): 1-9.

[3] Cui Xiao, Meng Xiangdong, Ye Lingping, et al. (2021) Airborne radar model simulation and effectiveness evaluation based on digital prototype [J]. Modern Radar,43 (5): 32-37.

[4] Liu Fei, Su Zhou, Jiang Siyu, CAI Zhicheng. Research on damage structure model of ground electronic countermeasures equipment [J]. Modern defense technology,2021,49(05):71-77.

[5] Du Zehai, Xue Lei. Multi-platform cooperative communication electronic warfare equipment training effect evaluation [J]. Journal of ship electronic countermeasures, 2015, 38 (4) : 102-105 + 110. DOI: 10.16426 / j.carol carroll nki JCDZDK. 2015.04.026.

[6] ZHAO Min. Research on Equivalent Simulation Method of Battlefield Electromagnetic Environment Construction [J]. Information Technology and Informatization,2020(06):160-161+165.

Anomaly detection based on temporal convolution Autoencoders

Lihao Wu*[a], Jiahui Liang[a]

[a]Guangzhou City University of Technology, Huadu District, Guangzhou, China, 510800;

[*] Corresponding author: wulh@gcu.edu.cn

ABSTRACT: Due to the rapid growth of the number of sensors in modern industrial society, detecting outliers in time series has become very important. And unsupervised detection of outliers in time series is a very challenging task. For most detection tasks of time series outliers, the autoencoder is one of the main choices. And the recursive network is usually used in the self-coding network structure. However, recent studies have shown that the network structure using dilated causal convolution performs better than the recursive network in all kinds of sequence modeling. In this paper, the encoder and decoder are network structures based on extended causal convolution The time series are reconstructed and then compared with the original data to calculate the distance between them, so as to identify the outliers. We use the Temporal convolution autoencoders to evaluate anomaly data sets in multiple time series. Our results show that the Temporal convolution autoencoders has better anomaly detection ability. In addition, we learned that the combination of Feature Engineering and super parameters will also have a great impact on the results, so ablation experiments need to be carried out carefully.

1. Introduction

In the research of time series anomaly detection, autoencoders are usually used as the standard model. And in autoencoders, RNN [1], LSTM [2] and other cyclic structure networks are used as the standard components. RNN and LSTM are typical sequence modeling network structures, which are famous for their effectiveness in sequence modeling. However, recent research shows that convolution structure can also achieve better accuracy than traditional methods in many sequence modeling tasks, which shows that to obtain the time characteristics of time series, it is not necessary to use a sequence model, and convolution network structure can be considered.

We explore this problem through systematic experiments on various time series anomaly detection data sets. We design an encoder-decoder architecture based on temporal convolution network (TCN) [3]. This architecture reconstructs errors by reconstructing time series, so as to achieve the purpose of anomaly detection. In this process, it maintains simplicity. Moreover, compare with traditional sequence modeling networks such as RNN and LSTM, it combines the best parameter adjustment by the practice of modern deep learning. We believe that the self-encoder can reconstruct the input very well. It should also create a useful and meaningful potential representation.

The results show that the encoder-decoder architecture based on the temporal convolution network is suitable for many different time series anomaly detection tasks, and the results are better than many baseline models. In order to further optimize the self-encoder, we also try to introduce the attention mechanism into TCN.

2. Related work

2.1 AutoEncoder
The autoencoders model is a kind of neural network. Its purpose is to learn a function without supervision, reconstruct the input and compress the data in the process, so as to find a more effective hidden space representation. The network can be seen as consisting of two parts: an encoder function and a reconstructed decoder as shown in figure 1 [4]. The encoder network aims to convert the original high-dimensional input into a potential low-dimensional space, and its input size is larger than the output size. The goal of the decoder network is to reconstruct data from this low dimensional space, and its output layer may become larger and larger. Therefore, the function of encoder network is dimension reduction. In addition, the automatic encoder can optimize the data reconstruction in the code. A good intermediate representation can not only capture potential variables, but also optimize the whole decompression process.

The detection method based on reconstruction error is introduced from the spectral anomaly detection technology, and it is pointed out that the method based on PCA belongs to this method. From the perspective of dimension reduction, it is introduced from the autoencoders, and the advantages are explained: better feature extraction and higher hidden layer can obtain some abstract features.

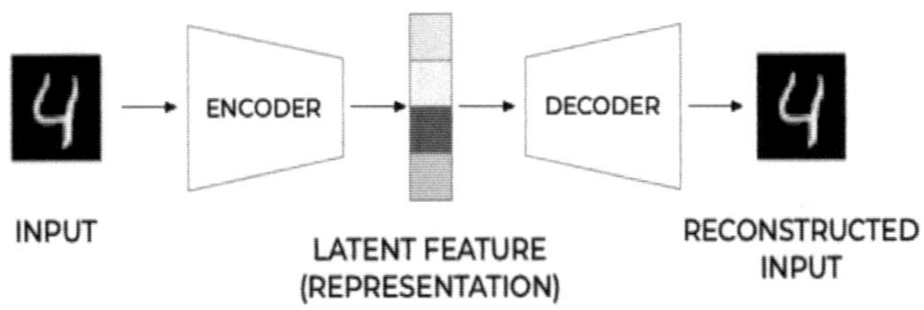

Figure 1. General structure of an autoencoders

2.2 Temporal convolution network
For time series, we can easily think of the following Markov model, where P represents the probability, y represents

$$P\left(y_k \middle| x_k, x_k - 1, \cdots, x_1\right) \tag{1}$$

the output value (label) at time k, and x represents the characteristic value at time k. RNN, LSTM, and other sequence models are designed to process this type of data. In fact, the convolution network can also operate time series data by using the extended causal convolution. TCN is basically a process of expanding causal convolution.

The temporal convolution network has the following advantages.

Firstly, TCN has many ways to change its acceptable field size. For example, stacking more dilated (causal) convolution layers, using a larger dilation factor, or increasing the size of the filter are all feasible options. Therefore, TCNs can better control the memory size of the model and easily adapt to different domains.

Secondly, it has a stable gradient. Different from the cyclic architecture, the anti-propagation path of TCN is different from the time direction of the sequence model. Therefore, TCN avoids the problem of explosion/disappearance gradient, which is a major problem of RNN.

Finally, it has Low memory requirements. Especially when the input sequence is long, LSTMs and GRU can easily occupy a large amount of memory to store some results of their multi-cell gates. However, in TCN, the filter is shared across layers, and the back propagation path only depends on the network depth. Therefore, it is found in practice that gated RNNs may use up to multiple times more memory than TCNs.

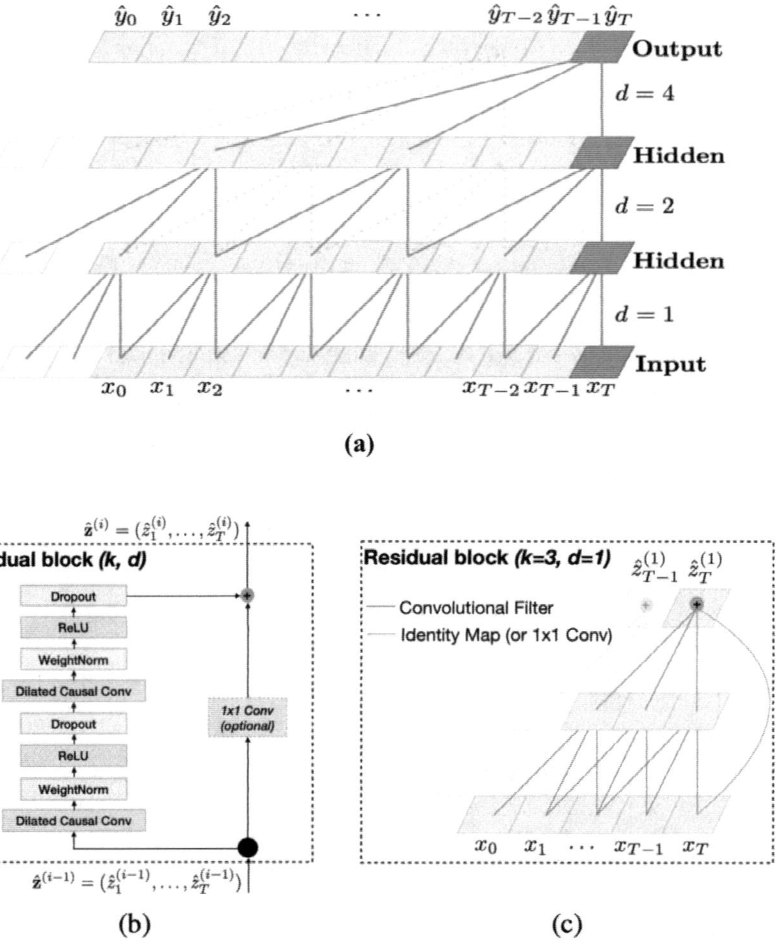

Figure 2. TCN Architectural . (a) All values in the input sequence can be sensed by the receptive field, and the expansion causal convolution, expansion factors d=1, 2, 4, and filter size k=3. (b) TCN residual block. When the remaining input and output dimensions are different, 1 x1 convolution kernel will be added (c) An example of residual connection in a TCN. The filter in the residual function is represented in blue, and the identification is mapped to green. [3]

3. Methodology

3.1 Encoder-Decoder

In most typical encoder and decoder structures, both encoder and decoder are neural networks [5]. The encoder proposed in this paper consists of three modules, each of which is composed of a time convolution network [6]. The decoder consists of two TCN modules, as shown in Figure 3. The encoder can be expressed by the equation (2).

$$h_i = g(X_i) \quad (2)$$

Where h represents the output from the encoder. Its function is to map the feature information into the implicit space, which can theoretically obtain more information in the time dimension. The decoder can be expressed by the equation (3).

$$z_i = f(h_i) = f\left(g(X_i)\right) \tag{3}$$

And the feature dimensions of Z and X are the same. The reconfiguration loss can be defined as equation (4).

$$LOSS = \frac{1}{N}\sum_{i=1}^{N}\left\|X_i - z_i\right\|_2^2 \tag{4}$$

3.2 Abnormal Score

We take the output reconstruction error as the final output, and then normalize it to judge the outliers. The smaller the value, the higher the degree of abnormality. In order to calculate this anomaly score, we use the cosinus similarity from the following the paper [7]， which is shown in equation (5).

$$Anomaly \quad score = 1 - \mathrm{Cos}\, Similarity(x, x) \tag{5}$$

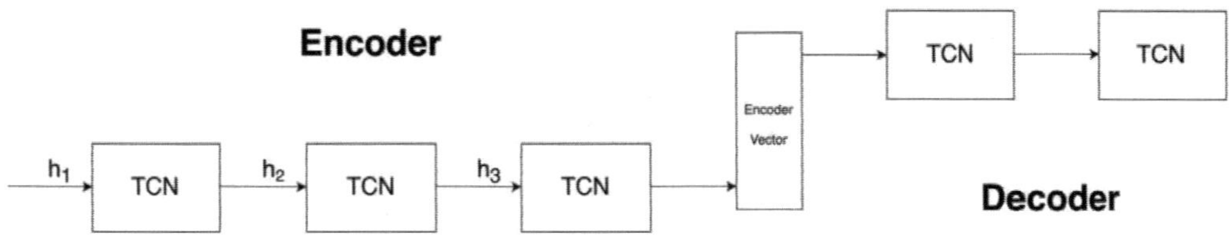

Figure 3. TCN AutoEncoder

4. Experiments

4.1 Experiment platform

The python library used is as follows.

NumPy: It is an open source code base for numerical calculation based on Python. This tool is mainly used for matrix operation. It is much more efficient than the built-in data structure of Python. It supports high-dimensional arrays and multiple matrix operations.

PyTorch: PyTorch is a python version of torch, and PyTorch is a deep learning framework developed by Facebook, which has been open source. Torch is a classic tensor library that operates on multidimensional matrix data, and is widely used in machine learning and other math-intensive applications.

The software and hardware configuration of the experimental environment is shown in the table 1.

Table 1. The software and hardware configuration

CPU	**graphics card**	**Memory**
AMD R9 5900X	3090Ti	32g
operating system	**Python**	**CUDA**
Ubuntu 20.04	3.8	11.3

4.2 Data Sources

We use the KPI exception data of Tsinghua NetMan lab, which is labeled with exception. Because the data set is sorted by time, we use the first 70% as the training set and the last 30% as the test set. And in the experiment, isolated forest, autoencoder based on RNN, autoencoder based on LSTM and autoencoder based on TCN were used to train and learn from data.

4.3 Experimental Design

This paper uses a single dimension of KPI traffic data, which has exception labels and timestamps, and is sorted according to the order of timestamps. The dataset used in this experiment is 3 million, of which outliers account for 2% of the total. And before the experiment, we first preprocess the data set. The preprocessing operations are as follows: (1) Scaling of the time series. (2) Extract the average value, minimum value, maximum value, peak value, offset value and other features.

We first adopt two methods to make the baseline model. The first one is to get a new time series by exponentially smoothing the data, and then use the medium absolute deviation (MAD) to detect anomalies. MAD is a method to detect outliers by calculating the sum of the distances between the observed values and the average values. The second is to use isolated forest to detect data anomalies. The effect of these two models is almost negligible, because the traditional methods consider the distance between data as the core idea, which will have great errors in time series data.

We selected Leaky ReLU, ReLU and Tanh as activation functions. The super parameters of encoder and decoder are selected as Table 2 and Table 3.

Table 2. Encoder parameters

	RNN-AE	**LSTM-AE**	**TCN-AE**
activation	Tanh, Relu	Tanh, Relu	Leaky ReLU, Tanh
hidden_size	300	300	0

Table 3. Decoder parameters

	RNN-AE	**LSTM-AE**	**TCN-AE**
activation	Tanh, Relu	Tanh, Relu	Leaky ReLU, Tanh
hidden_size	300	300	0

In the experimental comparison, we selected the autoencoders based on RNN and the autoencoders based on LSTM for comparison. The encoder part of the RNN based autoencoders is divided into three layers. The first layer is the RNN, the second layer is the full connection layer, the decoder part is symmetrical with the encoder part, and the same is true of the LSTM based autoencoder. The TCN based model is shown in the previous section.

In the training process, we used five-fold cross validation, which refers to dividing the data into five equal parts on average, taking one part of each experiment for testing, and the rest for training. The average value is obtained after five experiments.

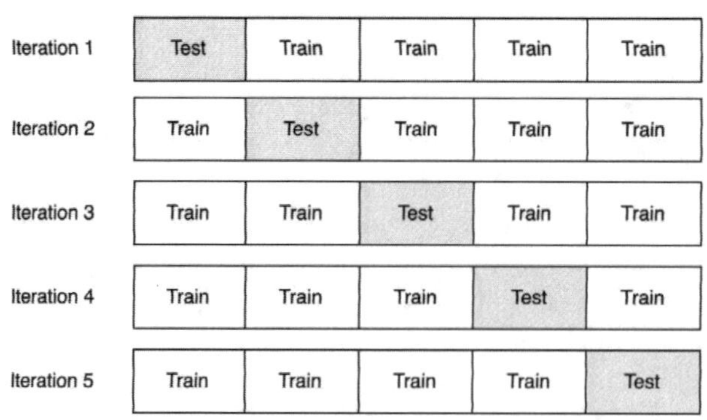

Figure 4. Five-fold cross validation design

Table 4. Comparison test results.

	RNN-AE	LSTM-AE	TCN-AE
precision	61.22%	67.63%	70.14%
recall	67.98%	73.29%	75.98%
F1	64.4%	70.34%	73%

The results show that TCN based autoencoders has better results than RNN or LSTM, especially in terms of precision, which indicates that recurrent neural network is not necessarily the first choice for processing time series data.

5. Conclusion

The topic of this paper is how to use time data more effectively for anomaly detection. The time series network structure is embedded into the encoder and decoder system to reconstruct the data, and then the reconstructed data is compared with the original data to calculate the reconstruction error, so the abnormal points are identified.

Firstly, this paper introduces some concepts of the temporary revolution network, focusing on the advantages of the temporary revolution network in temporal data. Then the general encoder and decoder architecture is introduced.

In order to make the experimental results more robust, we used five fold cross validation. In addition, we use the distance based on cosine to calculate the abnormal score, so as to avoid the error caused by Euclidean distance. After comparing the model based on the temporary revolution network proposed in this paper with the autoencoders based on RNN and LSTM, it is concluded that the model based on the temporary revolution network can better perform on time series data. In addition, we learned from the process of selecting parameters that the combination of Feature Engineering and super parameters will also have a great impact on the results, so ablation experiments need to be carried out carefully.

Acknowledgements

This work is supported by the Open Foundation of the Guangdong Provincial Key Laboratory of Electronic Information Products Reliability Technology, Climbing Program Special Funds(2021b0675), and Basic and Applied Basic Research Foundation of Guangdong Province (202002030228).

Jiahui Liang contributed equally to this study

References

[1] Zaremba W, Sutskever I, Vinyals O. Recurrent neural network regularization[J]. arXiv preprint arXiv:1409.2329, 2014.

[2] Hochreiter S, Schmidhuber J. Long short-term memory[J]. Neural computation, 1997, 9(8): 1735-1780.

[3] Bai S, Kolter J Z, Koltun V. An empirical evaluation of generic convolutional and recurrent networks for sequence modeling[J]. arXiv preprint arXiv:1803.01271, 2018.

[4] Michelucci U. An Introduction to Autoencoders[J]. arXiv preprint arXiv:2201.03898, 2022.

[5] Bao W, Yue J, Rao Y. A deep learning framework for financial time series using stacked autoencoders and long-short term memory[J]. PloS one, 2017, 12(7): e0180944.

[6] Kieu T, Yang B, Guo C, et al. Outlier Detection for Time Series with Recurrent Autoencoder Ensembles[C]//IJCAI. 2019: 2725-2732.

[7] Habler E, Shabtai A. Using LSTM encoder-decoder algorithm for detecting anomalous ADS-B messages[J]. Computers & Security, 2017, 78(sep.):155-173.

Research on Instrumentation Health Status Evaluation MethodBased on Neural Network

Wenrui Jiang[a], Yiyong Lin[*a], Lifen Zhao[a], Yi Wang[a]

[a]Xichang Satellite Launch Center, Sichuan, China, 615000

Corresponding author: 799729599@qq.com

ABSTRACT: Metrology support includes the management of measurement equipment and measurement process, which is an indispensable part of engineering construction. Aiming at the health status of instrumentation in use in enterprises metrological verification, BP and SOFM neural networks are introduced, the characteristics of the two neural networks are introduced, the sample data system and network algorithm are designed, and a set of scientific and complete evaluation methods are produced through algorithm practice and data analysis, which provides direction for the research on the health status of instrumentation in use.

1. introduction

Metrology support is to minimize the possible incorrect measurement results through the management of measurement equipment and measurement process[1]. The number of instrumentation in our enterprise is huge, and the corresponding metrology support work is very important. However, at present, we can only judge whether the equipment is available through periodic verification, and we can't find out its health status in use. Based on this, risk prevention and control can't be carried out. The construction of our project is faced with the problem of possible damage of instrumentation at any time. In recent years, neural network theory has developed rapidly, especially BP neural network has a wide range of applications in solving complex interrelated fields such as pattern recognition and expert system[2], while SOFM neural network is applied to a wide range of fields such as process system analysis, statistical pattern recognition, communication and image coding[3]. In China, Horizon Robotics Technology Co., Ltd. has applied artificial neural network algorithm to vehicle assisted driving system; In other countries, the TruthNorth neuron chip developed by IBM has opened up the precedent of "brain-like" computing[4]. To sum up, this paper introduces BP and SOFM neural networks, selects the pressure gauge as the sample and designs its data system, determines the health status baseline through the optimization algorithm, studies the threshold division based on the clustering characteristics, and finally provides a complete analysis method for the health status evaluation of instrumentation of enterprises.

2. introduction to two kinds of neural networks

2.1 BP neural network

BP neural network, the full name of error back-propagation neural network, belongs to a kind of forward network. The network is widely used in classification and recognition, approximation, regression, compression and other fields[5]. In practical application, about 80% of neural network models adopt BP network or its variation forms[6]. Its basic structure is shown in the figure below.

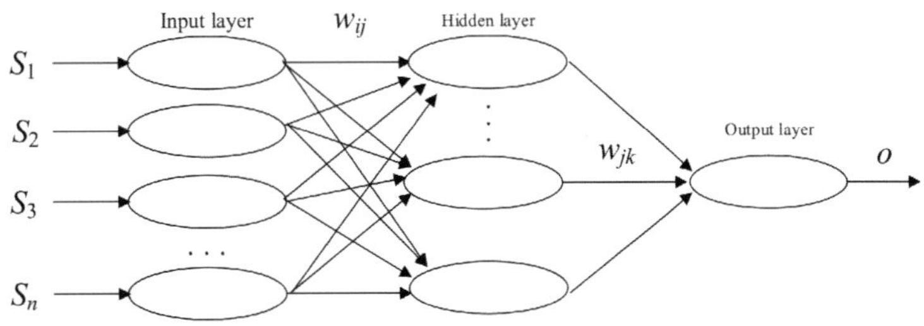

Figure 1. Basic structure of BP neural network

The basic algorithm is:
For sample set

$$S = \{(X_1, Y_1), (X_2, Y_2), \ldots, (X_n, Y_n)\} \quad (1)$$

The network calculates the actual output O_n and error measure E_n according to(X_n,Y_n), and makes the nth adjustment for $W^{(1)}$, $W^{(2)}$,..., $W^{(M)}$ respectively. This process is equivalent to a cycle of processing each sample in the sample set. This cycle needs to be repeated until the sum of error measures meets the requirements of the system for the whole sample set.

2.2 SOFM neural network

SOFM neural network is called self-organizing feature mapping network, which adopts the idea of competitive learning. The biological basis of this neural network is the phenomenon of lateral inhibition between neurons[7]. Its initial structure is as follows:

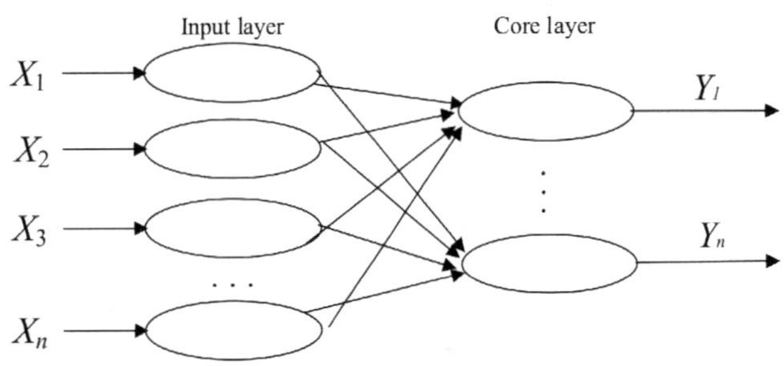

Figure 2. Initial structure of competitive neural network

The basic algorithm is summarized as follows: in a calculation, only one output neuron wins, the winning neuron is marked as 1, and the other neurons are marked as 0. The weights from the input layer to the core layer are given randomly, and finally there will be a neuron with the strongest excitation[8]. The neurons with the strongest excitation are further strengthened in the weight modulation. Through this competition, the distribution information of training samples, i.e. sample categories, can be obtained. When new samples are input, pattern classification can be carried out according to the excited neurons.

3. data system design

As an important part of instrumentation of our enterprise, general pressure gauges are prone to unqualified conditions in use. Therefore, taking the general pressure gauge (hereinafter referred to as pressure gauge) as an example, this paper introduces the process of data reconstruction, which brings universal design method to the data system of various instrumentation.

The verification items of the pressure gauge mainly include the indication error, hysterisis error and friction error under the two strokes of pressure rise and pressure drop. The indication data of pressure's rise and drop directly characterize the performance of the pressure gauge in operation. The record format is as follows:

14	2. Error of indication/hysteresis/friction:							Unit:	MPa
15									
16	Nominal	VOIEAT		PCBAT		Result			
17	Pressure	Boost	Depress	Boost	Depress	Project		Below 90% max range	At&above 90% max range
18	0	0.000	0.000	0.000	0.000	Indication Error	Permission	± 0.0096	± 0.0150
19	0.1	0.102	0.106	0.000	0.000		Max	0.006	0.002
20	0.2	0.202	0.206	0.000	0.000	Friction Error	Permission	0.0048	0.0075
21	0.3	0.302	0.306	0.000	0.000		Max	0.000	0.000
22	0.4	0.402	0.406	0.000	0.000	Hysterisis Error	Permission	0.0096	0.0150
23	0.5	0.502	0.506	0.000	0.000		Max	0.004	0.000
24	0.6	0.602	0.602	0.000	0.000	Result		√ Qualified\ ☐	Unqualified

Figure 3. Original record format of the pressure gauge

Considering the generalization ability of the research method, this paper needs to reconstruct the collected sample data. Pressure gauges in use have different ranges and accuracy levels, and their error levels cannot be directly compared. Therefore, a method for determining the error level is designed:

$$r = \frac{|e|}{|MPE|} \times 100\% \quad (2)$$

Where r represents the relative error based on the maximum allowable error limit, e represents the error value of a single data, and MPE represents the maximum allowable error. The algorithm of maximum allowable error is:

$$MPE = \pm(FS \times \text{LOA}\%) \quad (3)$$

Where FS refers to the value of full scale, which refers to the range of the pressure gauge; LOA indicates the accuracy level[9].

The relative error r can fully represent the error degree of the current pressure gauge, but if you expect to have a clearer performance in network training and result display, the training data samples need to be sorted in two-dimensional or three-dimensional form. Several data of each pressure gauge are summarized into two typical characteristic values, and the calculation method is as follows:

$$r_{11} = Max(r^{(2)}, r^{(3)}, \ldots, r^{(m)}) \quad (4)$$

$$r_{12} = \frac{r^{(2)} + r^{(3)} + \cdots + r^{(m)}}{m} \quad (5)$$

Where m represents the number of verification points. The zero point is removed from the value because the pressure gauge is often set with a stop pin, which makes the zero position error unable to accurately express the error level. r_{11} and r_{12} constitute an array to meet the basic description of the health status of the pressure gauge. r_{11} can reflect the linear error of a single pressure gauge or the nonlinear data jump caused by structural damage, and r_{12} can systematically describe the indication error level of the pressure gauge; The composition of r_{11} and r_{12} is not deeply processed to provide offset free information input for the neural network as much as possible. All data groups also performed preliminary normalization in the above reconstruction process, and the form is simplified as follows:

$$\text{R} = \{(r_{11}, r_{12}), (r_{21}, r_{22}), \ldots, (r_{n1}, r_{n2})\} \quad (6)$$

4. design and implementation of algorithms

4.1 Design and implementation of BP network algorithm

4.1.1 Algorithm design and implementation

A total of 100 samples are used in this round of algorithm, including 50 newly purchased pressure gauges (referred to as new gauges) and 50 old gauges in use. By default, the data of new gauges represent good health, while the other ones represent its health under actual working conditions. 25 groups are selected from the samples as the training data of BP neural network, and the remaining samples will be brought into the network for testing its classification performance.

Firstly, batch training is adopted to input all samples at the same time to calculate the overall error. The network flow is shown in Figure 4.

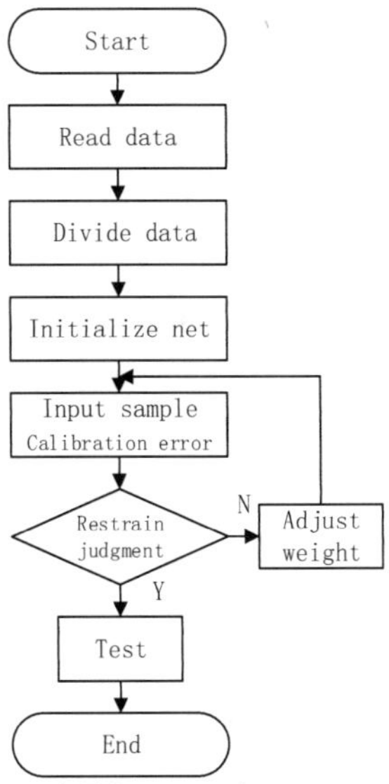

Figure 4. BP network process

1	Num	Attribute	Rn1	Rn2
2	21120254	New	0.375	0.1655
3	210100019	New	0.56	0.248
4	21120152	New	0.48	0.24
5	202202429	New	0.2	0.1175
6	202202449	New	0.5	0.1625
7	202111217	New	0.4	0.08
8	202111293	New	0.5	0.18
9	202202431	New	0.5	0.2125
10	202111288	New	0.5	0.13
11	202111213	New	0.2	0.07
12	202202441	New	0.3	0.06
13	202202444	New	0.5	0.17125
14	02202423[illegible]	New	0.2	0.05
15	202111285	New	0.5	0.23
16	202202456	New	0.4	0.1575
17	202202442	New	0.3	0.13
18	202202432	New	0.5	0.1575
19	202111212	New	0.4	0.175
20	202202448	New	0.1	0.0225
21	202202428	New	0.2	0.085
22	202202463	New	0.1	0.03
23	202202437	New	0.2	0.1075
24	202202433	New	0.4	0.1725
25	202111289	New	0.3	0.1275
26	202202436	New	0.3125	0.1325
27	1237147	Old	0.32	0.168
28	1237101	Old	0.32	0.264
29	1237136	Old	0.5	0.356
30	1237107	Old	0.8	0.58
31	11124330	Old	0.8	0.492
32	11107148	Old	0.64	0.4
33	1607048	Old	0.32	0.196
34	0298016	Old	0.5	0.2625
35	1607059	Old	0.96	0.504
36	YQXF004	Old	0.25	0.18125
37	04112171	Old	0.32	0.296
38	0612415	Old	0.96	0.528

Figure 5. Training data form

(1) Data reading and division. The data information is saved in an XLS format table, listing the number, attribute, r_{n1} and r_{n2} respectively, as shown in Figure 5. Use MATLAB to read the information and save it in data; 1 represents the new and 0 represents the old. Save the attribute in label; Random positive integers are generated by Randperm function to disperse the data, then divide it into two parts for training and testing.

(2) Initialize the BP network and calculate the sample error. The neural network contains a hidden layer, and the training method adopts the steepest descent method including momentum, which is carried out in batch. Since the output value of the output layer is not 0 or 1, the transfer functions of the hidden layer and the output layer use the Log-Sigmoid function:

$$logsig(n) = \frac{1}{1+e^{-n}} \tag{7}$$

Combine the threshold into the weight, write the function to construct the network, initialize the weight to a small random number, and set the number of hidden layer neurons to 3 in order to speed up the training;

In order to offset the possible relative error rate greater than 100% in data, the samples must be normalized again. Move the data to the center of the coordinate axis by subtracting the samples average, and then divide it by the sample standard deviation to standardize the variance. After completion, input the sample into the network and calculate the error.

(3) Judge whether the error converges and adjust the weight accordingly. Define an error tolerance, which is set to 0.01. When the sum of squares of sample errors is less than this tolerance, the algorithm converges; In addition, given a maximum number of iterations, set it to 2000 times, and stop the iteration when it reaches this number.

The weight is adjusted according to the following formula:

$$\Delta\omega_{ij}(n) = \eta\delta_j^i v_I^i(n) \tag{8}$$

Among them, δ^i_j is the local gradient, η is the learning rate. In addition, the steepest descent method with momentum factor is used here. Therefore, except for the first iteration, the weight modification of the previous iteration shall be considered in subsequent iterations:

$$\Delta\omega(n) = -\eta(1-\alpha)\nabla e(n) + \alpha\Delta\omega(n-1) \tag{9}$$

(4) Testing. The network is used for attribute recognition. The sample distribution, error classification, iteration times and accuracy are shown in the following figure:

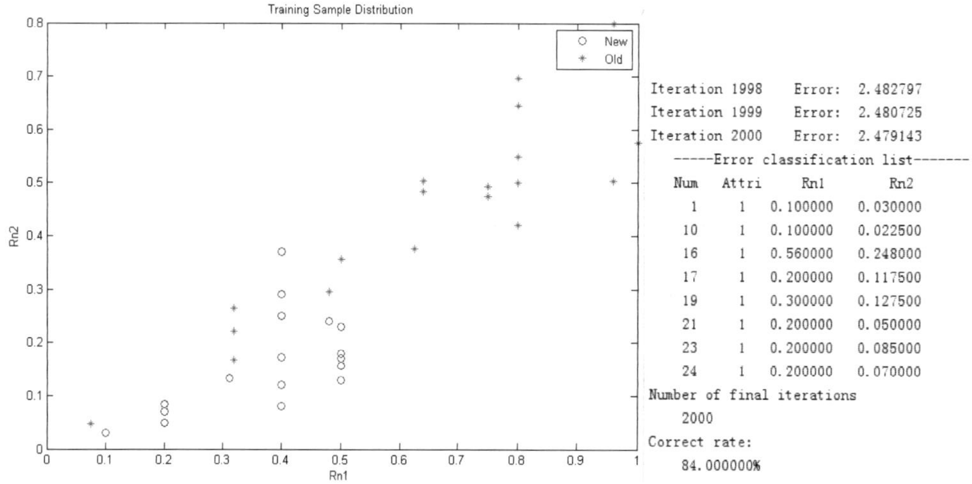

Figure 6. Batch algorithm results

Draw the error decline curve during training:

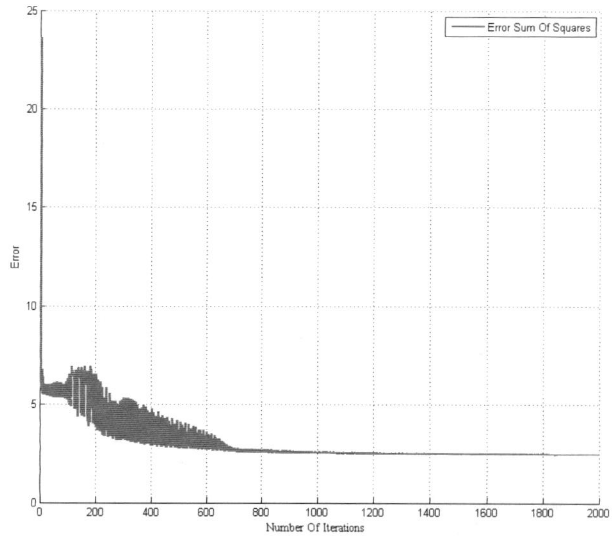

Figure 7. Iterative error decline curve

4.1.2 Algorithm optimization

The above method uses batch training method, which is prone to local optimal solution, resulting in no correction of training. The serial mode inputs the samples randomly one by one, which can avoid this problem to a certain extent, may also improve the division ability of training samples.

The design process of serial training method is similar, and the learning rate, momentum factor and iteration times need to be adjusted continuously in the test. Finally, when the learning rate is 0.1, the momentum factor is 0.2 and the number of iterations is 9000, the step-by-step advantage of the algorithm is brought into full play, and the classification accuracy is 92%.

At this time, the variation curve of training error is:

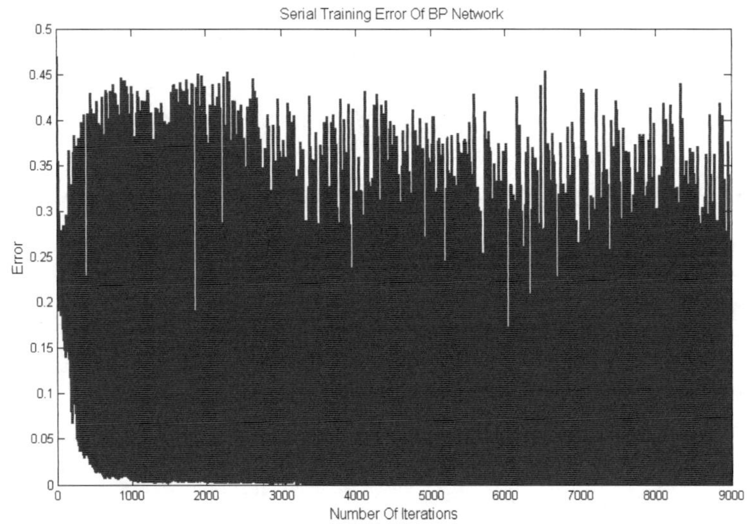

Figure 8. Variation curve of serial training error

So far, the algorithm has achieved 92% classification accuracy and reached a relatively excellent classification level.

4.1.3 Algorithm practice

A batch of pressure gauges submitted for inspection were selected as measured data, with a total of 183. The reconstructed data is brought into the serial algorithm training, and the partition accuracy is 37.7%; When the data is returned to the algorithm in the form of training for reiteration, it is found that the result fluctuates greatly, and the accuracy is directly related to the sampling data of the current old gauges.

4.2 SOFM neural network algorithm implementation

The custom SOFM algorithm still adopts the process of input network construction iterative update training classification, introduces the learning radius, and clusters by using the similarity and topology of the data itself. The specific design process will not be repeated.

Drawing the above 183 groups of data, it is found that the scale of the measured arrays r_{n1} and r_{n2} extends from 0 to nearly 15000 (if it is greater than 1, it refers to an unqualified pressure gauge), so that the image cannot be displayed correctly. Therefore, 15 groups of disqualified are eliminated. Draw the remaining data and cluster it with 8×8 topology network:

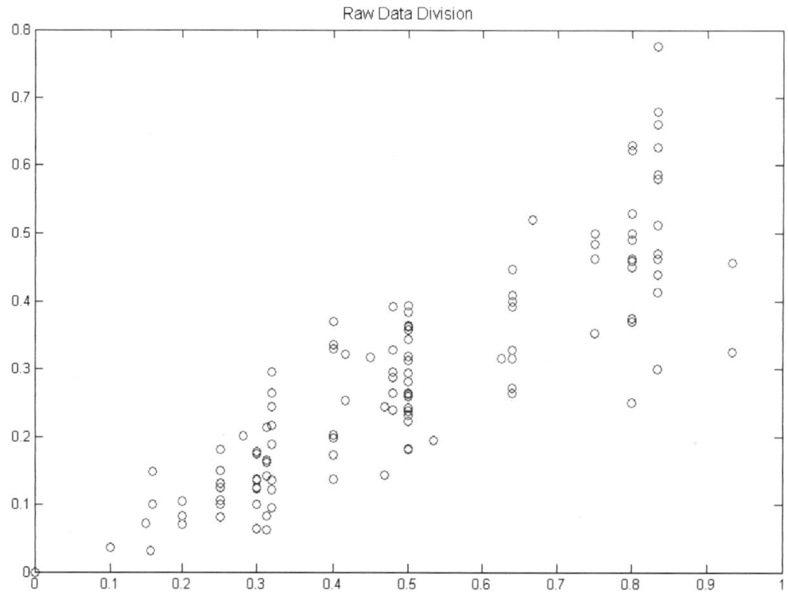

Figure 9-A. Preliminary drawing

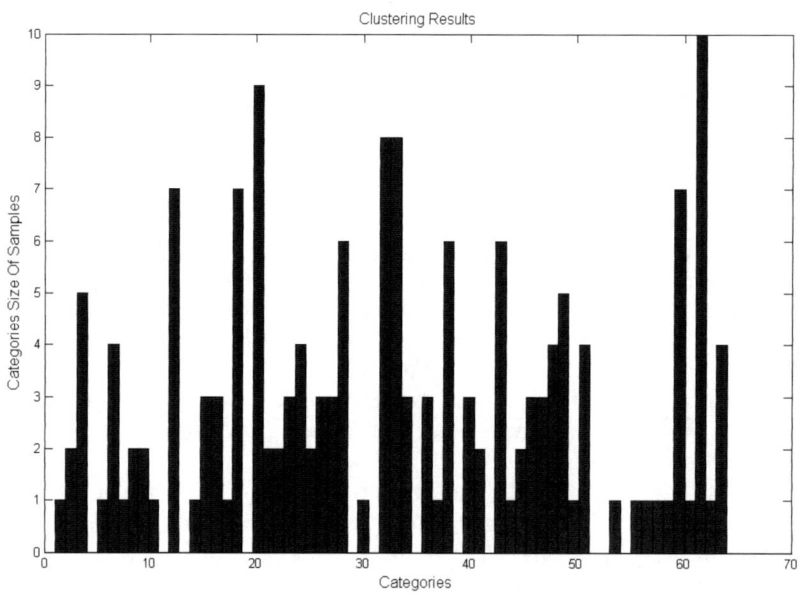

Figure 9-B. Clustering of data

Among the 64 classification frameworks in the figure above, the number of some categories is 1 or even 0, indicating that the dispersion of data does not reach this scale. Considering that the relative error rate scale has 5 areas in steps of 0.2 from 0~1, and $r_{n2} \leq r_{n1}$ can be deduced from formulas (4) and (5), combined with the above dispersion degree, reduce the network scale to 3×5, draw a more intuitive cellular topology, and draw the data center points as follows:

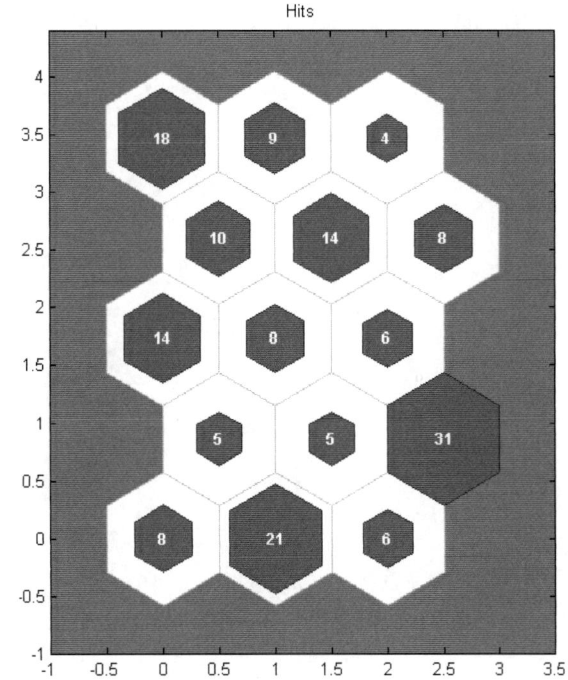

Figure 10-A. 3×5 cellular topology network clustering

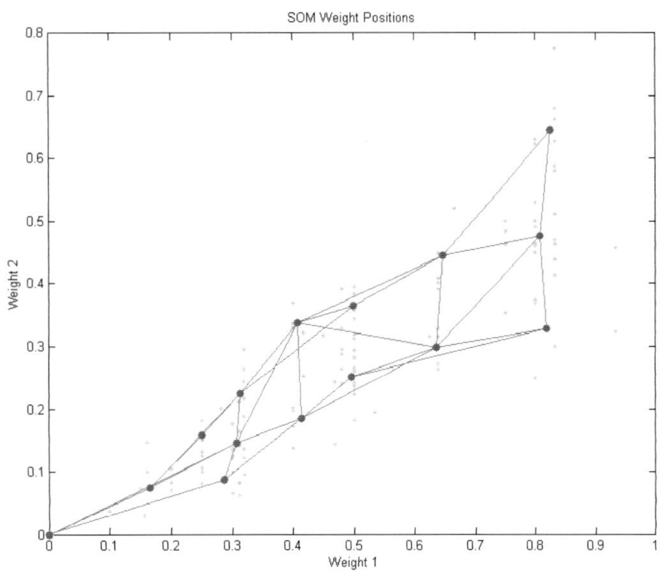

Figure 10-B. 3×5 cellular topology network center point drawing

The gradient formed by the connecting line of the central point does not reflect the health trend of a single pressure gauge, but several radial networks provide the threshold division radius of the health status of the pressure gauges in use.

Again, the competitive neural network algorithm is designed. According to the "New", "Old" and linear data forms defined when looking for the baseline, combined with the usual evaluation methods of "Good", "Medium" and "Bad", the threshold categories are divided into three corresponding ones. Finally, when the number of iterations is 2200, while the learning rate and learning radius decreasing adaptively with the number of iterations, the more rigorous clustering results are as follows:

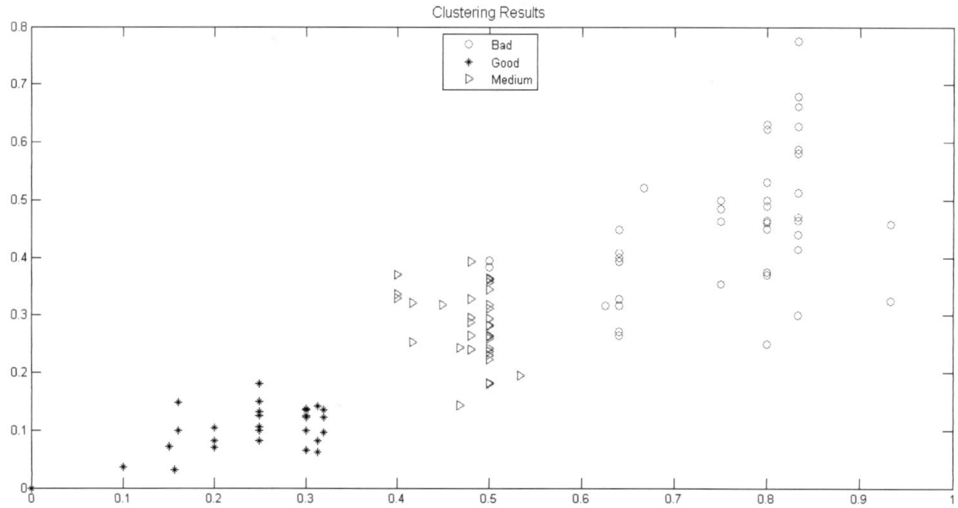

Figure 11. Clustering center point of competitive neural network

So far, the health threshold division of the pressure gauge has been completed.

4.3 Result analysis

Taking the pressure gauge as the starting point, this paper designs the sample data system and brings it into the BP neural network algorithm training to confirm the baseline of health status, and uses SOFM neural network to find the clustering and distribution characteristics. Finally, the algorithm is designed to realize the division of health threshold shown in Figure 11. The specific analysis process is as follows:

(1) The batch algorithm realizes the attribute differentiation of pressure gauge with 84% success rate, as shown in Figure 6; However, Figure 7 shows that with the iteration breaking through 1000 times, the sum of squares of errors tends to be stable, thus guess the algorithm is limited; By observing the distribution map, it can be seen that the data distribution of some old gauges intersects with the new gauges, guess that the linear indivisibility of the data is one of the factors affecting the accuracy;

(2) The optimization algorithm successfully achieves 92% accuracy, but figure 8 shows that it is difficult to continue to improve the algorithm to achieve 100% discrimination ability, the data samples are presumed to be the key reason;

(3) The classification accuracy of the algorithm practice is only 37.7%, but figure 9-A shows that the overall distribution of the test data is highly similar to the training data, and the "wrong" classified data is not a gross error; When the measured data is returned to the serial optimal algorithm as training data for reiteration, it is found that the result fluctuates greatly, and the accuracy is directly related to the sampling data of the current old gauges.

The above situation shows that with the increase of the measured number of pressure gauges in use, the linearly inseparable data is also increasing. It is inferred that there is no logic problem in the design and implementation of the algorithm itself, the accuracy depends on the good amount of data in the old gauges, and there is a baseline for the health status of the pressure gauges in use.

Therefore, SOFM network is introduced to draw the center point, as shown in Figure 10-A/B. The figures reflect the clustering of data quite well. At the same time, several radiation networks of the central points also indirectly prove that the health status can be measured by quantitative threshold. In order to fully demonstrate this assumption, SOFM algorithm is designed again, and the threshold is successfully divided into "Good", "Medium" and "Bad".

To sum up, it is concluded that the health status of the pressure gauge in use can be quantified, and different thresholds can be divided by algorithm to represent the health level.

5. conclusion

In this paper, the health status of the pressure gauges in use of our enterprise is quantified by introducing the neural network algorithm. The research results have excellent generalization ability, and can be nested and analyzed for other types of instruments in use. The research results can be directly used as a direction guide for evaluating the calibration cycle of pressure gauge, and also provide an excellent auxiliary means for the in-situ-calibration of instrumentation.

Of course, due to the incomplete sampling coverage in this paper, there is a certain deviation between the drawing of clustering threshold and the actual situation prediction, and it has not been extended to all kinds of instruments for actual measurement. In the next step, we will continue to explore the algorithm design, continuously enrich the simulation means, improve the accuracy of the algorithm, and make continuous in-depth breakthroughs in the metrology support work.

References

[1] ISO10012: Measurement Management System - Requirements for Measurement Process and Measurement Equipment [M] International Organization for Standardization. 2003 (4)
[2] Yi Yuan, Hongyu Zhao Application of BP Neural Network in Equipment Maintenance Support Capability Evaluation [J] Computer and Information Technology, 2009,17 (5): 28-31;
[3] Zhanhua Yang, Yan Yang A Document Clustering Algorithm Based on SOFM and K-Means [J] Computer Application Research, 2006,23 (5): 73-74
[4] Yuzhe Wang Research on Bionic Artificial Neural Network Simulator [D] Beijing: Tsinghua University, 2017
[5] Matlab Neural Network Principle and Example Refinement [M] Tsinghua University Press. 2020 (4);
[6] Zongli Jiang Introduction to Artificial Neural Networks [M] Higher Education Press. 2001 (8);
[7] Rumeng Yi Theory and Application of Artificial Neural Network [J] Science and Technology Economic Guide. 2015 (18)
[8] Xin Wen Et Al. Matlab Neural Network Simulation and Application [M]. Science Press, 2003 (7)
[9] JJG 52-2013 Elastic Element General Pressure Gauge, Pressure Vacuum Gauge and Vacuum Gauge [M] General Administration of Quality Supervision, Inspection and Quarantine. 2013 (12).

A Non-contact Heart Rate Measurement Approach Based on Modified EEMD Algorithm with Updated White-noise Iterating

Yingli Shi[1,2], Jian Qiu*[1,2], Li Peng[1,2], Peng Han[1,2], Kaiqing Luo[1,2], Dongmei Liu[1,2]

[1]School of Physics and Telecommunication Engineering, South China Normal University, Guangdong, Guangzhou 510006, China

[2]Qingyuan Institute of Science and Technology Innovation, South China Normal University, Guangdong, Qingyuan 511517, China

* Corresponding author: qiuj@scnu.edu.cn

Abstract. At present, Imaging Photoplethysmography (IPPG) has attracted extensive attention on heart rate measurement, because of its non-contact, non-invasive and safe characteristics. However, in reality, the IPPG signal is extremely feeble and susceptible to be submerged by ambient interference such as illumination. To this end, we propose a new IPPG signal processing approach to enhance the anti-interference capability of measuring heart rate through a modified EEMD algorithm with updated white-noise iterating. In the modified EEMD algorithm, based on the standard deviation of the input IPPG signal, zero-mean white noises are generated randomly for iterative ensemble averaging. Through this new method, the fault caused by the residual white noise of traditional EEMD is solved. Lastly, through the comparative experiments of our own acquisition statistics, the approach proposed by this paper has significantly improved the performance in terms of accuracy, timeliness and anti-interference under various scenarios. The mean absolute error reaches 1.4 beats per minute and the correlation coefficient r reaches 0.8089. In addition, the modified EEMD algorithm reduces the running time by more than 10 times compared to the original method, approximately 0.16s per time.

1. Introduction

As a result of the widespread prevalence of adverse lifestyles such as unhealthy diet, inactivity and cigarette smoking, the morbidity and mortality rates of cardiovascular diseases are escalating rapidly across the countries, with approximately 330 million patients currently suffering from cardiovascular diseases. The morbidity and mortality rates of cardiovascular disease in our country remain the highest, at 46.66% in rural areas and 43.81% in urban areas [1]. Therefore, daily monitoring and early prevention of cardiovascular disease are of great importance. Regular heart rate measurement can play an active role in the daily monitoring and diagnosis of cardiovascular diseases. Currently, the most widely used heart rate measurement methods include electrocardiography and photoplethysmography. However, they have to be in contact with the surface of the skin so as to measure the heart rate parameters of the subjects, which may bring about discomfort and skin damage to new-born babies, people with skin allergies and skin injuries. In view of the limitations of traditional contact methods, non-contact heart rate measurement has emerged. Of the available non-contact heart rate measurement approaches, imaging photoplethysmography (IPPG) offers the advantages of non-invasive, simplicity, and measuring security as well as comfortable, which makes it better applicable for routine heart rate monitoring than others. Moreover, the IPPG can also be applied to exceptional scenarios such as telemedicine, neonatal health monitoring, fatigue driving monitoring, emotion recognition, smart classroom, and detective polygraph [2-4].

In 2008, Verkruysse [5] et al. were the first to extract blood volume pulse signal (BVP) from facial images utilizing ambient illumination directly, and assumed that the green channel of the image was stronger than other channels in respect to the robustness of the BVP signal. However, they disregarded the environmental interference such that the impact of noisiness was not cancelled in the extraction proceeding, contributing to the inferior qualities of the eventually extracted signals. As the captured signal is extremely faint, external interference such as ambient luminance [6] is inevitable under conventional situations, where the BVP signal could be annihilated easily by noise. Hence, following the research of Verkruysse, the issue of minimizing exterior circumstantial interference and extracting the BVP signal accurately and speedily emerged as a hot topic for scholars investigating IPPG [7-8]. In 2010, Poh [9-10] et al. attempted to extract the heart rate source signal from the RGB signal of a facial image with the independent component analysis (ICA), which effectively averted the impact of ambient interference over its quality and strengthened the robustness of the source signal. In 2013, Haan [11] et al. instead proposed a chromaticity-based color space projection decomposition algorithm (CHROM) to construct the target heart rate signal. Those approaches yielded relatively positive performance under ideal illumination conditions, nevertheless, they did not address the inaccurate heart rate estimation due to illumination interference in a realistic scenario. A subsequent study has proposed using the background region as a reference to eliminate the noise caused by illumination variations in view of better measurement accuracies in realistic scenarios. In 2014, Li et al [12] proposed a framework to offset the impact of object locomotion and illumination variations utilizing facial tracking and normalized least mean square (NLMS) algorithm, which was validated on the public database MAHNOB-HCI. In 2018, Sang et al [13], then proposed to employ both background and facial region segmentation with NLMS algorithm to perform illumination rectification on IPPG signals. Subsequently, Cheng et al [14] in 2020, also taken the non-skin background region as a reference and proposed a joint blind source separation and ensemble empirical mode decomposition (EEMD) method to eliminate the illumination nuisance. Yet the above approaches with background region as reference still suffer from the restriction that the background and facial areas with inconsistent illumination circumstances likewise generate HR measuring mistakes. Besides, Chen [15] et al. proposed an EEMD-based heart rate measurement approach in 2014 in order to boost the immunity to ambient interference, demonstrating the feasibility of EEMD to mitigate ambient illumination interference. Subsequently, on the basis of the demonstration by Chen, in 2015, Khan et al [16] employed EEMD to eliminate the effects of noise such as motion and illumination, as well as increased robustness to deviation errors by recursive least squares filtration and temporal extraction approaches. In 2021, Yin et al [17] performed a logarithmic calculation on images with different ROI for each frame to segregate the reflection component and light noise so as to eliminate illumination interference, followed by EEMD decomposition of the reflection component to obtain the optimal IMF component as heart rate signal. Zheng et al [18] were able to obtain a purer heart rate signal by combining POS and EEMD processing on the RGB signal of the facial image after segmentation by Pixel2Pixel. Nevertheless, the original EEMD algorithm they used yielded discretization errors due to the fact that the white noise would not be fully averaged out by the iterations.

Thus, synthesizing the works above, in order to solve the problem of decomposition error generated by the white noise of the original EEMD algorithm, this paper proposes a new method of heart rate measurement based on the modified EEMD algorithm with updated white-noise iterating to improve the accuracy and speed of extracting heart rate signals under real conditions. In this paper, by adding zero-mean white noise generated according to the input signal at each data point and iterative ensemble averaging, the white noise at each data point is ensured to be zero, which not only eliminated the errors brought by the white noise of EEMD algorithm to a greater extent, getting intrinsic mode functions (IMF) with higher content of heart rate components, but also decreased the number of set iteration averaging and raised the efficiency of the algorithm. Lastly, the validity and feasibility of the modified approach were verified by the data collected by itself, which also demonstrated the apparent performance of the proposed approach in terms of its accuracy, temporal validity and anti-interference.

2. Methods

2.1 IPPG

The IPPG method captures periodic color variations in body skin due to the cyclic flow of capillary blood brought about by the pumping of the heart through an imaging equipment, thereby obtaining physiological information about the body such as heart rate. Figure 1 shows the flowchart of heart rate measurement by IPPG technology. Firstly, for the purpose of mitigating movement interference, the face image captured by the camera was adopted by the face detection algorithm so as to acquire the entire face region, and the cheek region was intercepted to be the region of interest (ROI). Next, the ROI region image was pixel averaged after green channel separation to obtain the original pulse wave signal. Finally, the original pulse wave after noise removal was subjected to frequency domain analysis for getting the final heart rate value. However, the variation of ambient illumination during the measurement directly affects the color change of the skin surface, leading to large errors in the measurement data. Thus, the original pulse wave requires filtration to eliminate noisy interference like illumination. Combining the current status of domestic and international studies, this paper utilizes a modified EEMD algorithm to eliminate the noise caused by ambient interference and extract a much more robust heart rate signal.

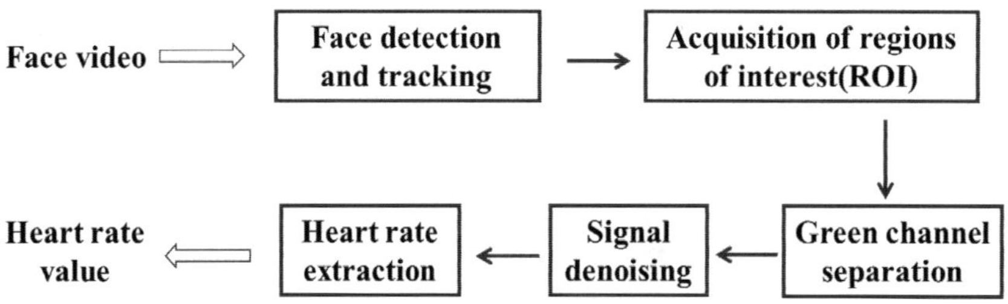

Figure 1. Flow chart of heart rate measurement by IPPG.

2.2 Regions of Interest Acquisition

Selecting appropriate ROI can effectively enhance the ruggedness of heart rate detection by IPPG [19-20]. The ROI ought to be capillary-rich, for instance, the frontal region of the face, the cheeks on the left and right, the lips and the chin, etc. However, the forehead is prone to be obscured by bangs and the brim of the hat, whereas the eyes and lips, etc., will generate apparent involuntary motion interference, adversely impacting the subsequent handling. Therefore, in this paper, after utilizing the 68 feature points of Dlib library face inspection algorithm to recognize and track the facial region of the face of the subjects, combined with the previous studies and experimental experience, the cheek region was integrated as the ROI. As shown in Fig. 2, the feature points in the left and right cheek regions were sequentially joined with the corresponding feature points in figure b in accordance with the serial number in figure a, forming a cheek polygon region and adopting it as the ultimate ROI.

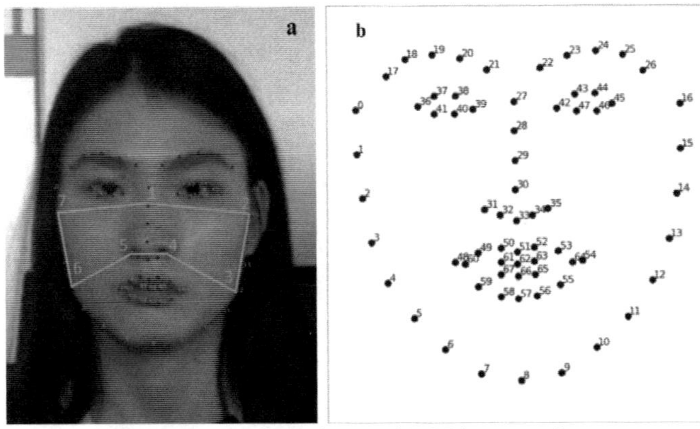

Figure 2. Face detection and ROI interception.

2.3 EEMD

Empirical mode decomposition (EMD) is an adaptation time-frequency dimension analysis approach, which overcomes the wavelet transform basis function maladjustment problem, and is now extensively applied to physiological signal analysis. It is decomposed into a series of intrinsic mode function (IMF) components with different frequency bands in accordance with stationary terms, and by removing the parts associated with interference, a much higher fraction of the objective component is acquired. Nonetheless, there is mode conflation in EMD algorithms. To address the mode conflation problem, Huang proposed the EEMD [21], which utilizes the zero-mean characteristic of white noise and appends Gaussian white noise with zero mean to the entire time series for the auxiliary analysis, for which the result of the ensemble mean is adopted as the ultimate output.

Prior to the decomposition of the signal by EEMD, it is essential to establish the ratio coefficient α of the added auxiliary white noise standard deviation (RSTD) to the input signal standard deviation (YSTD), as well as the number of white noise additions N. The α and N fulfil the statistical formula: $\varepsilon = \alpha/\sqrt{N}$, ε is the maximum relative error of the decomposition of the expectation signal. Usually, α = 0.2 with an ensemble average number N taken 100 times, the decomposition error ε is approximately 1%. the fundamental steps of the EEMD algorithm are listed below.

Step 1: Repeatedly adding N groups of Gaussian white noise sequences $w_{k(t)}$with a mean of 0 to the signal $X_{k(t)}$,$w_{k(t)} = \alpha * \text{YSTD} * \text{randn}(t)$, we get:

$$X_{k(t)} = x(t) + w_k(t)(k = 1,2,\cdots N) \tag{1}$$

Step 2: The signal added with white noise is taken as a whole, and then EMD is decomposed. where $I_{ik}(t)$、$C_{kn}(t)$ are expressed as IMF components and residual components after decomposition.

$$X_k(t) = \sum_{i=1}^{n} I_{ik}(t) + C_{kn}(t)(k = 1,2,\ldots,N) \tag{2}$$

Step 3: Repeat steps 1 and 2, adding a new Gaussian white noise sequence each time.

Step 4: The IMFs obtained each time are integrated and averaged as the final result.

$$I_i = \frac{1}{N}\sum_{k=1}^{N} I_{ik}\,, (\mathrm{i} = 1,2,\ldots,n) \tag{3}$$

2.4 Signal Extraction

Only a minority of the IMF components handled by the EEMD algorithm correlate closely with the BVP signal. In order to accentuate the signal characteristics, effective IMF components are necessitated to be elected. The majority of human heart rates can be captured in the range of 50-180 beats/min, which corresponds to the frequency interval of 0.8 HZ-3 HZ. Hence, by establishing the IMF component selection mechanism, the effective signal with strong components in the heart rate range can be acquired. At first, the cumulative total of spectral amplitudes in the frequency interval corresponding to the heart rate is computed. Second, the component with the maximum spectral area sum (IMFMax) along with the component larger than one-half of the maximum area sum (IMFhalf) are sifted out. Later, the total of the screened components is adopted as the reconstruction signal, in case there is no IMFhalf, the IMFMax is taken as the destination reconstruction signal. Eventually, the heart rate value corresponding to the reconstructed signal is calculated by Fourier transform. To further assure the accuracy of heart rate measurement, the average value of the first 20s is regarded as the heart rate value at the present moment of the paper.

Figure 3 shows all the IMFs and their corresponding spectra formed after the decomposition by EEMD of the original pulse signal. As shown in the figure, in accordance with the filter criteria of the IMF component selection mechanism, the IMF selection objects become IMF3 and IMF4, with the cumulative sum of the corresponding spectral amplitudes in the heart rate range of 0.2536 and 0.1396 (>0.5 × 0.2536). Consequently, the total sum of IMF3 and IMF4 signals was taken as the ultimate reconstructed signal. Under such circumstances, as shown in Figure 4, the maximum amplitude of the frequency of the reconstructed signal was 1.317 Hz, and its corresponding estimated heart rate was about 79.02 bpm.

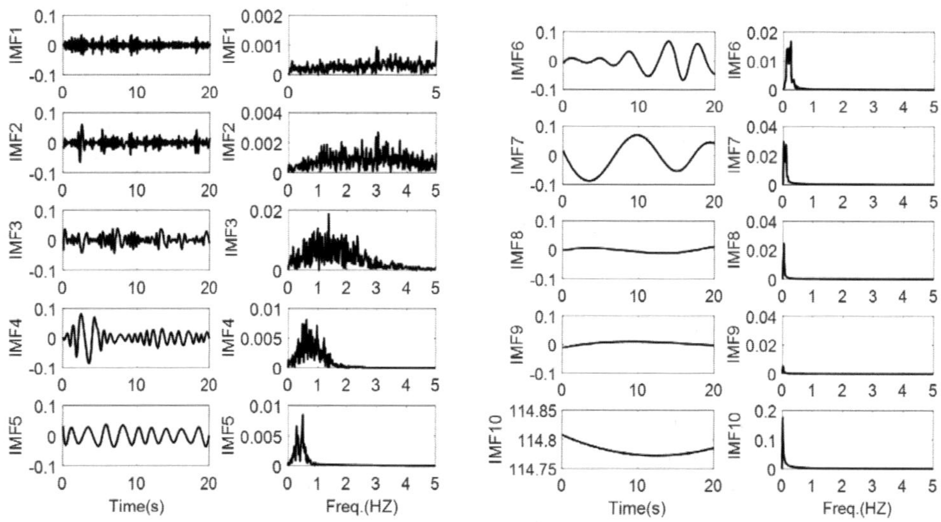

Figure 3. The figure shows all IMF components derived from the original pulse signal after decomposition by EEMD.

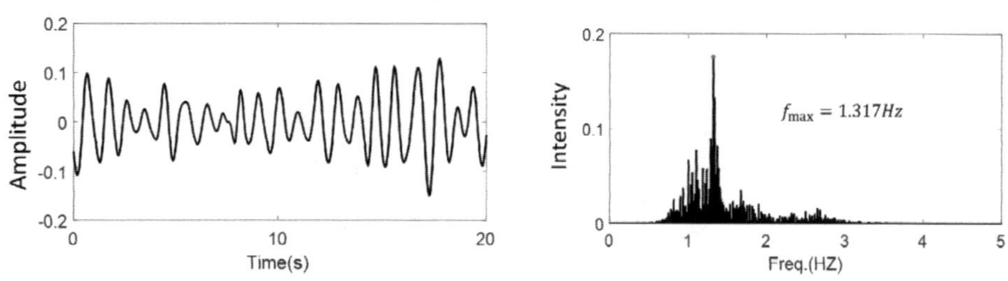

Figure 4. The figure shows the reconstructed signal formed by the EEMD decomposition of the original pulse signal and the corresponding spectrogram.

3. Modified EEMD

Since the EEMD algorithm adds white noise over the entire time series, the white noise at each data point grows as α increases, moreover, the white noise is not fully neutralized even though the number of ensemble averages continues to rise. Compared with the input ROI green channel signal, as shown in Figure 5, the signal after ensemble averaging by the original EEMD algorithm will have residual noise and inject additional energy.

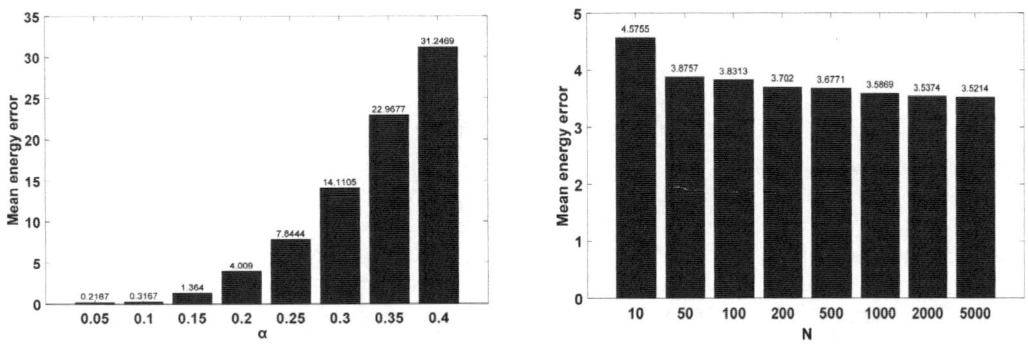

(a) (b)

Figure 5. Energy error before and after decomposition of the EEMD algorithm. Fig. (a) shows the reconstruction energy error corresponding to different α when the number of set iterations is 100, and Figure (b) shows the reconstruction energy error corresponding to different N when α is 0.2.

For the purpose of addressing the error problem brought by residual noise, firstly, different zero-mean white noise signals are generated according to the input signal standard deviation in our work, then the auxiliary white noise signal is appended to each data point for collective iterative averaging, which not only ensures that the white noise of the whole signal is 0 after iterative averaging, but also enables the white noise of each data point of the signal to be 0, reducing the residual noise and error and further improving the signal-to-noise ratio of the signal . Moreover, in general, the fixed choice of α = 0.2 and N = 100 for different signals will enlarge the actual decomposition error and computing time. As a consequence, under the condition that the maximum decomposition error ε is not more than 1%, it is demonstrated in the paper that the change in the size of the N parameter value of the modified method has no effect on the heart rate results through relevant experimental data, therefore, considering the accuracy and real-time, N inappropriate to be too large, which is why the value range of N is set to [1,10]. The processing flow of the modified EEMD algorithm in this study is illustrated as follows.

Step 1: Input the green channel signal in the ROI region and calculate the signal amplitude standard deviation YSTD.

Step 2: Set the maximum error of decomposition ε to 1%, consider the compromise by taking N to be 4, and obtain the size of α by combining $\varepsilon = \alpha/\sqrt{N}$.

Step 3: Calculate the standard deviation RSTD of the added auxiliary white noise according to α= RSTD/YSTD; as well as build a Gaussian random number set with mean 0 and standard deviation RSTD, where the number of random number sets is equal to the length T of the input green channel signal.

Step 4: One at a time, a randomly selected non-repeating value from the Gaussian data set is multiplied by a ratio factor α to form a Gaussian white noise $w_k(t)$.

Step 5: Add a white noise sequence $w_{k(t)}$ to each sampled data point in the original green channel pulse signal $X_G(t)$.

$$X_{G_{k(t)}} = X_G(t) + w_k(t), (k = 1,2, \cdots \mathrm{T}) \tag{4}$$

Step 6: Decompose the signal $X_{G_{k(t)}}$, which has been added with white noise, as a whole by EMD to obtain each IMF component and residual component.

$$X_{G_{k(t)}} = \sum_{i=1}^{n} I_{ik}(t) + C_{kn}(t), (k = 1,2, \dots, T) \tag{5}$$

Step 7: Repeat step 4, step 5 and step 6 for N times, adding a new white noise sequence each time, and finally take the IMF obtained each time as the final output result after the ensemble averaging.

$$I_i = \frac{1}{N}\sum_{j=1}^{N} I_{ik}\,,(\mathrm{i} = 1,2,\ldots,n) \quad (6)$$

4. Experiments and Discussion

4.1 Experimental setup

15 healthy adult subjects, 10 males and 5 females, with a mean age of 26.2 years and an age range of 18 to 46 years, were recruited for the test， to validate the feasibility and utility of the IPPG technique for individual surveillance of self-risk for cardiac activity. All participants gave informative consent to the test experiment. For the purpose of simulating complex real-life daily scenarios, the experimental data were captured in both daytime and night-time modes. Figure 6 shows the experimental setup with a double-armed lamp positioned in front of a table at a vertical height of approximately 1.5 m. The webcam (Logitech, C920 Pro HD) was utilized to acquire video data, which was mounted on a Lenovo Xiaoxin PRO 14 ACH laptop with automatic exposure and white balance disabled. All videos were recorded at a frame rate of 30 frames per second and at a resolution of 1920 × 1080. The subject was positioned approximately 1 m from the camera, blinking and breathing naturally. Simultaneous heart rate measurements were performed with a finger-clamp pulse oximeter (Beijing Chaos, MD300C) as an experimental reference while video was captured, totaling 45 videos were taken for experimental data manipulation.

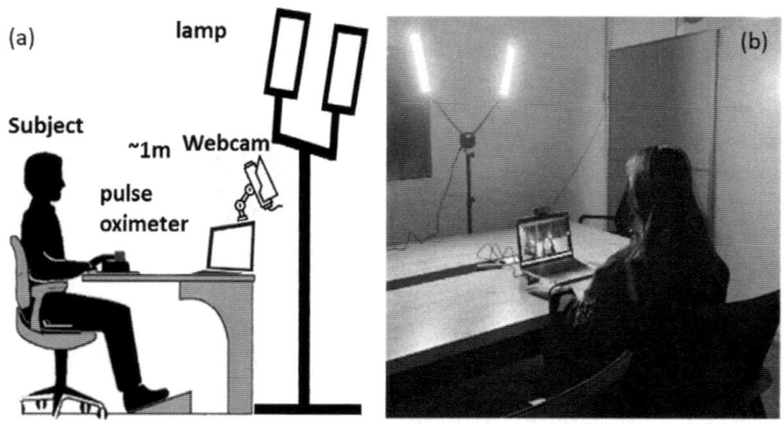

Figure 6. Diagram of the experimental measurement setup. Figure (b) shows an example of the experimental scenario in night mode.

4.2 Evaluation Indicators

Mean absolute error (MAE), mean absolute percentage error (MAPE), standard deviation (SD) and Pearson's correlation coefficient r were adopted as evaluation indexes to reflect the deviation of the measurement results from the reference heart rate and verify the accuracy of the measurement results. Furthermore, the signal-to-noise ratio (SNR), which can be briefly depicted as the ratio of the effective heart rate component to the noise component in the signal set, is also utilized to gauge the performance of the modified method. The relevant evaluation indexes are shown below.

$$SNR = 10\,log_{10}\left(\frac{\sum_{i=0}^{n}\hat{s}_i^2}{\sum_{i=0}^{i}(s_i-\hat{s}_i)^2}\right) \quad (7)$$

where s_i denotes the signal set and $\hat{s}_i$ denotes the heart rate active component.

Mean absolute error HR_{mae} is:

$$HR_{mae} = \frac{1}{n}\sum_{i=1}^{n}|x_i - y_i| \quad (8)$$

Where, x_i is the heart rate measured by the algorithm, y_i is the ground truth, and n is the number of test samples.

The mean absolute percentage error HR_{mape} and standard deviation HR_{sd} are:

$$HR_{mape} = \frac{1}{n}\sum_{i=1}^{n}\frac{|x_i - y_i|}{y_i} \times 100\% \tag{9}$$

$$HR_{sd} = \sqrt{\frac{1}{n}\sum_{i=1}^{n}(HR_e - \overline{HR_e})^2} \tag{10}$$

Where HR_e is the error between the measured heart rate and the ground truth, and $\overline{HR_e}$ is the mean value of the error.

$$r = \frac{\sum_{i=1}^{n}(x_i - \bar{X})(y_i - \bar{Y})}{\sqrt{\sum_{i=1}^{n}(x_i - \bar{X})^2}\sqrt{\sum_{i=1}^{n}(y_i - \bar{Y})^2}} \tag{11}$$

Among them, $\bar{X}$ represents the average heart rate measured by the algorithm, and $\bar{Y}$ represents the actual ref. heart rate average. The Pearson correlation coefficient r is used to reflect the degree of correlation between the algorithm's measured heart rate x_i and the reference heart rate y_i. In addition, we also used Bland-Altman analysis, a commonly used evaluation standard in clinical medicine, to evaluate the consistency between the measured heart rate and the actual ref. heart rate.

4.3 Discussion

As shown in Table 1, the modified EEMD algorithm effectively eliminates the effect of white noise and reduces the decomposition error of each data node in comparison with the algorithm of the original EEMD. And the results in Table 1 also demonstrate that the modified algorithm can enhance the signal-to-noise ratio and improve the signal more robustly than the original EEMD algorithm. Besides, the runtime of the modified EEMD algorithm is decreased by tens of times compared with that of the original EEMD algorithm, greatly raising the runtime. In summary, it indicates that the modified EEMD algorithm not only eliminates the decomposition error of the original EEMD algorithm which strengthens the signal quality, but also decreases the number of ensembles averaging and speeds up the algorithm.

Table 2 provides a comparison of our experimental results of video data acquired under different illumination scenarios utilizing different algorithms. From the table, we can reveal that the mean absolute error and the mean absolute percentage error of our proposed approach are smaller compared with other algorithms in any illumination scene, and the correlation degree r between the measured heart rate and the reference heart rate of our proposed approach is greater than that of other approaches, with extremely strong correlation, all of which indicate that our approach is more accurate and stable. The heart rate calculation using our approach in daytime and night-time light conditions is presented in Figure 7, respectively. The experimental results show much better performance in daytime natural light than in night-time indoor fluorescent light environment, probably due to the fact that daytime light is more abundant and less disturbed by ambient light than night-time light. Furthermore, the strobe of fluorescent tubes may have further increased the error of the measurement results.

Apart from that, it was also validated the consistency of the measured heart rate with the reference heart rate at different lighting scenarios of each tester by the Bland-Altman analysis in Figure 8, to assess the reliability of our approach compared to other algorithms for estimating heart rate. The horizontal and vertical coordinates in Figure 8 are the mean and difference between the estimated and reference heart rates, respectively. The solid line in the middle represents the mean of the two heart rates, the top and bottom dashed lines represent the 95% confidence interval, and the scatter is the sample standard deviation of the mean of the difference. The majority of sample points for all approaches in the figure are distributed within the 95% confidence interval, however, there are variations in the sample standard deviation of the differential means for each algorithm. Compared with other algorithms, overall, our approach has the lowest sample standard deviation and the strongest performance, followed by EEMD, while the ICA method has the highest error and the worst overall performance under less well-lit conditions. The experimental data demonstrate that the proposed approach can effectively eliminate the effects caused by external environmental interference in IPPG image signals and greatly improve the robustness of color signal fluctuations due to interference in real scenes.

Table 1. Performance comparison of EEMD before and after modification.

Method	Mean Energy Error	SNR/dB	Running Time/s
EEMD	0.3956	-1.3605	3.9821
Modified EEMD	0.0001	4.3653	0.1593

Table 2. Performance indicators of different methods for different scenarios.

Illuminated scenes	Evaluation Indicators	CHROM	ICA	EEMD	Ours
Daytime (natural light + computer screen light)	MAE	6.5102	5.4898	0.4245	**0.4082**
	MAPE	0.0946	0.0709	0.021	**0.0144**
	SD	2.3359	2.1791	1.7591	1.9436
	r	-0.5373	0.3335	0.6248	**0.8853**
Evening (fluorescent light + computer screen light)	MAE	13.675	25.8	8.55	**1.4**
	MAPE	0.1758	0.7472	0.1175	**0.0192**
	SD	5.0715	14.9724	2.2689	2.8572
	r	0.1201	-0.3726	-0.2436	**0.8089**

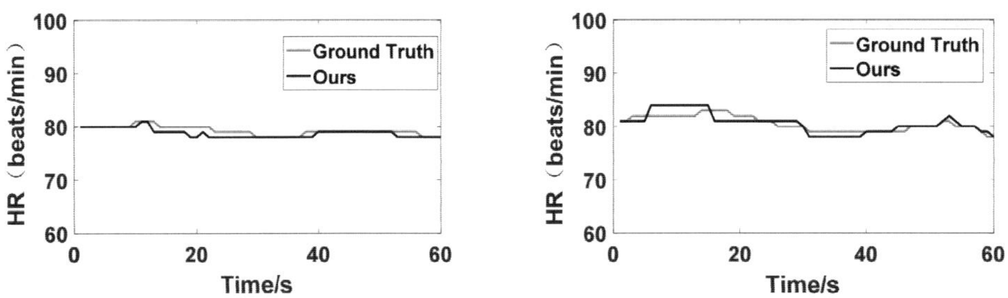

(a) (b)

Figure 7. Example of heart rate calculation using the approach proposed herein. Figure (a) shows the heart rate under daytime light conditions; Figure (b) shows the heart rate under nighttime light conditions. The black line represents the true ref. heart rate and the red solid line represents the estimated heart rate.

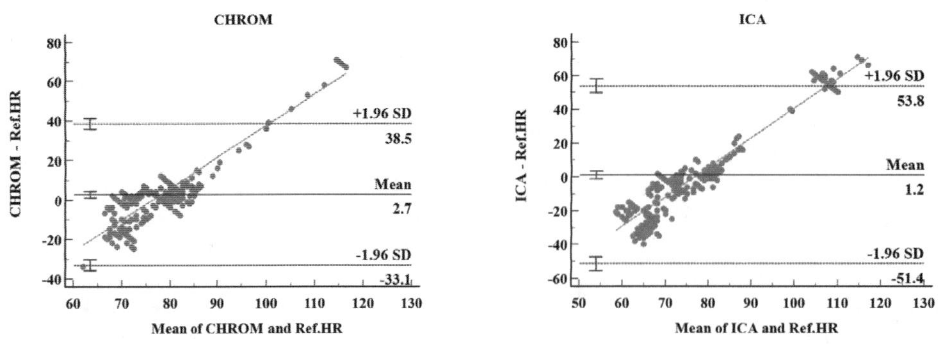

(a) (b)

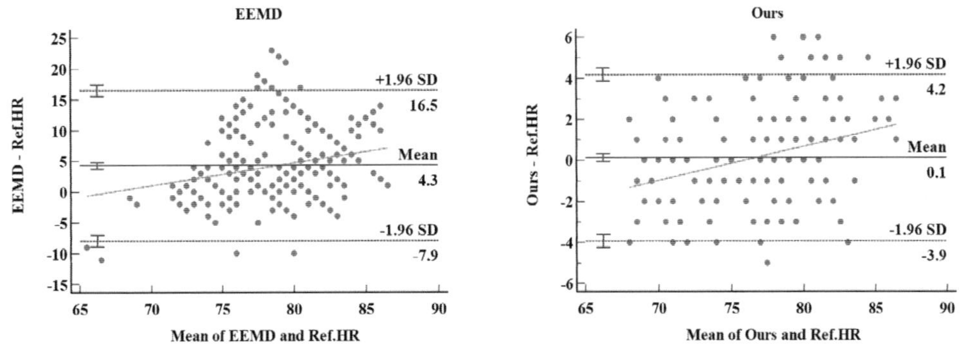

(c) (d)

Figure 8. Plots of Bland-Altman curves for various participants with different approaches under various light scenarios. Figure (a) shows CHROM and pulse oximetry; Figure (b) shows ICA and pulse oximetry, Figure (c) shows EEMD and pulse oximetry, and Figure (d) plots our proposed method and pulse oximetry.

5. Conclusion

On the basis of IPPG, a novel non-contact heart rate measurement approach with modified EEMD algorithm is proposed, which enhances the universality and practicality of a non-contact HR measurement in different scenarios. Meanwhile, to address the error problem brought by the residual white noise of the EEMD algorithm, a method is applied to eliminate the effect of residual white noise by aggregating and averaging the auxiliary Gaussian white noise at each sampling node and defining the white noise size according to the actual situation. Additionally, experiments on personal monitoring of heart health were conducted on several subjects in different real-life scenarios, to verify the feasibility and generalizability of the modified method through the experimental data acquired by itself, and the experimental results demonstrate that the performance of the proposed approach outperformed several other typical methods. Notably, the body of subject cannot be subjected to large movements, nor can the algorithm in the paper perform real-time heart rate monitoring. Consequently, how to enhance the stability and accuracy of the method for real-time usage for heart rate measurement in large-amplitude exercise scenarios is the focus of future study.

Acknowledgments

This paper was supported by: Funding. National Natural Science Foundation of China (61975058); Natural Science Foundation of Guangdong Province (2019A1515011401); the Science and Technology Program of Guangzhou (2019050001).

References

[1] CHINA CARDIOVASCULAR HEALTH AND DISEASE REPORT WRITING GROUP, HU SHENGSHOU. OVERVIEW OF CHINA CARDIOVASCULAR HEALTH AND DISEASE REPORT 2020[J]. CHINESE JOURNAL OF RECYCLING, 36(6):25.

[2] Kevat A C, Bullen D V R, Davis P G, et al. A systematic review of novel technology for monitoring infant and newborn heart rate[J]. Acta Paediatrica, 2017, 106(5): 710-720.

[3] Huang P W, Wu B J, Wu B F. A heart rate monitoring framework for real-world drivers using remote photoplethysmography[J]. IEEE journal of biomedical and health informatics, 2020, 25(5): 1397-1408.

[4] ZiaratNia S, Sripian P, Laohakangvalvit T, et al. End-To-End Deep Learning for pNN50 Estimation Using a Spatiotemporal Representation[C]//International Conference on Human-Computer Interaction. Springer, Cham, 2021: 588-593.

[5] Verkruysse W, Svaasand L O, Nelson J S. Remote plethysmographic imaging using ambient light[J]. Optics express, 2008, 16(26): 21434-21445.

[6] Ismail Shahid, Akram Usman, Siddiqi Imran. Heart rate tracking in photoplethy-smography signals affected by motion artifacts: a review[J]. EURASIP Journal on Advances in Signal Processing, 2021(1): 1-27
[7] Liu M, Wang C J, Wang N N, et al. An exploration of a heart rate sensing garment solution based on rPPG technology[J]. Journal of Physics: Conference Series, 2020, 1693(1): 012147.
[8] Niu X S, Shan S G, Han H, et al. RhythmNet: End-to-end Heart Rate Estimation from Face via Spatial-temporal Representation[J]. IEEE TRANSACTIONS ON IMAGE PROCESSING, 2020(29): 2409-2423
[9] Poh M Z, McDuff D J, Picard R W. Advancements in noncontact, multiparameter physiological measurements using a webcam[J]. IEEE transactions on biomedical engineering, 2010, 58(1): 7-11.
[10] Poh M Z, McDuff D J, Picard R W. Non-contact, automated cardiac pulse measurements using video imaging and blind source separation[J]. Optics express, 2010, 18(10): 10762-10774.
[11] De Haan G, Jeanne V. Robust pulse rate from chrominance-based rPPG[J]. IEEE Transactions on Biomedical Engineering, 2013, 60(10): 2878-2886.
[12] Li X B, Chen J, Zhao G Y, et al. Remote Heart Rate Measurement From Face Videos Under Realistic Situations[J]. 2014 IEEE CONFERENCE ON COMPUTER VISION AND PATTERN RECOGNITION (CVPR), 2014: 4264-4271
[13] Song Haifeng,Jin Zhiyi. Heart rate detection of local area of face based on light correction[J]. Journal of Computer Applications, 2018, 38(S2): 301-305
[14] Cheng J, Wang P, Song R, et al. Remote heart rate measurement from near-infrared videos based on joint blind source separation with delay-coordinate transformation[J]. IEEE Transactions on Instrumentation and Measurement, 2020, 70: 1-13.
[15] Chen D Y, Wang JJ, Lin K Y, et al. Image sensor-based heart rate evaluation from face reflectance using Hilbert–Huang transform[J]. IEEE Sensors Journal, 2014, 15(1): 618-627.
[16] Khan E, Al Hossain F, Uddin S Z, et al. A robust heart rate monitoring scheme using photoplethysmographic signals corrupted by intense motion artifacts[J]. IEEE Transactions on Biomedical engineering, 2015, 63(3): 550-562.
[17] Yin R N, Jia R S, Cui Z, et al. Heart rate estimation based on face video under unstable illumination[J]. Applied Intelligence, 2021, 51(8): 5388-5404.
[18] Wu, Zheng zhen, and Xin Li. "ROI Selected by Pixel2Pixel for Measuring Heart Rate Based on IPPG." 2021 40th Chinese Control Conference (CCC). IEEE, 2021.
[19] Kim D Y, Lee K, Sohn C B. Assessment of ROI Selection for Facial Video-Based rPPG[J]. Sensors, 2021, 21(23): 7923.
[20] Pirnar, Žan, Miha Finžgar, and Primož Podržaj. "Performance Evaluation of rPPG Approaches with and without the Region-of-Interest Localization Step." Applied Sciences 11.8 (2021): 3467.
[21] Wu Z, Huang N E. Ensemble empirical mode decomposition: a noise-assisted data analysis method[J]. Advances in adaptive data analysis, 2009, 1(01): 1-41.

A High-precision Digital Measurement System Design based on PSR

Jianyang Zhou*[a], Linhai Guan[a], Kai Zhang[a], Zhihua Zhang[a], Liqiang Guo[a], Yu Tao[a]

[a]Nanjing Glarun Defense Systems Co.,Ltd., Nanjing, China

* Corresponding author: jeanyangchou@qq.com

ABSTRACT: In engineering practice, a constant voltage output control system of synchronous rectifier primary side regulator (PSR) flyback converter is proposed to overcome the sampling defect of the existing digital approximation double line sampling scheme for primary side control synchronous rectifier structure. Regardless of the slope of the sampling waveform and the knee point, the single-output DAC midpoint sampling can achieve high sampling accuracy and good knee point following effect. At the same time, the neutral-point sampling error of the single-output DAC is analyzed, and a sampling compensation scheme which can simultaneously act in continuous current mode (CCM) and discontinuous current mode (DCM) is designed to further correct the sampling voltage obtained by neutral-point sampling, so as to compensate the voltage error of neutral-point sampling. Thus, the output voltage can be accurately constant in DCM and CCM.

1. Introduction

With the rapid development of the global electronic market, the demand for power management chips is growing in the future. With the increasing demand of consumers for thin and light products, the miniaturization of products is a major trend[1]. In addition, with the announcement of a new generation of Quick Charge 3.0 fast charging technology by Qualcomm, the charging power of mobile electronic devices has been further increased to a medium power level of 30 watts[2]. For the development trend of power supply technology, the demand for power will be greater and greater.

The traditional flyback converter uses diode rectifier. For the application of high-power power supply system, the loss of diode has become a bottleneck to improve the efficiency of the system[3]. In order to improve the efficiency and reduce the loss, synchronous rectification technology has become a necessary means[4]. Synchronous rectification technology is to use MOSFET with low on-resistance and high input impedance instead of diode to rectify. The synchronous rectifier tube has low on-resistance and low forward voltage drop, so the rectification loss is low and the efficiency of flyback switching power supply is greatly improved[5].

At present, the problem that hinders the application of digitally controlled synchronous rectification primary side feedback flyback converter in high power is that the mature digital approximation double-line knee point sampling scheme is not accurate in the case of synchronous rectification[6]. The invention relates to a double-line inflection point sampling scheme of a digital approximation type, which utilizes the particularity of the position of an inflection point. By utilize that different slopes of the sample waveforms at the inflection points, the difference of the gap time can be sampled, so that the double line are moved up and down and are always followed near the positions of the inflection points. However, in the synchronous rectification structure, the sampling waveform is not suitable for

 2366 (2022) 012044 doi:10.1088/1742-6596/2366/1/012044

the two-wire sampling scheme. In the synchronous rectification structure, the on-resistance of the synchronous rectifier tube is small, only a dozen to tens of milliohms. Therefore, in both CCM and DCM, there is a problem that the sampling waveform slope is small and steep after the inflection point.

In addition, in practice, the synchronous rectification is generally turned off before the inflection point. At this time, the parasitic diode DSR is turned on and the forward conduction voltage drops V_f is 1V. If it is inflection point sampling, the sampled voltage is $V_o + V_f$. At this time, the sampling accuracy will become worse due to the inflection point position, resulting in low sampling accuracy, and the effect of following the inflection point is poor, resulting in poor closed-loop stability and constant voltage accuracy. Therefore, it is necessary to propose a corresponding improved sampling solution for the problem.

Aiming at the sampling defect of the existing digital approximation double line sampling scheme to the primary side control synchronous rectification structure, a constant voltage output control system of synchronous rectifier primary side feedback flyback power supply is proposed.

2. Principle analysis

2.1. Principle introduction

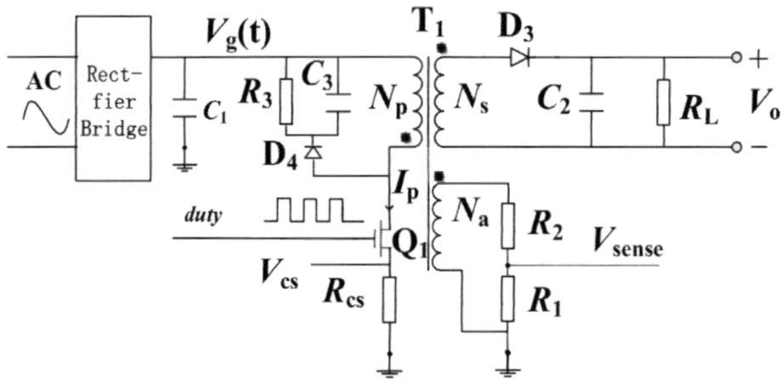

Figure 1. Circuit Diagram of Conventional PSR Flyback Converter Adopting a Diode rectification Mode

A circuit diagram of a PSR flyback converter adopting a diode rectification mode in the prior architecture is shown in Figure 1. The main topology of the PSR flyback converter mainly comprises an input rectifier filter part, a transformer part, an output rectifier filter part and a sampling part.

2.2. Cancellation model

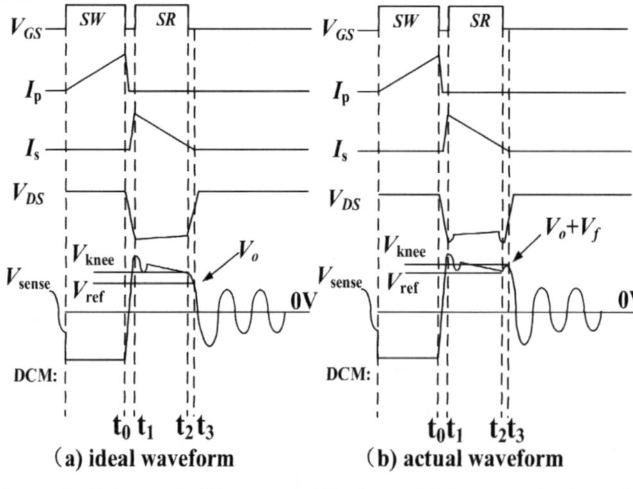

Figure 2. Schematic Diagram of Problem of Inaccurate Sampling

Figure 2 shows a schematic diagram of the problem of inaccurate sampling in the synchronous rectification structure of the digital approximation double-line inflection point sampling scheme. The invention relates to a double-line inflection point sampling scheme of a digital approximation type, which utilizes the particularity of inflection point positions, can sample the difference of gap time by setting a double-line gap pressure difference and utilizing different slopes of sampling waveforms at inflection points. Thereby moving the double line up and down always follows the vicinity of the inflection point position.

However, in the PSR flyback switching converter with synchronous rectification, the sampling waveform is not suitable for the two-wire sampling scheme. In that synchronous rectification structure, the on-resistance of the synchronous rectification tube is small and is only ten to dozens of milliohm, therefore, no matter in CCM or DCM, there is a problem that the slope of the sampling waveform is small and steep after the inflection point. At the same time, in the actual situation, at the moment of t_2, the synchronous rectification is generally turned off before the inflection point, and at this time, the parasitic diode DSR is turned on, and the forward conduction voltage drop V_f is 1V, in case of corner sampling, in the moment of t_3, the voltage sampled at the knee point may be $V_o + V_f$.

At this time, the sampling accuracy will become worse due to the inflection point position, resulting in low sampling accuracy, and the effect of following the inflection point is poor, resulting in poor closed-loop stability and voltage stabilization performance. Therefore, it is necessary to propose a corresponding improved sampling scheme for the problem.

3. Engineering realization

This paper proposes a constant voltage output control system of synchronous rectifier PSR flyback converter, aiming at the sampling defects of the existing digital approximation two-wire sampling scheme for the primary side control synchronous rectifier structure. The main topology of the flyback converter using synchronous rectification is taken as the control object, as shown in Figure 3.

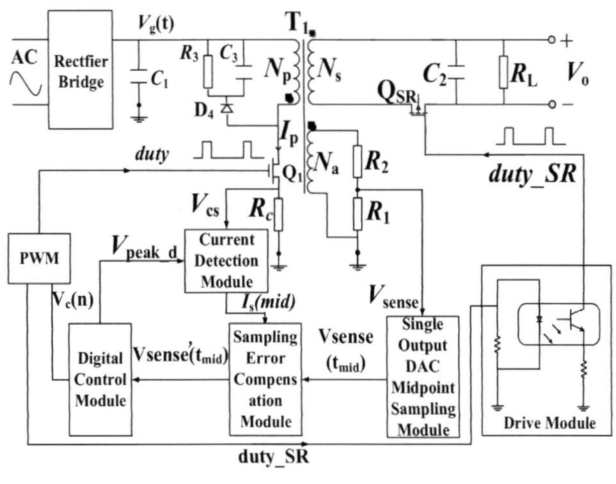

Figure 3. Circuit Diagram of PSR Converter Designed in this Paper

Compared with the traditional diode rectifier, the main topology of the flyback converter based on synchronous rectification is designed in this paper. The topology includes switches and transformers. The constant voltage control system includes a single output DAC midpoint sampling module, sampling error compensation module, current detection module, digital control module, PWM drive module and isolated drive module.

In this paper, the full digital control mode, the DAC sampling mechanism and the constant voltage control strategy of the single output DAC midpoint sampling scheme are designed. In the PSR main topology structure, the single-output DAC midpoint sampling module of the control system continuously compares Δt_{r_half} and $T_r/2$ according to each switching cycle through the internal feedback quantity calculation module. The starting voltage $V_{initial}$ of a single ramp digital wave is generated accurately and carefully. The digital value V_{ref_slope} is used for the triangular wave generator to generate a digital single ramp wave, and then is superimposed with the analog reference voltage generated by the DAC to form a digital wave for the single ramp. Finally, a comparator is used to compare the sampled ramp voltage with the voltage V_{sense} sampled by the auxiliary winding. The voltage V_{sense} can be accurately sampled on the auxiliary winding by comparing the voltage. Through a continuous comparison of Δt_{r_half} and $T_r/2$ in each switching cyclic, the feedback quantity calculation module precisely adjusts the starting voltage $V_{initial}$ of the single-slope digital wave.

In DCM, the current on the secondary winding will drop from peak I_{peak_s} linearity down to zero amps in the reset time T_r. This design uses a single-output DAC midpoint sampling scheme, so current $I_s(t_{mid})$ of secondary winding at the moment of midpoint t_{mid}, in fact, is the average current of the secondary side, which can be obtained. The size of $I_s(t_{mid})$ is $I_{peak\_s}/2$. According to the transformer principle, we can get I_{peak_s} and I_{peak_p}. As shown in the following formulas (1) and (2)[7].

$$I_{peak\_s} = \frac{N_p}{N_s} \cdot I_{peak\_p} \tag{1}$$

$$I_{peak\_p} = \frac{V_{peak}}{R_{cs}} \tag{2}$$

V_{peak} is the peak voltage of the primary current sampling resistor, and is the analog quantity converted by the digital-to-analog converter DAC of output voltage V_{peak_d} in digital control module. R_{cs} is a primary side current sampling resistor,I_{peak_p} is the primary winding inductance peak current, N_p is the number of primary winding turns of the transformer, N_s is the number of secondary winding turns of the transformer, which are all known quantity. After obtaining the average secondary current $I_s(t_{mid})$ in DCM, the reference voltage V_{sense} of the ideal waveform at the midpoint of is shown in Equation (3)[8].

$$V_{sense} = [V_o + I_s(t_{mid}) \cdot (R_S + R_{on})] \cdot \frac{R_1}{R_1+R_2} \cdot \frac{N_a}{N_S} = \left[V_o + \frac{N_P}{N_S} \cdot \frac{I_{peak\_p}}{2} \cdot (R_S + R_{on})\right] \cdot \frac{R_1}{R_1+R_2} \cdot \frac{N_a}{N_S} \tag{3}$$

Sampling error V_{err_dcm} of single-output DAC midpoint sampling scheme in DCM can also be obtained, as shown in formula (4).

$$V_{err\_dcm} = V_{sense} - V_{sense}(t_{mid}) = \left[V_o + \frac{N_P}{N_S} \cdot \frac{I_{peak\_p}}{2} \cdot (R_S + R_{on})\right] \cdot \frac{R_1}{R_1+R_2} \cdot \frac{N_a}{N_S} - V_{sense}(t_{mid}) \tag{4}$$

In CCM, the current on the secondary winding does not fall to zero. I_{peak_p}, the peak current of primary winding inductance, is a known quantity. I_{n_p}, the initial current of the primary winding inductance in CCM, is the unknown quantity. At the moment midpoint t_{mid}, current of secondary winding $I_s(t_{mid})$ is still the secondary side average current, but due to the unknown of I_{n_p}, the average secondary current cannot be obtained. Thus, a new reference current I_{ref} is introduced. Borrowing the time interval t_a of primary current rise from the initial value I_{n_p} to I_{ref} in the case of CCM to calculate the average current of the primary side, and using the proportional relationship of the primary and secondary side currents of the transformer to obtain the average current of the secondary side $I_s(t_{mid})$.

Let the average primary current be I_{mid_p}, the average secondary side current be I_{mid_s}. In CCM, the peak current I_{peak_p} of the primary winding inductance, initial current I_{n_p} of primary winding inductance, the average primary current I_{mid_p} and reference current I_{ref} can be obtained from the proportional relationship of Formula (5)[9].

$$\begin{cases} \frac{t_a}{t_a+t_b} = \frac{I_{ref}-I_{n\_p}}{I_{peak\_p}-I_{n\_p}} \\ I_{mid\_p} = \frac{1}{2} \cdot (I_{n\_p} + I_{peak\_p}) \end{cases} \tag{5}$$

Equation (5) can be derived:

$$I_{mid\_p} = \frac{t_b - t_a}{2t_b} \cdot I_{peak\_p} + \frac{t_b + t_a}{2t_b} \cdot I_{ref} \quad (6)$$

According to the transformer principle, the average current of the secondary side $I_s(t_{mid})$ is in a proportional relationship with the average primary current I_{mid_p}, the formula (7) is deduced[10]:

$$I_s(t_{mid}) = \frac{N_P}{N_S} \cdot I_{mid_p} = \frac{N_P}{N_S} \cdot \left(\frac{t_b - t_a}{2t_b} \cdot I_{peak\_p} + \frac{t_b + t_a}{2t_b} \cdot I_{ref}\right) \quad (7)$$

When the reference current I_{ref} take the value $k \cdot I_{peak\_p}$ $(0 < k < 1)$, the expression can be simplified to Equation (8).

$$I_s(t_{mid}) = \frac{N_P}{N_S} \cdot I_{mid_p} = \frac{N_P}{N_S} \cdot \frac{(1+k)t_b - (1-k)t_a}{2t_b} \cdot I_{peak\_p} \quad (8)$$

After the obtaining of average secondary current $I_s(t_{mid})$ in CCM, the reference voltage V_{sense} of ideal waveform at the midpoint of is shown in Equation (9).

$$V_{sense} = [V_o + I_s(t_{mid}) \cdot (R_S + R_{on})] \cdot \frac{R_1}{R_1 + R_2} \cdot \frac{N_a}{N_s} = \left[V_o + \frac{N_P}{N_S} \cdot \frac{(1+k)t_b - (1-k)t_a}{2t_b} \cdot I_{peak\_p} \cdot (R_S + R_{on})\right] \cdot \frac{R_1}{R_1 + R_2} \cdot \frac{N_a}{N_s} \quad (9)$$

Sampling error V_{err_ccm} of single-output DAC midpoint sampling scheme in CCM can also be obtained, as shown in formula (10):

$$V_{err\_ccm} = V_{sense} - V_{sense}(t_{mid}) = \left[V_o + \frac{N_P}{N_S} \cdot \frac{(1+k)t_b - (1-k)t_a}{2t_b} \cdot I_{peak\_p} \cdot (R_S + R_{on})\right] \cdot \frac{R_1}{R_1 + R_2} \cdot \frac{N_a}{N_s} - V_{sense}(t_{mid}) \quad (10)$$

In summary, equation 4 and equation 10 give the error V_{err_dcm} and V_{err_ccm} between actual midpoint waveforms and ideal waveforms for DCM and CCM respectively. Then using the actual waveform $V_{sense}(t_{mid})$ adaptively and rapidly compensate V_{err_dcm} and V_{err_ccm}, the compensated midpoint voltage $V_{sense}'(t_{mid})$ is obtained. The actual waveform can be made to reach the ideal position and output to the digital control module for negative feedback adjustment to make the output voltage$V_o$to be constant. Voltage precision of the synchronous rectification primary side feedback flyback converter under any load is highly improved.

Sampling error V_{err_dcm} of Single-Output DAC Midpoint Sampling Scheme in DCM is obtained from the following formula.

$$V_{err\_dcm} = V_{sense} - V_{sense}(t_{mid}) = \left[V_o + \frac{N_P}{N_S} \cdot \frac{I_{peak\_p}}{2} \cdot (R_S + R_{on})\right] \cdot \frac{R_1}{R_1 + R_2} \cdot \frac{N_a}{N_s} - V_{sense}(t_{mid}) \quad (11)$$

$V_{sense}(t_{mid})$ is the voltage on the auxiliary winding at the time of $T_r/2$, R_1 and R_2 are sampling divider resistors at both ends of the auxiliary winding, N_p , N_a and N_s are the number of turns of the primary, auxiliary and secondary windings of the transformer respectively.R_S is the ohmic resistance of the secondary winding of the transformer, R_{on} is the on-resistance of the synchronous rectifier tube, V_o is the voltage at the output end of the secondary side.

Sampling error V_{err_ccm} of single-output DAC midpoint sampling scheme in CCM is obtained from the following formula.

$$V_{err\_ccm} = V_{sense} - V_{sense}(t_{mid}) = \left[V_o + \frac{N_P}{N_S} \cdot \frac{(1+k)t_b - (1-k)t_a}{2t_b} \cdot I_{peak\_p} \cdot (R_S + R_{on})\right] \cdot \frac{R_1}{R_1 + R_2} \cdot \frac{N_a}{N_s} - V_{sense}(t_{mid}) \quad (12)$$

I_{peak_p} is the peak current of the primary winding inductor, and k is still the analog voltage across the primary current sampling resistor. V_{peak_half} is k times the primary peak current, i.e. $V_{peak\_half} = k \cdot I_{peak\_p}$, $0<k<1$.

Through the above two formulas, the error V_{err_dcm} and V_{err_ccm} between actual waveforms $V_{sense}(t_{mid})$ and ideal waveforms of midpoint in DCM mode and CCM mode are obtained respectively. Adaptively compensating V_{err_dcm} and V_{err_ccm} , corrected midpoint voltage signal $V_{sense}'(t_{mid})$ can be obtained.

4. Engineering verification

The algorithm is tested and verified in engineering. Figure 4 is a waveform diagram of the mid-point sampling scheme for a single output DAC. When V_{ref_slope} aligns exactly at the midpoint, the time interval of intersection point of $V_{ref_slope}$and V_{sense} and intersection point of and the zero-voltage and V_{sense} is shown in fig. 4 (a). When V_{ref_slope} is above the midpoint, the time interval of intersection point of V_{ref_slope} and V_{sense} and intersection point of zero-voltage and V_{sense} is shown in fig. 4 (b). WhenV_{ref_slope} is below the midpoint, the time interval of intersection point of V_{ref_slope} and V_{sense} and intersection point of zero-voltage and V_{sense} is shown in Figure 4 (c).

In case of DCM, reset time T_r is the time for the secondary current dropping from peak to zero. In case of CCM, the secondary side current does not fall to 0, reset times T_r indicates how long the primary switch is turned off. In steady state, reset time T_r remains the same either in DCM or CCM. Through a continuous comparison of Δt_{r_half} and $T_r/2$ in each switching cyclic, and continuously accurately adjusting the starting voltage $V_{initial}$ of the single-slope digital wave up and down, the accurate starting position of single ramp digital wave voltage signal V_{ref_slope} is guaranteed. Which always ensure the starting position follow the voltage signal V_{sense} on the auxiliary winding under different load condition, and obtain better mid-point following effect under steady state and dynamic conditions.

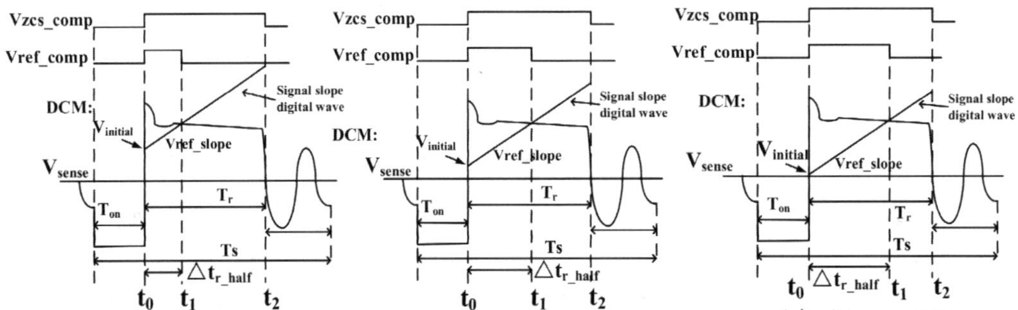

(a) $\Delta t_{r\_half} < T_r/2$ (b) $\Delta t_{r\_half} = T_r/2$ (c) $\Delta t_{r\_half} > T_r/2$

Figure 4. Waveform Diagram of Digital Wave Operation for Single-Output DAC Mid-Point Sampling Scheme

Figure 5 is the comparison chart of the constant voltage accuracy test of three sampling schemes, namely, the digital approximation double-line inflection point sampling scheme, the single-output DAC midpoint sampling scheme without the compensation strategy, and the single-output DAC midpoint sampling scheme with the compensation strategy. The theoretical analysis of the invention is verified, and the test conditions are as follows: the load size is 1 to 5 A, and the output rated voltage is 20 V. The test results are basically similar to the theoretical analysis. Compared with the scheme without error compensation, the scheme with error compensation in the prior art has poor constant voltage accuracy under light load and good constant voltage accuracy under heavy load, while the mid-point sampling scheme with compensation strategy in this design has the best constant voltage accuracy in both light load and heavy load.

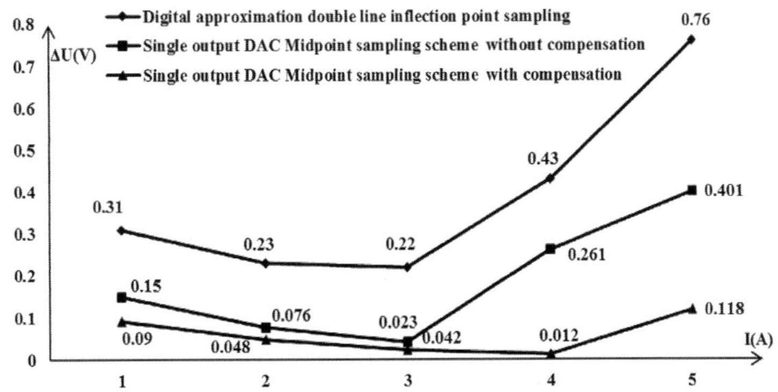

Figure 5. Comparison Chart of Three Sampling Schemes

5. Conclusions

The advantage and effect of this paper are significant. For synchronous rectification structure, the sampling waveform slope is small, and steep after the inflection point, and the sampling precision is low due to voltage from V_o turns $V_o + V_f$ at inflection point.

This paper proposes the single-output DAC midpoint sampling scheme. A single-slope digital wave V_{ref_slope} is generated and sampling the midpoint voltage signal $V_{sense}(t_{mid})$ on the auxiliary winding, avoiding the problem that the sampling accuracy of the existing sampling scheme is affected by the position of the inflection point and becomes poor. And a compensation algorithm based on the mid-point sampling error calculated by the current detection module is proposed to correct the sampling voltage $V_{sense}'(t_{mid})$ on the auxiliary winding at the midpoint. It quickly compensates for the output voltage error of a single-output DAC midpoint sampling scheme in a single cycle, enabling accurate constant voltage. Through the simulation and verification, it is found that the constant voltage accuracy of ± 1% can be obtained, which greatly improves the constant voltage characteristics of the PSR flyback converter under the synchronous rectification structure.

References

[1] Chen, C., Wa Ng, L., Chang, C., Han, X., "An adaptive multi-mode PWM control PSR flyback converter," Journal of Circuits, Systems and Computers. Papers 13(4), 67-94 (2020).

[2] Xu, S., Kou, X., Wang, C., Sun, W., Shi, L., "A digital control scheme for PSR flyback converter in CCM and DCM," IEEE Journal of Emerging and Selected Topics in Power Electronics. Papers 99, 1-1 (2020).

[3] Xu, S., Kou, X., Wang, C., Sun, W., Shi, L., "New digital control method for improving dynamic response of synchronous rectified PSR flyback converter with CCM and DCM modes," IEEE Transactions on Power Electronics. Papers 99, 3-8 (2020).

[4] Huang, C. S., Wang, S. S., "Modeling and design of cable compensation for a primary side regulation (PSR) flyback converter," Future Energy Electronics Conference, IEEE. Papers 804, 697-703 (2013).

[5] Zhao, C., Xie, X., Dong, H., Liu, S., "Improved synchronous rectifier driving strategy for primary-side regulated (PSR) flyback converter in light-load mode," IEEE Transactions on Power Electronics. Papers 29(12), 6506-6517 (2014).

[6] Wang, Z., Lai, X., Qiang, W., "A PSR CC/CV flyback converter with accurate CC control and optimized CV regulation strategy," IEEE Transactions on Power Electronics. Papers 32, 7045-7055 (2016).

[7] Lei, T., Li, Q., Wang, W., "Research of the adaptive control on modulation factor for PSR fly-back PSM converter," Etri Journal. Papers 41, 124-132 (2018).

[8] Zhang, H., Chan, P. K., Tan, M. T., "A high PSR voltage reference for DC-to-DC converter applications," IEEE International Symposium on Circuits & Systems, IEEE. Papers 816-819 (2009).
[9] Lim, S. F., Khambadkone, A. M., "A simple digital DCM control scheme for boost PFC operating in both CCM and DCM," Energy Conversion Congress & Exposition, IEEE. Papers 1218-1225 (2010).
[10] Li, Y., Zhu, Z., "A constant current control scheme for primary-side controlled flyback controller operating in DCM and CCM," IEEE Transactions on Power Electronics. Papers 35(9), 9462-9470 (2020).

ADC Clock Jitter Measurement Based on Simple Coherent Sampling Algorithm

Shuoyan Zhang[1], Qinghe Sun[1], Xiaolong Chen[1]*, Bocheng Wang[1], Yifan Li[1]

[1]Xi'an Key Laboratory of Intelligent Instrument and Packaging Test, Xidian University, Shaanxi, China

*E-mail: xlchen@mail.xidian.edu.cn

Abstract. In a high-speed sampling system, the clock jitter of analog-to-digital converters (ADCs) will greatly affect its sampling accuracy, leading to the reduction of the signal-to-noise ratio (SNR) of the system output. Therefore, it is necessary to compensate the sampling results to reduce the sampling error caused by clock jitter by measuring the distribution sequence of jitter. In this paper, the influence of clock jitter on the ADC sampling process is analyzed, and an ADC clock jitter measurement scheme based on a simple coherent sampling algorithm is investigated. This scheme can accurately measure the distribution sequence of clock jitter, and has the characteristics of low computational complexity and high precision. The simulation results show that this algorithm can accurately measure the clock jitter sequence with root-mean-square (RMS) greater than 5ps when the amplitude noise of the input signal is greater than 35dB, and the relative error is less than 5%.

1. Introduction

In a low-speed sampling system, the jitter of clock signal is far less than the sampling period, because of which it has little influence on the sampling result. However, with the continuous improvement of analog-to-digital converter (ADC) sampling rate, in a high-speed sampling system, the size of the clock jitter relative to the sampling period is no longer a negligible "tiny amount". The jitter of the sampling clock will cause non-uniform sampling of the input signal, and reduce the signal-to-noise ratio (SNR) of the system output [1] . To compensate for the sampling error caused by clock jitter, it is necessary to measure the real-time distribution sequence of the jitter, so as to improve the influence of clock jitter, improve the SNR, and achieve noise suppression. Therefore, the measurement of clock jitter has become an important technology in high-speed ADC.

In the early stage, the influence of clock jitter on SNR of sampling sinusoidal signal in high-speed sampling system was studied [2] . Later, some scholars proposed a high-resolution clock jitter measurement method using ADC sampling [3] , from which the basic idea of "coherent sampling algorithm" was derived. In recent years, some scholars have proposed an on-chip clock jitter digital measurement scheme for high performance microprocessors [4] . In [5] , a new approach based on the calculation of a cyclostationary variance to measure the jitter at the output of a sample-and-hold circuit is presented. Some other scholars present a technique which makes use of existing circuits in a phase interpolator-based clock and data recovery to measure random jitter [6] . In [7] , a new method based on the evaluation of the cyclic spectrum of the acquired signal for measuring the power spectral density (PSD) of the jitter contributions is proposed. [8] investigates a method of using $\Delta\Sigma$ sampling to suppress clock jitter errors in high-performance sampling and holding systems. Some other scholars have analyzed the theory of jitter and skew measurement, and proposed a fully on-chip measurement scheme based on incoherent subsampling for high-speed clock jitter and skew [9] . Among these methods, "coherent sampling algorithm" can measure the size and distribution sequence of clock jitter, which is in line with the requirements of compensating for the error caused by clock jitter. At the same time, it has the characteristics of low computational complexity and high precision. Aiming at the measurement problem of distributed sequence of clock jitter in high-speed ADC, a clock jitter measurement scheme based on a simple signal coherence method is studied in this paper. In this method, the distributed sequence of jitter can be measured by processing and calculating the sampled signal, which provides an important basis for compensating for the sampling error caused by clock jitter.

The rest of this paper is organized as follows. Section 2 analyzes the influence of clock jitter on sampling and the error between ideal sampling result and actual sampling result. Section 3 introduces the basic principle of the clock jitter measurement scheme based on a simple coherent sampling algorithm. Simulation results are shown in section 4. Section 5 concludes the paper.

2. Impact of clock jitter on ADC

2.1. Impact of clock jitter on sampling

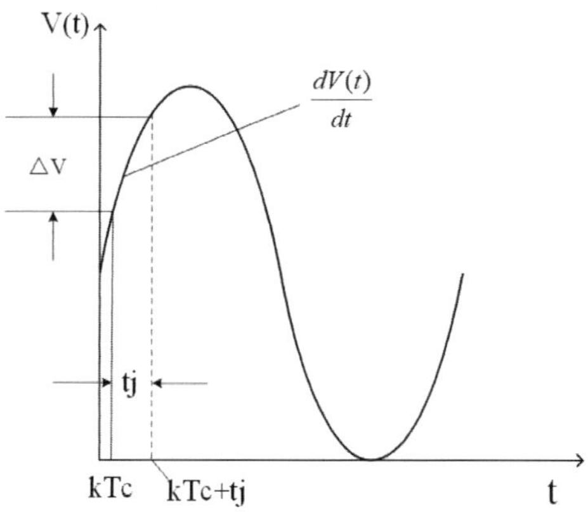

Figure 1. Sampling with jitter.

Suppose the input is a sinusoidal signal: $V(t) = A\sin(2\pi f_{in} t)$, as shown in Figure 1. As the clock jitter value is very small, the error curve can be approximated as a straight line, and the sampling error is:

$$\Delta V = t_j \frac{dV}{dt} = 2\pi f_{in} A t_j \cos(2\pi f_{in} t) \tag{1}$$

Where t_j is the jitter of the clock signal, f_{in} and A are the frequency and amplitude of the input sinusoidal signal, respectively. Generates maximum error at zero crossing (maximum slope):

$$V_{err} = \Delta V\big|_{t=0} = 2\pi f_{in} A t_j \tag{2}$$

In an ADC, ΔV must be less than *LSB*/2 in order not to cause errors in the final output of the data conversion, i.e.:

$$\Delta V < \frac{1}{2} LSB = \frac{1}{2} \cdot \frac{2A}{2^n} = \frac{A}{2^n} \tag{3}$$

Where *LSB* and *n* are the least significant bit and significant bits of an ADC respectively. From (2) and (3) can be obtained:

$$f_{in} < \frac{1}{2^{n+1}\pi t_j} \tag{4}$$

According to (4), with the increase of clock jitter, the frequency of analog input signals that can be effectively processed by ADC decreases continuously. By (2), the *SNR* of the system considering only clock jitter is:

$$SNR_{jitter} = 20\log\frac{A}{V_{err}} = -20\log(2\pi f_{in} t_j) \tag{5}$$

According to (5), the larger the clock jitter is, the smaller the SNR of ADC is. For a higher frequency input signal, clock jitter will make the SNR of the system output smaller.

2.2. Sampling error caused by clock jitter

This part deduces the expression of sampling error caused by clock jitter from the point of mathematical expression. Suppose the input is a sinusoidal signal: $x(t) = A\sin(2\pi f_{in} t + \varphi)$, the ideal sampling results are as follows:

$$x(n) = x(nT_s) = A\sin(2\pi f_{in} nT_s + \varphi) \tag{6}$$

Where f_{in}, A and φ are the frequency, amplitude and initial phase of the input sinusoidal signal, respectively, T_s is the sampling interval determined by the ideal clock signal. The actual sampling result with clock jitter is as follows:

$$x_j(n) = x(nT_s + t_j) = A\sin(2\pi f_{in} nT_s + \varphi + 2\pi f_{in} t_j) \tag{7}$$

Where t_j is the jitter of the clock signal. Then, the sampling error is:

$$\begin{aligned} e(n) &= x_j(n) - x(n) \\ &= A\sin(2\pi f_{in} nT_s + \varphi + 2\pi f_{in} t_j) - A\sin(2\pi f_{in} nT_s + \varphi) \\ &= A\sin(2\pi f_{in} nT_s + \varphi)\cos(2\pi f_{in} t_j) + A\cos(2\pi f_{in} nT_s + \varphi)\sin(2\pi f_{in} t_j) \\ &\quad - A\sin(2\pi f_{in} nT_s + \varphi) \end{aligned} \tag{8}$$

Because the clock jitter t_j is generally on the order of ps, namely 10^{-12}, it can be approximated as follows: $\cos(2\pi f_{in} t_j) \approx 1, \quad \sin(2\pi f_{in} t_j) \approx 2\pi f_{in} t_j$

$$\begin{aligned} \therefore e(n) &\approx A\sin(2\pi f_{in} nT_s + \varphi) + A 2\pi f_{in} t_j \cos(2\pi f_{in} nT_s + \varphi) \\ &\quad - A\sin(2\pi f_{in} nT_s + \varphi) \\ &= 2\pi A f_{in} t_j \cos(2\pi f_{in} nT_s + \varphi) \end{aligned} \tag{9}$$

According to Equation (9), as the frequency of input signal increases, the maximum of error sequence caused by clock jitter on sampling results increases. Therefore, it can be explained that clock jitter has a greater impact on higher frequency input signals.

3. Simple coherent sampling algorithm measurement principle

The coherent sampling algorithm refers to the sinusoidal signal whose input frequency meets a certain relationship with clock frequency, calculates the jitter of the clock signal after sampling, then obtains the jitter sequence, and finally obtains the specific distribution of jitter.

The simple coherent sampling algorithm refers to the method used to calculate clock jitter by sinusoidal signal whose input frequency is an integer multiple of clock frequency, namely:

$$f_{in} = Nf_c \tag{10}$$

Where f_{in} is the frequency of the input sinusoidal signal, N is a positive integer, f_c is the frequency of the clock signal.

The sampling process of the simple coherent sampling algorithm is shown in Figure 2. Let the input sinusoidal signal be: $V(t) = A\sin(2\pi f_{\text{in}} t + \varphi_0)$. Since the input sinusoidal signal frequency is an integer multiple of the clock frequency, under ideal sampling conditions, the sampling result is a straight line, namely:

$$\begin{aligned} V(k) = V(kT_c) = A\sin(2\pi f_{in} kT_c + \varphi_0) = A\sin(2\pi f_{in} k\frac{1}{f_c} + \varphi_0) \\ = A\sin(2\pi kN + \varphi_0) = A\sin(\varphi_0) \end{aligned} \tag{11}$$

Where A, f_{in} and φ_0 are the frequency, amplitude and initial phase of the input sinusoidal signal, respectively, T_c is the sampling interval determined by the ideal clock signal, which is defined as $T_c = 1/f_c$.

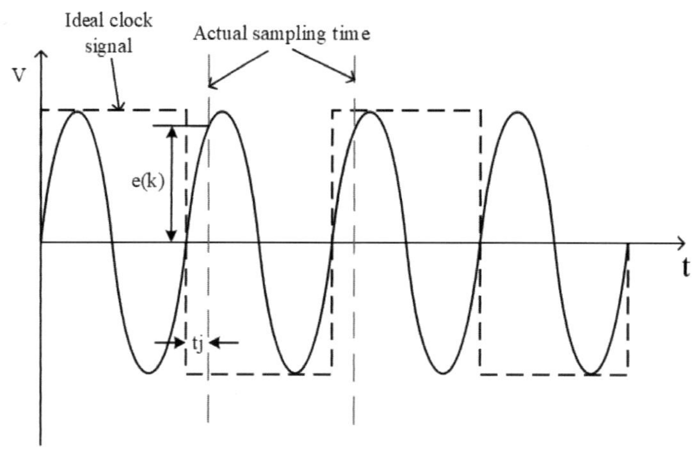

Figure 2. Sampling process of simple coherent sampling algorithm.

In actual sampling, clock jitter $t_j(k)$ exists at the sampling time, namely:

$$t_k = kT_c + t_j(k) \tag{12}$$

Then the sampling result is:

$$\begin{aligned} V_j(k) = V(t_k) &= A\sin(2\pi f_{in} t_k + \varphi_0) \\ &= A\sin(2\pi f_{in} kT_c + 2\pi f_{in} t_j(k) + \varphi_0) \\ &= A\sin(2\pi f_{in} t_j(k) + \varphi_0) \\ &= A\sin(2\pi f_{in} t_j(k))\cos(\varphi_0) + A\cos(2\pi f_{in} t_j(k))\sin(\varphi_0) \end{aligned} \tag{13}$$

Since the size of the general clock jitter is on the order of ps, it can be approximated as: $\cos(2\pi f_c t_j) \approx 1, \quad \sin(2\pi f_c t_j) \approx 2\pi f_c t_j$. Thus, Equation (13) can be simplified as:

$$V_j(k) \approx 2\pi A f_{in} t_j(k)\cos(\varphi_0) + A\sin(\varphi_0) \tag{14}$$

As the time jitter is generally distributed with an average value of 0, the average value of the above formula is:

$$\begin{aligned}\overline{V_j(k)} &\approx \overline{2\pi A f_{in} t_j(k)\cos(\varphi_0)} + \overline{A\sin(\varphi_0)} = \overline{A\sin(\varphi_0)} \\ &= A\sin(\varphi_0) = V(k)\end{aligned} \tag{15}$$

Therefore, the ideal sampling result can be approximated by averaging the actual sampling result, and the error between the actual sampling result and the ideal sampling result can be denoted as $e(k)$, then:

$$\begin{aligned}e(k) &= V_j(k) - V(k) \\ &= t_j \frac{dV(t)}{dt}\Big|_{t=kT_c} = 2\pi f_{in} A t_j(k)\cos(2\pi f_{in} k T_c + \varphi_0) \\ &= 2\pi f_{in} A t_j(k)\cos(\varphi_0) \\ &= 2\pi f_{in} t_j(k)\sqrt{A^2 - (A\sin\varphi_0)^2} \\ &\approx 2\pi f_{in} t_j(k)\sqrt{A^2 - (\overline{V_j(k)})^2}\end{aligned} \tag{16}$$

The ideal sampling results in the above formula are approximately replaced with the average value of actual sampling results, and the following results can be obtained after sorting:

$$\begin{aligned}t_j(k) &= \frac{V_j(k) - V(k)}{2\pi f_{in}\sqrt{A^2 - (\overline{V_j(k)})^2}} \\ &\approx \frac{V_j(k) - \overline{V_j(k)}}{2\pi f_{in}\sqrt{A^2 - (\overline{V_j(k)})^2}}\end{aligned} \tag{17}$$

The distribution sequence of clock jitter can be calculated by equation (17).

4. Simulation

In this section, we use MATLAB for simulation. The simulation process is realized by programming in MATLAB file editor. In the simulation, we set the following parameters: the significant bits of an ADC BITS=12bit, clock frequency f_c=50MHz, input sinusoidal signal frequency is 4 times of clock frequency, namely $f_{in}=4f_c$=200MHz, initial phase $\varphi_0=\pi/6$, amplitude A=10. The number of sampling points is 10000. A sinusoidal signal with amplitude noise is input, and the amplitude noise follows Gaussian distribution with mean a value of 0. After jitter is added, the input sinusoidal signal is sampled, and then the sequence is quantized by $y(k) = round(2^{BITS-1}) \cdot y(k)$. Jitter is measured by this algorithm. Because this algorithm needs to measure the distribution of jitter, histogram is used to display.

4.1. Jitter measurement under different SNR conditions

The SNR of amplitude noise was set as 30dB~80dB, and the jitter was random jitter with Gaussian distribution with mean value of 0. The jitter with root mean square of 0.1ps, 1ps, 5ps, and 10ps were measured. The measurement performance is expressed by relative error, namely:

$$\varepsilon(k) = \frac{t_j'(k) - t_j(k)}{t_j(k)} \times 100\% \tag{18}$$

Where $t_j'(k)$ is the root-mean-square (RMS) of the jitter sequence measured by this algorithm, $t_j(k)$ is the RMS of the jitter sequence of injection. The simulation results are shown in Figure 3.

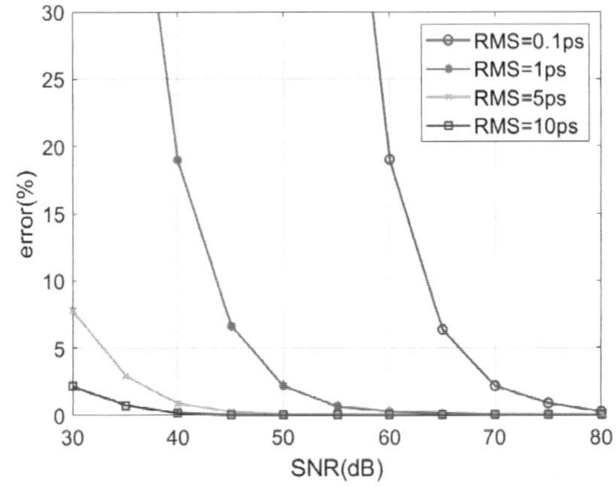

Figure 3. Jitter measurement error at different SNR.

It can be seen from Figure 3. that as the SNR of the input sinusoidal signal increases, the clock jitter can be measured more accurately by the simple coherent sampling algorithm. When the SNR is greater than or equal to 35dB, the algorithm can accurately measure the distribution sequence of clock jitter with root mean square of 5ps and 10ps, and the relative error is less than 5%. To accurately measure the distribution sequence of clock jitter with root mean square of 1ps and 0.1ps, the SNR of the input sinusoidal signal needs to be greater than 50dB and 70dB, respectively.

4.2. Jitter measurement of different distributions

The SNR of amplitude noise was set as 50dB, and the jitter was random jitter with the mean value of 0 and the root mean square of the clock. The jitter of Gaussian distribution, triangular distribution, uniform distribution, and periodic distribution were measured, respectively. Simulation results are shown in Figure 4, Figure 5, Figure 6 and Figure 7.

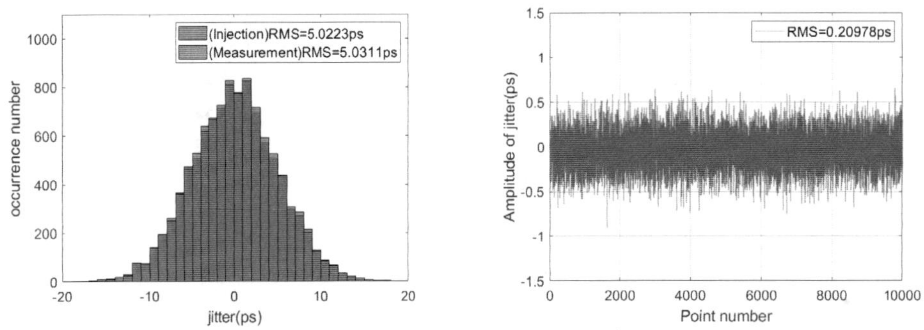

(a) Comparison between injection and measurement (b) Error sequence

Figure 4. Jitter measurement of Gaussian distribution.

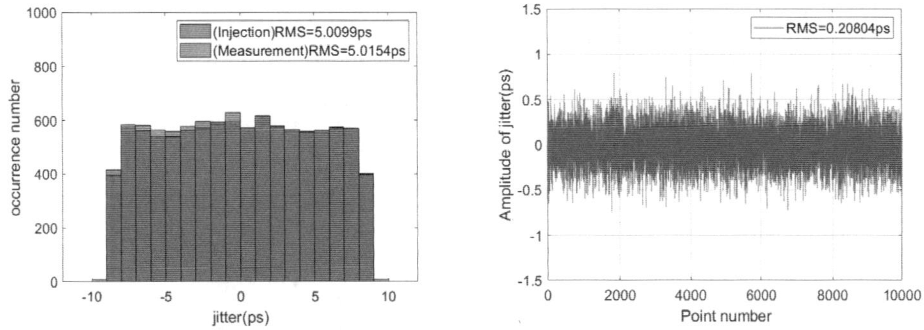

(a) Comparison between injection and measurement (b) Error sequence

Figure 5. Jitter measurement of uniform distribution.

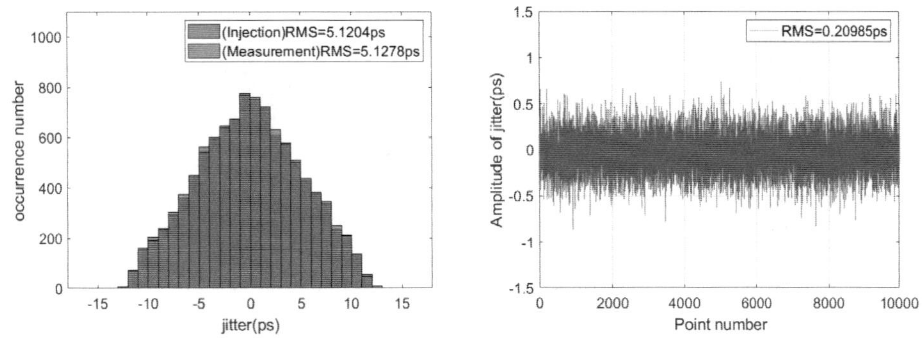

(a) Comparison between injection and measurement (b) Error sequence

Figure 6. Jitter measurement of triangular distribution.

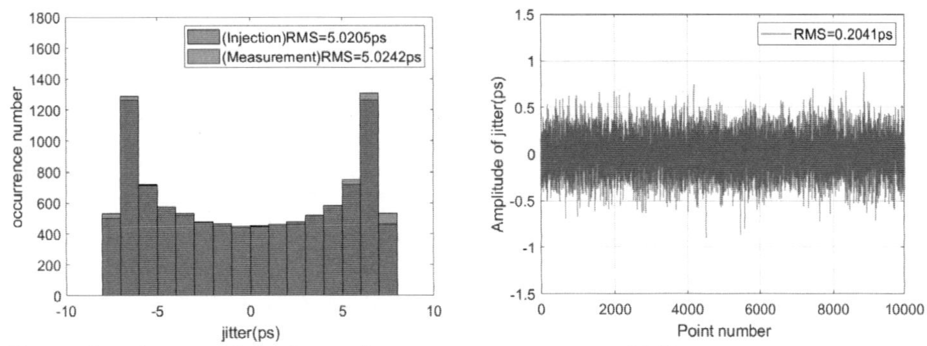

(a) Comparison between injection and measurement (b) Error sequence

Figure 7. Jitter measurement of period distribution (sinusoidal distribution).

According to Equation (18), it can be calculated that the relative errors of the measurement results under these four distributions are 0.1752%, 0.1098%, 0.1445% and 0.0737%, respectively.

According to the above results, when the SNR of the amplitude noise of the input sinusoidal signal is 50dB, the simple coherent sampling algorithm can accurately measure the clock jitter which is subject to the Gaussian distribution, triangular distribution, uniform distribution, and periodic distribution. When the RMS of the clock jitter is 5ps, the relative error is less than 0.2%, and the error sequence of jitter injected and measured is within the range of plus or minus 0.5ps.

4.3. Measurement of total jitter

According to the classification of jitter, jitter can be divided into deterministic jitter and stochastic jitter. Therefore, to simulate the actual situation, we set the clock jitter to total jitter with both random jitter and periodic jitter (periodic jitter is a type of deterministic jitter). Based on the above simulation conditions, the jitter distribution is measured by simple coherent sampling algorithm. Simulation results are shown in Figure 8.

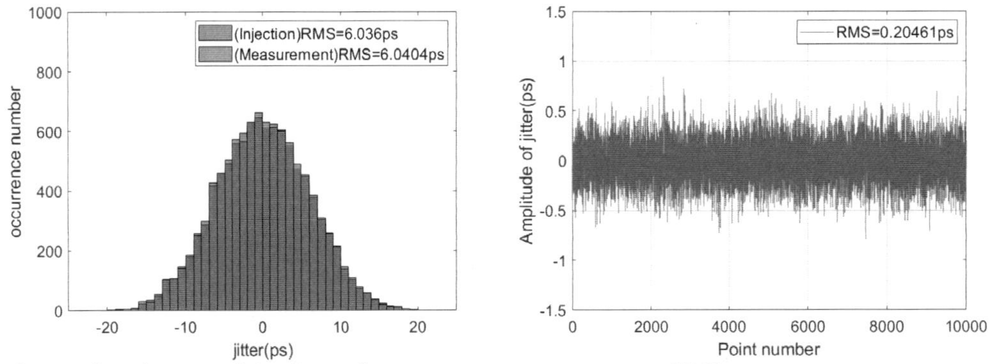

(a) Comparison between injection and measurement (b) Error sequence

Figure 8. Measurement of total jitter.

As can be seen from Figure 8(a), when the root mean square is 6ps and the amplitude noise of the input sinusoidal signal is 50dB, the algorithm can accurately measure the distribution sequence and size of the total jitter including random jitter and periodic jitter. The RMS of the error sequence of the measurement results is 0.2ps, as shown in Figure 8(b).

5. Conclusion

In this paper, the influence of clock jitter on ADC sampling process is analyzed, and it is explained from two aspects that clock jitter has more influence on higher frequency input signal. On this basis, an ADC clock jitter measurement scheme based on a simple coherent sampling algorithm is investigated. The algorithm can measure the distribution sequence of clock jitter by calculating and processing the sampled signal. The experimental results show that when the amplitude noise of the input sinusoidal signal is greater than 35dB, this algorithm can accurately measure the magnitude and distribution sequence of clock jitter with RMS greater than 5ps, and the relative error is less than 5%. The clock jitter sequence with Gaussian distribution, triangular distribution, uniform distribution and periodic distribution can be accurately measured by this method. When the SNR of amplitude noise is 50dB and the RMS of the clock jitter is 5ps, the relative errors of the measured results are within 0.2%, and the error sequence of jitter injected and measured is within the range of plus or minus 0.5ps. Therefore, this method can be applied to the precise measurement of ps level clock jitter in high-speed ADC, and then provide an important basis for further research on sampling error compensation algorithm caused by clock jitter.

Acknowledgments

This work has been financially supported by the National Natural Science Foundation of China with Grant No. 61727804.

References

[1] Da Dalt N. (2004) Effect of jitter on asynchronous sampling with finite number of samples. IEEE Transactions on Circuits and Systems II: Express Briefs, VOL. 51, NO. 12, pp. 660-664.

[2] Shinagawa M, Akazawa Y and Wakimoto T. (1990) Jitter analysis of high-speed sampling systems. IEEE Journal of Solid-State Circuits, VOL. 25, NO. 1, pp. 220-224.

[3] Cherubal S and Chatterjee A. 2001. A high-resolution jitter measurement technique using ADC sampling. In: International Test Conference 2001 Proceedings. Baltimore, MD, United states, pp. 838-847.
[4] Omaña M, Rossi D, Giaffreda D, Metra C, Mak T M, Rahman A and Tam S. (2015) Low-Cost On-Chip Clock Jitter Measurement Scheme. IEEE Transactions on Very Large Scale Integration (VLSI) Systems, VOL. 23, NO. 3, pp. 435-443.
[5] Schweiger T, Krüger B and Wiedmann F. (2017) Cyclostationary Stochastic Jitter Measurement Method With Uncertainty Prediction. IEEE Transactions on Instrumentation and Measurement, VOL. 66, NO. 9, pp. 2371-2379.
[6] Liang J, Sheikholeslami A, Tamura H and Yamaguchi H. (2018) On-Chip Jitter Measurement Using Jitter Injection in a 28 Gb/s PI-Based CDR. IEEE Journal of Solid-State Circuits, VOL. 53, NO. 3, pp. 750-761.
[7] Daponte P, De Vito L, Picariello F and Rapuano S. (2018). A Cyclostationarity-Based Method for Jitter Measurement in Waveform Recorders. IEEE Transactions on Instrumentation and Measurement, VOL. 67, NO. 8, pp. 1786-1794.
[8] Jamali-Zavareh S and Harjani R. (2020) Jitter Suppression Techniques for High-Speed Sample-and-Hold Circuits. IEEE Transactions on Circuits and Systems I: Regular Papers, VOL. 67, NO. 1, pp. 1-11.
[9] Lyu Y, Feng J, Zhu K, Ye H, Yu D. (2020). Fully on-chip clock jitter and skew measurement scheme via incoherent subsampling. Microelectronics Journal, VOL.96.

A new type of vertical wall-climbing robot

Jinfan Lin[1], Zeming Ren[1], Xiaobin Hong[1]*

[1]School of Mechanical & Automotive Engineering, South China University of Technology. Guangzhou 510640, China.

* Corresponding author: scut_hongxiaobin@126.com

ABSTRACT: Using robots to replace human beings to achieve large-scale automatic operations has become an inevitable trend. Wall-climbing robots have important application value in the field of aerial work. For the existing wall-climbing robots have limitation on movement flexibility and overcome own gravity by adsorption friction, this paper innovatively proposes a wall-climbing robot with suspension traction, back push adsorption and McNum wheel driving. Meanwhile, an electronic control system and control algorithm are designed. Experiments results show that the wall-climbing robot has advantages of high relative load, omnidirectional movement, strong continuity of movement and smooth operation.

1. Introduction

Wall-climbing robot is a kind of special robot, which can climb and crawl on vertical walls and other occasions, and is usually equipped with corresponding equipment to realize various functions such as cleaning, flaw detection, and detection of the wall. Automatic wall-climbing robots are widely used in various fields with rapidly development. In recent years, a large number of automatic wall-climbing robot solutions have emerged. The automatic wall-climbing robot needs to drive and stick to the wall, to overcome the self-gravity in working. Therefore, the key technologies of automatic wall-climbing robot are the adsorption device and running gear. The main function of the adsorption device is to provide a force perpendicular to the wall. The common adsorption methods include negative pressure adsorption, suction cup adsorption, magnetic adsorption, etc. [1]. The negative pressure adsorption is to draw out the air in the closed cavity between wall surface and the robot to form a partial vacuum, which can provide adsorption force by the difference of internal and external pressure [2]. The principle of suction cup adsorption is similar to that of negative pressure adsorption, but the suction cups for adsorption also can in the form of foot-like structure [3]. Magnetic adsorption is to use the gravitational effect between permanent magnet or an electromagnet and a ferromagnetic wall to generate adsorption force [4]. The running gear is the key component to drives the automatic wall-climbing robot. The running gear is usually combined with the adsorption device, which can be divide into the legged type, the crawler type and the wheeled type. The legged type imitates the walking mode of dog to drive the robot, which is often combined with the suction cup adsorption [5]. The crawler-type uses two crawler tracks to achieve movement and steering, which can be combined with suction cup adsorption and magnetic adsorption [6]. The wheeled type is usually used as a driving mode of negative pressure adsorption or magnetic adsorption.

In the above scheme, the gravity of the wall-climbing robot is completely borne by the friction between the robot and the wall, which cannot adapt to some high-load situations, and the continuous movement ability is weak. Therefore, a novel wall-climbing robot with suspension traction, back push adsorption and McNum wheel driving is proposed.

2. Mechanical Structure Design of Wall-Climbing Robot

The wall-climbing robot mainly consists of suspension traction device, Mecanum wheels, ducted fans and electronically controlled platform, as shown in Figure 1. The suspension traction device installed a movable servo spool which is mounted on the horizontal rail. The suspension traction device has two ropes to pull the wall-climbing robot to overcome self-gravity and assist the vertical movement. The movement speed, accuracy and direction of the wall-climbing robot have great influence on the detection quality. In order to control the posture of the wall-climbing robot in real time, the chassis of the wall-climbing robot is mount with four Mecanum wheels which can be realized omnidirectional movement of the robot. A number of small rollers are evenly arranged on the surface of the Mecanum wheel and fixed on the hub. The axis of the small rollers forms a certain angle with the rotation axis of the hub, and the angle is usually 45°. According to the installation direction of the small rollers, the Mecanum wheel can be divided into left-handed wheel and right-handed wheel, as shown in Figure 2. A ducted fan is composed of a fan and a brushless permanent magnet DC motor. The motor is driven and controlled by a special electronic governor. Four ducted fans are mounted on the chassis with circular air intake openings, which exhausts backwards to generate thrust force so the wall-climbing robot can attach to the wall, and provide positive pressure for the Mecanum wheels to drive the wall-climbing robot. The single-axis electronically controlled platform can be rotated 360°, which can install with detection devices or other functional equipment.

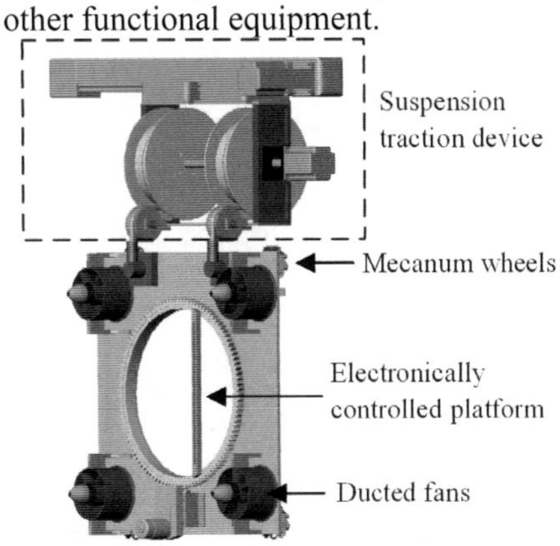

Figure 1. The three-dimensional model of the wall-climbing robot.

(a) Right-handed wheel (b) Left-handed wheel

Figure 2. The Mecanum wheel

3. Control System Design of Wall-Climbing Robot

3.1 Electric control system design

The framework of the electric control system of the wall-climbing robot is shown in Figure 3. The wireless remote controller sends commands to the receiver and then sends to the microcontroller, which controls the McNum wheel and the suspension device according to the remote controller commands and the feedback signals from the gyroscope. The radio remote controller is consisted of a ten-channel RadioLink AT9S model remote controller and a R9DS receiver. The model of microcontroller is STM32F405RGT6. The model of gyroscope is MPU6050.

The signal output mode of the receiver adopts uses SBUS-PWM mode, and the other two independent PWM channels directly control the front and rear ducted fans motor speed respectively. The microcontroller receives multiple PWM signals causing high burden of the hardware performance. Using SBUS serial communication to receives the remote controller command can greatly reduce the microcontroller burden. The microcontroller controls and drives the motor by through L298N. The motor rotation direction, suspension or braking are controlled by changing two IO levels, and the motor speed is controlled by changing the duty cycle of the PWM signal. The output shaft of motor connects to the wheel and motor encoder. The motor speed is measured by the M method and then feedbacks to the microcontroller. The gyroscope is used to obtain the robot pose. The microcontroller controls the speed of the two motors of the suspension traction device through the ATK-LORA-01 wireless serial communication.

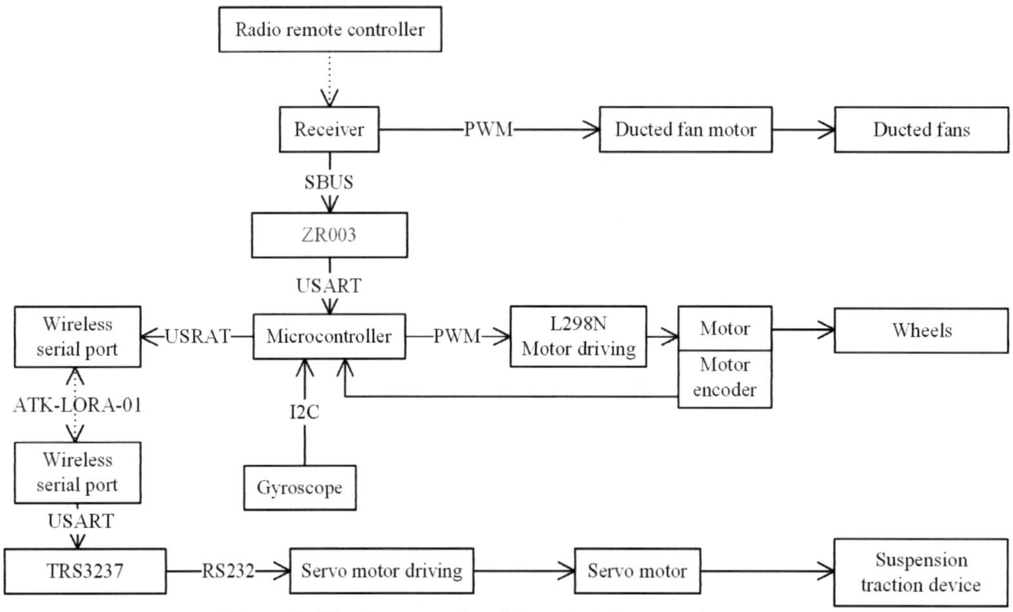

Figure 3. The framework of the electric control system.

The energy power for the whole machine is provided by lithium batteries. The power of the wall climbing robot is large, and the peak current of the ducted motor can reach 80A. Therefore, the wheels motor adopts 11.7v3s lithium battery as power supply, while the ducted motor adopts ESC used 22.2V lithium battery as independent power supply, as shown in Figure 4. The ducted motor ESC has a built-in buck module to supply power to the remote controller receiver, and the L298N has a built-in buck module to supply power to the microcontroller. All the electronic control original parts share a common electrical ground.

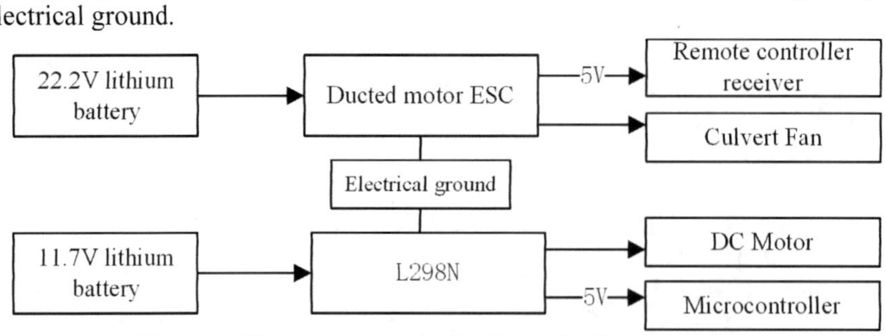

Figure 4. The power supply for the wall-climbing robot.

3.2 Kinematic analysis and PID control design

The relationship between the chassis motion and the McNum wheels motion is analyzed by establishing the kinematic model. Firstly, the motion state of an individual McNum wheel is studied, and the velocity decomposition is carried out for the right front wheel as an example, as shown in Figure 5. $\boldsymbol{V}_{wa}$ is the linear velocity at the center of a wheel, $\boldsymbol{V}_{wa1}$ is the component of $\boldsymbol{V}_{wa}$ along the axis direction of the small roller, and $\boldsymbol{V}_{wa2}$ is the vertical component, which makes the small roller do spin motion rotate around the axis and will not affect the overall motion of the wall climbing device. Therefore, the $\boldsymbol{V}_{wa2}$ is not considered in the chassis motion analyzing.

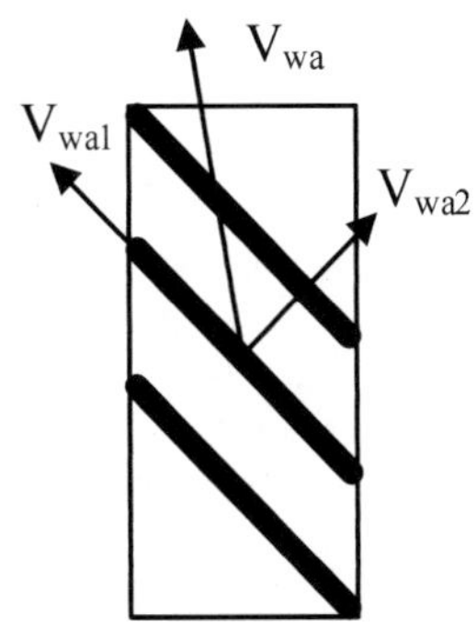

Figure 5. The velocity decomposition diagram of the right front wheel.

Then analyze the rotation of the McNum wheel. The radius and angular velocity of the McNum wheel are r and ω_i, respectively. The linear velocity of McNum wheel landing point rotation v_{wi} can be calculated as follows:

$$v_{wi} = \omega_i r \tag{1}$$

The angle between the small rollers of the McNum wheel and the rim is 45°. The relationship between the component $\boldsymbol{V}_{wi1}$ of the $\boldsymbol{V}_{wi}$ which along the axis direction of the small rollers can be calculated as follows:

$$|\boldsymbol{V}_{wi1}| = v_{iw1} = v_{wi} \cdot \cos 45^{\circ} \tag{2}$$

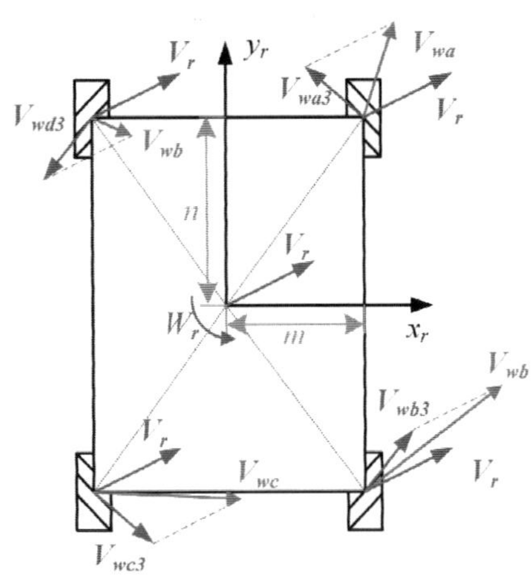

Figure 6. The velocity analysis of the wall-climbing device.

The overall velocity analysis of the wall-climbing device is shown in Figure 6. The x_r-o-y_r is the coordinate system of the wall-climbing device, which origin is the center of the rectangle formed by the four wheels. The motion state of the climbing robot can be expressed as $\boldsymbol{V} = [\boldsymbol{V}_r \quad \boldsymbol{W}_r]^T$, where $\boldsymbol{V}_r$

is linear velocity vector of wall climbing robot the and $\boldsymbol{W}_r$ is the rotation. The linear velocity $\boldsymbol{V}_{wi}$ of the McNum wheel is provided by the wall-climbing device, which can also be considered as a superposition of the wall-climbing device linear velocity $\boldsymbol{V}_r$ and the rotational angular velocity on the McNum wheel linear velocity $\boldsymbol{V}_{wi3}$, as shown in equation (3).

$$\boldsymbol{V}_{wi} = \boldsymbol{V}_r + \boldsymbol{V}_{wi3} = \boldsymbol{V}_r + \boldsymbol{W}_r \times \boldsymbol{R}_i \tag{3}$$

where $\boldsymbol{R}_i$ is the vector of the chassis center of the wall climbing robot to the McNum wheel center.

According to equations (1-3), the relationship between the rotation angular velocity of four-wheels and the velocity vector of device movement under ideal conditions can be calculated as follows:

$$\boldsymbol{V} = \begin{bmatrix} \boldsymbol{V}_r \\ \boldsymbol{W}_r \end{bmatrix} = r \begin{bmatrix} -1 & 1 & -1 & 1 \\ 1 & 1 & 1 & 1 \\ \dfrac{1}{m+n} & \dfrac{-1}{m+n} & \dfrac{-1}{m+n} & \dfrac{1}{m+n} \end{bmatrix} \times \begin{bmatrix} \omega_a \\ \omega_b \\ \omega_c \\ \omega_d \end{bmatrix} \tag{4}$$

where m and n are the distance from origin to wheels in x-axis and y-axis, respectively.

A cascade PID control algorithm is designed according to the actual demand, as shown in Figure 7. The angular displacement PID controller of wall-climbing robot chassis is the first-PID controller, and the motor speed PID controller is the second-PID controller. The angular displacement signal sent by the remote controller is the expectation of the first-PID controller. The the absolute value of angular displacement which is measured by MPU6050 is the feedback of the first-PID controller. The first-PID controller outputs the angular velocity $\boldsymbol{W}_r$, and the remote controller sends the linear velocity expectation $\boldsymbol{V}_r$. The $\boldsymbol{V}$ is calculated by kinematic solution, which is the input of the second-PID controller, and the motor encoder increment is the feedback of the second-PID controller.

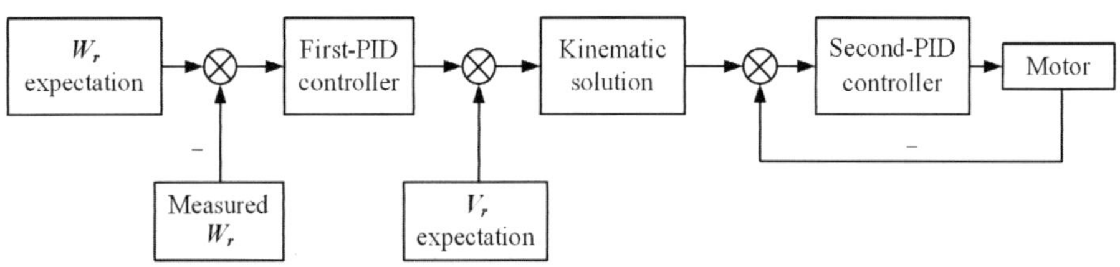

Figure 7. The schematic diagram of the cascade PID controller.

In the digital system, the PID controllers use the incremental PID algorithm, which is calculated using the increment of the control quantity, as follows:

$$u(k) = u(k-1) + (K_P + K_I + \frac{K_D}{T})e(k) - (K_P + \frac{2K_D}{T})e(k-1) + \frac{K_D}{T}e(k-2) \tag{5}$$

Where $e(k)$ is the difference between measurement and the expectation of k-th calculation; K_P, K_I and K_D are the coefficients of proportionality, integration and differentiation, respectively; $u(k)$ and $u(k-1)$ are the k-th and (k-1)-th controlled quantities, respectively.

4. Experimental results and analysis

The prototype of the wall-climbing robot was fabricated according to the design scheme, as shown in Figure 8. The size of wall-climbing robot body is 550 × 550 mm. The rated speed of the ducted fan is 51060 rpm. The robustness of the control algorithm indicates the stability of motion state under different motion commands, which is determined by the PID control parameters. The dual-loop PID parameters were adjusted through repeated experiments and finally determined as Eq.(6).

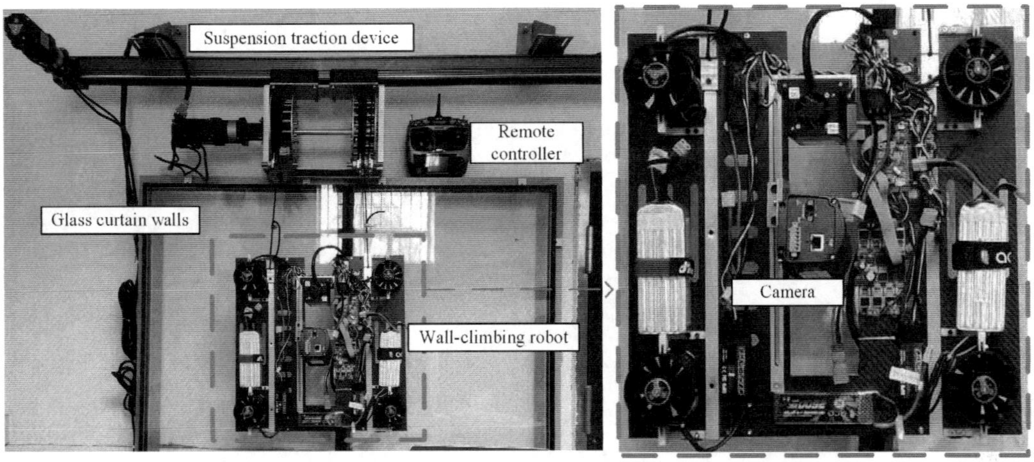

Figure 8. The physical prototype of the wall-climbing robot.

$$\begin{aligned} K_{P1} = 40, K_{I1} = 5, K_{D1} = 40 \\ K_{P2} = 40, K_{I2} = 0, K_{D2} = 20 \end{aligned} \tag{6}$$

Where K_{P1}, K_{I1}, K_{D1} are the inner-loop proportional, integral and differential coefficients respectively; K_{P2}, K_{I2}, K_{D2} are the inner-loop proportional, integral and differential coefficients respectively.

In the paper, the wall-climbing robot is used to detect to damage of glass curtain walls by installing a camera. In detection operation, the shouting fan is turned on and the thrust magnitude is adjusted to make the wall-climbing robot stably adsorbed on the glass curtain walls. The wall-climbing robot is perpendicular to the ground, controlled by a rocker offset. The running speed in the vertical and transverse directions are controlled by the other two rocker offsets. During the robot's motion, its heading remains constant and the suspension traction device is able to move in concert with the robot to overcome the robot's own gravity. By comparing the remote controller command with the McNum wheel speed, the response speed of the motion command does not exceed 0.5s when the wall-climbing device is running, and the maximum overshoot does not exceed 5%. The performance of this control system can meet the inspection demand.

5. Conclusion

In this paper, a wall-climbing robot with suspension traction device, Mecanum wheels, ducted fans and electronically controlled platform is designed and fabricated, which has advantages of high relative load, omnidirectional movement, strong motion continuity and smooth operation. Moreover, targeted electronic control system and control algorithm are designed. The experiment verifies that the wall-climbing robot can achieve all-round smooth operation under remote control, which can be equipped with corresponding functional devices to realize the detection and cleaning of walls, especially glass curtain walls.

References

[1] Da C. "Study on the dynamics analysis and navigation control of tracked wall-climbing robot". Changchun: Jilin University (2021).

[2] Zhao J, Li X, Bai J. "Experimental study of vortex suction unit-based wall-climbing robot on walls with various surface conditions". Proceedings of the Institution of Mechanical Engineers, Part C: Journal of Mechanical Engineering Science, 232(21), 3977-3991 (2018).

[3] Ruihua J. "CFD analysis and optimization research on the adsorption device of tracked wall climbing robot". Changchun: Jilin University (2021).

[4] Fan J, Xu T, Fang Q, Zhao J and Zhu Y. "A novel style design of a permanent-magnetic adsorption mechanism for a wall-climbing robot". Journal of Mechanisms and Robotics,

12(3), 035001 (2020).
[5] Huo P, Xu Y, Yu J and Wang, Y. "Design and analysis of a new frame-foot wall-climbing robot". Proceedings of the Institution of Mechanical Engineers, Part D: Journal of Automobile Engineering, 236(7), 1460-1473 (2022).
[6] Yoshida Y, Ma S. "Design of a wall-climbing robot with passive suction cups". 2010 IEEE International Conference on Robotics and Biomimetics. IEEE. 1513-1518 (2010).

The Application of Deep Learning on Room Temperature Conductivity of $Li_{10}GeP_2S_{12}$ Type Solid State Electrolyte

Jie Peng[1], Jianwen Du[2] and Liping Wang[1*]

[1] Electronic and Electric Engineering Department, Zhaoqing University, Zhaoqing, China, 526061

[2] Electronic Engineering Department, Jinan University, Guangzhou, China, 610632

[*]982685359@qq.com

Abstract. Room temperature ionic conductivity of the solid-state electrolyte is the essencial element of the commercialization of solid-state lithium-ion batteries. First-principles method is a good tool as it simulates the atomic level dynamics, and provides accurate insides of the ion moving velocity, path, and the microscopic environment. However, the first-principles method is time and computational resources-consuming. It is usually applied to systems with hundreds of atoms and the simulation time is less than 100 picoseconds. The classical molecular dynamics method applies to systems containing thousands of atoms and the simulation time can reach the scale of nanoseconds. However, it is based on the experimental force-field parameters and can only be applied to limited systems with experimentally-proofed good parameters. The method used in this paper is a combination of the first-principles simulation and classical dynamics. Deep-learning tool Deepmd-kit is used to train the first-principles simulation data of the targeted system into a tailor-made force-field model. The verified model for the specific system will then be used in the classical molecular dynamics simulation. With the first-principles precise and classical efficiency, the molecular dynamics simulation of ion movement at room temperature is achieved. The results of $Li_{10}GeP_2S_{12}$ and $Li_{10}SiP_2S_{12}$ showed good agreement with the experiments.

1.Introduction

As a next-generation lithium battery technology, all-solid-state lithium batteries show great application potential in terms of safety, energy density and operating temperature range compared with the traditional organic liquid lithium batteries [1]. However, among the known superionic conductor materials, the lithium ionic conductivity rarely reaches 1 mS/cm or higher at room temperature, which is believed to be the lowest requirement for applicable charging efficiency. [1-3] First-principles simulation method has been applied to obtain the detail information of the atomic structure and ion moving path to improve the ion moving inside the electrolyte. Ab initio molecular dynamics simulation has been conducted for the $Li_{10}GeP_2S_{12}$ (LGPS) system by Ceder and fellow scientists [4]. However, due to the time and computational resource consuming nature of the ab initio molecular dynamics(MD) simulation, the lithium ions were hardly move within the simulation time scale. To obtain the visible movement of ion, the temperature of the system was set to 600K, 800K, 1000K, 1200K and 1400K respectively. The room temperature ion conductivity was obtained by the Arrhenius relationship between the temperature and the diffusion coefficient. The limitation of this method is the assumption of the linearity of the Arrhenius relationship. However, Arrhenius relationship is not valid in the extended range of temperature in most of solid-state systems. It is not precise to achieve the room temperature ionic conductivity through the linear extension from the high temperature results.

To simulate the diffusion coefficient at room temperature, we propose a combination method of the ab initio and classical molecular dynamics through the state of art deep learning method. Two LGPS-type materials were under investigation. The potential energy functional models were trained and verified before being applied to the classical molecular simulation process. The ab initio accuracy and the classical efficiency are achieved at the same time. The simulation results of the two systems have good agreement with the experiments.

2.Methods

2.1.Deep learning applied to Molecular Dynamics

In recent years, machine learning has emerged as a promising tool for addressing the difficulty of high-dimensional function representation, with heuristic implications for constructing complex physical models. As an important branch of machine learning, deep learning has an excellent performance in computer vision [5], speech recognition [6], and natural language processing [7]. The key element is a deep neural network (DNN), with multiple hidden layers consisting of nonlinear functions. In 2007, Behler and Parrinello [8] constructed a potential energy surface model (BPNN) trained with the density functional theory (DFT) simulation data DFT and extended the model to investigate a chemical reaction process. Based on the BPNN model, Boes et al. studied the vacancy diffusion in defected gold bulk [9], and Li et al. studied the lithium-ion diffusion coefficient of amorphous Li3PO4 [10]. The applications demonstrated that the BPNN approach has good commonality and flexibility.

2.2.The tool Deepmd-Kit

Deep potential molecular dynamics Kit (DeePMDkit) is an open source tool to describe the potential energy of interatomic interactions. This concept was first proposed by Zhang Linfeng, Wang Han, and co-workers in 2018 [11]. The tool has built interfaces with several popular first-principles packages, for example, the first-principles packages vasp[12] and quantum-espresso[13] and the classical molecular dynamics package lammps[14]. DeepMD-kit is trained by the neural network, and the goal is to complete the mapping of atomic coordinates to energy:

$$E = E(r_1, r_2, \dots\dots, r_n) \approx DNN(r_1, r_2, \dots\dots, r_n) \tag{1}$$

Wherer_n is the coordinate of the atom number n. DNN represents the deep learning neural network model. The total Energy E defined in DeepMD-kit contains not only the energy of each atom but also the potential energy generated by the interaction of the atom with surrounding adjacent atoms. Atoms close to the targeted atom contribute a large interaction force; atoms far away contribute a small interaction force. Truncation radius is a tuneable parameter. Atoms outside of the circle is approximated given zero contribution.

3.Models construction and verification

3.1.First-principles simulation of structures $Li_{10}GeP_2S_{12}$ and $Li_{10}SiP_2S_{12}$

Two structures, $Li_{10}GeP_2S_{12}$ and $Li_{10}SiP_2S_{12}$ were under investigation. They are identical except for the Ge/Si atoms. Both structures were optimized in Density Functional Theory (DFT) with Perdew-Burke-Ernzerhof generalized gradient approximation (PBE-GGA) functionals in the Quantum Espresso package. The DZP basis and an energy cutoff of 80 Hartree were also applied. The Brillouin zone was sampled with 2×2×2 Monkhorst and Pack k-points. This setup is carried out in the DFT simulation throughout the paper. The optimized structure was given an initial temperature of 100 K and a Boltzmann distribution before being heated up to 5 different temperatures in 2 ps. After that, a canonical ensemble pre-equilibration process (particle number N, volume V and the simulated temperature T remain unchanged) was operated for 5ps before the 60 picoseconds ab initio molecular dynamics (AIMD) simulation.

3.2. The formation and verification of the training sets

Each material was simulated with a step size of 2 fs for a total of 30,000 steps of 60 ps at each temperature. In total 30,000 frames of data were obtained after the AIMD simulation. Each frame was a snapshot on an MD trajectory, which contained information on atom coordination, total energy, and forces applied to each atom. To reduce the data redundancy and the correlation between adjacent frames, the valid data sets were selected every 100 MD steps. The total number of frames obtained for each material was 1500. The 300 frames obtained at a temperature of 600K were selected as the testing set and the rest of the frames were used to train the model. The reason for choosing 600K as the testing set was to check if the model trained by high-temperature data was applicable at a lower temperature. The potential energy surface model was constructed using the smoothing model of the DeePMD-kit [12]. The Adam stochastic gradient descent method [13] built into the TensorFlow [14] framework is used to update the neural network hyperparameters. The number of layers and sizes of neurons in the embedding network and fitting network were set to (25, 50, 200) and (240, 240, 240), respectively. The cutoff radius was set to 6 Å, and the edge smooth transition range was set to 0.5 Å. The initial learning rate was set to 10^{-3}, and the final drop ends at 3.5×10^{-8}. The loss function does not consider the virial term ($\mathcal{P}G = 0$), the initial energy loss factor $\mathcal{P}1$ was set to 0.02; the final change was set to 2. The initial stress loss factor $\mathcal{P}7$ was set to 1000, and the final change was set to 1.

3.3. Verification of models

Figures 1 and 2 are the loss function of the LGPS and LSiPS potential energy models vs the number of training steps in logarithmic coordinates. The errors in the legend refer to the root mean square error (RMSE) of each component, the top orange and blue curves are the total errors, the middle brown and purple curves are the force errors, and the bottom red and green curves are the error in energy. It can be seen that the RMSE of each component of the validation set and the training set of the two models almost coincides, and the changing trend is consistent, indicating that the model trained with the training set data of high-temperature (excluding 600 K) AIMD can well predict low-temperature behavior, the model has temperature adaptability.

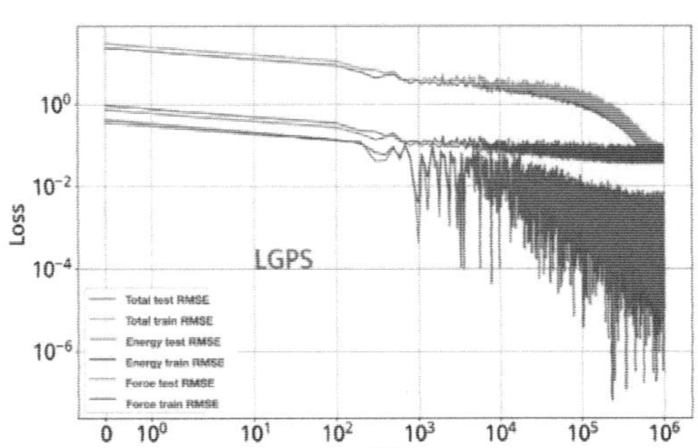

Figure 1. loss function of various parameters of system LGPS.

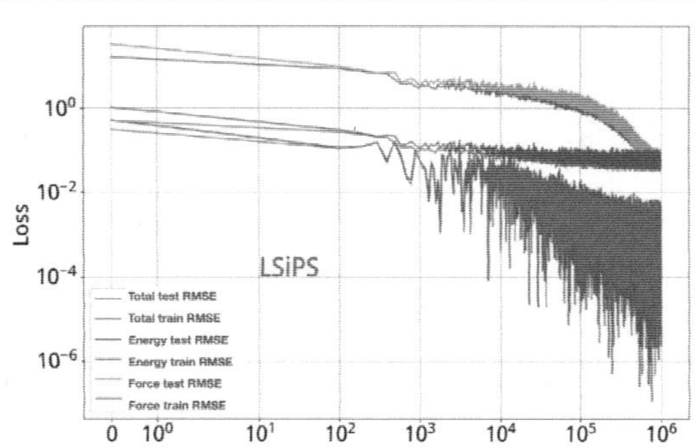

Figure 2. loss function of various parameters of system LSiPS.

The RMSE of energies were much easier to converge comparing with the forces applied on each atom. That's because the total energy had only one data set in one frame and forces had the dimension of atom number times 3 directions. Adding force as an evaluation criterion for the total loss error helps to prevent the model from overfitting. Table 1 listed the RMSE of energy/atom and force/atom.

Table 1. RMSE of energy/atom and force/atom.

	LGPS	LsiPS
Energy/atom(meV)	1.3	5.3
Force/atom(meV/Angstrom)	48.7	48.0

Figures 3 and 4 are the prediction and verification of the energy and force of the two models for different materials at 600 K. The DFT energy and forces were taken from the AIMD data at 600 K, and the DPMD energy and force were calculated through the potential energy function model. The black solid lines (y=x) is for guiding the eyes. The red data dots distributed close the straight line, indicating that the DPMD data and DFT data predicted by the three models are close to the relationship of y=x, indicating the model was successful.

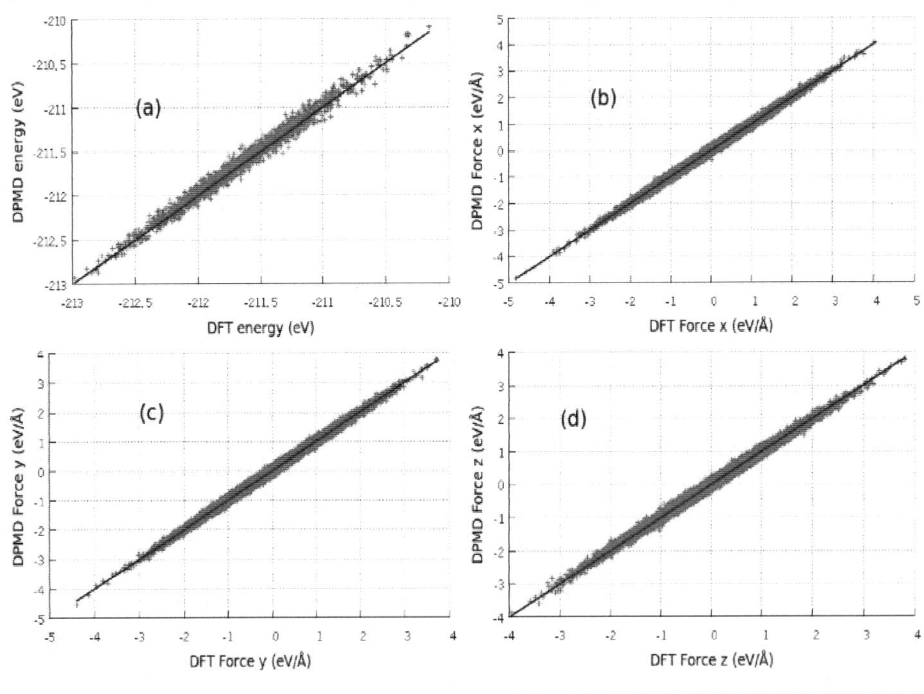

Figure 3. DFT energy and forces comparing with the DPMD energy and forces of system LGPS

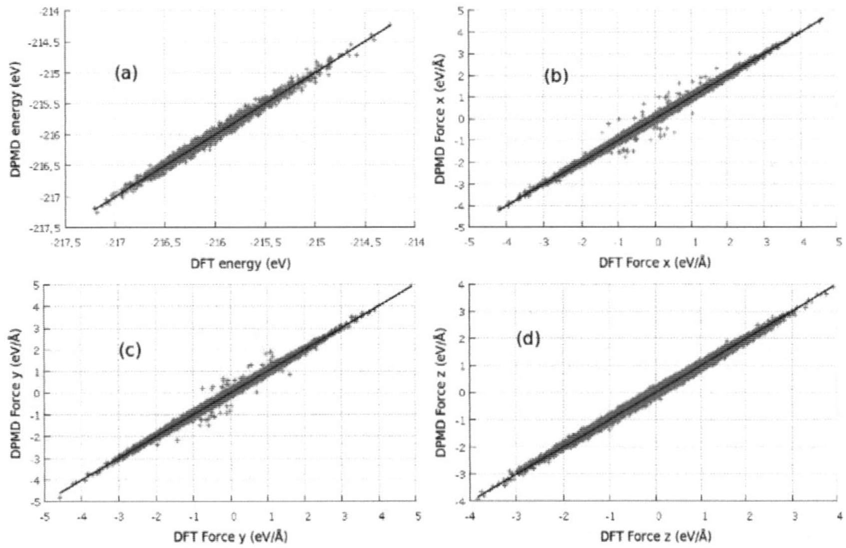

Figure 4. DFT energy and forces comparing with the DPMD energy and forces of system LSiPS

4.Room temperature ionic conductivity simulation

The package lammps was applied to run the room temperature molecular dynamics simulation with the verified potential model obtained above. The time step was set to 0.5fs, and the total simulation time was set to160 ps. Figures 5 and 6 displaced the variation of the mean square displacements (MSDs) of lithium ions in all directions at room temperature. Both of the materials have a tetragonal structure with isotropic diffusion properties in directions a and b and an anisotropic diffusion property in direction c. Ions moved much faster at the c direction, which was in agreement with the previous the DFT simulation[4].

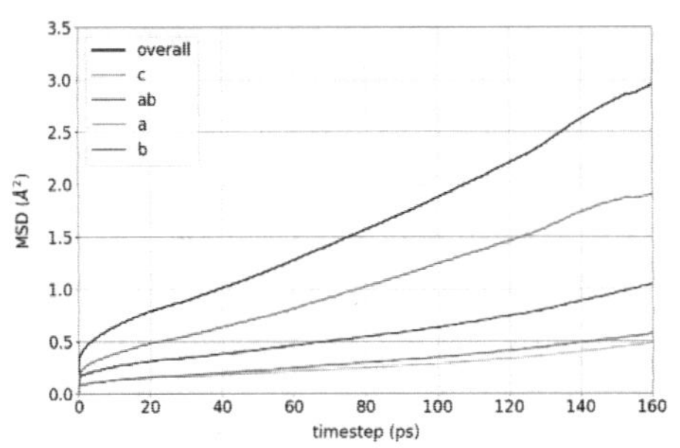

Figure 5. MSDs of lithium ions in all directions at room temperature (LGPS).

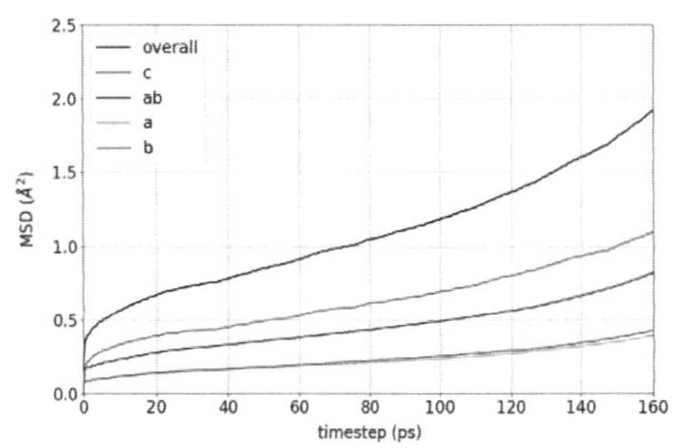

Figure 6. MSDs of lithium ions in all directions at room temperature (LSiPS).

The simulated and experimental ionic conductivities were listed in Table 2. The simulation results show good agreement with experiments.

Table 2. Ionic conductivities from simulation and experiments(mS/cm)

material	simulation	experiment
LGPS	30.9	2-40
LSiPS	14.6	0.4-6

5.Conclusion

In summary, the deep learning method has been successfully applied to the simulation of ionic conductivities in solid state electrolyte systems. The potential models trained by the AIMD data obtained at the temperatures of 800K, 1000K, 1200K, and 1400K were verified by the AIMD data obtained at the temperature of 600K, indicating that model trained by high temperature is applicable in low temperature. The standard criterion of RMSE of force/atom in the machine learning potential model is below 100meV/Angstrom. In both LGPS and LSiPS models, the RMSE of force/atom was lower than 50meV/Angstrom. Applying the potential model to the classical molecular dynamics, room temperature simulation for the ionic conductivity of solid state electrolyte became feasible. The calculated results were in good agreement with the experiments.

Acknowledgments

This paper has been funded by Guangdong Science and Technology Commissioner Project through the grant GDKTP2021024100 and the PhD start-up fund from Zhaoqing University.

References

[1] Hu Y S 2016 Batteries: Getting solid. (Nature Energy) 1(4): 16042.

[2] Kamaya N et al. 2011 *A lithium superionic conductor* (Nature Materials) 2011, 10(9): 682-686.

[3] Kato Y, Hori, Saito T 2016 *High-power all-solid-state batteries using sulfide superionic conductors* (Nature Energy) 1(4): 16030.

[4] Mo Y, Ong S P, and Ceder G 2012 *First Principles Study of the Li10GeP2S12 Lithium Super Ionic Conductor Material* (Chem. Mater.) 24, 1, 15–17

[5] ZHANG L. 2020 *Machine learning for multi-scale molecular modeling: theories, algorithms, and applications*. (Applied and Computational Mathematics, Academic dissertations Ph.D.)

[6] Krizhevsky A, Sutskever I, Hinton G E 2017 *ImageNet Classification with Deep Convolutional Neural Networks* (Communications of the ACM) Volume 60 Issue 6 p 84–90

[7] Hinton G, Deng L, Yu D 2017 *Deep Neural Networks for Acoustic Modeling in Speech Recognition : The Shared Views of Four Research Groups* (IEEE Signal Processing Magazine) vol. 29, no. 6, pp. 82-97

[8] Behler J, Parrinello M 2007 *Generalized Neural-Network Representation of High-Dimensional Potential-Energy Surfaces* (Physical Review Letters) 98(14): 146401.

[9] Boes J R, Groenenboom M C, Keith J A 2016 *Neural network and ReaxFF comparison for Au properties* (International Journal of Quantum Chemistry) 116(13): 979-987.

[10] Li W, Ando Y, Minamitani E. 2017 *Study of Li atom diffusion in amorphous Li3PO4 with neural network potential* (The Journal of Chemical Physics) 147(21): 214106.

[11] Zhang L, Han J, Wang H 2018 *Deep Potential Molecular Dynamics: A Scalable Model with the Accuracy of Quantum Mechanics* (Physical Review Letters) 120(14): 143001.\

[12] G. Kresse and D. Joubert, 1999, *From ultrasoft pseudopotentials to the projector augmented-wave method* (Phys. Rev.) 59 , 1758.

[13] P. Giannozzi et al. 2009 *Quantum Espresso: A Modular and Open-Source Solfware Project for Quantum* (J.Phys.: Condens.Matter) 21, 395502

[14] S. Plimpton 1995 *Fast Parallel Algorithms for Short-Range Molecular Dynamics* (Comp Phys) 17, 1-19
[15] Zhang L, Han J, Han H, Saidi W, Car R, Weinan E 2018 *End-to-end symmetry preserving inter-atomic potential energy model for finite and extended systems* (NIPS'18: Proceedings of the 32nd International Conference on Neural Information Processing Systems) p 4441–4451
[16] Kingma D P, Adam BA J 2022 *A Method for Stochastic Optimization* (arXiv) :1412.6980
[17] Abadi M, Agarwal A, Bbrham P. 2016 *TensorFlow: Large-Scale Machine Learning on Heterogeneous Distributed System* (OSDI'16: Proceedings of the 12th USENIX conference on Operating Systems Design and Implementation) p 265–283

Matlab/Simulation of Brushless DC motor vector control

Zi-xi TAN[1,*], Li-bo Ding[1]

[1]School of Mechanical Engineering, Nanjing University of Science and Technology, Nanjing, China 210014

[*]Corresponding author: Zi-xi TAN 1074733279@qq.com

ABSTRACT: Based on improving the dynamic and static performance of the brushless DC motor control system, a speed current closed-loop brushless DC motor control scheme is designed by using SVPWM. In the light of the environment of matlab/simulink, the vector control model of BLDCM is established by using the control strategy of id=0 and the idea of modularization. The simulation results show that the system has good dynamic and static performance and stable operation, which proves the reliability and effectiveness of the simulation model.

1. Introduction

With the rapid development of current power electronics technology, microelectronics technology and motor control theory, motor vector control technology is becoming more and more mature. The research on vector control of Brushless DC motor is also widely used in the industrial field, such as rocket actuator control, UAV control, industrial robot and so on. Compared with other motors, brushless DC motor maintains the excellent speed regulation performance of DC brush motor, overcomes a series of shortcomings brought by the converter composed of rotating mechanical commutator and static brush, and becomes the main power element in motor energy conversion and motion control[1]. The brushless DC motor is usually driven by square wave, and the control method is simple, but it is difficult to be used in high-precision and low-noise occasions because of the shortcomings of torque ripple and commutation noise; Through the sine wave drive control of the brushless motor, the sine wave voltage is applied to the motor winding, and the control output torque is obtained by controlling the amplitude and phase of the sine wave voltage. Compared with traditional square wave control, vector control technology can greatly improve the smoothness of motor torque, and meet the requirements of high precision, fast and stable operation[2]. Under the environment of Matlab /Simulink, this paper simulates the brushless DC motor vector control system and analyzes the simulation results to obtain the position, current, speed and torque response characteristics of the control system.

1.1 Principle of Brushless DC motor

Using electronic switch circuit to replace the mechanical commutator of Brushless DC motor is the design idea of Brushless DC motor. The brushless DC motor cancels the brush in order to realize no mechanical contact commutation. Therefore, the brushless DC motor has a commutation device composed of rotor position sensor, control circuit and power inverter bridge[3]. Its basic structure consists of motor body, power drivert and position sensor.

The body of the brushless DC motor includes the stator with armature winding and the rotor with permanent magnet. The stator winding is divided into three phases a, B and C, which are star connected, and the phase difference is 120 °. The power driving circuit is a full bridge driving circuit, which controls the on and off of the three-phase winding current of the motor according to the logic signal of the hall element. Generally, it is a three-way conduction mode, and the phase is changed every one sixth cycle[4]. The power logic switch unit turns on and off with a certain logic distribution relationship, and distributes the current to the three-phase windings on the motor stator, so that the motor can produce continuous torque. The purpose of position sensor is to detect the position of rotor magnetic pole, provide correct commutation information, and drive the rotor to rotate. Hall sensor is often used as position sensor in Brushless DC motor control system.

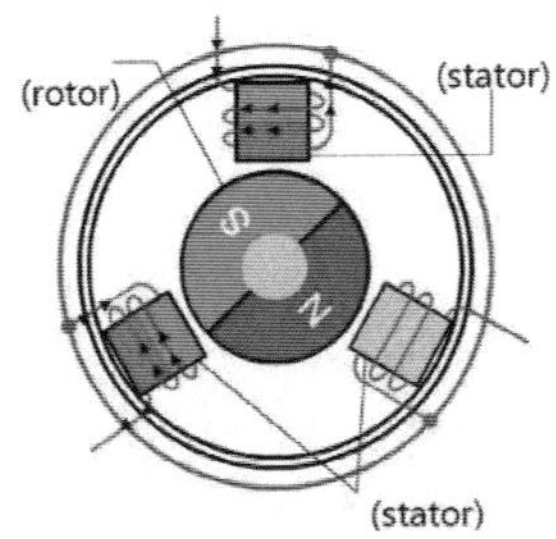

Figure 1. working principle of Brushless DC motor

2 Establishment of simulation modules of Brushless DC motor based on matlab/Simulink vector control

According to the principle of vector control, the simulation model of vector control system is modeled in a modular way, and the current speed closed-loop control system is designed, which can well meet the requirements of vector control. The sub modules of the control system mainly include: SVPWM module, coordinate transformation module, inverter module, etc.

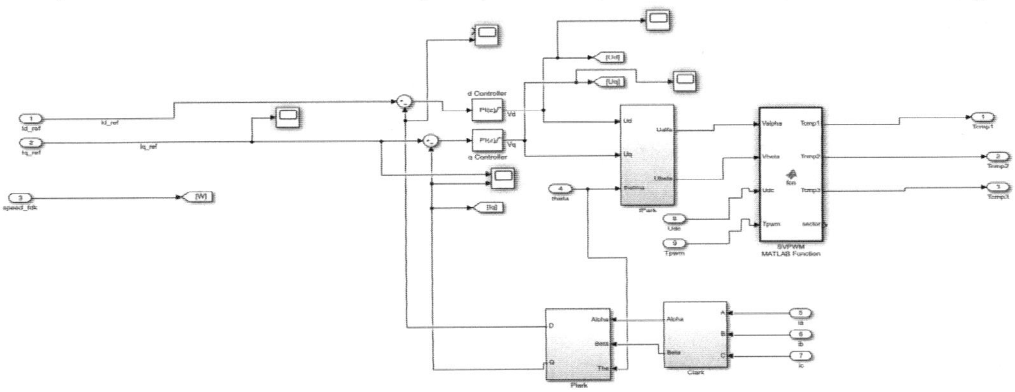

Figure 2. vector control model

The functions of each functional module are as follows:

2.1 SVPWM module

SVPWM adopts the principle of mean value equivalence, judges the sector interval according to the different voltage vectors, and continuously synthesizes the required voltage vector by using the basic vector of the corresponding sector. SVPWM outputs time-varying PWM wave through six power tubes of the three-phase inverter[5], simulates the voltage vector circle formed by three-phase symmetrical sinusoidal current, and generates a nearly circular flux path to achieve the effect of space vector pulse width modulation. Figure 3 shows a typical three-phase inverter drive circuit.

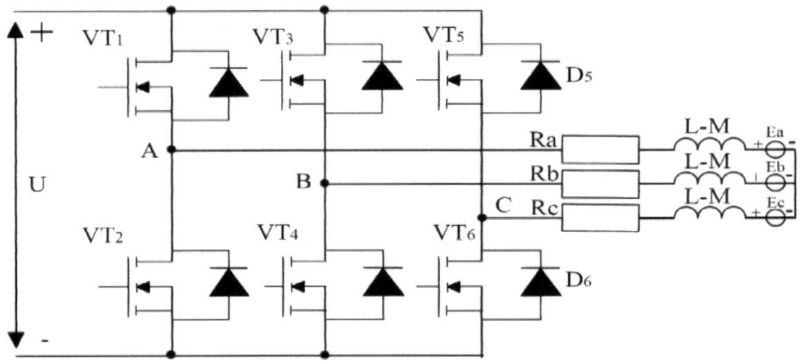

Figure 3. the driving circuit with three-phase reverse converte

To judge the sector of the vector, it is necessary to obtain the components of the two axes of the α - β two-phase stationary coordinate system, Calculate the sector of the composite vector according to U_α , U_β .If U_β >0,then A=1,otherwise A=0;If $\sqrt{3}U_\alpha - U_\beta$ >0, then B=1,otherwise B=0;If $-\sqrt{3}U_\alpha - U_\beta$>0, then C=1,otherwise C=0. Let N=A+2B+4C, calculate the N value through different combinations of A, B and C, and it is very convenient to judge the sector where the vector is located.

The size and direction of the voltage rotation synthetic vector at any time can be composed of other vectors, including the size of the basic voltage vector and the action time of the zero vector[6]. The size and rotation direction of the synthetic vector are determined by the basic voltage vector, and the purpose of the zero vector is to improve the degree of the rotating magnetic field approaching the circle.

2.2 Coordinate transformation module

The vector control process includes Clarke transform, park transform and inverse coordinate transform. Coordinate transformation aims to solve the complex current relationship. The purpose of coordinate transformation is to decouple the complex current relationship. Clarke transformation converts the current to α - β axis through the conversion between three phase stationary coordinates and two phase static coordinates [7]. Park transformation is to convert the two phase stationary coordinates and the d-q coordinate system rotating with the magnetic field to obtain i_d, i_q .

2.3 Inverter module

SVPWM adopts inverter mode and generates corresponding space voltage vector through the switch mode combination of MOSFET or IGBT to approximate the reference circular magnetomotive force generated by three-phase symmetrical sinusoidal current[8]. In the matlab/simulink vector control simulation model, the module drives the brushless DC motor by outputting three-phase voltage. The specific principle is that the six IGBT power switching devices generate the disconnection respectively according to the six groups of control signals generated by the SVPWM module to form the three-phase PWM wave of the driving motor.

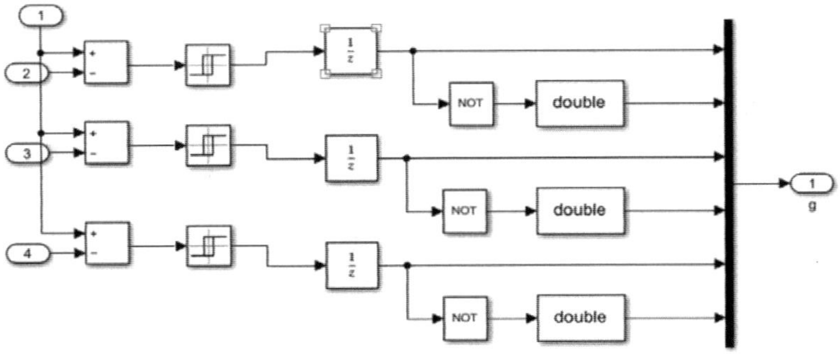

Figure 4. inverter module model

3 Simulation research and Analysis on vector control model of BLDC

In order to verify the correctness and accuracy of the vector control model, the simulation experiment is carried out. The three closed loop vector control model of Brushless DC motor is as follows:

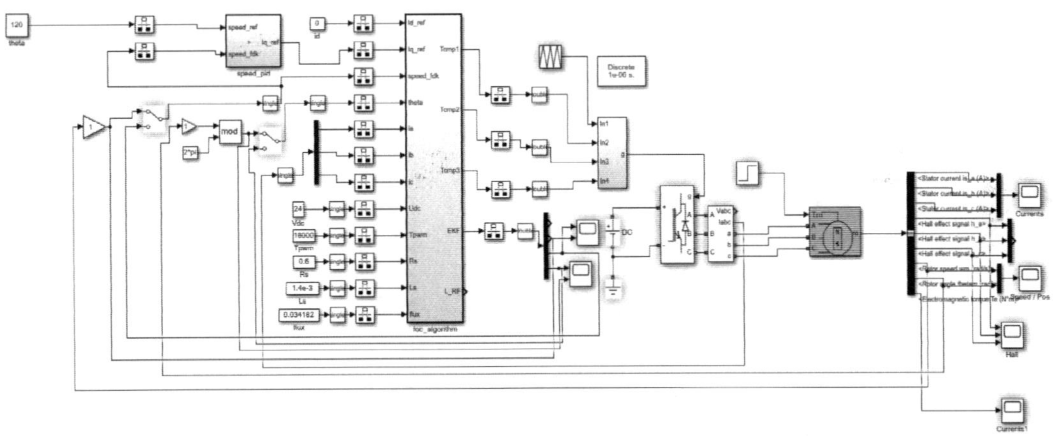

Figure 5. three closed loop vector control system model of Brushless DC motor

In matlab/simulink environment, according to the established simulation model of Brushless DC motor vector control, the motor parameters adopted are as follows: DC voltage V_d =24V,Stator winding resistance R =0.6Ω,Winding self inductance L_σ =0.0014H,Moment of inertia $J=1.1e^{-5}$ kg×m²,Polar logarithm is 1, Rotor flux linkage ψ_f =0.034182Wb.

The system speed is given at 120rad/s and the Load Torque is 0.03 N·m, at this time, the speed response, rotor position response waveform, the electromagnetic twist moment wave shape and stator three phase current wave shape are obtained as shown in the following experiments.

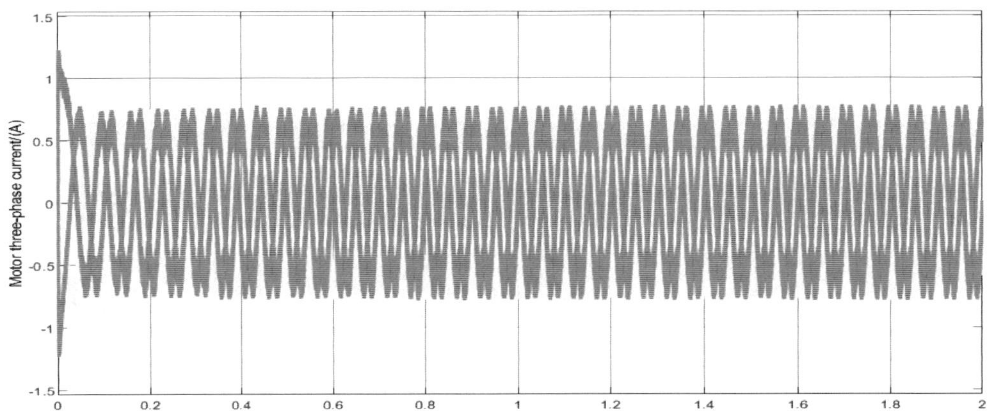

Figure 6. three phase current waveform of motor

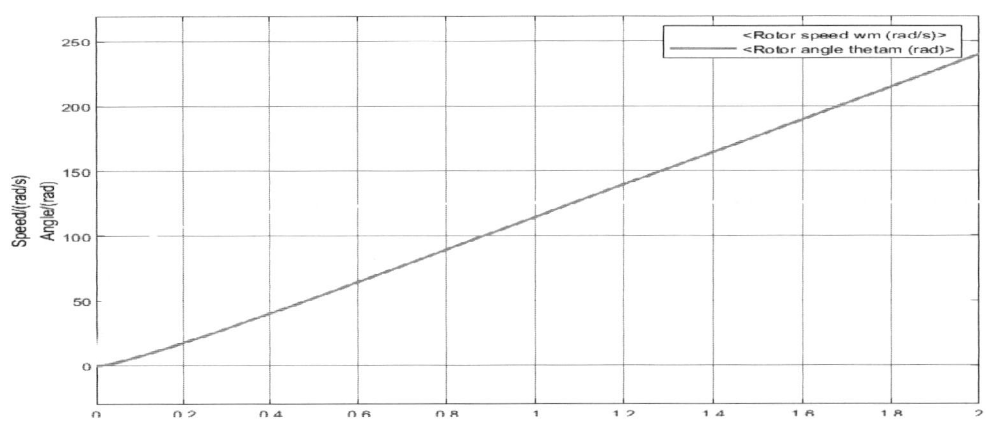

Figure 7. motor speed and position waveform

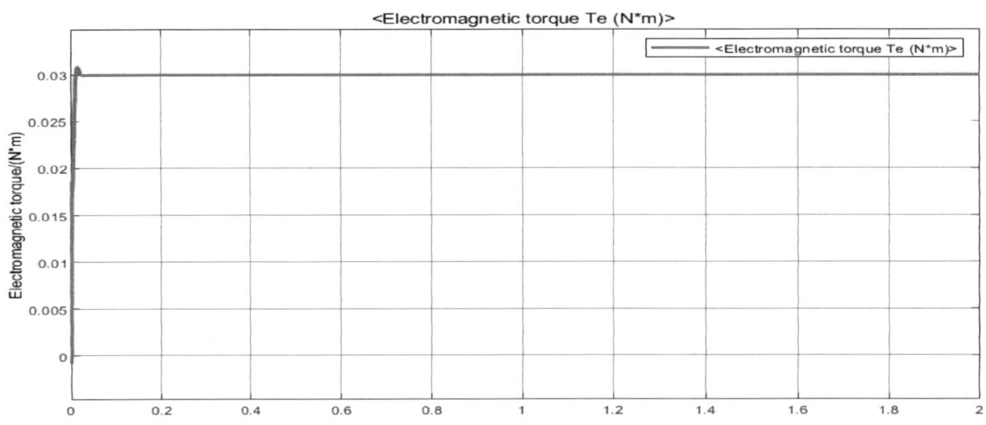

Figure 8. motor torque waveform

According to the result waveform, the starting time of the vector control system is short, the speed rises rapidly, and the motor quickly reaches the given speed of the system. The torque fluctuates greatly in the initial starting time, but quickly returns to 0.03 N·m, in a very short time, and tend towards less volatility; The sinusoidal characteristic of the three phase current of the motor stator is obvious and the operation is stable.

4 Conclusion

According to the principle of vector control, under the environment of Matlab/Simulink, the speed and current closed loop control system of Brushless DC motor is established, and the specific design of each sub module is carried out. According to the experimental simulation results, it can be seen that the control system has good dynamic and static performance and stable operation, which verifies the reliability of the experimental system. It provides a good theoretical basis for the Design and debugging of the actual DC brushless rotary motor control system.

References

[1] Guo qing ding, zhao xi mei. Principle and technical application of Brushless DC motor [M]. China Electric Power Press, 2008:23-24.
[2] Jing xiao bo Simple sine wave control of Brushless DC motor based on xc866 [J]. World electronic components, 2010, (09): 51-54.
[3] Zhang Gang.SVPWM simulation of permanent magnet synchronous motor based on PR control[J]. Automation technology and application, 2019,38 (6): 6-10.
[4] McDonagh James, Palumbo Nunzio, Cherukunnath Neeraj et al. Modelling a permanent magnet synchronous motor in FEniCSx for parallel high-performance simulations[J]. Finite Elements in Analysis & Design, 2022, 204.
[5] QU L M,WANG C,SUN W,et al.Design and simulation of three-level SVPWM based on FPGA[J].IOP Conference Series:Earth.
[6] Chen jian qing. Research and implementation of asynchronous motor vector control system based on DSP [d]. Kunming: Kunming University of technology, 2017.
[7] Chikh H K,et al. PMSM Vector Control Performance Improvement by Using Pulse with Modulation and Anti-Windup PI Controller. Proceedings of International Conference on Multimedia Computing and Systems, Chengdu, China.August 1-3,2011:1-7.
[8] Li SH, Liu ZG. Adaptive Speed Control for Permanent-Magnet Syn-chronous Motor System With Variations of load Inertia [J]. IEEE Transactions on Industrial Electronics,2009,56（8）:3050-3059.

Tool life prediction based on multi-source feature PSO-SVR neural network

Ji Wang[1], Jian Zhou[1], Wen-an Mo[1], Chao Liang[1], Li-jun Sun*[1], Chun-bo Wen[1]

[1] China Electronic Product Reliability and Environmental Testing Research Institute Guangzhou 510610, China.

*Corresponding author:sunlj@ceprei.com.

Abstract. With the continuous improvement of modern manufacturing automation and process intensification, the complexity of tools and machined parts has greatly increased, and their performance directly affects the quality of workpieces and production efficiency. Accurate prediction of tool life is conducive to improving production efficiency and reducing enterprise costs. In this paper, a tool life prediction method based on multi-source feature PSO-SVR neural network is proposed. By monitoring and collecting the current and vibration signals of the tool during the machining process of CNC machine tools, the time-frequency eigenvalues are extracted. The effective features are extracted by feature selection technology as the input of support vector regression (SVR) neural network, and the parameters of the network are optimized by particle swarm optimization (PSO), so as to improve the accuracy of the predict, and finally predict the remaining useful life(rul) of the tool.

1. Introduction

In modern machining, tool life will have an impact on tool scheduling and replacement, processing cost, process parameter formulation and processing quality. Due to the complex factors affecting tool life, simply relying on existing machining experience or related theoretical formulas to calculate tool life is inefficient and inaccurate, making it difficult to guide production. Accurate and effective prediction of tool life can not only improve the quality and efficiency of workpiece processing, but also reduce the additional processing cost caused by tool failure. Therefore, the application research of tool life prediction has been widely recognized in machining[1]. Direct tool life prediction based on raw signals in the machining environment has many problems, such as many types of data, large differences in collection methods, and unclear internal correlation characteristics between data, resulting in difficult prediction and low accuracy[2].

Many well-known scholars or authoritative institutions have done a lot of work in this direction and have made major breakthroughs. Suheng et al. The feature extraction is carried out on the vibration signal by wavelet transform, which is used as the input of the neural network parameters to predict the rul of the tool.[3].Shen Zhi proposed to apply artificial intelligence technology to the field of tool condition monitoring, which solved the multiple correspondence between various signals and tool wear levels, and provided a new way for research in this field. Cao Deng used the cutting force signal to evaluate the influence of the parameter characteristics on the smoothness of the workpiece, and predicted the wear degree of the tool based on machine learning. [4].

In this paper, a multi-source feature-based PSO-SVR neural network tool life prediction method is proposed. By monitoring and collecting current and vibration signals during CNC machining, its time-frequency eigenvalues are extracted. The effective features are extracted by the feature selection technology as the input of the model, and the rul of the tool is predicted. The basic flow of this method is shown in Figure 1.

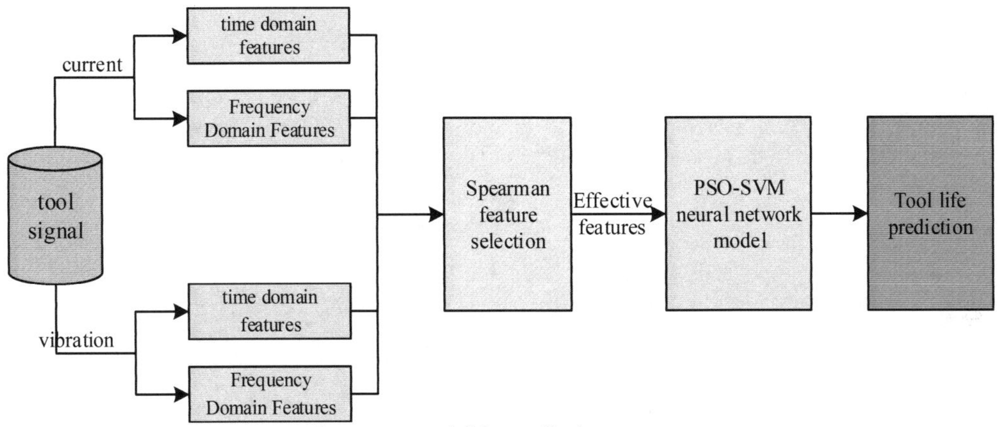

Figure 1. Tool life prediction process.

2. PSO-SVR model construction

2.1. Feature selection

By monitoring the current, vibration, temperature and other physical signals that characterize the state of the tool, the degree of tool wear can be inferred, so as to predict the life of the tool, considering the characteristics of different sensor signals in practical applications and the complexity of the cutting process, multi-sensor signals Fusion has become a popular monitoring method at present, and scholars at home and abroad have conducted in-depth research on sensor configuration, signal processing and feature fusion. In this paper, the time-frequency domain feature extraction is carried out by monitoring the tool current and vibration signal. This paper extracts 22 vibration time and frequency domain features, 37 current time and frequency domain features, and a total of 59 features[5].

Since different features have different physical meanings and different degrees of sensitivity to tool wear, feature selection can not only reduce the number of features, reduce the interference of irrelevant features on the model, but also improve the generalization ability of the model to prevent overfitting. Spearman's correlation coefficient can reflect the dependence between variables. By evaluating the correlation between variables, assigning different weights to each feature to represent the importance of the feature, and sorting the features according to the weights.

In order to characterize the correlation between the fault feature and the number of cuts, it is assumed that the number of cuts is represented by X, a feature extracted earlier is represented by Y, and the i-th ($1<=i<=N$) values are represented by X_i and Y_i respectively. Sort the sets X_i and Y_i respectively (both in ascending or descending order), then x_i and y_i are the serial numbers of X_i and Y_i, then the ranking difference set $d=x_i-y_i$, then the Spearman of the feature correlation coefficient ρ_s can be calculated by the following formula[6]:

$$\rho_s = 1 - \frac{6\sum d^2}{N(N^2-1)} \tag{1}$$

Among them, d represents the level difference of different features, and N represents the number of features. When the obtained ρ_s is greater than 0.8, it indicates that the feature has a deep connection with the original signal and can be used as an effective fault feature

2.2. Support Vector Regression (SVR)

ε-SVR is the most common SVR algorithm. For the training sample set , it is necessary to find a model in the sample space, making it as close as possible to formula (2),

$$f(x) = w^T \mathrm{x} + b \tag{2}$$

As shown in **Figure 2.** SVR Regression Principle, svr is equivalent to constructing a spacer with width ε centered on . On this basis, formula (2) can be transformed into:

$$\begin{cases} \mathrm{y}_i - wx_i - b \le \varepsilon \\ wx_i + b - \mathrm{y}_i \le \varepsilon \end{cases} (i = 1, 2, \cdots, \mathrm{n}) \tag{3}$$

Then the optimization objective is

$$\min \frac{1}{2}\|w^2\| \tag{4}$$

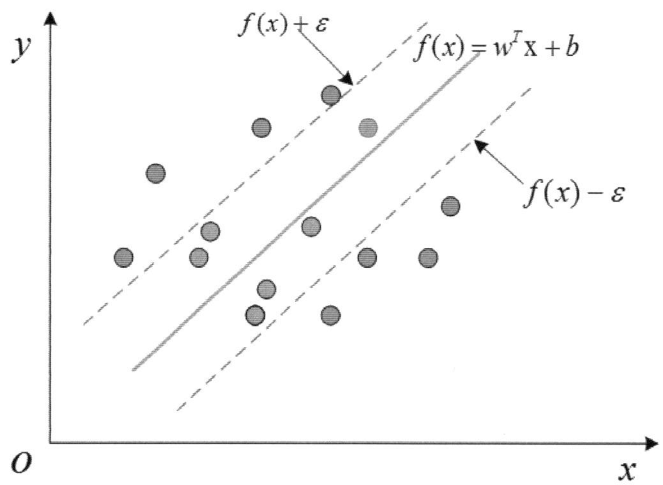

Figure 2. SVR Regression Principle

In the case of allowing the fitting error, the slack variables $\xi_i \ge 0$ and $\xi_i^* \ge 0$ are introduced, then the formula (2) changes to:

$$\begin{cases} \mathrm{y}_i - wx_i - b \le \varepsilon + \xi_i \\ wx_i + b - \mathrm{y}_i \le \varepsilon + \xi_i^* \end{cases} (i = 1, 2, \cdots, \mathrm{n}) \tag{5}$$

Then the optimization objective is:

$$\min \left\{ \frac{1}{2}\|w^2\| + C\sum_{i=1}^{n} \xi_i + \xi_i^* \right\} \tag{6}$$

In the formula, the first item on the right is to improve the generalization ability of learning, and the second item is to reduce the error. The penalty coefficient $C > 0$ makes a compromise between the two, indicating the degree of penalty for the samples that exceed the error ε . Building the Lagrange function

$$L(\mathrm{w},\xi_i,\xi_i^*)=\frac{1}{2}\left\|w^2\right\|+C\sum_{i=1}^{n}(\xi_i+\xi_i^*)-\sum_{i=1}^{n}\alpha_i^*\left[\xi_i^*+\varepsilon+y_i-wx_i-b\right]$$
$$-\sum_{i=1}^{n}\alpha_i^*\left[\xi_i+\varepsilon-y_i+wx_i+b\right]-\sum_{i=1}^{n}(\xi_i^*\gamma_i^*+\xi_i\gamma_i) \quad (7)$$

Taking partial derivatives for w,b, ξ_i and ξ_i^*, we can get:

$$L(\mathrm{w},\xi_i,\xi_i^*)=-\varepsilon\sum_{i=1}^{n}(\alpha_i^*+\alpha_i)+\sum_{i=1}^{n}\gamma_i(\alpha_i^*-\alpha_i)$$
$$-\frac{1}{2}\sum_{i=1}^{n}\sum_{j=1}^{n}(\alpha_i^*-\alpha_i)(\alpha_j^*-\alpha_j)(x_ix_j) \quad (8)$$

Therefore, according to the definition of Wolf duality, under KKT conditions, the dual form of Lagrange is obtained as:

$$\max W(\alpha^*,\alpha)=-\varepsilon\sum_{i=1}^{n}(\alpha_i^*+\alpha_i)+\sum_{i=1}^{n}\gamma_i(\alpha_i^*-\alpha_i)$$
$$-\frac{1}{2}\sum_{i=1}^{n}\sum_{j=1}^{n}(\alpha_i^*-\alpha_i)(\alpha_j^*-\alpha_j)(x_ix_j) \quad (9)$$
$$\sum_{i=1}^{n}(\alpha_i-\alpha_i^*)=0,(0\le\alpha_i,\alpha_i^*\le C,i=1,2,\cdots,n)$$

The regression function is obtained as:

$$f(x)=wx+b=\sum_{i=1}^{n}(\alpha_i^*-\alpha_i)(x_ix)+b \quad (10)$$

Using the kernel function $K(x_i,x)$ to replace the original inner product operation (x_ix) to achieve nonlinear function fitting:

$$f(x)=wx+b=\sum_{i=1}^{n}(\alpha_i^*-\alpha_i)K(x_ix)+b \quad (11)$$

Through comparative analysis, a more appropriate radial basis kernel function is selected:

$$K(x_i,x_j)=\exp(-\frac{\left\|x_i-x_j\right\|^2}{\sigma^2})=\exp(-g\left\|x_i-x_j\right\|^2),g=\frac{1}{\sigma^2} \quad (12)$$

2.3. *Particle Swarm Optimization (PSO)*

In the SVR model, the model penalty function coefficient C and kernel function coefficient g determine the complexity and accuracy of the model. Based on this, this paper adopts the PSO algorithm to optimize the parameters of the SVR neural network according to its characteristics and properties.

The PSP algorithm is used to optimize the PSO-SVM neural network C and g with the highest diagnostic accuracy as the optimization goal. As shown in Figure 3, the model parameter optimization process is as follows[7]:

- Initialize particle group Xi(C,g),including initial position and velocity, Vmax=2, Vmin=-2.
- According to the fitness function, evaluate the fitness of each particle Xi(C,g);

- Find the local optimum point, compare the current fitness value corresponding to each particle (that is, the diagnostic accuracy) with the diagnostic accuracy of the individual best position, pick the current position of the particle with a higher fitness value ;
- Find the global optimum point, compare the current fitness value (the diagnostic accuracy) corresponding to each particle with the diagnostic accuracy of the global best position, pick the current position of the particle with a higher fitness value;
- The velocity and position updates of particles are given by:

$$V_{id} = \omega V_{id} + C_1 random(0,1)(P_{id} - X_{id}) + C_2 random(0,1)(P_{gd} - X_{id}) \quad (13)$$

$$X_{id} = X_{id} + V_{id} \quad (14)$$

Among them, ω is called the inertia factor, C_1 and C_2 are called acceleration constants. Usually set $C_1 = C_2 = 2$, but not necessarily equal to 2, generally take $C_1 = C_2 \in [0, 4]$.

- The algorithm stops when the number of iterations reaches 20 or the prediction accuracy of the validation set reaches more than 90%.

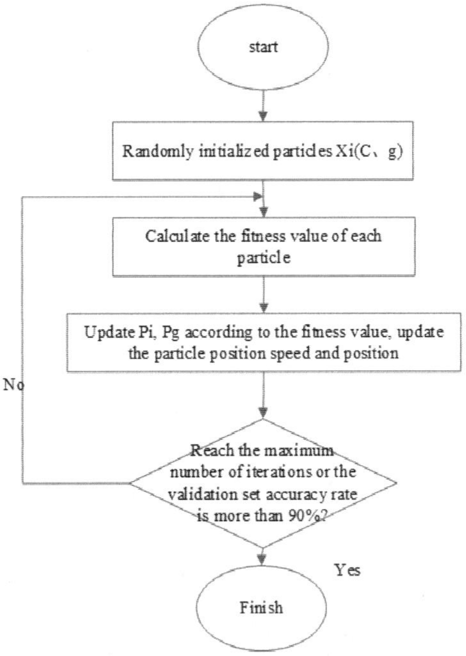

Figure 3.PSO algorithm flow

The above is the algorithm flow of PSO. The hyperparameters in the support vector regression are regarded as the optimization object particles of PSO, and the life prediction accuracy is regarded as the optimization goal. Through the process of the above-mentioned PSO optimization algorithm, the SVR model is continuously optimized, and the most suitable model parameters are finally obtained. The prediction is then optimized to improve the accuracy of tool life prediction[8].

3. Application

Taking a CNC machine tool as the test object, as shown in the Figure 4, the spindle motor is an AC asynchronous motor with a rated current of 26A, a rated speed of 1500rpm, a rated power of 11kW, and 2 pairs of poles. The tool machined the workpiece until it was worn out. As shown in Figure 5 before and after the workpiece was machined, the tool life was measured by the number of remaining cuts.

Figure 4. tool object

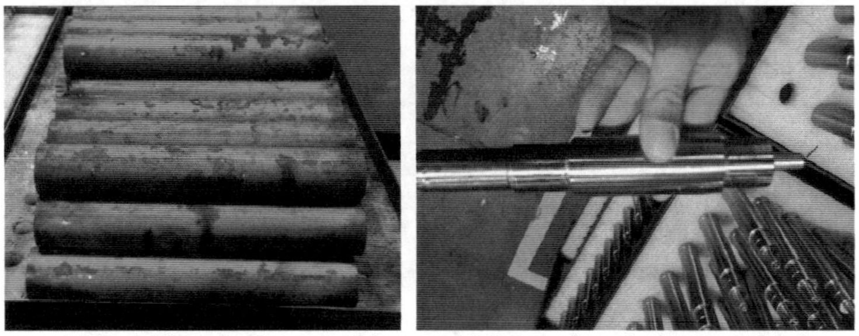

a) Workpiece before machin b) Workpiece before machining

Figure 5. Workpiece before and after machining

The set speed of the frequency converter is 2363rpm, and the acquisition is triggered every time in the first round of cutting, the data is stored for 10s, and the sampling rate is 80kHz. The monitoring points are shown in the **Figure 6**:

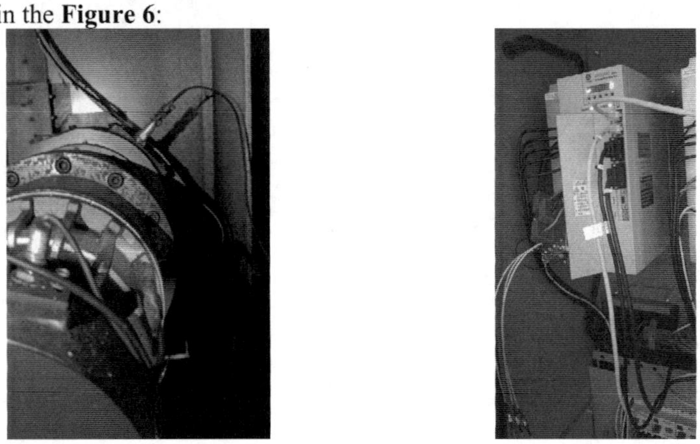

a) Spindle vibration measuring point b) Spindle three-phase current measuring point

Figure 6. Sensor measuring point layout

The vibration and current signals of a certain sampling are shown in Figure 7, and 22 time-domain and frequency-domain features of the vibration signal and 37 time-domain and frequency-domain features of the current signal are extracted for a total of 59 features. The Spearman correlation coefficient is used to screen the features with strong correlation with time, and finally 10 features that are more sensitive to tool wear are obtained. The three-dimensional Figure 8 is shown in the figure.

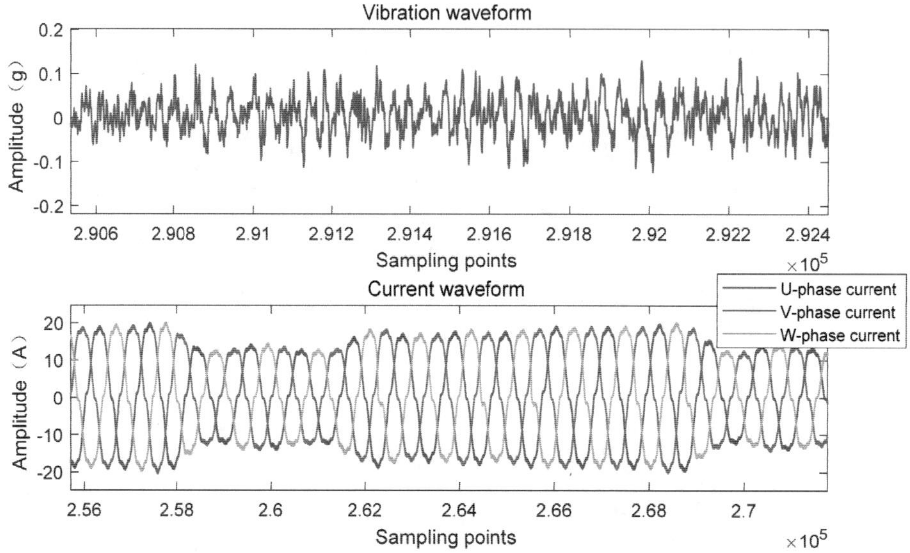

Figure 7. Vibration signal and original current signal

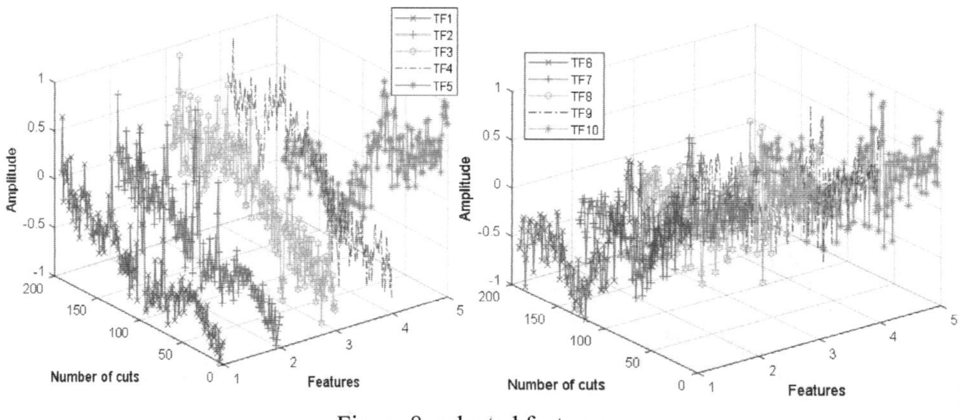

Figure 8. selected features

The screened 10 wear-sensitive features are used as the input of the PSO-SVR neural network, 100 sampling samples are used as the training set, 50 validation sets, and 42 test sets. Through the continuous optimization of the neural network by the PSO algorithm, the optimal network parameters are obtained. The obtained validation set prediction results are shown in Figure 9. The validation set prediction accuracy rate reaches 93%, and the test set results are shown in Figure 10.

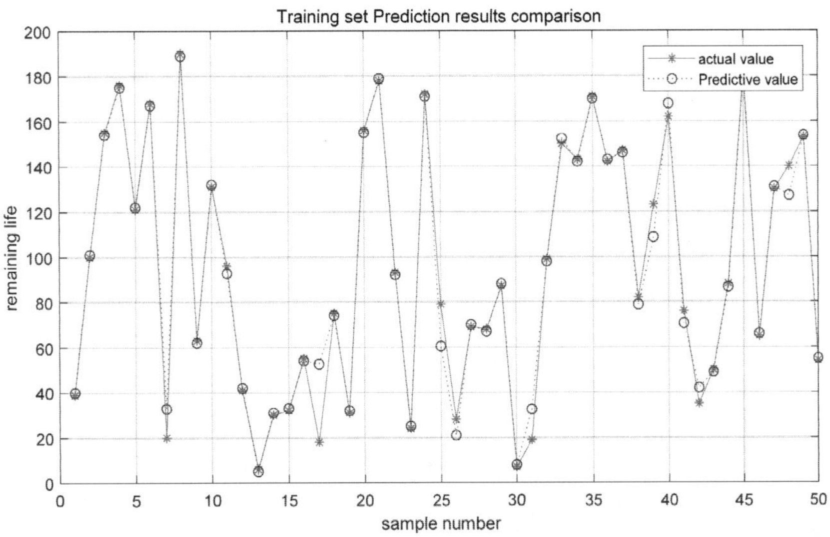

Figure 9. Validation set prediction results

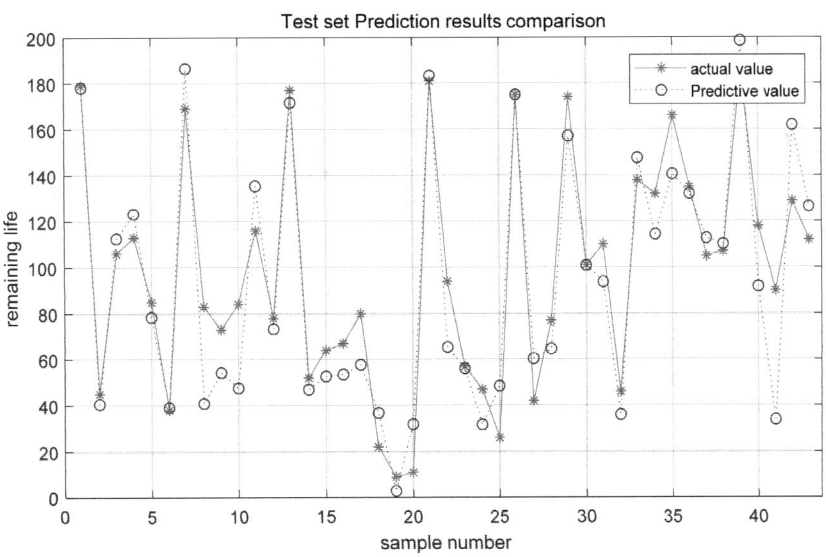

Figure 10. Test set prediction results

Table 1. Comparison of prediction results of different methods

	RMSE	MAE	Accuracy
BP	32.1	23.9	57.8%
SVM	23.5	20.2	78.5%
PSO-SVM	18.7	14.4	85.7%

Table 1 shows the prediction results of methods such as BP, SVM and PSO-SVM. PSO-SVM neural network prediction The method is superior to other neural network methods in tool life prediction, the prediction accuracy is significantly improved, and the danger of falling into local minima is reduced at the same time, and it has excellent global search ability.As shown.

4. Conclusions

Tool life is a vital basis and means to evaluate tool quality, formulate tool demand plans and optimize cutting parameters. Aiming at the problems of the difficulty in selecting tool fault features, the difficulty of determining the network structure in the neural network prediction method of tool life, and the tendency to fall into local extreme values, the tool life prediction method based on multi-source feature PSO-SVM neural network was proposed. The application example shows that the established tool life prediction model based on SVR machine has strong generalization ability and high prediction accuracy, and the absolute error, average error and accuracy of tool life of test samples are better than other neural networks The method has high theoretical significance and practical value.

Acknowledgments

This project is supported by the Key Research and Development Program of Shaanxi Province（Grant No. 2021GY-335）and the Science and Technology Program of Guangzhou (Grant No. 202201010298).

References

[1] Bagga P.J.,Bajaj K.S.,Makhesana M.A.,Patel K.M.. An online tool life prediction system for CNC turning using computer vision techniques[J]. Materials Today: Proceedings,2022,62(P6).
[2] Business - Manufacturing; Findings on Manufacturing Reported by Investigators at Tianjin University (Tool Life Prediction Based On Gauss Importance Resampling Particle Filter)[J]. Journal of Technology,2019.
[3] A.V. Antsev. Cutting tool life prediction in case of rough machining by the fracture model[J]. Materials Today: Proceedings,2019,19(Pt 5).
[4] Hua An,Guofeng Wang,Yi Dong,Kai Yang,Lingling Sang. Tool life prediction based on Gauss importance resampling particle filter[J]. The International Journal of Advanced Manufacturing Technology,2019,103(9-12).
[5] Jinsong Yu,Shuang Liang,Diyin Tang,Hao Liu. A weighted hidden Markov model approach for continuous-state tool wear monitoring and tool life prediction[J]. The International Journal of Advanced Manufacturing Technology,2017,91(1-4).
[6] Qin Guohua, Xie Wenbin, Wang Huamin. Tool wear detection and control based on neural network and genetic algorithm [J]. Optical Precision Engineering, 2015, 23(05): 1314-1321.
[7] He Yan, Ling Junjie, Wang Yulin, Li Yufeng, Wu Pengcheng, Xiao Zhen. Online monitoring model of tool wear based on long-short-term memory convolutional neural network [J]. China Mechanical Engineering, 2020, 31(16): 1959-1967 .
[8] Luo Huan, Zhang Dinghua, Luo Ming. Research progress on the prediction of cutting tool wear and remaining life for aviation difficult-to-machine materials [J]. China Mechanical Engineering, 2021,32(22):2647-2666.

Research on Fusion Algorithm of Beidou and AIS Based on Improved Federated Filter

Jingjun Jia[a], Feijiang Huang*[ab], Yujun Cao[b], Hui Cao[b], Ming Bai[b], Bin Feng[b], Zhong Chen[b]

[a] School of Electronic Engineering, Guangxi Normal University, No. 15 Yucai Road, Guilin, Guangxi, CHN 541004

[b]School of Information and Communication Engineering, Guangzhou Maritime University, No. 101 Hongshan 3rd Road, Guangzhou, Guangdong, CHN 510725

Corresponding author: 458655369@qq.com

ABSTRACT: In order to solve the problem of insufficient accuracy and reliability of ship position display in AIS(Automatic Identification System) system, a federated Kalman filter is designed, an open algorithm architecture is built, AIS and BDS positioning information sources are selected for fusion, and the information fusion part of the federated Kalman filter framework is improved by IMMF algorithm and covariance crossover algorithm to obtain more accurate and reliable target position information and ship motion state. Simulation experiments are carried out, and the experimental results show that, compared with GPS positioning or BDS(Bei Dou Navigation Satellite) positioning in AIS alone, AIS/BDS combined positioning based on federated Kalman filter structure has higher system fault tolerance, positioning accuracy and stability, and improves the safety of ship navigation.

1. Introduction

AIS can realize the main functions of automatic information exchange and automatic identification between ships. The display of ship position is mainly based on the GPS. Due to the existence of various errors, the accuracy of ship position display by AIS system is slightly insufficient. Secondly, the reliability of single use of AIS equipment based on GPS system is insufficient, and the ship's AIS track often shows large blank points[1].This might be the result of accidental technical issues, radio interference, attenuation, or genuine device sabotage [2]. A vulnerability to national security may arise from the ease with which terrorists or other parties seeking to conceal their whereabouts might trick and manipulate AIS signals [3]. Since Beidou Satellite Navigation System is an independent creation and operation of BDS, it offers very accurate positioning, navigation, and timing services. but Beidou navigation system is rarely used in our ships, and the replacement and elimination of GPS positioning AIS equipment in a large area will cause unnecessary economic losses. Beidou navigation's unique accurate timing function can be indispensable in ship navigation. Beidou and AIS carry out information fusion. It can also greatly improve the accuracy of fusion. Therefore, how to integrate Beidou navigation and AIS track information in order to improve the accuracy and reliability of the target position and maintain the accurate timing function of Beidou is of great research value.

At present, in the aspect of domestic navigation information fusion, Ref. [4] proposes a marine navigation system fusion algorithm based on factor graph, but the improvement of fusion accuracy is limited. Ref. [5] studies the robust information fusion method in SINS/DVL/AST underwater integrated navigation, and proposes a federated robust Kalman filtering algorithm based on Mahalanobis distance algorithm, which can achieve higher precision and more stable integrated navigation under non-Gaussian conditions. A data fusion method based on IMM-Kalman in integrated navigation system is proposed in Ref. [6], which can realize the dynamic optimal estimation of system error and sensor zero bias. In the aspect of foreign research, JungMinPak proposed a hybrid interactive multi-model filter to improve the reliability of the radar front collision early warning system [7]. The FIR filter is used to reset and restore the divergence in the IMM collision PDA filter. Therefore, the use of integrated navigation information fusion to improve positioning accuracy is a hot research topic.

2. Target Motion and Observation Model

Assuming that the measurement is incomplete, the linear discrete signal model and the linear observation model for a single target in Beidou and AIS systems are respectively:

$$\dot{X}(t) = F(t)X(t) + Q(t) \tag{1}$$

$$Z_i(t) = H_i(t)X(t) + v_i(t)(i = 1,2) \tag{2}$$

$X(t)$ is the state vector of the system at t time,$F(t)$ is the predictive state transition matrix.$Q(t)$ is the state noise of the system, $Z_i(t)$ is the observation vector of Beidou and AIS, $H_i(t)$ is the observation matrix of Beidou and AIS, $v_i(t)$ is the observation noise, $v_1 \sim N(10,1)$,$v_2 \sim N(5,1)$, $Q(t)$ and $v_i(t)$ are independent of each other.

3. Bds/Ais Track Fusion Algorithm Based on Improved Nr Federated Kalman Filter

3.1 Information Distribution

In this study, the NR federated filter with strong fault tolerance is used. The federated filter of this structure has no information allocation, and the output of the main filter is determined by time update.

3.2 Information Updates

3.2.1 Time Updates

The time update process is carried out independently between each sub-filter and the main filter. The filtering algorithm for time update is

$$\hat{X}_i(k+1|k) = \phi(k+1|k)\hat{X}_i(k|k) \tag{3}$$

$$P_i(k+1|k) = \phi(k+1|k)P_i(k|k) \tag{4}$$

$$\phi^T(k+1|k) = \Gamma(k)Q_i(k)\Gamma^T(k) \tag{5}$$

Where ϕ is the state transition matrix of the system and Γ is the process noise distribution matrix.

3.2.2 Measurement Updates-IMM-Kalman Algorithm

The time and measurement update of the original federated Kalman filter uses the Kalman filter algorithm, which is improved in this paper. Based on the traditional Kalman filter, the interactive multi-model filtering algorithm (IMM) is adopted. The interactive multi-model data fusion algorithm is a commonly used multi-model estimation method for target tracking [9-10]. The adaptive switching between different models can be realized through the transition probability matrix [11]. It has been widely used in integrated navigation system, on the one hand, it is used to describe the uncertainty of system model and observation noise [12-14], on the other hand, it interacts with a variety of fusion methods, in order to improve the accuracy of system data fusion. The advantages of using IMM algorithm in Beidou and AIS information fusion algorithm are as follows: first, the algorithm model contains multiple motion models, which can be added or subtracted according to the actual situation, which not only greatly improves the scope of application of the algorithm, but also adapts to the transformation of target motion mode. Secondly, the algorithm introduces the concept of model transition probability matrix, which can switch between models independently and has a certain degree of adaptability. Finally, each model in the algorithm has a corresponding filter. According to the actual motion model of the target, the appropriate filtering algorithm can be selected for modular programming, which has the characteristics of flexible use. It can have a strong adaptability to the complex environment at sea.

The schematic diagram of the data fusion algorithm based on IMM-Kalman is shown in figure 1, which mainly includes four steps: input interaction, filtering calculation, model probability update and output fusion. From these four steps, this paper will introduce in detail the track information of Beidou and AIS using IMM-Kalman algorithm to filter to produce more accurate position information.

(1) input interaction

According to prior system state estimation and covariance estimation, the measurement Z（k）of Beidou or AIS at the present time is inferred. The model is then re-initialized, in which the new starting value is acquired by the Markov operation matrix between various models. The initial model of this article is set to two, and the optimal state estimation value and estimate covariance matrix of model KMur1 are set $X_j(k-1|k-1)$ and $p_j(k-1|k-1)$ as follows. The beginning conditions of the model following interaction are then calculated, and these are the CA model and CV model, respectively.

$$X^{0j}(k-1|k-1) = \sum_{i}^{2} X^j(k-1|k-1)\, u_{ij}(k-1|k-1)\ (j=1,2)$$

$$p^{0j}(k-1|k-1) = \sum_i^2 u_{ij}(k-1|k-1)\{ p^j(k-1|k-1) + [x^j(k-1|k-1) - x^{0j}(k-1|k-1)][(k-1|k-1) - x^{0j}(k-1|k-1)]^T\} \quad (6)$$

Where:

$$u_{ij}(k-1|k-1) = \frac{p_{ij}u_i(k-1)}{\bar{c_j}} \quad (7)$$

$$\bar{c_j} = p_{ij}u_i(k-1) \quad (8)$$

(2) filtering of filter

In the IMM algorithm, the Kalman filter algorithm is used to filter and predict the target, and the input of the j model at time k-1 is the optimal estimation and update covariance. The Kalman filtering process of the IMM algorithm is as follows:

the Kalman filter primary prediction equation and prediction covariance matrix are

$$X^j(k|k-1) = A_j X^{0j}(k-1|k-1) \quad (9)$$

$$p^j(k|k-1) = A_j p^{0j}(k-1|k-1) A_j^T + Q_j \quad (10)$$

The Kalman gain is

$$K_j(k) = p^j(k|k-1) H^T [H p^j(k|k-1) H^T + R_j]^{-1} \quad (11)$$

The system state estimation and covariance estimation corresponding to the output of a single filter at k time are respectively

$$X^j(k|k) = X^j(k|k-1) + K_j(k)[Z(k) - HX^j(k|k-1)] \quad (12)$$

$$p^j(k|k) = [I - K_j(k)H] p^j(k|k-1) \quad (13)$$

(3) the update probability of the model

The IMM algorithm's effectiveness is directly impacted by model updates. The algorithm updates the model using the maximum likelihood function approach, which determines the weight of the most appropriate tracking model by comparing it to the current state of the moving target, or the ratio of the CA model to the CV model. The maximum likelihood function at k time that most closely matches the j model is

$$\Lambda_j(k) = \frac{1}{\sqrt{2\pi|S_j(k)|}}\exp\{-\frac{1}{2}d_j^T(k)S_j^{-1}(k)d_j(k)\} \tag{14}$$

Where $S_j(k) = Hp^j(k|k-1)H^T + R, d_j(k) = Z_j(k) - H_jX_j(k|k-1)$.

The update probability of model j can be obtained as follows:

$$u_j(k) = \frac{1}{c}\Lambda_j(k)\sum_{i=1}^{2} p_{ij}u_i(k-1) \tag{15}$$

Where c is the normalized constant, that is:

$$c = \sum_{i=1}^{2} \Lambda_j(k)\overline{c_j} \tag{16}$$

(4) New converged data

The final output result of k-time interactive data is supplied and is as follows, taking into account the tracking outcomes calculated independently by each model and the weight of model matching:

Overall estimates:

$$p(k|k) = \sum_i^2 u_i(k|k)\{ p_i(k|k) + [x_i(k|k) - x(k|k)][x_i(k|k) - x(k|k))]^T\} \tag{17}$$

The final results can not only effectively reflect the motion state of the ship, but also obtain more accurate position information, which is higher in accuracy than the traditional Kalman filter and has real-time tracking.

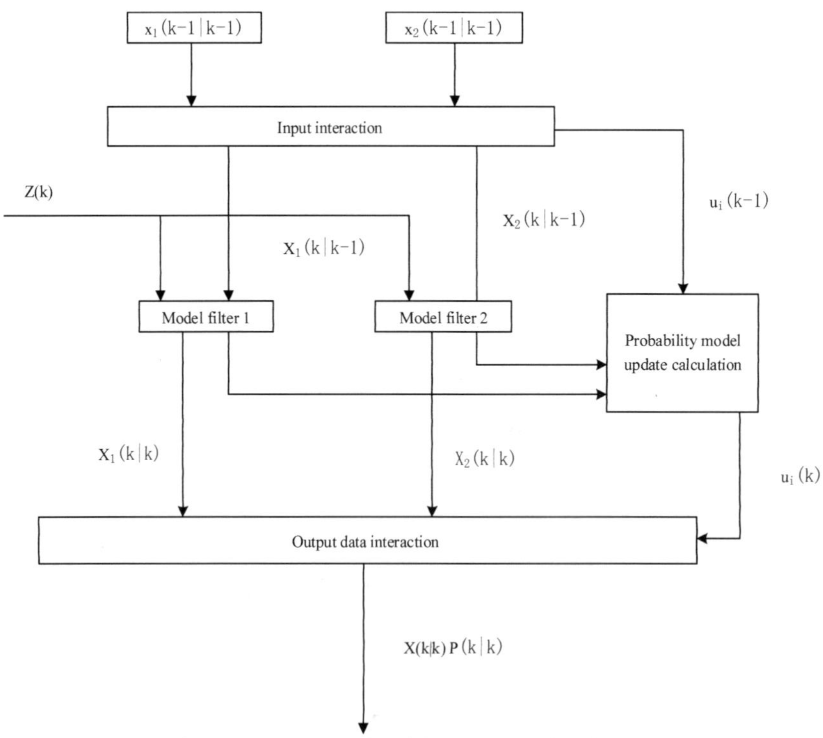

Figure 1. IMM Algorithm Schematic Diagram

3.3 Information Fusion-Covariance Crossover Information Fusion Algorithm

In the federated filter main filter, the local track estimation information obtained by IMM-calman filtering algorithm in each local filter is fused to obtain the global optimal estimation. In this study, the covariance cross fusion algorithm is used for information fusion, and the optimal fusion estimation is obtained. The covariance cross fusion algorithm has the following advantages: first, the covariance of Beidou track and AIS track is very complex. If we ignore the correlation and use the formula under the independent condition for fusion estimation, the error variance estimation of the fusion estimation will be too low, which will further lead to the divergence of the fusion estimation algorithm. The covariance cross estimation fusion algorithm can completely avoid calculating the correlation between the two and reduce the computational burden. Secondly, the fusion accuracy is higher than that of the local filter and has excellent performance. Finally, multi-dimensional model filtering can be realized, which is not limited to Beidou and AIS navigation equipment, and can be flexibly used for positioning. Therefore, it is very necessary to use covariance crossover fusion algorithm for track information fusion of Beidou and AIS.

The covariance cross fusion algorithm of Beidou and AIS track information can be simply expressed as

$$P^{g}_{k+1|k+1}{}^{-1} = \sum_{j=1}^{2} \omega_j P^{j}_{k+1|k+1}{}^{-1} \tag{18}$$

$$\hat{x}^{g}_{k+1|k+1} = P^{g}_{k+1|k+1}\left(\sum_{j=1}^{2} \omega_j P^{j}_{k+1|k+1}{}^{-1} \hat{x}^{j}_{k+1|k+1}\right) \tag{19}$$

$$\omega_1 + \omega_2 = 1$$

In the formula, $P^{g}_{k+1|k+1}$ is the estimation matrix of the covariance matrix, $P^{j}_{k+1|k+1}$ is the covariance matrix of the j th sensor, $\omega_i \in [0,1]$. The optimality of the estimation is determined by ω, and ω can be obtained by optimization method according to the criterion of minimum norm of C.

4.Simulation Experiment And Analysis

The algorithm is simulated and verified, and a group of target trajectories are generated, and the motion states are as follows: uniform motion 10s, uniform acceleration 10s, uniform deceleration 10s, uniform acceleration 10s, uniform deceleration 20s, uniform deceleration 20s, uniform acceleration 20s, uniform motion 100s, sampling period 1s, simulation time 200s.

The track information generated by a single positioning system is filtered by IMM algorithm and KF filter respectively, and the mean values of relative position and velocity errors are obtained. 52 Monte Carlo simulations are carried out to get figures 5 and 6. As can be seen from figures 5 and 6, the relative position error and relative velocity error obtained by 52 simulations of IMM algorithm are much less than that of KF filter algorithm. Therefore, the accuracy of the algorithm can be significantly improved by using IMM algorithm instead of traditional Kalman filter algorithm, and relatively accurate acceleration information can be obtained by using IMM algorithm.

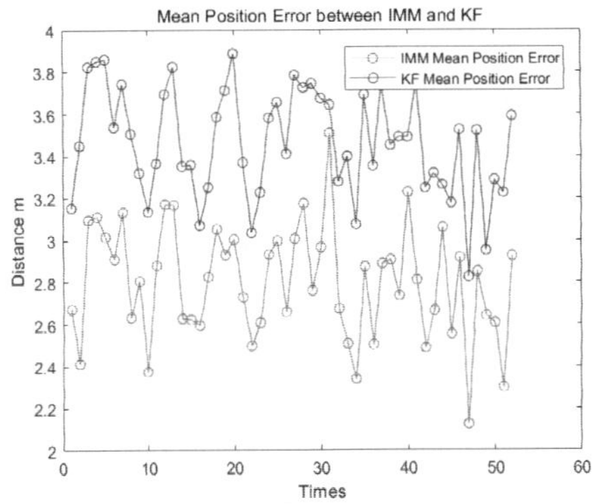

Figure 2. Mean Position Error between IMM and KF

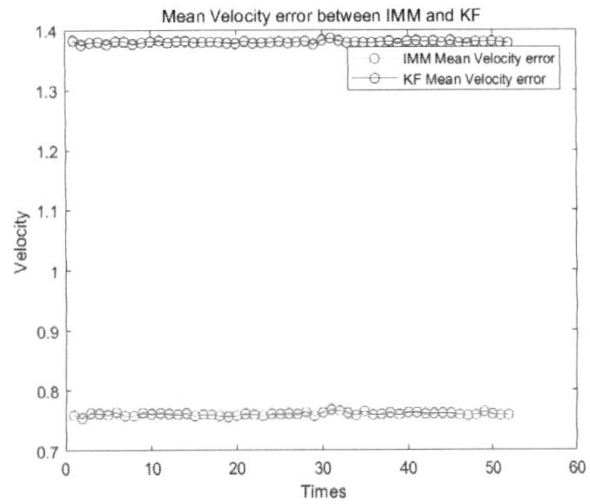

Figure 3. Mean Velocity error between IMM and KF

The data filtered by IMM algorithm is sent to the fusion center and processed by covariance crossover algorithm. The value of 0.618 is taken to get the information of position, velocity and acceleration after fusion. 52 Monte Carlo simulations are carried out to get the average error of relative position, velocity and acceleration, as shown in figures 7, 8 and 9, compared with the track filtered by IMM-Kalman. The track fused by the covariance crossover fusion algorithm is closer to the actual route, the position accuracy is more obvious, the acceleration and velocity accuracy is slightly improved, it is easier to avoid danger in practical application, and it is easier to grasp the navigation condition of the hull.

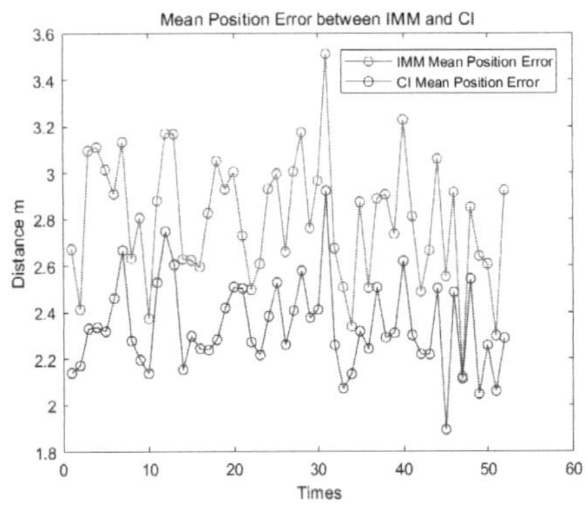

Figure 4. Mean Position Error between IMM and CI

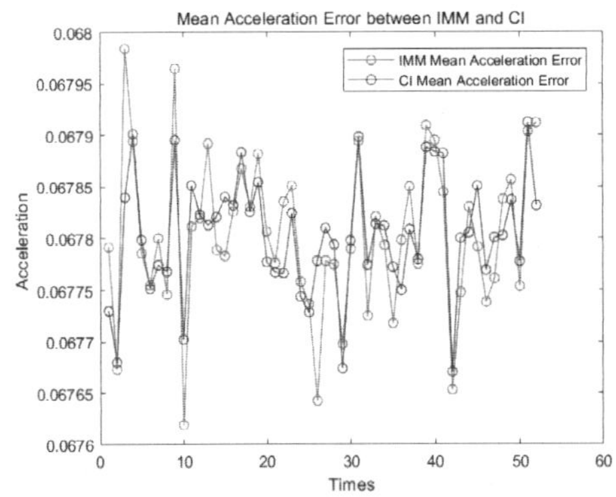

Figure 5. Mean Acceleration Error between IMM and CI

Through 50 Monte Carlo simulations, the simulation results are compared from three aspects: root mean square error (RMSE), mean absolute error (MAE) and standard deviation (SD). As shown in Table 1, Table 2 and Table 3, the effectiveness of the algorithm is further proved.

Table 1. RMSE、MAE、SD on Position table

	AIS Observation	BDS Observation	AIS-IMM Filtering algorithm	BDS-IMM Filtering algorithm	CI Algorithm fusion
RMSE	4.4200	4.0306	2.8108	2.9393	2.3471
MAE	4.4147	4.0247	2.7977	2.9279	2.3393
SD	0.2198	0.2193	0.2608	0.2733	0.1928

As shown in Table 1, the position accuracy of the observation data is greatly improved after the position information of the observation data is successively processed by IMM filtering algorithm and CI fusion algorithm. The IMM filtering algorithm is relatively unstable, but after the fusion of the CI algorithm, the stability is also significantly improved, and the position accuracy is improved by about 40% and 50% compared with the observed initial data.

Table 2. RMSE、MAE、SD on Speed table

	AIS-IMM Filtering algorithm	BDS-IMM Filtering algorithm	CI Algorithm fusion
RMSE	0.8720	0.8722	0.8714
MAE	0.8720	0.8722	0.8714
SD	0.0015	0.0015	0.0010

Table 3 RMSE、MAE、SD on Acceleration table

	AIS-IMM Filtering algorithm	BDS-IMM Filtering algorithm	CI Algorithm fusion
RMSE	0.2604	0.2605	0.2604
MAE	0.2604	0.2605	0.2604
SD	0.1477×10^{-3}	0.1709×10^{-3}	0.1112×10^{-3}

As shown in Table 2, the CI fusion algorithm slightly improves the accuracy and stability of speed measurements based on IMM filtering. The increase in acceleration is small, as shown in Table 3.

To sum up, using the improved federal Kalman filter AIS/BDS information fusion, positioning accuracy is higher, and more stable, can provide more accurate velocity and acceleration information, easier to grasp the navigation state of the ship.

5. Conclusion

Based on the AIS/BDS information fusion of the improved federated Kalman filter, the Beidou navigation and positioning system is used to assist the AIS positioning system, which overcomes the lack of position information caused by the system failure caused by signal interruption when AIS is used alone. The federated Kalman filter technology combined with IMM algorithm and CI algorithm is used to fuse AIS/BDS data. The system achieves the best overall performance and has high accuracy of position, velocity and acceleration, which is conducive to real-time tracking of the state of ship navigation. The AIS and BDS subsystems run independently and run in parallel, which makes the whole system have stronger fault tolerance and real-time computing ability.

Acknowledgements

This study was supported by the Characteristic Innovation Project of Department of Education of Guangdong Province (No.2021KTSCX098), the Key Construction Discipline Scientific Research Capability Improvement Project of Guangdong Province (No.2021ZDJS059), the Special Projects in Key Fields of General College of Department of Education of Guangdong Province (No. 2020ZDZX3072), the Special Research Project on Building a National Science and Technology Think Library in Guangzhou(No.SXK20220201035), the Education Science Planning Project of Guangzhou(No. 202012589)

References

[1] Mazzarella F, Vespe M, Alessandrini A , et al. A novel anomaly detection approach to identify intentional AIS on-off switching[J]. Expert Systems with Applications, 2017, 78(JUL.):110-123.
[2] Iphar C, Ray C, Napoli A.Uses and Misuses of the Automatic Identification System[C]. OCEANS 2019 - Marseille. 2019.
[3] Balduzzi M , Pasta A , Wilhoit K . A security evaluation of AIS automated identification system[C]. Computer Security Applications Conference. 2014:436-445.
[4] Zhang Chongmeng, Deng Fujian, Yang Tao, Lan Xiaoyang. Information fusion algorithm of marine navigation system based on factor graph [J]. Chinese Journal of Inertial Technology, 2020, 28 (04): 448-455.
[5] Zhu Bing, Chang Guobin, he Hongyang, Xu Jiangning.Robust Information Fusion method for SINS/DVL/AST underwater Integrated Navigation[J].Journal of National University of Defense Science and Technology,2020,42(05):107-114.
[6] Tian Yi, Yan Yuepeng, Zhong Yanqing, Li Jixiu, Meng Zhen. A data fusion method based on IMM-Kalman in integrated navigation system [J/OL]. Journal of Harbin University of Engineering: 1-6 [2022-05-30].
[7] Pak, J.M. Hybrid Interacting Multiple Model Filtering for Improving the Reliability of Radar-Based Forward Collision Warning Systems[J]. Sensors 2022, 22, 875.

[8] Han Chongzhao, Zhu Hongyan, Duan Sheng. Multi-source information fusion [M]. Beijing: Tsinghua University Press, 2006.
[9] Luo X , Jiu B , Chen S , et al. ML estimation of transition probabilities for an unknown maneuvering emitter tracking[J]. Signal Processing, 2015, 109(apr.):248-260.
[10] Jo K , Chu K , Sunwoo M . Interacting Multiple Model Filter-Based Sensor Fusion of GPS With In-Vehicle Sensors for Real-Time Vehicle Positioning[J]. IEEE Transactions on Intelligent Transportation Systems, 2012, 13(1):329-343.

Research on flight control system hardware in the loop simulation system based on NI hardware platform for Unmanned Aerial Vehicle

Helong Wu[a,b,c], **Mingxi Cai***[a,b,c], **Zongke He**[a,b,c]

[a] China Electronic Product Reliability and Environment Testing Research Institute Guangzhou, China;

[b] Guangdong Provincial Key Laboratory of Electronic Information Products Reliability Technology

Guangzhou, China

[c] Guangdong Provincial Engineering Technology Research Center of UAV Reliability and Safety, Guangzhou, 510610, China

* Corresponding author: 461636930@qq.com

ABSTRACT: HIL(Hardware in the loop) simulation of UAV(Unmanned Aerial Vehicle) plays an important role in flight control sysrtem development and algorithm verification. Taking the open source flight control system Pixhawk as the research object, this paper establishes a HIL system based on NI(National Instrument) hardware platform. The dynamic model and kinematic model are modeled through Simulink, the online management of simulation model is realized through Veristand, and the online display of model parameters and fault injection are realized through Labview. The simulation system can test the functional performance, reliability and safety of flight control system. The simulation results show that the simulation platform works reliably and has good real-time performance. It can effectively and reliably carry out the HIL of flight control system for UAV.

1. Introduction

HIL simulation refers to replacing a part of the simulation system with a real object and connecting it to the loop of the whole simulation system[1-3]. HIL simulation belongs to system simulation science and technology. It is the first and most successful application of the combination of simulation technology and physical object in the field of UAV[4-5]. Compared with digital simulation, HIL simulation has higher reliability, can be closer to the actual system and environment, and can get more accurate and reliable data[6-7]. Compared with physical simulation, HIL simulation can connect the hardware in the simulation system to verify the operation of the physical and the overall performance of the system, which can not only reduce the research cost, but also reduce the research time.

This paper takes the open source flight control pixhawk as the research object and establishes a HIL simulation system based on NI hardware platform. The HIL simulation system for flight control system is constructed based on the combination of Simulink, Veristand and Labview. The dynamic model and kinematic model are modeled through Simulink, the online management of simulation model is realized through Veristand, and the online display of model parameters and fault injection are realized through Labview. The simulation system can test the functional performance, reliability and safety of flight control system.

2. Design of HIL simulation for flight control system

2.1 Overall scheme

The HIL simulation system for flight control system is composed of NI hardware platform, flight control system and the upper computer control software. Send commands (such as throttle, roll angle, pitch angle and flight mode) to the flight control system through the remote control. The PWM signal output by flight control system is transmitted to the UAV model to control the rotation of the motor model, which controls the movement of the UAV model. The UAV model output the data information of gyroscope, Euler angle, speed and position, which are converted into gyroscope, accelerometer, magnetometer and barometer data and transmitted to the flight control system through the data interface model. The PWM signal output by the flight control system is transmitted to the UAV model through the NI data acquisition board, and the output of the UAV model is displayed on the data display and control interface. The flight state of UAV model serves as the judgment basis for normal operation of flight control system. The fault simulation program injects the preset fault mode into the UAV model to test whether the flight control system can correctly and safely respond to the fault. The overall scheme of HIL simulation is shown in Figure 1.

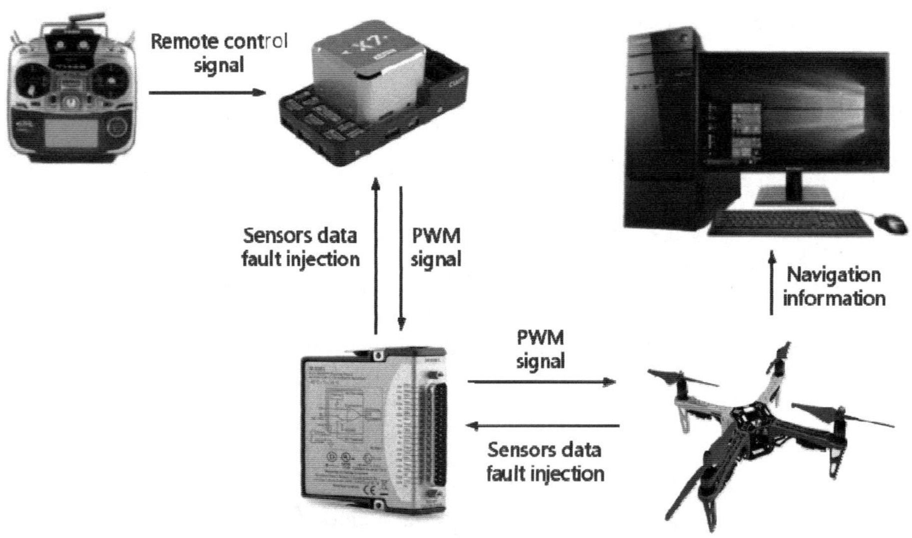

Figure 1. The overall scheme of HIL simulation

2.2 Modeling of UAV based on Simulink

The UAV model simulates a real multi-rotor UAV, with the PWM signals of the motor as input, the status parameters and sensor information of the UAV as the output. The UAV model includes seven main modules: motor module, force and torque module, rigid body motion dynamic module, environment module, mass module, battery module and output interface module. In this paper, Simulink graphical tools are used to build the UAV model, and RTW code tools are used to automatically generate the embedded code of the simulation platform, which is loaded into the model debugging system. The UAV model developed by Simulink is compiled into DLL (Dynamic Link Library) file through NI Veristand software. Use veristand blocks NI in and NI out on Simulink to replace the input and output of UAV model, as shown in the figure 2. Generated DLL file is imported into veristand. The input of UAV model is 4-channel PWM signal, and the output of 6-channel signals are position posE, velocity velE, attitude angle AngEuler, angular rate AngRateB, motor speed MotorRPMS and rotation matrix DCM.

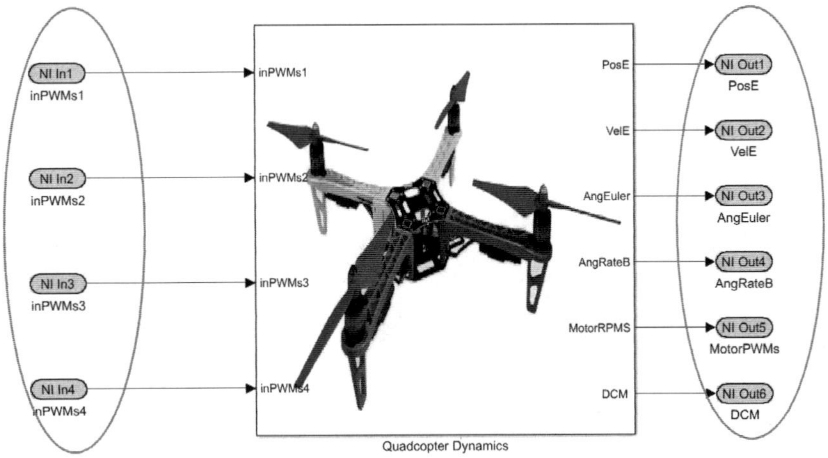

Figure 2. UAV model for dll file

2.3 Establishment of data interface model

The function of the data interface model is to input the sensor data output from the UAV model into the flight control system according to a certain protocol, and input the 4-channel PWM signal output from the flight control system into the UAV model. The sensor data necessary for UAV attitude estimation include gyroscope, accelerometer, magnetometer and barometer data. The gyroscope signal corresponds to the AngRateB output from the UAV model and can be directly input to flight control system. The accelerometer, magnetometer and barometer data need to be obtained by mathematical calculation through the output value of the UAV model.

The accelerometer measures the acceleration value. In the North-East-Earth coordinate system, The value of Z axis of the accelerometer placed horizontally is –g, and the value of X axis and Y axis is 0, which can be recorded as (0,0, -g). When the UAV rotates a certain attitude, the gravitational acceleration will generate corresponding components on the three axes of the accelerometer, which is essentially the transformation from (0,0, -g) in the geodetic coordinate system to the UAV coordinate system. There is a rotation matrix DCM difference between these two coordinate systems, the accelerometer value (a_x, a_y, a_z) under the UAV coordinate system is:

$$\begin{bmatrix} a_x \\ a_y \\ a_z \end{bmatrix} = \begin{bmatrix} \cos\varphi\cos\theta & \cos\varphi\sin\phi\sin\theta - \cos\phi\sin\varphi & \sin\phi\sin\varphi + \cos\phi\cos\varphi\sin\theta \\ \cos\theta\sin\varphi & \cos\phi\cos\varphi + \sin\phi\sin\varphi\sin\theta & \cos\phi\sin\varphi\sin\theta - \cos\varphi\sin\phi \\ -\sin\theta & \cos\theta\sin\phi & \cos\phi\cos\theta \end{bmatrix} \cdot \begin{bmatrix} 0 \\ 0 \\ -g \end{bmatrix} \quad (1)$$

$$= -g \cdot \begin{bmatrix} \sin\phi\sin\varphi + \cos\phi\cos\varphi\sin\theta \\ \cos\phi\sin\varphi\sin\theta - \cos\varphi\sin\phi \\ \cos\phi\cos\theta \end{bmatrix}$$

Where θ is pitch angle, ϕ is roll angle, φ is yaw angle.

The Magnetometer measures the magnetic field strength of the body. When the UAV is stationary and the X axis faces north, only the X axis can measure the magnetic field strength, and the value of Y axis and Z axis is zero. Assuming that the value of magnetic field is B, the triaxial magnetic field strength is (B, 0, 0). When the UAV rotates a certain attitude, the magnetic field strength will generate corresponding components on the three axes of the Magnetometer, which is essentially the transformation from (B,0, 0) in the geodetic coordinate system to the UAV coordinate system. There is a rotation matrix DCM difference between these two coordinate systems, the Magnetometer value (M_x, M_y, M_z) under the UAV coordinate system is:

$$\begin{bmatrix} M_x \\ M_y \\ M_z \end{bmatrix} = \begin{bmatrix} \cos\varphi\cos\theta & \cos\varphi\sin\phi\sin\theta - \cos\phi\sin\varphi & \sin\phi\sin\varphi + \cos\phi\cos\varphi\sin\theta \\ \cos\theta\sin\varphi & \cos\phi\cos\varphi + \sin\phi\sin\varphi\sin\theta & \cos\phi\sin\varphi\sin\theta - \cos\varphi\sin\phi \\ -\sin\theta & \cos\theta\sin\phi & \cos\phi\cos\theta \end{bmatrix} \cdot \begin{bmatrix} B \\ 0 \\ 0 \end{bmatrix} \quad (2)$$
$$= B \cdot \begin{bmatrix} \cos\varphi\cos\theta \\ \cos\theta\sin\varphi \\ -\sin\theta \end{bmatrix}$$

The barometer measures the height of an object by using the pressure of air. Its function is to measure the height of the plane where the UAV is located. Therefore, the atmospheric pressure at the current height can be calculated according to the flight height of the UAV.

$$P = 1013.25 - 0.11H(mbar) \quad (3)$$

It can also calculate the current temperature value according to the height:

$$T = 25 - 0.065H(°C) \quad (4)$$

The UAV model and flight control system communicate with each other to transmit virtual sensor signals and PWM signal. Because the amount of sensing data is small, and the data flow is only about 5kb/ s, the requirements for bandwidth can be met by using serial communication.

2.4 Software design of upper computer

The upper computer software is written by labview 2020, which mainly displays some data of the hardware in the loop simulation and sends control instructions to the control system. Its basic functions are as follows:

(1) It has the corresponding control function of sending and receiving data of common serial port assistant.

(2) It has fault injection function and can inject the preset fault mode into the UAV model.

(3) It has data display function, including data and curve display of PWM signal,angular rate, speed, position, Euler angle and motor speed. In addition, it is also required to display the data of four sensors: gyroscope, accelerometer, magnetometer and barometer.

3. HIL simulation experiment for the flight control system

3.1 System composition

The HIL simulation system mainly includes UAV model, remote controller, flight control system, data acquisition card and the upper computer software, as shown in Figure 3. The hardware model selected for HIL simulation are shown in the Table 1.

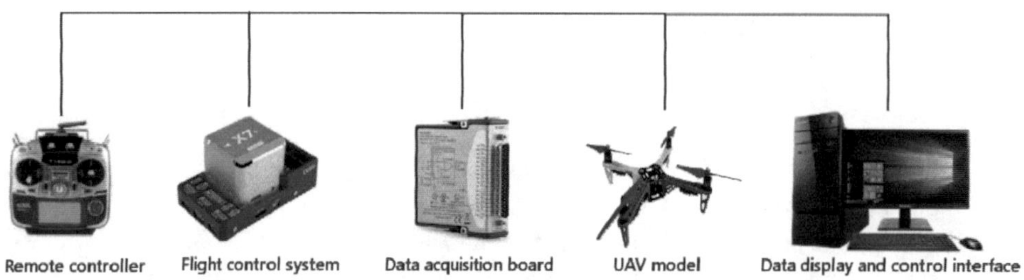

Figure 3. Hardware type selected for HIL simulation

Table 1. The hardware model selected for HIL simulation

Hardware type	**Hardware model**
Flight control system	CUAV-X7
Remote control / receiver	FUTABA T14FG/7008
NI software and hardware platform	PXIe-1085
data acquisition card	NIcDAQ9147/NI9361

3.2 Hardware in the loop simulation

First, carry out the communication test of the acquisition card. In order to ensure that the 9361 board can realize the data transmission function, the NI Max upper computer is used to verify the data of the acquisition channel. Select channel 0-3, the mode adopts edge counting, and the counter value is 734, indicating that the 9361 board can realize the data communication function, as shown in Figure 4.

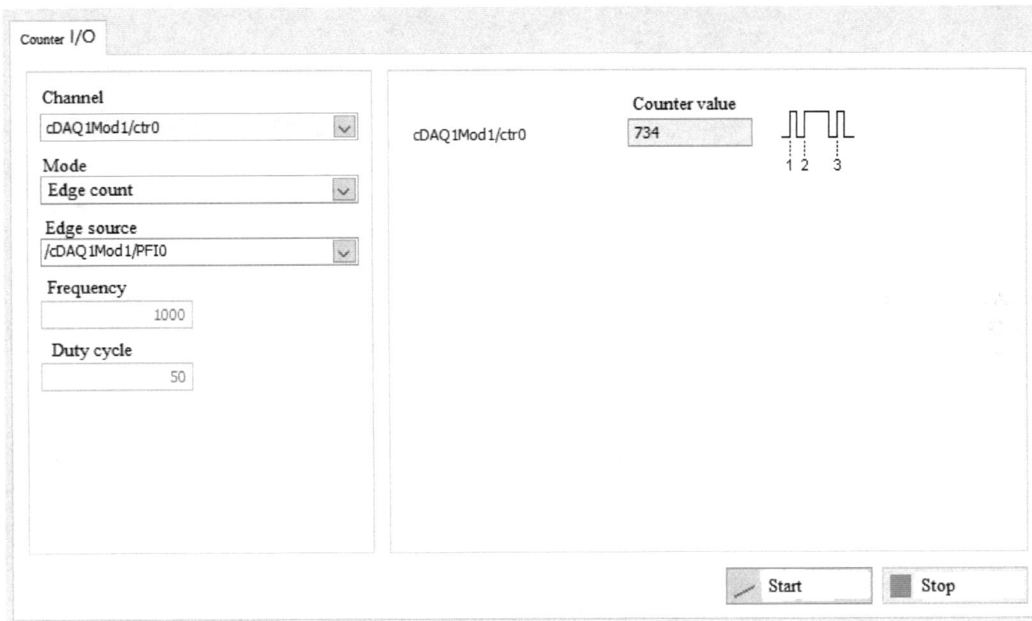

Figure 4. Data acquisition board test

After the channel data acquisition test is completed, the HIL simulation for flight control system should be conducted. Test unlocking status, flight status and fault injection status in HIL for the flight control system. In time 0s to 100s, when the flight control system is unlocked, the remote controller does not give command information, and the motor is in idle state, it can be seen that the four command channels are 1100, as shown in Figure 5. In time 100s to 800s, when the UAV was flight, the remote controller gives commands, and output the position posE, velocity velE, attitude angle AngEuler, angular rate AngRateB, motor speed MotorRPMS. Inject the fault at 1200 seconds, modify the PWM value to simulate the fault, and the response of each channel is as shown in the Figure 6.

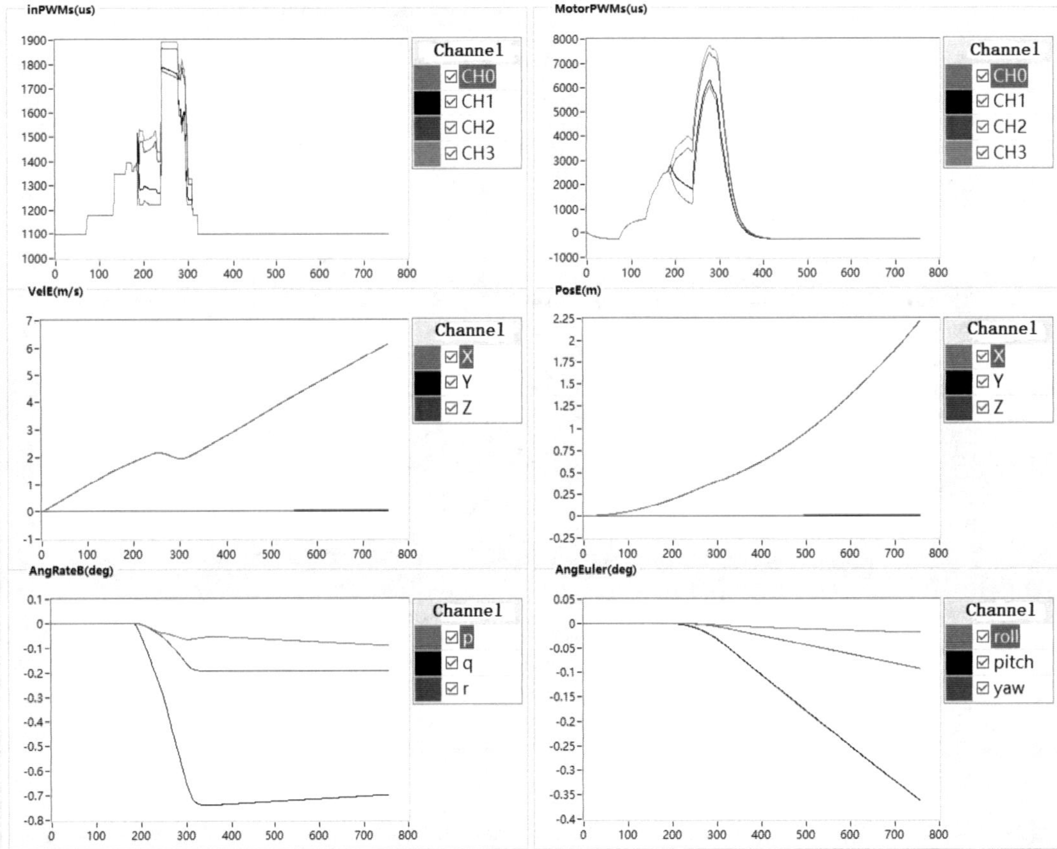

Figure 5. Hardware in the loop simulation for flight control system

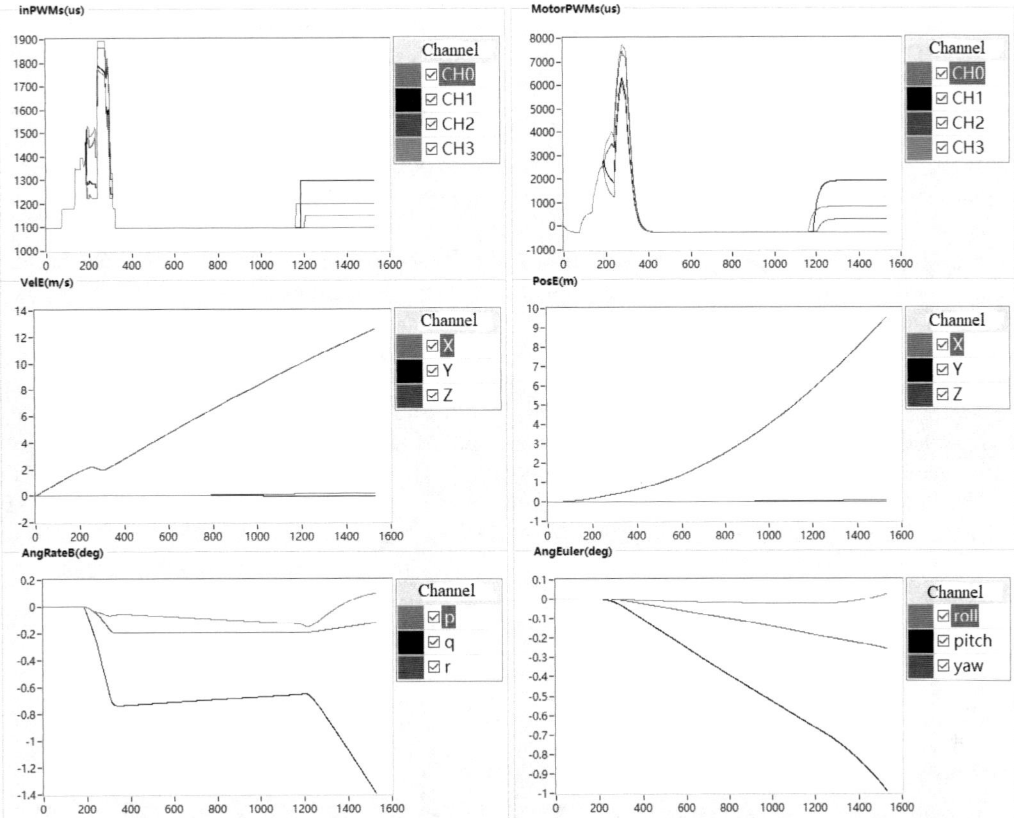

Figure 6. Fault injection test for flight control system

The data interface model calculates the values of accelerometer, gyroscope, compass and barometer, as shown in Figure 7. The sensor data can be accurately simulated, which shows the effectiveness of HIL simulation.

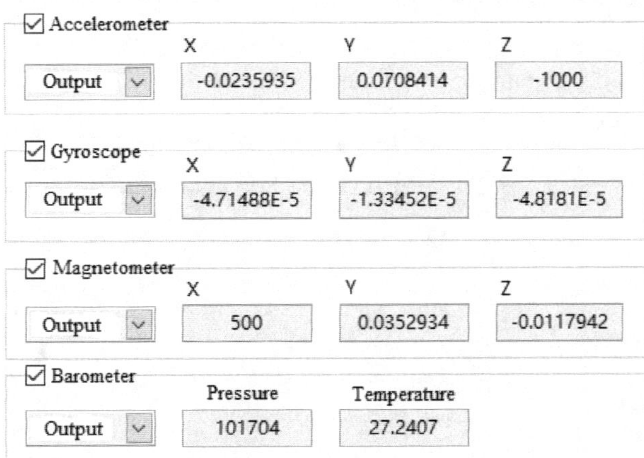

Figure 7. Analog sensor data test

4. Conclusion

In this paper, the HIL simulation system for flight control system based on NI hardware platform is established. Firstly, the PWM signal output by the flight control system is input to the UAV model in NI Verisand, and then the gyroscope, accelerometer, magnetometer and barometer data estimated by the UAV model are input to the flight control system through the serial port to realize the closed loop simulation. All simulation data can be displayed on the data display and control interface. The fault simulation program injects the preset fault mode into the UAV model to test whether the flight control system can correctly and safely respond to the fault. The simulation results show that the simulation platform works reliably and has good real-time performance. It can effectively and reliably carry out the HIL simulation of flight control system for UAV.

Acknowledgment

The authors are supported by the Key-Area Research and Development Program of Guangdong Province under No.2020B0404020002, the Special Fund of Guangdong Province for promoting high-quality economic development under No. JQR217000010, the Development fund for the Fifth Electronics Research Institute of Ministry of Industry and Information Technology under No.19Z14, Ministry of industry and information technology project No.TC210804R-1.

References

[1] Tomlin C, Jang J. Autopilot Design for the Stanford DragonFly UAV: Validation through Hardware-in-the-Loop Simulation[C].American: American institute of aeronautics and astronautics paper,2001.
[2] Holger Spangenberg, Holger Friehmelt. Hardware-In-the-Loop Simulation with Flight Control Actuators. AIAA Modeling and Simulation Technologies Conference and Exhibit,2005:15-18.
[3] T. L. NG, P. Krishnamurthy, F. Khorrami, et al. Autonomous Flight Control and Hardware-in-the-Loop Simulator for a Small Helicopter[C]. The 16th IFAC World Congress. 2005:1531-1536.
[4] Jung D,Tsiotras P. Modeling and Hardware-in-the-Loop Simulation for a Small Unmanned Aerial Vehicle[J].Georgia Institute of Technonlogy,2007:07-27.
[5] Saunders J., Beard R. (2009) UAS Flight Simulation with Hardware-in-the-loop Testing and Vision Generation. In: Valavanis K.P., Beard R., Oh P., Ollero A., Piegl L.A., Shim H. (eds) Selected papers from the 2nd International Symposium on UAVs, Reno, Nevada, U.S.A. June 8–10, 2009.
[6] Guowei Cai, Ben M. Chen, Tong H. Lee, Miaobo Dong, Design and implementation of a hardware-in-the-loop simulation system for small-scale UAV helicopters, Mechatronics, Volume 19, Issue 7, 2009, Pages 1057-1066.
[7] Khaligh, S.P., Martinez, A., Fahimi, F. et al. A HIL Testbed for Initial Controller Gain Tuning of a Small Unmanned Helicopter. J Intell Robot Syst 73, 2014: 289–308.

AUTHOR INDEX